W9-BJR-828

Introduction to
MASS
COMMUNICATION

Introduction to
MASS
COMMUNICATION

Second Edition

JAY BLACK
University of Alabama

FREDERICK C. WHITNEY
San Diego State University, Emeritus

wcb
Wm. C. Brown Publishers
Dubuque, Iowa

Book Team

Editor *Stan Stoga*
Developmental Editor *Kathy Law Laube*
Production Editor *C. Jeanne Patterson*
Designer *Carol S. Joslin*
Photo Research Editor *Shirley Charley*
Visuals Processor *Joyce E. Watters*
Product Manager *Marcia H. Stout*

wcb group

Chairman of the Board *Wm. C. Brown*
President and Chief Executive Officer *Mark C. Falb*

wcb

Wm. C. Brown Publishers, College Division

Executive Vice-President, General Manager *G. Franklin Lewis*
Editor in Chief *George Wm. Bergquist*
Director of Production *Beverly Kolz*
National Sales Manager *Bob McLaughlin*
Marketing Research Manager *Craig S. Marty*
Production Editorial Manager *Colleen A. Yonda*
Manager of Design *Marilyn A. Phelps*
Photo Research Manager *Faye M. Schilling*

Cover photograph © Cam Chapman

The credits section for this book begins on page C1, and is considered an extension of the copyright page.

Library of Congress Catalog Card Number: 87-070283

ISBN 0-697-00478-3

Printed in the United States of America by Wm. C. Brown Publishers
2460 Kerper Boulevard, Dubuque, IA 52001

10 9 8 7 6 5 4 3

Contents

Part Two
The Print Media 76

Part Three

The Electronic Media 200

Chapter 6

Radio 203

Chapter 7

Recordings 245

Chapter 8

Television 268

Chapter 9

Film 327

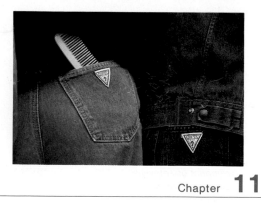

Preface

In writing the second edition of *Introduction to Mass Communication,* we continue the main thrust found in the first edition. We have attempted to analyze the mass communication empires on their own terms, to recognize them for what they are—large, sometimes monolithic industries established to earn a profit as well as to serve the interests of their customers. The media are studied here within the broad contexts in which they operate. Technological, economic, political, philosophical, and sociological factors are considered throughout. We try to ask and then carefully answer such questions as:

How does each of the media operate; what technological, economic, and human factors are involved in the production of newspapers, magazines, books, recordings, radio, television, film, and new electronic media?

What are the effects of media upon their audiences? As they fulfill their basic functions of persuading, informing, entertaining, and transmitting culture, to what extent are the media beneficial or harmful to individuals and to society?

What is the nature of media auxiliaries—the indirect but enormously influential advertising, public relations, news, and feature services?

Why have the media developed as they have—into pervasive, profit-oriented concerns, often in conflict with both the government and the governed?

What might the alternatives be to the present mass communications network? Is the future likely to bring more of the same, or are major changes at hand?

What forces in government, society, and within the media themselves serve to put a check on media excesses? Left to their own devices, would profit-oriented media be responsible to society?

Answers to these and many other questions are sought through a balanced study of the overall mass communication system. They cannot be found without some attention to detail, but our approach is not encyclopedic. Names, dates, and other details are cited, but not at the expense of the broader picture.

Mass communication is approached in this text as the central nervous system of society—a concept that suggests a deep, complex interrelationship between society and its means of communication. This approach, we feel, is consistent with audience reactions to the media, as well as with the inherent structures of the media themselves.

The media are often damned on the one hand as pollutants, filling heads with specious trivia, beclouding perception, and feeding confusion in the ranks through rising decibels of noise. On the other hand, media are often heralded as catalysts of the social organism—offering ever-changing, ever-multiplying views of both fantasy and reality—that enrich our lives and bring us to the brink of truth, or self-understanding. The mass media are regarded as both guardians of the status quo and radical vehicles of change.

This paradox emanates, in part, from the composition of the individual media. Because of their diverse, often corporate nature, the media cannot be appropriately perceived from within any one academic pigeonhole. They cut across journalism with their information content and media forms. Sociology and social psychology come into play insofar as the media's involvements with and effects upon large and small audiences. Psychology encompasses the media's effects upon the individual. Political science and economics are involved because media are instruments of political socialization and are at the same time profit-motivated. Advertising and public relations are closely related to dollars and votes. Economics plays an additional role in the relative affluence upon which mass media are so dependent, and in the expanding technology responsible for this development. Most recently, mathematics and engineering have influenced media and communications, not only because of computerization but also because of instantaneous transmission and feedback of individualized, localized, and global information.

Thus, what we offer here is a thoroughly updated and integrated approach to mass communication that distills pertinent contributions from many disciplines. *Introduction to Mass Communication* examines each mass medium in light of its historical development, its relationship to other media, its effects on audiences, and its probable future. The media are also discussed in terms of the broad social functions and their individual characteristics, which are surprisingly complementary.

Helping students understand media has been the foremost consideration of the authors and the book team. For this reason, attention to pedagogical techniques should be evident to even the casual reader. Several hundred pages of carefully researched and written text have been divided into five major parts according to function or theme. Each part is introduced by a short overview aimed at helping students understand that part's place within the broader context of the "social organism." Each chapter begins with an outline of the headings within the chapter, plus a brief introduction, to alert students to the topics covered therein. In addition, capsule commentary statements along the margins of the text highlight events, concepts, and concerns in the ever-evolving media scene. Subheads, illustrations, boxed quotes, chapter summaries, and reference lists have been carefully designed to render each chapter as accessible as possible. A glossary of terms is included at the end of the book to define and clarify terms used in the text. And for those students who desire additional reading resources, the bibliography will be a useful tool, for it is from many of these sources that we have gleaned much of our information when writing this text.

As practice of what we preach concerning service to clients, we request any feedback you care to give, so that, like other mass media producers, we can adjust our messages to suit our audience and to make ourselves heard. Comments may be sent in care of: Speech and Journalism Editor, Wm. C. Brown Publishers, 2460 Kerper Boulevard, Dubuque, IA 52001.

We would like to thank the tens of thousands of students and hundreds of instructors who used the first edition of *Introduction to Mass Communication*. Many of the suggestions they have made have been incorporated into this revision. Especially we would like to thank the following reviewers for their help with the manuscripts of both the first and second editions: Ralph Barney at Brigham Young University; ElDean Bennett at Arizona State University; Carol Burnett at California State University in Sacramento; Penny Byrne at Utah State University; Robert Carrell at the University of Oklahoma; Raymond Carroll at the University of Alabama; Mary Cassata at the University of Buffalo; Edgar Eaton at Green River Community College; Michael Emery at California State University in Northridge; David Gordon at Emerson College; Earl Grow at the University of Wisconsin in Milwaukee; Milton Hollstein at the University of Utah; James Hoyt at the University of Wisconsin in Madison; Donald Jugenheimer at Louisiana State University; Val Limburg at Washington State University; Kelly Leiter at the University of Tennessee; James T. Lull at the University of California in Santa Barbara; Lawrence Mason, Jr., at Syracuse University; John Merrill at Louisiana State University; Sharon Murphy at Marquette University; Carole Oukrop at Kansas State University; William Porter at the University of Michigan; Keith Sanders at the University of Missouri; Robert O. Shipman at Mankato State University; Gerald Stone at Memphis State University; John Wittig at the University of Alabama, Birmingham; and Alan Zaremba at Northeastern University.

A very special thank you goes to William Oates, University of Miami at Coral Gables and one of the nation's leading researchers on new electronic communications systems, for his work in researching and writing the bulk of chapter 10, "The New Electronics."

Introduction to
MASS
COMMUNICATION

Part One

The mass media (newspapers, magazines, books, radio, recordings, television, film, and new electronic communications systems) and their auxiliaries (advertising, public relations, and news and feature services) are significant institutions in today's world. They can be looked at in terms of how they perform in society. Why do they do what they do; how do they reflect and mold the priorities of society; what have they achieved in the past; and what will they be capable of achieving in the future?

The main purpose of the introductory chapters is to come to grips with the basic nature of the mass media and their audiences.

We begin by considering the fundamental character of communication. Without an appreciation of how meaning and information are exchanged among individuals, we cannot fully appreciate the more sophisticated process of mass communication, or communication through complex media to large, anonymous, and heterogeneous audiences.

Our focus is on the four primary functions of communication: (1) information; (2) entertainment; (3) persuasion; and (4) transmission of the culture. These functions are seldom performed singly; rather, they are performed in varying combinations, often with contradictory impacts. We investigate the nature of mass media as social and economic institutions, reliant upon consumer acceptance for their own existence. Media are ever changing to reflect changes in their audiences and levels of economic support. Those media which yesterday appealed to massive conglomerates of audiences are being supplanted by more specialized media, about which so much of this book is devoted.

Overview

In this overview section we look at audiences as consisting of far more than the lowest common denominator of individuals into whom media fare is indiscriminately poured. As individuals, we attend to the media, reacting as individuals seeking to gratify our own special interests and needs. If hundreds, thousands, or even millions of us choose to react similarly, it is still basically an individual decision, even though it appears in the form of mass behavior.

Numerous mysteries surround the process and effects of mass communication. Social scientists and philosophers offer us many explanations of how and why the mass media operate as they do, and what effects they have on society. We devote a chapter to a historical tour of these mysteries. In bygone days there were simple, all-encompassing explanations of the media-societal interface. At mid-century researchers became tentative, recognizing the complexity and individuality of audience members. Recently, evidence is growing that the media do indeed have powerful effects, that they are influential factors in making life meaningful, giving us a sense of self, and shaping our agendas. Such effects are elusive and not easily measured in the laboratory, but that is no reason not to pursue them.

As one media scholar has explained, we shape our tools and then our tools shape us. Our hope in writing this first section of the text is that students understand their media environments well enough so they are not unwittingly controlled by those tools of communication.

Communication
and Audiences

Figure 1.1
The basic communication model.

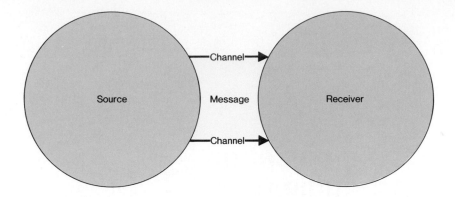

Introduction

The mass media are newspapers, magazines, books, radio, film, and television and its related technologies. They are generally divided into print media and electronic media. The print media are older, having developed over the last 500 years, while the electronic media are products of the twentieth century.

What do the mass media do in society? They give us baseball scores and tell us about the Middle East; they explain inflation, and they interpret current events. The media sell goods, services, candidates, and opinions. They make us laugh, they create drama, and they bring music into our lives. In short, they communicate.

In order to understand *mass* media and *mass* communication, we need an elementary understanding of the communication process—the events that define communication. That is the subject of the first part of this chapter. The model we will use (fig. 1.1) is later expanded to account for the elements of mass communication. In the final sections we will look at the functions of mass communications and the nature of audiences. ("Communication," without the "s," refers to the theory or theoretical process; "communications," with an "s," refers to the mechanical means by which communication occurs. Thus "mass media" and "mass communications" are synonymous.)

Elements of Communication

Communication has a variety of definitions.

No single definition of communication is agreed upon by all scholars interested in the subject; diversity abounds. Sociologists, psychologists, anthropologists, linguists, and speech communication specialists all offer definitions, some of which follow. Communication:

is the process of transmitting meaning between individuals;
is the process by which an individual (the communicator) transmits stimuli (usually verbal symbols) to modify the behavior of other individuals (communicatees);
occurs whenever information is passed from one place to another;

Communication—a
human stimulus-response
process by which we
inform, persuade, and
entertain.

is not simply the verbal, explicit, and intentional transmission of
    messages; it includes all those processes by which people influence
    one another;
occurs when person *A* communicates message *B* through channel *C* to
    person *D* with effect *E*. Each of these letters is an unknown to some
    extent, and the process can be solved for any one of them or any
    combination.

      For our purposes we will describe communication in simple terms as
s-r for the stimulus-response process, or the interaction between a source
and receiver. (Thus we have two referents for "s" and two for "r" in this

definition, which our examples and models will attempt to make clear.) Although scholars debate whether communication occurs between other species or even inanimate entities, we will limit our discussion to the idiosyncracies of human communication in general and its complex subcategory, mass communication.

Communication takes at least two entities: an event outside the individual (the stimulus) and the individual reacting (the response). Thunder and lightning striking fear in a young child constitute communication. The thunder and lightning are outside the child, and fear is the child's reaction. From this simplest example, another factor in the communication process can be inferred. The child is changed. The thunder and lightning have been programmed into his or her mental computer, and will remain there as a part of the child's experience. This is the effect of communication, and all communication has some effect, if only that of total boredom.

The two principal factors in the communication process are a *source* and a *receiver,* and both are required. A thunderstorm in the wilderness out of sight and earshot constitutes no communication by our definition (there is no receiver, no human effect).

Further, this simple s-r model points up that all communication is an individual, even a personal process. Thunder and lightning over a city strike fear into many hearts, but into each one separately, without reference to any other. Each child will react differently—some sobbing, crying, cowering, or even curiously watching out the window. Some, sensing no fear at all, will thrill at the sounds and sights of the natural fireworks. In this sense, the thunderstorm becomes a form of mass communication.

Individuals' reactions to different stimuli vary. The fire alarm means different things to the firefighter and the theatergoer. Feminism evokes diverse reactions from Gloria Steinem and Phyllis Schlafly. Human communication involves a *message*—fire or feminism—that is carried over a *channel,* whether alarm bell or magazine article. Human communication has a purpose; it does not rise out of thin air. In these instances, the purpose is to call attention to a fire (information) or to voice an opinion on a controversial topic (persuasion). That these messages were received differently is the crux of communication. Reception and, consequently, reaction and effect differ according to each individual's orientation. Orientation, in turn, depends on many factors that can be reduced to the individual's experience—the sum total of all that has gone before. Since a good deal of that experience has been communication, the complexities of the process and its circularity become apparent.

The Communication Model

Emerging from all this is a model of the communication process that consists of a *source* or *encoder* sending a *message* over a *channel* to a *receiver* or *decoder.* Communication shorthand for this is shown in figure 1.1.

The source has a purpose in trying to communicate: to inform, to persuade, to entertain. In order for communication to occur, there must be

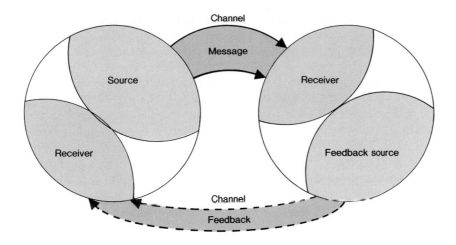

Figure 1.2
The interactive model of
communication.

some kind of effect on the receiver: a change in cognitive (thinking), af-
fective (feeling), or behavioral (acting) processes. This response or reaction
is seen as *feedback* and constitutes another element in the basic communication model. This *feedback* is basically another phase in the same process of transferring messages between sources and receivers. We see in figure
1.2 that the roles of sender and receiver have been reversed and a communication cycle begun.

The simplest illustration, of course, is an ordinary conversation. A
woman says, "Good morning." She is the source; her purpose is to establish
contact. The message is "Good morning," the channel is speech, and the
man to whom the message is sent is the receiver. This may seem a bit complicated for such a simple transaction, but if the sequence and ingredients
of this simple interpersonal communication are understood at the outset, it
will enormously simplify the investigation of the much more complicated
mass communication process.

Once the man has heard the greeting and responds, communication
has taken place. The effect is one of warmth and reciprocity, and he responds, smiling and saying, "How are you?" This is his reaction; it constitutes feedback to the woman and completes the simple communication
process.

Feedback Note, however, that both the smile and response of the man were
a part of the feedback. Note further that in replying, the man became a
secondary sender and the woman a receiver; thus, the model reverses itself
with his reply. Note finally that the man's answer partially determines what
the woman will say next. "How are you?" demands an answer: "Fine, thank
you," or "Terrible." The principle here is that feedback conditions the course
of future communication between the two by limiting the options available
to them.

Feedback completes the
communication cycle.

Interference There remains one more element in the basic communication process—interference. The technical term for interference is *noise,* and it consists of two types: channel noise and semantic noise. *Channel noise* is interference within, or exterior to, the channel or medium. If, for instance, the woman was seized with a fit of coughing when she said "Good morning," the man would have difficulty understanding her. The woman might have to repeat her greeting; in any event, there was channel interference in her speech. Alternately, if she said "Good morning" in a New York subway during rush hour, it is doubtful that anyone could hear her.

Channel noise can be corrected in two ways. First, the woman may try not to cough or sneeze, to use clear enunciation, and to speak loudly enough for the man to hear; in short, she may perfect the channel—speech. The other means is by repetition; if the man didn't hear her the first time, he may the second or third time.

Semantic noise, on the other hand, is more complex; it is interference within the communication process itself. If the woman greeted the man in a Chinese dialect, the man probably would not understand her if he spoke only English. This language barrier is the simplest example of semantic noise, but there are more prevalent and subtle forms indicated by differences in education, socioeconomic status, residency, occupation, age, experience, and interest.

Often we hear someone say "I simply can't talk to him." This is an example of semantic noise. These two individuals may be poles apart in their orientation, interests, backgrounds, and habits. "The generation gap" is a glib description of semantic noise provoked by age differences.

The solutions to semantic noise, as in the case of channel noise, are incumbent upon the sender. After all, her original purpose was to communicate. One solution is to try to communicate on the level of the receiver. Kindergarten teachers use simple words, short sentences, and brief lessons because they know their students' vocabularies are limited and their attention spans are short. These teachers are trying to eliminate semantic noise. In briefest summary, the solution to semantic noise is to appeal in terms of the receivers' interests.

Recognizing the semantic problems that can arise in one-to-one encounters, and the degree of interference that surrounds the simplest conversation, the scope of interference that is inevitable becomes apparent when such situations are multiplied by many in mass communication.

Mass communication occurs
when mass-communicated
messages are transmitted to
large, anonymous, and
heterogeneous masses of
receivers.

Mass Communication For our purposes a simple definition will suffice: mass communication is a process whereby mass-produced messages are transmitted to large, anonymous, and heterogeneous masses of receivers. By "large" we mean a larger mass than could reasonably be assembled in a single place in physical proximity; by "anonymous" we mean that the individuals receiving the messages tend to be strangers to one another and to the sources of those messages; by "heterogeneous" we mean that the messages have been sent "to whom it may concern"—to people from all walks of life, with unique characteristics, and not necessarily a single homogeneous type of audience. In this and subsequent chapters, we shall see that this definition, while useful, is somewhat oversimplified. Recent advances in mass communications technology, and highly sophisticated means of understanding and reaching out to special interest audiences, mean that the audiences/receivers are somewhat less anonymous and heterogeneous than in the past. Basically, the discrepancy is so vast between the single one-to-one situation and contemporary mass communications, often involving tens of millions of receivers, that the numerical differences become differences in kind. Further, these differences stretch across the entire model, showing various changes in the source, channel, receiver, and feedback, as well as in noise.

The basic model of communication also applies to mass communication. A sender uses a channel to reach receivers, which prompts feedback to the sender. The main difference between mass communication and the interpersonal model is, of course, the matter of multiple receivers. Sometimes they receive simultaneously, immediately, as in network television; other times they receive individually over longer periods as with a movie, or even over centuries, as with some books, such as the Bible.

    To make this distinction in the communication model, the sender is called the *source,* and the multiple receivers are called the *audience.* The channels, whether they be television, radio, newspapers, magazines, books, or movies, are known as *media.*

It should be obvious that audiences come in different sizes, from the forty million or so of a network television program, to the several thousands of an average book, to the few hundreds of a scholarly journal. Regardless of size, it is crucial to remember that each audience is composed of that many individual persons, each one a separate thinking machine reacting to the medium's message in a different fashion, viewing the message through his or her own separate lens, ground from personal experience and orientation. This individuality of audiences belies the concept of a single mass reacting as so many automatons.

    Multiple receivers of a mass medium may also react with one another. The members of a family make comments to one another about a television show. Scholars discuss articles from academic journals. "Did you see . . . ? Have you read. . . ? The papers said. . . . Have you heard. . . ? They say. . . ." The contents of mass media constantly become topics of conversation in daily life, and thus media influence is extended; indirect or secondary audiences may, in many cases, be far larger than the original audiences.

Elements of Mass Communication

Audiences

Audiences consist of individuals, not automatons.

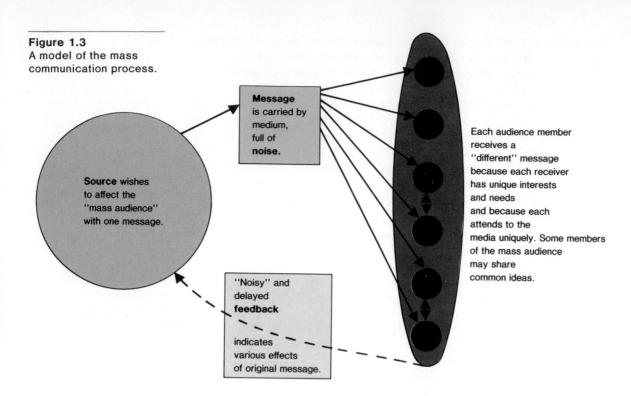

Figure 1.3
A model of the mass communication process.

Source wishes to affect the "mass audience" with one message.

Message is carried by medium, full of **noise.**

"Noisy" and delayed **feedback** indicates various effects of original message.

Each audience member receives a "different" message because each receiver has unique interests and needs and because each attends to the media uniquely. Some members of the mass audience may share common ideas.

Thus, it appears that the effect of mass media reaches far beyond the initial audience. This is a significant point, for it illustrates the catalytic nature of mass communication in triggering individual reactions. However, for now it is enough to adapt the basic communication model to accommodate these additional factors, as can be seen in figure 1.3.

Delayed Feedback

Delayed feedback is indigenous to mass communications.

The new model indicates some subtle changes. First, feedback in mass communication is rarely instantaneous and direct, as it is in face-to-face conversation. Rather, feedback becomes an aggregate ingredient reflected to the source after a considerable lag in time, often from great distance, and frequently in a different nature. Feedback from the issues and rhetoric of a political campaign will be reflected, often many weeks later, at the ballot box on election day. The appeal of a television commercial or a magazine advertisement will be known at the sponsor's cash register. The popularity of a movie can only be measured in dollars at the box office, and the success of a book generally by over-the-counter sales, both of which may involve a wait of more than a year. Delayed feedback is indigenous to mass communication.

As a result of technological advances in computerization over the last couple of decades, this delay has been shortened by cultivating some new forms of feedback. In a political campaign, often costing millions of dollars, it is tactically unwise to await the verdict at the polls, by which time strategies cannot be corrected. Public opinion polling has proven able to offer an indication of election day results to perceptive candidates as a guide in their campaigns.

Similarly, when advertisers spend hundreds of thousands of dollars for prime-time television commercials, they need to know far in advance whether this kind of investment will pay off at the cash register. This need has led to television and radio ratings. There is an empirical correlation between audience size and the subsequent sales of consumer products. Consequently, the audience size of television shows, as reflected in ratings, gives a reasonable clue to the relative success of programs and commercials.

Public opinion polling as a feedback device has another use, too. By indicating what is acceptable and unacceptable to different audiences, polls and ratings tend to condition the kind of campaign or the kinds of programs that will be offered in the future. The candidate who finds his or her corruption-at-city-hall issue falling flat will abandon it. The advertiser who finds a science fiction TV series less appealing than a dramatic program will try to switch to a "Dallas" or "The Cosby Show."

Channel Noise

Noise in mass communication is a mammoth aggravation. Within the media, channel noise consists of such things as typographical errors, misspellings, scrambled words, or omitted paragraphs in the newspaper. It is the fuzzy picture on the tube, static on the radio, or missing pages in a magazine. It is also a broken television set, a dead battery in the transistor radio, the Sunday paper in a mud puddle outside the door, or the magazine subscription that doesn't arrive. Obviously, the more technologically complex society becomes, the greater the opportunity is for this kind of channel noise. As the numbers, varieties, and complexities of media increase, the greater the likelihood is that we will be exposed to mounting noise.

Mass communication inevitably involves channel noise, or outside interference.

Since channel noise also includes outside interference, it encompasses such things as kids fighting during a television program, or visitors interrupting our reading. Other such interferences may be the persistent ringing of the telephone as we watch television, a teenager's stereo at full volume while parents are trying to read, competing programs scheduled in the same time slot, or a variety of magazines and media from which to choose. These examples show evidence that in many cases the media interfere with one another and constitute a considerable part of their own noise. As more media develop and become available, the problem will get worse.

One of the solutions for channel noise is repetition, and it is in use constantly in mass communications, especially in advertising. Disc jockeys repeat phone numbers; television commercials reappear during a program; and department stores advertise daily with multiple pages in both morning and evening papers. Repetition employs the law of averages. If the message was interrupted the first time—by the doorbell or by conversation—chances

Channel noise is lessened by repetition of messages, and attention to mechanics.

are it will not be the second or third time. Repetition in broadcast media offers an opportunity to reach those who tune in late. However, repetition operates on a law of diminishing return. There comes a point in repetition when the receiver, as an individual, will tune the message out. When multiplied by many individuals, the message is lost. Repetition must be used with discretion.

Another cure for channel noise is perfecting the channel performance. This includes avoiding static on the radio, prolonging the life of transistor batteries, proofing the typos and scrambled paragraphs in the newspaper, and cleaning up the fuzzy picture on the tube. These are rather obvious solutions, but accomplishing them leads in several directions.

Removing static on the radio, for instance, may require increasing the wattage of the station, which calls for approval by the Federal Communications Commission (FCC) and demands considerable capital investment. The long-life transistor battery stems directly from improved technology—a constant probing of new frontiers. Such technology already has developed transistors (in lieu of bulky vacuum tubes), printed circuits, and miniaturization, making the radio a personal, portable mass communications tool.

A fuzzy picture may demand a new picture tube at the owner's expense, but it also may require new cameras at the studio or a new control console or better engineers. Perfecting the channel on television can run all the way from improving the transmitter's output via intricate engineering through the quality of engineers, to the talent of directors, and the diction of the announcer.

The typos and misspellings in the newspaper demand better copy editors and proofreaders, of course, but they may also demand better computer operators, updated typesetting machines, or better-trained printing personnel. In the distribution system employed by newspapers, a part of channel noise will depend on the working condition of the newscarrier's bicycle.

Semantic Noise

Semantic noise is inevitable in mass communication. It is found in both sources and receivers.

Channel noise may be omnipresent and aggravating in mass communication, but no more so than semantic or psychological noise. As described earlier, semantic noise is interference within the communication process itself, within the human sources and receivers of the communication. In mass communication, this type of interference is inevitable, if only because audiences, individually and collectively, have so many different expectations of their mass media. Likewise, from the sources' perspectives, semantic noise occurs whenever the audiences do not appear to be receiving and responding to carefully crafted messages the way the senders had hoped.

Earlier, we listed some reasons for semantic noise in interpersonal communication: language barriers, differences in education, socioeconomic status, residency, occupation, age, experience, and interest. Semantic noise in mass communication differs both quantitatively and qualitatively from that in interpersonal communication. With so many people composing a mass audience, it is impossible to pinpoint a given message toward any specific individual's complete set of values, interests, needs, expectations, moods, life experiences, and language ability. However, mass media do try. That they succeed at all is because they employ simplicity and commonality in their messages, aiming at the lowest common denominators of audiences' values, interests, and so forth.

Consider news, for instance. "Journalese," the style in which the daily newspaper is written, is an example of simplicity. The most significant and most captivating facts are placed in the lead or top paragraph as attention getters. The "who, what, when, where, why, and how" (the five Ws and H) are addressed early in the story so a time-pressed reader can skim without missing the really pertinent data. The balance of the article is composed of additional data in decreasing order of significance, in an inverted pyramid of presumed interest and newsworthiness. By and large, the words used are simple and denotative, the sentence structure uncomplicated, and the paragraphs brief. A simple style aimed at a mass audience is designed to convey maximum information requiring minimal reader effort.

Broadcast media use a similar formula to minimize semantic noise. Television anchors headline their stories with simplistic, attention-grabbing devices ("Five area residents are dead following a blazing car crash; we'll give you the details after these commercial messages"). They then "back into" their story, using fewer words than the newspaper writer, but relying heavily upon graphic video to convey the total message. Frequently, reporters on the scene will wrap up their thirty- to ninety-second news item with a didactic "kicker," letting viewers in on the real significance of the story. The graphic video, the pace and tone of the report, and the minor editorializing at the conclusion are standard operating procedure, and are therefore part of the presumed contract between broadcast news producers and most consumers. The producers simplify the world's complexities, and package them so the news consumers' common interests and tastes are fulfilled. Indeed, the process has become so institutionalized that for many years America's best-known anchor, Walter Cronkite, concluded his CBS network newscast with a sonorous announcement that "And that's the way it is," followed by the day's date. Americans could then turn to their affairs, comforted by the revelation that the significant events of their day had been neatly packaged and delivered to them, somehow the nation had coped, and tomorrow would be another day. That, of course, was a myth, as could be attested by anyone who, in any given day, had experienced firsthand anything of significance—or heard or read anything secondhand—that had somehow been ignored by Cronkite and his CBS crews.

Other techniques of simplicity and commonality are found in mass communication. The clearest example is seen on network television, although it is by no means restricted to this medium. This is the lowest

Simplicity and commonality are attempts to overcome semantic noise.

common denominator, wherein television's programming content is theoretically aimed at that audience intelligence or interest level that will attract and hold the greatest number of viewers. Much of the criticism of television programming is directed toward a bland and sometimes mindless array of situation comedies, prime-time soap operas, law enforcement dramas, and celebrity specials, all interspersed with simplistic, catchy commercials. Whether this video diet is all that bad is speculatory, but it certainly is an example of purposeful appeal to the presumed lowest common denominator of a vast and invisible audience in an attempt to overcome semantic noise. Consider the semantic noise occurring when we're captured by a fictional dramatic story about nuclear war, AIDS, or teenage suicide, interspersed by commercial messages convincing us that the world's more serious problems include whether we can cope with spots on our china, those embarrassing rings around the collar, or toilet paper that is not squeezably soft. To confuse the reality of nuclear war, AIDS, and suicide with dramatic portrayals is one thing; to confuse minor household problems with items of substance is quite another.

The sophisticated means by which the mass media package reality, entertainment, and persuasion with simplicity and commonality would seem to break down semantic noise. That may be true if media satisfy the largest common collectivity of audience interests. But, as we shall see again and again throughout this book, media efforts to be all things to all people inevitably result in, at best, an incomplete and superficial delivery of the goods— i.e., news goods, entertainment goods, advertising goods, and cultural goods. Our discussions of news, entertainment, advertising, and media ethics are replete with examples of semantic noise—how readily communication backfires, or short-circuits, despite all efforts to the contrary. With this breakdown in communication, we can see what is known in the science of thermodynamics as *entropy:* the tendency of a system to move from a state of order to one of disorder or chaos.

Having said that semantic noise and entropy are inevitable in mass communication, it might seem appropriate for us to give up and condemn the media for arrogantly pretending they are successfully serving society's needs. We would do better to attempt to understand the numerous and subtle ways in which such breakdowns occur, and the means by which intelligent consumers and dedicated communicators can work to overcome—or at least diminish—the inevitable. As a start in that direction, we should consider the role and influence of media *gatekeepers,* those individuals who sift and sort through the available media fare and then package it for consumption.

Gatekeepers

Gatekeepers determine what audiences will read, see, or hear.

Gatekeeping has generally been associated with the news, specifically with newspapers. The editors are the gatekeepers of the newspaper. They determine what the public reads, or at least what is available for them to read. The events they bypass are events that never happened as far as the public is concerned. Society's exposure to the day's reality and fantasy is in the gatekeepers' hands. Theirs is a prime responsibility.

Editors constantly have an eye on the audience as they sort through the day's events. They tend to place emphasis on the unusual, the sensational, and the spectacular, as well as on the criminal and the deviant. These types of stories historically make good reading; subscribers like them. Within the severe space limitation in which editors operate, they sometimes find they must forego a story on zoning controls in favor of a gory three-car accident, or pass up a scientific breakthrough for an axe murder. This is because the number of pages available for news is determined by the amount of advertising that has been sold. Typically, only 40 percent of the paper can be devoted to news.

Before an editor gets a story, the reporter has already exercised a form of gatekeeping in the selection and presentation of facts. No matter how objective the reporter has tried to be, something of that individual and his or her orientation has crept into the story. No two reporters will write the same story; more broadly, no two observers will see the same thing. Thus, the public's view of an event will be colored to a degree by the kind of fact-finding glasses the reporter wore.

Editorial policy is also a form of gatekeeping. Different newspapers have different values. Two examples that come readily to mind are the *New York Times* and the *New York Daily News*. The *New York Times* prides itself on completeness and detail in its substantive reporting. It plays down the sensational and deviant in the interest of propriety, taste, and thoroughness. On the other hand, the *New York Daily News,* as a matter of policy, emphasizes the sensational, the odd, the different. Both approaches constitute forms of gatekeeping. While each newspaper deprives its audience of something, together they achieve a kind of imperfect balance for the residents of New York.

These examples illustrate how media tend to specialize in order to reach selective audiences. Gatekeeping activities reflect this specialization. Magazines and radio stations further emphasize this principle. Each has developed a format of appeal to a specific audience, and audiences differ.

Some stations play rock 'n' roll, some classical music, some "oldies but goodies." These stations are not free to depart from their format, except at the risk of losing their established audiences.

Magazines also have equally well defined formats to appeal to specific audiences: *Vogue* for the fashion-conscious, *Field and Stream* for the outdoor buff, *Cosmopolitan* for the unmarried working woman. Such audiences have grown to expect a certain point of view from these periodicals, and editors screen and prune all the available material to come up with the exact contents their readerships expect. There are no articles on fly-fishing in *Mademoiselle,* no economic forecasts in *Popular Mechanics.* Magazines and radio are as selective in what they present as the selective audiences they serve, and this is a form of gatekeeping.

Programmers of prime-time television wrestle with what to air and what kind of a balance to strive for. From hundreds of potential shows— serials, specials, dramas, situation comedies, police shows, mysteries, and news magazine shows—they must each screen out a dozen or so for evening viewing.

In television news, producers are affected by the limitations on time in the same way newspaper editors are affected by space. The result is highly fragmentary. Newscasters have little time for headline news representing perhaps no more than 2 or 3 percent of the total news of the day.

Even book publishers and movie producers who cater to a self-selective, numerically unpredictable audience have their gatekeeping problems. From hundreds of manuscripts, screenplays, and scenarios that come to their attention from authors and agents, solicited and unsolicited, they must select those to be published or produced. The public will never know of the remaining hundreds, the thousands in a year, that failed to be approved. For audiences, they never existed, dying before birth.

From all this two things become apparent concerning the gatekeeping function. First, it is limiting in that it restricts what the public is exposed to as an audience, whether in news, television programs, movies, books, or radio. Due to the marvelous diversity of the media, however, a certain balance of exposure is achieved in the aggregate from so many media catering to so many different audiences. Second, the gatekeeping function is subjective, personal. It is the judgment of a surrogate substituted for that of the audience, and is basically a professionally educated guess as to what the public will like or react to.

Functions of Mass Communication

Media inform, entertain, persuade, and transmit culture.

The mass media can be described as having four primary functions: (1) information, (2) entertainment, (3) persuasion, and (4) transmission of the culture. We refer to them here as the primary or most important functions, but there are, of course, others. As we shall see in chapter 2, and in later descriptions of individual media and the regulatory and ethical components of mass communication, several other types of media functions could be considered. For instance, the media serve an adversary/symbiotic function with government; they serve to help people cope with or escape from

The mass media have four primary functions: information, entertainment, persuasion, and transmission of culture.

their environment; they create new values and channel old ones; and they serve an important economic function by keeping the wheels of commerce turning. For each of these positive and intended functions, there may be an opposing or negative "dysfunction." Some examples of this are war nerves or unrest among citizens who are constantly told of crises among governments, or anti-social, greedy, or self-destructive behavior among some who have experienced entertainment programming or commercial messages originally intended to bring diversion or pleasure.

In individual communication, functional gears can be shifted at will. We plead, assert, instruct, joke, question—moving from one mode to another as situation and inclination demand. In mass communication, altering the primary functional thrust of a medium is somewhat more difficult.

This is because the highly organized, institutionalized nature of the mass media creates a ponderous inertia resistant to change. Like living organisms, the systems making up the media are in delicate balance, and even slight adjustments can have quite an impact.

Audiences have grown to expect one format—or a certain combination of formats—from each medium, and a departure deceives audience expectations and threatens profit. The thrust of the *National Enquirer* is basically entertainment, providing sensation, spice, and escapism. The concern of the *New York Times* is information, to be "the newspaper of record." Drastic changes in these—or any—media's basic functional mix are seen as indications of serious identity crises. In such cases, the death of such a medium is more likely than metamorphosis or transmutation.

We will take up each basic function in turn, pointing out some overlapping of the functions and problems that arise when the functions are confused in the minds of the gatekeepers and media consumers.

Information

When we think of the various functions of the mass media, the information function frequently comes to mind first. Information is the easiest of the functions to identify, because it comprises a part of each, and the most prominent form of information is news.

The emphasis on news has camouflaged the fact that 60 percent of the average newspaper is advertising and that a considerable portion of what is left over after the ads have been inserted is entertainment of one sort or another, starting with the comics and ranging from selected features, editorials, and columnists to a variety of sensational and human-interest stories. However, the basic thrust of the newspaper remains informational. That is what its audience expects of it. The *New York Daily News* satisfies its information function, for example, by finding a high quotient of sensational material—such as violence and sex—and presenting it with proper news leads, written in a punchy style. Whether or not this constitutes the responsible exercise of the information function will be discussed more fully in later chapters.

Print media have a strong informational thrust.

As adjuncts to mass media, the wire services (particularly the Associated Press [AP] and United Press International [UPI]) have the highest information content. Their business is selling information, specifically news of current events gathered worldwide. Further, their customers or clients—the newspapers, television networks, and individual television and radio stations across the nation—represent such a broad spectrum of approaches, interests, formats, and editorial policies that the wire services tend to rely on straight news—an objective, unadulterated informational approach that avoids all attempts to color the news. The wire services leave it to their clients to choose specific items from those offered and to season those items as they see fit.

Textbooks, making up about half of the book publishing industry, are expected by their publishers and readers to consist primarily of information. Their voluntary readership is slight, and they are generally read at the direction of an instructor. For this reason publishers frequently attempt to interject humor (entertainment) as a relief from the heavy dose of information. Not incidentally, publishers also hope to sell a lot of books in order to make a profit. The other branches of book publishing—trade books, fiction, and nonfiction—are freer to depart from pure information as their thrust, but their information function is still relatively high, covering a wide variety of topics from which readers select and choose at their own discretion.

Although television's primary function is entertainment, it does include some information. There are regularly scheduled newscasts that tend to take on entertainment overtones. News commentators are not so much in competition with each other as they are with other prime-time personalities. Their formats are doctored to move quickly and dramatically regardless of the significance of the information they are presenting.

Documentaries also take on a dramatic quality and sometimes, but not always, develop a point of view that: (1) is designed to appeal to the presumed taste of their massive audience, and (2) is not necessarily objective in its analysis. Award-winning and highly rated "60 Minutes" is probably the most obvious example. The CBS team uses such dubious techniques as that of accosting unsuspecting news sources in the street with cameras rolling and choosing controversial story topics that are intended to raise ratings.

Television cannot be entirely blamed for cursory treatment of serious topics. Profit is the motivating force. Television's ratings indicate massive regular departures of viewers whenever a documentary appears. For instance, would you be more interested in watching an hour-long program on industrial waste or a network showing of a Hollywood blockbuster?

Entertainment

The broadcast media—radio, television, and film—have a basic, although by no means exclusive, entertainment thrust. Film is included in this category because, although there are differences, film is such a large part of contemporary television that one cannot realistically be considered without the other.

That the broadcast media are intensely, purposefully, and enthusiastically entertainment oriented is fairly obvious. Also apparent is the persuasive, commercial aspect of television. Anyone who has had the misfortune of being hospitalized for a week or so can testify to the unremitting and highly imaginative diet of entertainment and persuasive fare that daytime television offers, with its fantasies of soap operas, game shows, old movies, and reruns. Prime time (during the evening) is a wonderland of scheduled police situations, private eyes, hospitals, situation comedies, personalities, serials, and premiere movies that move inexorably in the direction of sex, violence, and deviance.

The Public Broadcasting Service (PBS) is confirmation of broadcast's need to entertain. PBS—supported by government funding, private donations, and corporate underwriting—has no commercial requirements

Broadcast media tend to be entertainment oriented.

to make money or show a profit. Devoid of these commercial pressures, it is also free to offer programs of consequence (culture, education, documentaries), and it presents these to almost always smaller audiences than the commercial networks would tolerate. There is a serious question about how much incentive exists either to produce or to watch a non-commercial medium in a commercial society.

The largest portion of radio is the same mixture of entertainment and commercials that television offers. Lacking television's video quality, radio must concentrate on what it does best—appeal to the ear—and this generally means music. The spectrum of music offered by radio is impressive. Station by station, radio has selectively carved out a segment of the audience to which it appeals, and station by station, it continues to move goods to these audiences: components to stereo enthusiasts, blemish ointments to teenagers, and annuities to the affluent. Radio's audiences are far more diversified than television's, and its production costs are far less. This permits it to specialize toward selective audiences, and such specialization may take a form other than music. Some radio stations have found their niche in broadcasting a series of constantly updated news bulletins, emphasizing information in a largely entertainment medium.

Film has shown a gradual metamorphosis in recent years. In its heyday during the 1930s and early 1940s, it was essentially a mass entertainment medium, playing in baroque palaces to large audiences. Now it has moved under television's competitive pressure to be a more expressive medium, freed from the tyranny of appealing to the tastes of the lowest common denominator. Further changes are starting to occur with the growth of the home video market.

Although most of film still has a basic entertainment thrust, particularly that portion serving the television industry, two other facets of filmmaking are becoming evident: (1) a distinctly persuasive-informative orientation in industrial or commercial films, and (2) a self-conscious role as social critic in the so-called popular movies that typically play to smaller, more selective audiences.

While the broadcast media and film have been identified as the basic purveyors of entertainment, this function is always intermixed with others. A high degree of entertainment pervades all mass media, often serving as the vehicle for more serious functions.

The entertainment function of the media requires an extraordinarily affluent society to support the level inherent in American mass communication. When considering that mass communication is time-consuming, and that a substantial portion of it is pure entertainment, as evidenced by television, we become aware of the prerequisite leisure and affluence of our society that enables us to spend so much of our time unproductively. But this makes a certain amount of sense in our democratic society when we figure that the average consumer of mass media is consciously seeking to fulfill personal needs and interests.

The persuasive function of the media in contemporary society is as significant as the information and entertainment functions. Advertising, of course, is its most apparent form, but there are other more subtle manifestations of persuasion that are likely to have lasting effects on the future of mass communication. The one-hundred billion dollars spent annually on advertising is only a portion of the amount spent on American mass media persuasion. Public-relations activities, special promotional events, and blatant as well as subtle efforts at image manipulation and public-opinion formation pervade the media environment. Editorials, letters to the editor, and opinion columns are obvious examples of overt persuasion; the subtle ways in which individual, social, and economic values are reflected in news columns, cartoons, music videos and other entertainment programs are something else. Because we recognize advertisements and editorials for what they are—blatant persuasion—we give them their proper due. But how do we cope with subtly implanted commercial products such as brand name beer, candy, cereal, or household products in our Hollywood films? And do we tell our children that their favorite Saturday morning cartoons are nothing but lures to bring them into the stores to buy the commercial products those cartoon characters represent?

Much of the persuasion in mass communications is concealed. Any public-relations practitioner can testify to the fact that a considerable portion of what passes for news in the media has a persuasive origin and an ulterior purpose. Much of what the public reads, hears, or watches in all the media is designed to influence in one way or another.

Political campaigns, which periodically command vast attention in the mass media, are almost pure persuasion. Much of governmental news at all levels has a propaganda base as government seeks to declare or justify its actions in a democratic society. A good part of business and financial news is advocacy. In today's environmental and consumer-conscious society, business is increasingly under attack and seeks to utilize the mass media in defense.

The doses of persuasion masquerading as information in mass communication are huge, and inevitably the functions have tended to merge, obliterating distinctions. This in turn leads to a credibility gap, because what appears to be bona fide news repeatedly turns out to be political or commercial advocacy, or is flavored by newspaper bias or television distortion.

Most of American mass media are supported by advertising in one way or another; commercial radio and television are 100 percent so. Newspapers and magazines in varying degrees rely heavily upon advertising revenues. The price of a newspaper does little more than cover distribution costs, leaving all editorial costs, all production costs, and all profit to be borne by paid advertising. Magazines vary widely in their use of advertising, but generally at least half their revenue, and hence, all their profit, comes from advertising.

Advertising is an obvious form of persuasion, but much of the persuasion in mass media is concealed and subtle.

It is obvious that television is not all entertainment. The schedule is regularly interspersed during daytime and evening hours with a mosaic of commercials touting used cars, beers, cosmetics, household sprays, fast-food restaurants, intimate hygiene products, major appliances, and ball-point pens. Some of the commercials, we might note subjectively, appear better than the surrounding programming. That advertising should conform to or even surpass the format of the medium is not at all surprising.

In fact, a good case can be made that the role of advertising agencies in mass communication industries is to inject entertainment into commercial persuasion, lest the public's attention, subjected to unrelenting exposure to so many sales pitches, begins to wane and thus defeat the advertiser's purpose. In any event, it becomes apparent that the purpose of television programming is to provide a vehicle for the commercials, to deliver customers to the advertisers.

It is also significant that, although only 10 percent or less of the information or news that is available to media reporters and editors eventually appears in the news medium or newscast, nearly all of the available advertising is published or broadcast (with only minor exceptions, such as ads that are obscene, for illegal or overly controversial products, or that simply cannot be squeezed into the available time and space). This may offer a commentary on the relative values placed on advertising and information in a commercial society.

We generally think of entertaining feature movies when we consider films, but of the sixteen thousand films produced each year in the United States, only 150 or so are for theater distribution. A highly profitable aspect of filmmaking is concerned with persuasion and information. Commercials are an obvious example. In addition, there are training, educational, and institutional films: travelogues, and driver-training, how-to-do-it, and sales-orientation films. Like textbooks, these films have limited appeal and are shown for specific reasons before captive audiences. They constitute a large portion of filmmaking, easily the greatest number of new films each year. But they lack the public exposure of either paid-admission films or television's "nights at the movies."

Film today shows considerable persuasive and informational content. It begins to lay serious claim as a prime medium of cultural transmission, recording and playing for inspection the triumphs, failures, and foibles of society. A good deal of social concern has proven profitable, and movies appear to have capitalized on this.

Transmission of the Culture

Cultural transmission is a widespread but little understood function of the mass media.

Cultural transmission is one of the most widespread but least understood functions of mass communication. Cultural transmission is inevitable, always present, for any communication has an effect on the individual recipient. Thus, any communication becomes, if ever so slightly, a part of the individual's experience, knowledge, and accumulated learning. Through individuals, communication becomes a part of the collective experience of groups, publics, audiences of all kinds, and the masses of which each individual is a part. It is this collective experience reflected back through communications forms, not merely in the mass media, but also in the arts and sciences, that paints a picture of the culture, of an age, of a society. Heritage, then, is the cumulative effect of previous cultures and societies that have become a part of humanity's birthright and being. It is transmitted by individuals, parents, peers, primary and secondary groups, and the educational process. This cultural communication is constantly modified by new experience.

Thus, cultural transmission takes place at two levels: the contemporary and the historical. These two levels are unseparated and constantly interweaving. Furthermore, the mass media are major tools in the transmission of the culture on both levels. On the contemporary level, media constantly reinforce the consensus of society's values, while continually introducing the seeds of change. It is this factor that leads to the enigma surrounding mass media; they are simultaneously the conservator of the status quo and the vehicle of change. Television, for instance, is both mirror and molder of the times. As television programs and original movies increasingly show previously taboo themes, such as nudity and sex, they reflect a change in the social structure—a change that television itself may be partially responsible for causing. The process is no less true of other media messages, even those primarily informational or persuasive.

Some records left by the mass media are conscious and intentional; others seem incidental but are nevertheless important. The market ads, for example, and white goods sales in daily newspapers are good indicators of contemporary living standards and of the tastes and values of society. Anthropologists of the twenty-second century, in searching for a full record of twentieth-century culture, would do well to study the cultural records transmitted in the catalogs issued by Sears, J. C. Penney, and Montgomery Ward. Although some may question the accuracy of television's portrayals of everyday life, this medium too collates and transmits today's cultural values. It must, because it is a commercial medium.

To understand the process of cultural transmission, indeed of communication itself, it is worthwhile to peer back as far as possible into prehistory.

Two things about human communication are unique. The first is what, in his theory of general semantics from *Science and Sanity,* Alfred Korzybski called humans' "time-binding" ability based in memory.[1] *Homo sapiens* alone, of all the creatures on earth, has been able to consciously store its experiences and pass them along from one generation to the next. Thus, the progress of the species has been more or less constant. This ability has led to cultural transmission as a function of the media, and to the entire institution of education, so much a part of this function. Nor should it be forgotten, particularly in today's deafening competition for attention, that only a part of this education is a formal, in-the-classroom, from-the-textbook education; an enormous amount of it is acquired willy-nilly from the mass media (most often, television) as they transmit their version of contemporary culture.

Historically the human race has been able to continually draw on the past and add new experience from the present to guide the future. Not only have humans been able to accumulate experience, but they have proven themselves able to sort and sift among these memories, discarding the un-needed and ordering the rest for ease in transmission both to their fellows and to posterity. It is this process that prunes knowledge from raw experience. With other species that lack the time-binding ability, each new generation starts more or less where its predecessor did and finishes at roughly the same state of development that all previous generations did, subject only to the ponderous process of biological evolution. In other words, all other species are somewhat static in time, whereas humans collectively and consciously determine their own future. Elephants of the twenty-first century, for all their intelligence and longevity, will be about the same as they are presently. However, it is safe to predict that humans will be substantially changed—socially, politically, economically, and technologically—and that the mass communications network serving the third-millenium culture as its central nervous system will be radically different both in cause and effect from that of today.

Related to all this, possibly a cause of it, is humanity's other unique distinction—its ability to deal in abstractions, to let symbols stand for things, thoughts, events, states of mind, and even for emotions, for very complex processes indeed. For example, human communication itself is an abstraction, requiring tools for the transfer of meaning. Sets of symbols, called "codes," or more simply, "languages," are employed. Most familiar to us are verbal symbols, such as the spoken language. However, they probably are no more important to successful communication than are the entire range of nonverbal symbols such as body English, eye contact, mime, music, art, and graphics.

Words are symbols of things, thoughts, and emotions. No symbol is exact; it is, after all, only the attempted portrayal of reality, and "the map is not the territory," as general semanticists say. We all know what "flower" stands for—or do we? Since people's experiences with them differ, which flower does "flower" bring to mind—a rose, a violet, daisies in the field? Even narrowing the categories, does "rose" bring to mind the flower picked by young lovers, the one sent by an individual seeking forgiveness, or the one on a casket?

People think in words, verbally, and words mean special things to different people. But some words are more distinctive than others. To describe this relative difference, consider denotative and connotative words. Denotative words mean pretty much what they say they do. They are fairly explicit and have a general commonality, with little variation in meaning from

Humans are unique in their ability to "bind time" and use abstractions.

person to person. "Bookcase" is a good example; "blackboard" is another. Wars are not likely to start over different interpretations of such words.

Connotative words, on the other hand, are less explicit; they imply rather than denote. They are far more abstract, referring rarely to things but rather to thoughts and abstract concepts such as justice, patriotism, love, beauty, truth, freedom, and courage. There is little commonality of acceptance in the meaning of these words. What is strategic defense? What is a peacekeeper missile? What is equal opportunity? What is the right to life? What is newer than new, whiter than white, fresher than fresh? The answers to these questions depend not only upon whom we ask, but the way in which we ask and the conditions under which we ask. Yet this is the stuff of which mass communication and public opinion are made, and, as we well know, the stuff over which battles are fought. Language and abstractions—the uniquely human tools mankind has designed to separate itself from other species—divide individuals and societies just as readily as they bind them together. As noted earlier, this semantic noise is an inevitable component of the mass communication process, just as it is a prime ingredient in cultural transmission.

Manipulation of language—verbal and non-verbal—in transmitting culture is the media's stock-in-trade. Such manipulation is not limited to the persuasive arts of advertising and public relations. It may be more readily observed in persuasion, but is no less influential when employed in informational and entertainment contexts. Two prime ways media use to bind culture together is through use of stereotypes and myths.

Manipulation of verbal and nonverbal language in transmitting culture is the media's stock-in-trade.

Stereotypes and Myths In this complex society, with its mounting competition for attention, rising decibels of noise, confusion, and accelerating pace, individuals live vicariously and, for the most part, experience their world only indirectly through the mass media. Reality, in large part, is what the media say reality is. In this situation stereotypes and myths provide some perceptual shortcuts to understanding. They are an economic means of ordering confusion, saving both time and labor; they are a useful mental filing system permitting an individual to sort and store experience with minimal effort.

Stereotypes and myths provide some perceptual shortcuts to understanding.

Consider the nature of stereotypes. When some people hear "yuppies," it immediately invokes media-enhanced images of upwardly mobile baby-boomer professionals who dress casually but expensively, eat at fine restaurants, drink vintage wines, and talk about little other than work, money, leisure, and people. The stereotype also works in reverse. When those people see an under-forty man or woman driving a sporty car or engaging in any of the leisure time activities typically enjoyed by "yuppies," they automatically think "yuppy," and rightly or wrongly, that individual has been tucked into a pigeonhole of understanding. People carry all sorts of stereotypes around with them pertaining to politicians, absentminded professors, hookers, bankers, ditchdiggers, and rednecks.

One of the problems with stereotypes is that they may be in error. For decades, Stepin Fetchit was the stereotype in white America of the black male: a shiftless, subservient, comically ignorant black man of uncertain age. It was wrong, but it was convenient and it fitted well with the prevalent racial viewpoint. Thus, stereotypes, for all their usefulness and economy, must be used with great caution lest they help paint an erroneous picture of the world, and inhibit rather than assist understanding.

Myths are related to stereotypes; indeed, they are institutionalized stereotypes. They generally refer to beliefs and situations rather than to people. There is the myth of the power of the press, or the myth of women's superior intuition, both still prevalent. One thing noteworthy about stereotypes and myths is that generally there is a sufficient modicum of truth in them to make them believable. This credibility, once established, holds them over long after the original model passes away.

Like language itself, stereotypes and myths exist by consensus because a sufficiently large portion of society finds them to be a convenient shortcut to deeper, more analytical thought that also suits its particular world view. For the same reason, mass communication is replete with myths and stereotypes. In a thirty-second commercial, the little pictures of the good life in a viewer's head are invaluable aids in creating wants and desires. Network television and popular movies are full of white hats and black hats, good and bad personified for the audience in the quaint morality plays that compose most television series and many popular films. In a sociological and psychological sense, there really is little difference between the stereotyping and mythmaking of "Dallas," "Falcon Crest," "Days of Our Lives," "The Cosby Show," "Miami Vice," *Rambo,* or *Rocky.*

News reporters similarly use stereotypes and myths as a kind of shorthand for their readers. The headline "Yuppies Influence Election" creates a legend of impressions in only three words without saying much at all about what happened. Yet somehow people are deceived into thinking that they know all about it, enough at least to repeat it to a neighbor ("Say, did you see in the paper where the yuppies have had a major effect on the outcome of elections all over the country?"), and the stereotype passes on through cultural transmission.

As shortcuts to emotion, as well as shortcuts to understanding, myths and stereotypes are put in the hands of mass media gatekeepers for exploitation, whether to further a cause, to attract attention, or to appeal to the known prejudices of their particular audiences. Most likely, the use of stereotypes and myths may simply reflect the subconscious attitudes of a particular gatekeeper, or reflect the gatekeeper's views of audience orientation. They are not used perniciously to manipulate the unsuspecting public. Again and again in this text we will come back to the point that a successful media system in a democratic society depends upon audiences' selection of media fare that is consistent with their own values, beliefs, needs, interests—and, indeed, stereotypes and myths.

A good case can be made for the fact that it is the degree to which the mass media fulfill the mythology of their particular audiences that determines the media's relative success in the marketplace. Cumulatively, the process results in cultural transmission, the most generic function of mass communication.

Understanding Audiences

In many discussions of mass communication, as John Merrill and Ralph Lowenstein point out in *Media, Messages and Men,* scant attention is paid to the audiences of the mass media except to acknowledge their presence.[2] The mass media are tangible, audible, visible, and treatable—the institutionalized product of a corporate society. Consequently, it has been far easier to examine them in operational terms than it has been to probe audiences in terms of reaction and effect. Audiences are vague will-o'-the-wisp conglomerates composed of different individuals at different times.

Though hard to define, audiences do exist in the plural. The best way to describe the range of possible audiences is to visualize a giant funnel with the largest audiences at one end. These huge audiences are served best, although not completely, by national network television, a medium frequently criticized for its natural tendency to program for the lowest common denominator. The funnel narrows as it includes in turn various publics, groups, and associations, all of them collectivities of individuals defined in geographic, demographic, or functional terms. Finally the mass media funnel culminates at the individual. This "single" audience is, of course, served by telephone and the mails and, most importantly, by face-to-face encounters with other individuals.

We often overlook the simple fact that audiences are comprised of individuals.

It is the individual who composes different audiences. It is this single individual, not a crowd or group, who casts a ballot in the sanctity of the polling place, who passes dollars across the counter, who buys an admission to the stadium or the theater, and who holds opinions and beliefs that, with others, will be translated into some sort of temporary and incomplete consensus called public opinion. Further, it is the individual who subscribes to a newspaper, buys a magazine, reads a book, laughs at a TV comic, or listens to the news on the way to work. In this technological age and with a social emphasis on size and quantitative measurement, attention has perhaps been too preoccupied with mass audiences.

Mass media audiences are not necessarily permanent. Ordinarily they ebb and flow, changing as individuals move in and out of them. Audiences are comprised of individuals, each a unique member of society. Membership is voluntary for the most part, and an individual joins these different media audiences to satisfy personal needs for information, entertainment, professional advancement, diversion, and so forth. It is the individual who makes the choice, and in recent years the media have come to understand and accept that reality.

How Audiences Affect Media

In addition to size, anonymity, and heterogeneity, there are other characteristics of audiences that affect how the media carry out their functions. Among them are longevity, specialization, and audience expectations.

Audiences come and go, and media effects vary accordingly.

Longevity Television's prime-time audience on any given evening will come and go over a three-hour period, while the audience for any given program will last only an hour or so. On the other hand, a popular movie may take a year to reach its audience in the thousands of theaters spread throughout the land, and a best-seller may take a couple of years or more to sell its quota. In effect, the more durable the audience, the more concrete the effect of the medium. While the impact of a book or film can be substantial and sustained, the specific effect of radio and television programs is usually short-lived.

Specialization For a person to say that he or she is a Republican is no longer descriptive enough. There is an Eastern liberal wing and a conservative wing, there are moderates, and there are members of the John Birch Society. What used to be considered a cohesive women's audience is now broken into at least four subgroups: the young working woman, the traditional housewife, the professional, and the feminist. Large, monolithic audiences are becoming more fragmented for several reasons, one of which is an increase in knowledge.

Increased knowledge has made general practitioners of medicine almost a thing of the past. Their place has been taken by specialists—neurosurgeons, radiologists, opthalmologists, and so on—unheard of a generation or so ago. In profession after profession, increasing specialization to meet the demands of society has evolved. As these professions specialize, they tend to develop their own languages, techniques, and body of knowledge, which not only demand their attention but require media to feed them.

A second reason for specialization is increased leisure time. Shorter work weeks give Americans the opportunity to explore new interests and hobbies—from cabinetmaking to boating. Others find satisfaction in matters of the mind that may be satisfied by a magazine such as *Psychology Today,* which caters to an audience of educated young liberals who have discovered fascination in the world of human behavior.

Personal mobility is a third reason. The growth of suburbs following World War II focused interest in suburban affairs and led to their own newspapers and even some FM radio stations. The huge metropolitan dailies lacked the physical resources, personnel, and space to adequately cover the civic, social, and governmental activities of the communities on their periphery. Nor could local merchants afford the space and time rates for the larger metropolitan media; these rates were based on wide-area audiences, only a fraction of which a local merchant could ever hope to serve. The answer from both an information and advertising standpoint was smaller, localized media serving a specialized audience.

Audience Expectations Audiences, regardless of size, are not simply passive receivers of media; rather, they are active participants in the dynamic process of mass communication. Each audience determines within certain limits exactly what kind of material it will consume. Media gatekeepers attend to audience expectations by devising their programs' content mix, or specific titles that, in their professional judgment, will best appeal to the largest numbers of the particular audience they are trying to reach.

This occurs in mass communications in the same way that it does in merchandising. The variety of clothing, furniture, housewares, and sporting goods available in the local shopping center is prescribed by the people who shop there. The mass media are like the stores on the mall, from the enormous chain stores to the tiny boutiques. Each store caters to certain customers in groups of various sizes, and is available to customers who wander in and out more or less as the mood strikes them. Like mass media, shopping centers are social institutions whose gatekeepers, called buyers, are guided by management policy as to what public they will cater to, and by that public as to what they will buy. Sometimes the gatekeepers are successful and sometimes they are not, but if they are to stay in business, they have to be responsive to the interests, needs, and expectations of their customers.

Increased knowledge, leisure time, and mobility have brought about specialized media.

Audiences are active, not passive, participants in the mass communication process. Media are like shopping malls, and audiences are their customers.

Search for Specific Audiences

Media reach homogeneous target audiences by either specializing in whole or in part.

Throughout much of the present century, many media have survived by seeking smaller and more specific audiences. Some media have specialized in whole—an entire magazine or broadcast station gearing itself to reach a singularly homogeneous audience. Others have recognized the value of partial or internal specialization—sections of a magazine or newspaper, or specific programs on the broadcast station—attempting to reach an audience that is homogeneous on both demographic (age, sex, income, education, etc.) and psychographic (values, needs, beliefs, interests, etc.) levels.

For example, radio, which could have died when television was born, not only survived, but prospered. Today it is far healthier and wealthier than it was when it was America's preeminent mass medium during the depression and World War II.

Television took radio's evening prime time and usurped radio's national advertisers and network operation. Like radio, television was enjoyed in the home, but it had the visual dimension of pictures and motion. On the basis of its ability to command the involvement of more senses for a more comprehensive communication experience, television should have killed off radio.

However, radio found a new home and a new prime time, which television couldn't match—in the automobile when America goes to and from work each day. It also found new local markets, and it found music and teenagers and fed one to the other in a true symbiotic relationship. Under technological pressure, radio became portable, and under economic pressure, it became local. It created the disc jockey as a personality. It diversified, with each station picking a segment of the total audience for its own. It fragmented, concentrated, specialized, and in so doing built up new audiences with transistors to their ears.

Radio's survival is a fascinating case study, but no more so than that of the magazine industry. A generation ago, general interest magazines with circulations approaching ten million set the standard. When television came along and stole their national advertisers, the behemoths collapsed, but like the phoenix, reemerged in a new and more vigorous form. The specialized magazine of today is the standard of the industry, with its capacity to pinpoint audience expectations and marry those expectations to the sales pitches of eager advertisers. Nowhere else has the de-massification of media been as obvious as in the special interest magazine, which has narrowed its audiences, researched them more carefully, and pandered to their interests as defined by demographic and psychographic surveys.

Radio, magazines, and, to a lesser extent, film and cable television have begun to move away from trying to be all things to all people. The "shopping center" analogy referred to earlier may be buttressed by a major chain store or two (in terms of media, the chain stores are "network television," or "best-selling paperback book," or "block-buster motion picture"). However, all the shops around that major chain store are increasingly the tiny boutiques of special format radio stations or customized magazines, carefully targeting their customers and positioning themselves for greatest impact on the market. In instance after instance, specialization under pressure has restored obsolescent media as they concentrate on that unique combination of media functions they are best able to perform. In the structure of American society each media form finds its own place where its particular formula of characteristics ideally makes it suitable to a particular grouping of people—its audience.

In this process, today's mass media audiences are slowly but significantly becoming smaller than before, less anonymous to the producers and to each other than before, and less heterogeneous than before. It may mean that there is getting to be less systematic and universal transmission of culture, sharing of specific values, information, entertainment, and persuasion. Indeed, the time may be fast approaching when a new definition of the mass communication process and the nature of the audience will be called for.

As audiences grow smaller, less anonymous, and less heterogeneous, it may be time to redefine the mass communication process.

Summary

Human communication requires a source and a receiver. The source may be human or it may be inanimate, but the receiver must be human, and if communication takes place, the receiver will be changed in some way. This can be described as a stimulus-response relationship between the source-encoder and receiver-decoder. The response of the receiver is called feedback; it will affect the course of future communication.

Interference, known as noise, will be either channel noise or semantic noise. Channel noise is interference within or external to the medium, such as the screeching sounds of the subway that prevent conversation. Semantic or psychological noise occurs when the meaning of the sender's message is not clear to the receiver, or when communicators do not share values, interests, attention levels, or the like.

Mass communication can be defined as a means whereby mass-produced messages are transmitted to large, anonymous, and heterogeneous masses of people. (*Mass communication* refers to the theoretical processes, whereas the term *mass communications* is used interchangeably with *mass media,* or the vehicles conveying that communication.) The model used for mass communication is similar to the model for basic communication, except that the receivers are plural and the feedback is delayed. Channel and semantic noise are particularly serious problems in mass communication. Mass media personnel constantly try to improve upon channels used to mechanically transmit messages, and gatekeepers utilize simplicity and commonality to reach the lowest common denominator of audiences. Despite their best efforts, and the efforts of audiences, entropy occurs as the communication system tends to move from order to disorder. Although gatekeepers are not illustrated on the model (fig. 1.3), they serve the important role of determining what will be printed, broadcast, or produced and made available for consumption. Editors, reporters, television programmers, and movie producers are examples of gatekeepers.

The four primary functions of the mass media are (1) information, (2) entertainment, (3) persuasion, and (4) transmission of the culture. These functions overlap at times, but each medium tends to emphasize one over the others. The format and functional thrust of each medium is somewhat resistant to change. The print media tend to emphasize information, the broadcast and film media, entertainment. Advertising is the most blatant form of persuasion, but much of the persuasion in mass media is concealed, and some is passed off as news. Finally, the media transmit the culture by preserving a record of events and by noting changes in the social structure. They bind time, frequently manipulating language and using stereotypes and myths in the process.

Audiences for media are not monolithic. Audiences change and affect media by their longevity, their tendency to become specialized, and their collective expectations. Although media measure audiences in terms of size, all audiences are made up of individuals. Media have survived by specializing and seeking more specific audiences.

Notes

1. Alfred Korzybski, *Science and Sanity,* 4th ed. (Lakeville, Conn.: Institute of General Semantics, 1962).
2. John C. Merrill and Ralph Lowenstein. *Media, Messages and Men: New Perspectives in Communication,* 2d ed. (New York: Longman, 1979).

Mass Communication Theory

2

Introduction

Theory is the end result of
long, careful analysis of
variables. But theories are not
universal laws.

In the first chapter we were introduced to some basic characteristics of the communication process and the nature of mass media and their audiences. It is appropriate to turn our attention to some important theoretical issues centering on the effects of mass media in society. In order to fully appreciate how social scientists have come to understand what the media do to us and what we do to the media, we will take a historical journey for the next several pages, observing the development and refinement of mass communication theory throughout the current century.

Theory doesn't develop full-blown from the pens of casual observers. It is one of the end results of long, careful analysis of many variables. Theories are not universal laws explaining in complete and unquestionable detail all the causes and effects of phenomena being studied. Rather, theories are systematically related generalizations, calling for further analysis of phenomena and variables. In the case of mass communication scholarship, theory emerges from social scientists who have coupled the tools used in such fields as psychology, sociology, anthropology, history, and economics. Intellectual kleptomaniacs abound in mass communication research, as scholars borrow sometimes indiscriminately from this wide variety of disciplines. This has been both an asset and a liability. It is an asset in that mass communication theory is broadly based and exciting; a liability in that it just might be too easy for sloppy, overly eclectic scholarship to serve as the foundation for what passes as substantive theory. For those reasons, our trip through some of the significant theories about mass communication will be tentative, as we point to highlights and milestones along a path whose ultimate destination is still a mystery to us.

The chapter is divided into three sections, reflecting the three dominant thrusts social scientists have taken. First, we discuss the early theories whose arguments were that the mass media were enormously influential forces in society—theories weakly based on research, but strongly based on somewhat naive assumptions about powerful propagandistic messages and malleable audiences. Next is the mid-century's collection of studies about the minutiae of forces, influences, noise, and audience traits culminating in conclusions that were, at best, highly tentative—theories maintaining that the media must not have very much direct influence because there were too many contaminating variables to take into consideration. Finally, we see a recent series of speculations and theories maintaining that the media are powerful but subtle influences on increasingly discriminatory audiences—theories emerging from long-term and sophisticated studies and somewhat different viewpoints on humanity.

This chapter is not offered as the last word on the subject, but as a cursory overview of a highly significant aspect of media studies. Throughout the book, practical applications of several of the theories will be presented, as we attempt to gain further insights into how the mass media operate within and upon this complex society.

As this edition of *Introduction to Mass Communication* was being prepared, several intriguing issues were grabbing headlines in the national news media. All of them relate directly to this chapter and to this entire text, because they all say something about the assumptions we have made concerning the effects of media upon society and of society upon media. As you read these scenarios, ask yourself what assumptions or theories you have in mind. How powerful are the media in triggering reactions from individuals and influencing the general public, and what defenses do individuals and society have against such assaults? As we will see during our journey through mass communication theory, the same questions have been asked—and only tentatively answered—again and again throughout this century. The scenarios:

Causes and Effects

What assumptions and theories about mass media do we hold to as we evaluate current events?

A "mild-mannered" electronics specialist gained international attention after taking out a handgun and shooting four youths who had confronted him in a New York City subway. The media did not hesitate to dub him the "Death Wish Vigilante," comparing him with the "hero" of a popular Charles Bronson film. Debates were waged over whether he was the manifestation of a new American self-reliance, or whether he was just one more example of an impressionable individual who had taken a movie role too seriously, who had confused illusion with reality.

In Los Angeles, police arrested a drifter—"The Night Stalker"—suspected of killing sixteen people in a rampage that terrorized California. At one murder and rape site authorities had found a baseball cap bearing the logo of the heavy metal rock band, AC-DC. A friend of Richard Ramirez, the man arrested for the killings, told reporters Ramirez was particularly fond of the rock group, whose album *Highway to Hell* contains a song, "Night Prowler," which says in part, "Was that a noise outside your window? What's that shadow on the blind? As you lie there naked like a body in a tomb, suspended animation as I slip into your room."

AC-DC's lyrics and album covers, along with those of Prince, W.A.S.P., Motley Crue, Frankie Goes to Hollywood, Twisted Sister—and the whole gamut of rock 'n' roll, punk, funk, new wave, rap, and pop music and music videos—were coming under scrutiny for their sexual explicitness, violence, or drug and alcohol messages. The National Parent-Teacher Association and the Parents Music Resource Center, representing both liberal and conservative groups, pushed for warning labels to be placed on offensive album covers. The music, they said, was pandering to youth who were unaware of how powerful the anti-social messages were on their psyches and behaviors. As hearings progressed in Washington, D.C., we were

Students of public
opinion and mass
communication continue
to debate the mass
media's influence on
society. (a) Bernhard
Goetz, the so-called
"Death Wish Vigilante,"
is arrested after shooting
four youths on a New
York City subway.
(b) Farrah Fawcett
portrays a battered wife
in the provocative TV
movie, *The Burning Bed*.
(c) Rock musicians raise
millions of dollars for the
world's hungry and
needy.

(a)

(b)

(c)

reminded that Aristotle believed all music for the young should be regulated by law; that Confucius disapproved of foreign performers in China because he did not know what kind of music they would play; that devotees of Hinduism use mantras or chants to raise consciousness, to heal, and to become one with God; that the negative and insidious effects of rock music are attributable to the fact that the rock beat works directly against the diastolic/systolic rhythm of the body; and even that plants thrive or die depending on the type of music to which they are exposed.

Millions of television families viewed with varying degrees of interest a powerful dramatization called *The Burning Bed,* in which Farrah Fawcett portrayed a physically abused wife whose final desperate act was to set her husband afire after he had fallen asleep in a drunken stupor. For at least one viewer, the film showed a solution to his own family problem; he was arrested after setting his own wife ablaze. For tens if not hundreds of thousands of others, the film motivated phone calls to local spouse abuse and crisis centers in search of help for domestic problems they had heretofore been reluctant to share. Several questions arose: was television the direct cause of the "copycat" crime? What role did it play in awakening the consciousness of thousands of others, who may not have recognized the enormity of—and solutions to—their own problems? Was television indirectly responsible for the dissolution of many marriages that might have been salvageable? If television had done all of the above, what were the trade-offs between the death of one person, the termination of many potentially salvageable relationships, and the liberation of many previously abused spouses?

In Springfield, Pennsylvania, a woman well known in the area as "Ms. Rambo," due to her surly demeanor and custom of dressing in green combat fatigues like those worn by Sylvester Stallone in 1985's most popular movie, was charged with spraying a suburban shopping mall with semiautomatic gunfire, killing two people and wounding eight others. Rambo, the hero of a fictional story about a one-man army who returned to Vietnam to free MIAs, was seen by many social commentators as both a reflection of America's new national pride and an indication of some seriously confused social values. In the post-Olympic jingoism and the delayed acceptance of Vietnam veterans, flag-waving and individualism were the order of the day. Scores of American tourists, released from Lebanon after being hijacked while on a commercial jetliner, received heroes' welcomes upon their return home.

President Ronald Reagan, when praising American pilots for intercepting four Palestinians who had hijacked the cruise ship *Achille Lauro* and killed an elderly American in a wheelchair, claimed, "There's a new patriotism alive in our country." He was referring to a public mood that was being captured not only in films like *Rambo,* but also in popular music, literature, and in the headlines of newspapers and broadcast media in the mid-1980s. It was a time of seemingly unprecedented public selflessness—popular musical groups were raising millions of dollars for starving Ethiopians and embattled American farmers—and anti-social trends—vigilantes, serial killers, violence, and sex dominated our information and entertainment media.

Students of public opinion and mass communication seemed confused over whether the mass media were the cause or the effect in all of these scenarios. (As is true of most such debates, there is a bit of truth in both points of view.) These scholars can be excused for their confusion, for the history of mass media's influence on society, and society's influence on mass media, demonstrates a good deal of misunderstanding. As we work our way through some of that history, investigating how we have come to make assumptions about the media in society, we might bear in mind that no matter how drastically times and situations may have changed, some of the basic debates over mass media seem to go on forever. Theory building, as we pointed out earlier, is a difficult and tenuous business.

The Mass Communication Process

In the preceding chapter, some basic distinctions between interpersonal and mass communication were discussed. It was noted that the main difference between the types of communication resided in the nature of single versus multiple receivers, and immediate versus delayed feedback.

We can better appreciate the complexity of the mass communication process by applying it to a framework constructed in the late 1940s by a political scientist, Harold Lasswell. In order to understand the complete process, Lasswell said, answers must be found for each of the following questions: *Who, Says What, In What Channel, To Whom, With What Effect?*[1]

Each of these five elements can be pictured as it would apply to the interpersonal communication example cited in chapter 1, in which a woman and man exchange greetings. The "who" is the sender, the "says what" is the verbal message, "in what channel" is the air waves separating the sender and receiver, the "to whom" is the receiver, and "with what effect" in this case would be the simple, clearcut and appropriate (from the sender's perspective) response. In this simplest sense, each of the five variables is basic, easily identified, and therefore readily controlled or predictable. It is the basic stimulus-response, source-receiver model of the process. Even if contaminating variables of channel and semantic noise were to be added, the simple two-person communication environment is readily adapted or improved, as the noise factors themselves are identifiable and, to a degree, controllable.

Look what happens when we expand this to the mass media situation. The "who" or source of information cannot be readily identified and understood, because he or she is part of a complex organization, complete with institutional, financial, social, and individual pressures. In order to understand the performance of the source, researchers would have to undertake *gatekeeper* studies—studies of those factors which motivate the journalist, movie producer, cartoonist, advertiser, public relations practitioner, or other mass media decision-maker to function as he or she does. What is important to remember is that while there appears to be only one source or gatekeeper functioning, for all practical purposes there are as many sources at any given time as there are audience members. If Dan Rather on CBS News and Michael Jackson on MTV are being viewed by millions of Americans on a particular evening, each member of the audience is perceiving Rather and Jackson through his or her own eyes, projecting personal values, biases, attention or inattention levels to the message being presented by the newscaster and singer at any given moment. There is not "one" Dan Rather or Michael Jackson, but millions of them.

It follows that the "says what"—the message—is equally complex. For each receiver there is a slightly different message, regardless of the fact that the sender is transmitting a single message. Just as it is impossible to appreciate the totality of audience attitudes toward the sender, so it is impossible to comprehend its understanding of the message. But researchers undertake *content analyses* of the messages in order to ascertain patterns of the sender's verbal and nonverbal behavior. In isolation, the content analysis is no better means of gaining a full understanding of the complete process than is a gatekeeper study. Both have to be done in conjunction with studies of the other factors in Lasswell's communication matrix, or we have incomplete insights into the process.

"In what channel" refers to the physical medium being used to transmit the message from senders to receivers. Because it is a physical medium, it is often studied by mathematicians, electricians, physicists, engineers, or other scientists and technicians trained to deal with mechanical instruments and their problems. Extending our example of a Dan Rather newscast or a Michael Jackson music video, a full comprehension of the total effectiveness of the shows would have to include physical studies of the network's and local stations' transmission capabilities, atmospheric disturbances, cable capacities, and quality of individual television sets. When analyzing the effectiveness of print media—words and pictorial images in newspapers, magazines, or books—researchers are concerned with everything from whether typefaces are legible to whether the ink smears on

"Who, says what, in what channel, to whom, with what effect"—the Lasswell model—is far more complex in mass communication than in interpersonal communication.

readers' hands. This type of study is referred to as *media* or *channel analysis,* and, as implied above, should be considered only one part of the complete mass communication analysis.

Naturally, Lasswell's "to whom" refers to the audiences. It is not a mass of people behaving as one unit; as outlined in chapter 1, it is millions of individuals with their unique demographic and psychographic characteristics. The large, anonymous, and heterogeneous audiences, as described by sociologist Charles Wright, are more numerous than the source can possibly interact with or address at any given time on a face-to-face basis. Their members generally are unknown by the source; and they are made up of the broad spectrum including ages, educational levels, races, sexes, religions, interests, and attitudes.[2] Given these characteristics and the inevitable factors of semantic noise, researchers conducting surveys or other audience analyses began to realize several decades ago that at best they can only approximate the entire audience reaction to a mass-mediated message. In short, there are things going on in the media-societal mix that the basic stimulus-response model of the mass communication process cannot possibly account for.

Hypodermic or Bullet Theory

According to the hypodermic or bullet theory, mass media have direct and powerful influences upon audiences.

To today's student of mass media, it should be obvious that media gatekeepers go to great lengths to cater to their customers. But it has not always been that way. Actually, for most of their history, mass media have appeared to be convinced that the whims, values, and intentions of the senders were of greater importance than those of the receivers.

Until the 1930s and 1940s, the prevalent view of mass communication held that every media message was a direct and powerful stimulus that would elicit an immediate and predictable response. Audiences, members of "mass society," were presumed to share uniform characteristics, and to be motivated by biological and environmental factors over which they had little control. No interference was seen between the messages and the receivers, so a clear-cut, simplistic message would have a clear-cut, simplistic response. A model of this process, which looked just like our earlier figure 1.1, went by the name of the *hypodermic theory* or *bullet theory,* because the contents of the needle or rifle (in this case, the messages) were thought to get transferred lock, stock, barrel, and bullet directly into the receiver. The theories were based more on intuition than on research evidence; social science research was rather rustic, and little empirical evidence of mass media effects had been accumulated. Besides, massive propaganda efforts during the First World War, coupled with numerous examples of successful advertising campaigns, seemed to be highly effective in mobilizing public opinion and consumer behavior.

It may be significant that this basic stimulus-response model is still accepted by many critics of the mass media, who assume the media to be extremely powerful institutions, and media consumers to be woefully naive and malleable. Such critics seem to believe that audiences are ruled by instincts, "human nature," and irrational forces over which they have little control, while media gatekeepers are seen as inherently manipulative and somehow much more clever than their audiences. Consider, for instance, some of the accusations about the media cited earlier. We looked at copycat criminals, social misfits apparently driven by misguided senses of loyalty, justice, and heroism, and adolescents supposedly influenced by antisocial popular music. In addition, media critics want ready explanations of why twenty-nine Americans died "playing" Russian roulette after viewing *The Deer Hunter* on television, or why several children in Batman costumes fell from their rooftops, or why John W. Hinckley, Jr., tried to gain the affection of his *Taxi Driver* movie heroine, Jodie Foster, by assassinating the nation's best-known public official. Contemporary media, especially television, are accused of all of these insidious influences, along with shortening the public's attention span, rushing children into adulthood, and even reducing all public discourse to fluffy, trivialized amusement that no one—particularly those weaned on entertainment-laden television—can take seriously.[3] Those who subscribe to this view of how mass communication works may be serious scholars, but they seem to be avoiding the dominant thrust of research and theory over approximately the past forty years.

To believe in the bullet theory is to maintain that the audience is made up of an enormous, undifferentiated mass of humanity—that there are no differences among the individuals who constitute that audience. In the extreme view, each member of that audience would have to react identically to the mass-mediated messages. In the narrower view, biological or environmental factors would at the very least bring about a largely undifferentiated response to the media. In other words, thousands or millions of people who viewed *Death Wish* would actually become vigilantes, a like number who viewed *The Burning Bed* would set their spouses ablaze, and most preteenagers and teenagers exposed to AC-DC's *Highway to Hell* would become night prowlers (or paranoid potential victims). The fact that all of this did not happen should cause us to have second thoughts about the viability of the hypodermic or bullet theory.

By the 1930s psychologists and social psychologists were having serious doubts about the degree to which instincts and basic "human nature" accounted for reactions to media messages. Study after study had concluded that there were some *intervening variables* at work; something or some things besides the message and the sender's intentions accounted for differing reactions to media. Researchers were learning that individuals possessed widely divergent psychological mechanisms. Media content, while capable of activating these mechanisms, did not do so indiscriminately. For instance, as illustrated in figure 2.1, it made a difference how motivated the audience members were; how predisposed to accept or reject a given message; how much native intelligence and formal schooling they had; how sensitive, moody, fearful, prejudiced, or generally perceptive they were.

Individual Differences Theory

By the 1930s social scientists had begun to appreciate the fact that individuals possess widely divergent psychological mechanisms.

Figure 2.1
Individual differences
model of the mass
communication process.

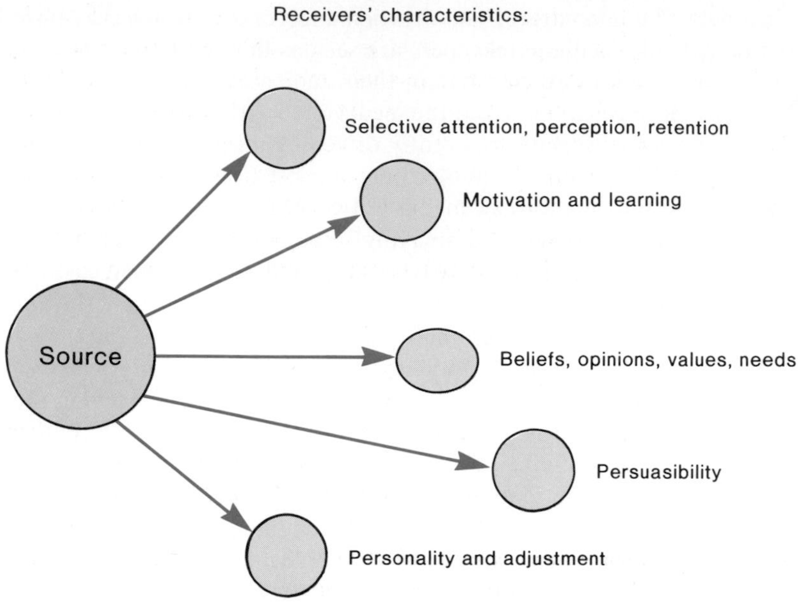

In order to investigate psychological differences among media audience members, social psychologists in laboratory settings carefully manipulated messages and media environments on different "types" of receivers. The research literature came to be filled with fascinating but tentative conclusions about communication effects. For instance, research on persuasion concentrated on what types of people are most likely to be influenced by what kinds of messages. Researchers raised such questions as, whether one side or both sides of the argument should be presented, whether the messages should be presented clearly and explicitly or vaguely and implicitly, whether fear rather than rational arguments should be used, and whether the climactic argument should be given at the beginning or at the end of the presentation. Persuasion research also included careful studies on perception, or audiences' tendencies to misconstrue arguments to suit their own prior beliefs or stereotypes. Other studies dealt with source credibility, attempting to discover what type of person was most likely to be highly influential in persuading people. To cite conclusions from these studies would fill volumes, but a couple of generalizations can help us to appreciate the work of social psychologists from this period.

Selective Perception

One of the most intriguing findings of the post-World War I social scientists was that different personality variables resulted in different reactions to the same stimuli. At the core of this is the concept of *perception:* how our values, needs, beliefs, and attitudes influence which stimuli we will select from the environment and how we will interpret those stimuli.

Selective attention or *exposure, selective perception,* and *selective retention* are highly interrelated psychological characteristics, often lumped under the single term, selective perception, and explain how people confront the content of the mass media. As DeFleur and Ball-Rokeach concluded, they are all intervening psychological mechanisms that have entered into the stimulus-response model of mass communication. From a multiplicity of available content, individual members of the audience selectively utilize messages, particularly if they are "related to their interests, consistent with their attitudes, congruent with their beliefs, and supportive of their values."[4]

Selective exposure, selective perception, and selective retention help explain how people confront the content of the mass media.

Selective Attention People tend to expose themselves to various messages or stimuli in accordance with their existing opinions and interests, and to avoid communications not in accordance with such opinions and interests. Democrats are far more likely than Republicans to attend Democratic rallies, read Democratic literature, and discuss Democratic policies. People who are predisposed toward sports and religion are far more likely to watch sports or religious programming than are those not so predisposed. Producers of *Rambo* or *Bambi,* of *Rocky* or *The Muppets,* of *The Burning Bed* or *The Sound of Music* have a pretty good idea what effects their media fare will have, if only because they know a bit about the kinds of audiences they will attract. Occasionally, of course, people expose themselves to belief-discrepant viewpoints (viewpoints conflicting with their predispositions). The pacifist may go to *Rambo,* the insensitive sadomasochist to *Bambi.* Usually this kind of selective attention occurs because people are consciously or subconsciously analyzing the weaknesses of belief-discrepant viewpoints in order to strengthen their basic beliefs, arguments, or even values. This kind of exposure is not normal, and not without some pain. True open-mindedness, despite the lip service played to it in a democratic, libertarian society, is extremely hard work for most people.

Selective Perception Once individuals have consciously sought out those media which support their predispositions, or even if they have accidentally exposed themselves to general information which does not support those predispositions, those audience members "read into" the messages whatever suits their needs. It is the "I told you so" proposition, as confirmation of any and all beliefs, opinions, and values is readily forthcoming.

Selective perception also describes the tendencies to misperceive and misinterpret persuasive messages in accordance with the receivers' predispositions, by distorting the message in directions favorable to those inclinations. Classic examples of this tendency were found in the "Mr. Biggot experiments," in which prejudiced people twisted the meaning of antiprejudice propaganda so it ended up reinforcing their existing biases. More recently, studies of the television program "All in the Family" have shown

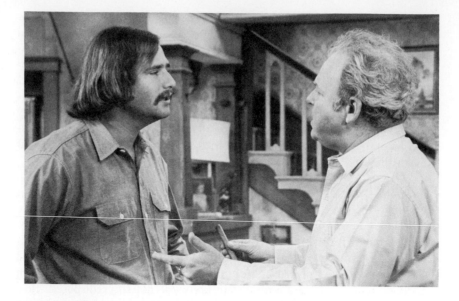

Studies of the television program "All in the Family" have shown that authoritarian audience members perceive Archie Bunker as justifying their own prejudices.

that inherently authoritarian audience members perceive Archie Bunker as fully justifying their own tendencies to discriminate against minorities and to see the world in simplistic, black-and-white stereotypes.

Selective Retention   Sometimes it is difficult to delineate between selective perception and selective retention or recall, because people are far more likely to retain those messages they consciously perceived than those they consciously rejected. Factors influencing selective retention include the saliency or importance of the message for later utility, the extent to which the message coincided with predispositions, the intensity of the message, the means by which it was transmitted, and sometimes even the extent to which receivers strongly disagree with the message. For instance, we tend to remember lecture materials if our instructors mention that we will be accountable for them on the next quiz; we remember the finer points of an argument advocating liberal spending policies if we have long been in favor of such policies; and we remember a message better if it has been presented in a basic, readily graspable, and emphatic manner. In addition, selective recall also involves the distortion of what is remembered. Research on televised presidential debates showed that over time, viewers associated arguments with their favorite candidate, regardless of who originally made the argument.

Inevitably we are exposed to information contrary to our basic opinions. "Accidental exposure" influences us. Television entertainment shows, movies, our regular newspapers, and magazines all contain information that is peripheral to our basic interests, but which nevertheless influences us. For instance, when Bill Cosby and Robert Culp costarred in a television series called "I Spy" in the 1960s, they may have done more to advance interracial understandings—in long-run, unmeasurable fashion—than any number of speeches, sit-ins, and court decisions at the time. But the individual-differences psychologies of the mid-1900s were incapable of measuring or accounting for such influences.

At about the same time psychologists were demonstrating fascination with the individual psychological differences of audience members, sociologists were emerging with an entirely different picture of why mass communication did what it did to certain people. Building upon the stimulus-response model, they too sought to discover what variables intervened between media and audience. As sociologists, they looked first at the common characteristics shared by various segments of the audiences, while the psychologists were looking primarily at the nature of the individual in the audience.

This *social categories* perspective was based largely on a collection of stereotypes held by social scientists. Researchers assumed that persons sharing particular demographic characteristics would have more or less uniform reactions to a given set of stimuli. Demographic variables—such as age, sex, income level, educational attainment, rural-urban residence, or religious affiliation—were seen as the basic distinctions between and guides to the type of communication content a given individual would or would not select from the available media.[5]

Whatever unit of analysis they studied—whether it was a public, mass, or audience—sociologists conceived of its membership as a dispersed group of individuals who never meet together and whose interaction must take place indirectly, through media. Therefore, public opinion or mass behavior was seen as merely the summation of all the private decisions made and acted upon by individuals acting on their own.[6] Naturally, audience analysis, surveys, and demographic variables received the lion's share of these sociologists' attentions. They compiled study after study of the ways in which different categories of individuals utilized, learned from, and were persuaded by the mass media. Political scientists, intrigued by this research perspective, contributed volumes of voter-behavior studies and established numerous opinion-research centers throughout the nation during the 1930s and 1940s.

Figure 2.2 shows that, according to the social categories model of mass communication, audience members behave quite as Charles Wright has described them; they are anonymous to one another and to the source of communication. It also may be worth noting that social categories research produced a great deal of quantitative and descriptive data about what categories of people respond to what kinds of media content, but it paid very little attention to why people respond to the media the way they do. In essence, the generalizations about who reads, sees, or listens to what, never became actual theories, but remained simple descriptive models of the process.

Social Categories

Theories of social categories emerged from sociologists who were studying audience demographics.

Figure 2.2
Social categories model
of the mass
communication process.

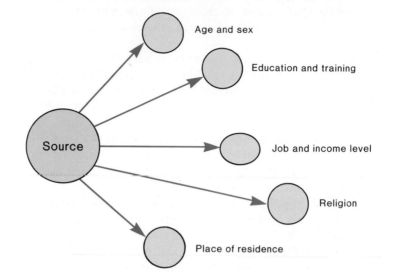

Receivers' characteristics:

Age and sex

Education and training

Source

Job and income level

Religion

Place of residence

Personal Influence

Election studies in 1940
showed the importance of
personal influence, and led to
the two-step model of mass
communication.

By 1940 scholars had come a long way from the hypodermic needle or bullet model of media effects. Even though their investigations into psychological and sociological characteristics as intervening variables in the mass communication process were admirable, such investigations remained inconclusive. For one thing, neither the individual differences nor social categories approaches even hinted at the possibility that anything other than specific traits of audiences served as barriers or interventions between messages and effects. Basically, the models were still single-step views of causality.

The 1940 presidential election changed much of that. Almost by accident, an elaborate research project, which set out to demonstrate how the mass media influenced voters, ended up demonstrating that while the media had little direct bearing on voters, interpersonal relationships had an enormous influence. In stage after stage of the study, the researchers found respondents indicating that political discussions with other people constituted a more relevant portion of their decision-making than did their direct use of the mass media.[7]

This finding was significant because it broke the back of the model of an atomistic mass audience. People did talk to one another, despite sociologists' pictures of an urbanized, indifferent, and defensively segregated humanity. "Informal communications" studies were born, and it was found that these informal communications networks, in which audiences talked to one another and sought out advice from opinion leaders, necessitated a new model of the influence of mass media.

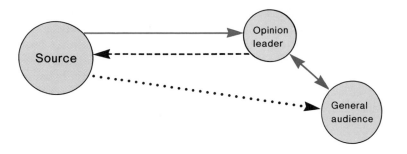

Figure 2.3
Two-step flow model of
the mass communication
process.

The 1940 study concluded that many voters had limited exposure to the mass media, but obtained most of their information from other people who had received it firsthand. The stages of progression were therefore from the media sources directly to the opinion leaders, who passed it along to others (figure 2.3). Not only were opinion leaders passing along information, they were adding their own interpretations to it. But it was acceptable to the ultimate receivers, who actually sought out the ready-made conclusions.

Two-Step Flow

Sociologists therefore began to pay very close attention to the nature of the interpersonal influences. Why did certain kinds of people emerge as opinion leaders, what kinds of people relied on what kinds of opinion leaders for what kinds of information and conclusions, etc.? It was found that opinion leaders are not necessarily city council members, bank executives, chamber of commerce presidents, and the like. They exist everywhere, in the classroom, in the shop, on the assembly line, in the ranks. They are those to whom others defer, possibly because of perceived expertise, certainly because of a confidence in them. They may be only vaguely recognized as opinion leaders. In their 1940 presidential election study, Lazarsfeld and Berelson cited the example of a cafe waitress who accepted an unknown customer's opinion overheard in a conversation because "he looked like he knew what he was talking about."[8] Thus, it is perceived expertise that matters. With reference to civic leaders and the like, whose prestige is great and whose opinions are sought, it is more likely that they hold their positions of prestige because they are, in effect, "natural" opinion leaders. This concept also goes a long way toward explaining the popular appeal of certain charismatic leaders, to whom others defer in business, government, in society as a whole, and at the ballot box because "they look (and act) like they know what they're doing."

The greatest shortcoming of the two-step theory appears to be that if pursued too diligently, it discredits the considerable original, direct influence of the mass media. While the dimensions of this influence are not entirely known, it is safest to hypothesize that the real function of the mass media in molding public opinion lies in some combination of the two-step and other theories.

Two-Step Critique

The two-step model tends to ignore the direct impact of mass media. Therefore, a multi-step model is suggested.

There are other shortcomings to the two-step theory. It evolved from the political realm, which, because of its polarized nature in society, is relatively easy to study. Consequently, it omits many other factors that exist in broader issues involving the mass media—demographics and psychographics, matters of taste, importance of the topics, and so forth. Because the studies used were concerned with presidential elections, they were necessarily concentrated in time. Public opinion formation is not so restricted. The studies were also conducted decades ago and have not been thoroughly updated on the same scale as the originals. Because the 1940 research had been conducted in an atmosphere devoid of television, a medium whose full impact, while unknown, is enormous, its direct applicability to contemporary political science is questionable.

Once opinion leadership was studied, it was found not only that opinion leaders came in all sizes and shapes, but that their intervention between the media and audiences worked in more than one direction. Their influence was not always "downwards," as when they interpreted the media messages for audiences. Sometimes it was "upwards" or back toward the media sources, when they helped tell gatekeepers how to do their job. At other times their influence was "sidewards," sharing insights with other opinion leaders. Indeed, the two-step flow was soon recognized as a *multi-step flow,* because the social relations of audiences (and opinion leaders) are generally complex and not unidirectional or unidimensional (figure 2.4). Surveys, previously used to study audiences, were found to be inadequate in giving an understanding of these complex social relations.[9]

Social Influence

Group norms—shared opinions, attitudes, and values—form a social reality capable of withstanding mass-mediated messages which oppose them.

In addition, further studies on interpersonal influences demonstrated that individuals' sense of belongingness to a group, club, organization, or mass strongly affects the kinds of information sought and utilized. Group norms, whether informally agreed upon or spelled out in codes of behavior, are significant variables. Shared opinions, attitudes, and values constitute a type of "social reality" that remains quite impervious to mass-mediated messages to the contrary. (Taking a group of Quakers to *Rocky* or *The Exterminator* is not very likely to convince them that violence is the solution to social ills.) Naturally, media produced by one's own group serve to maintain the status quo, as members utilize the media to reinforce their own individual and collective prejudices and values.

All in all, a wide body of research in social psychology has shown how groups and other individuals influence audience members' reactions to mass media messages. Much of that research has been underwritten by advertisers interested in the effects of persuasion. Increasingly, however, the research is being done by entertainment and news researchers whose primary concern is getting individual and collective members of audiences to pay attention and remember what they are being shown; such researchers also have come to appreciate the importance of social influences upon individual audience members.

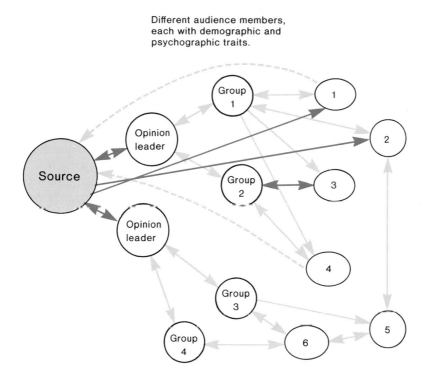

Different audience members, each with demographic and psychographic traits.

The Theories Modernize

Early models and theories of mass communication inadequately accounted for subtle nuances in the mass media's physical, social, and semantic environments. Complex interrelationships of source, message, medium, and receiver became more readily apparent when looked at through contemporary lenses focusing on such elements as feedback, noise, and competition for audience attention. Most fascinating of all seems to be the conclusion reached by many current media observers that much of the impact and effect of the mass media reside not in the hands of omnipotent media sources, but in the hands of the media users. As one generally accepted modern theory puts it, it is not just a case of what media do to people; it is a case of what people do with media. The recent emphasis on audiences, on the consumer-as-producer, may be the most significant finding by mass communication researchers in the past two or three decades.

Current media researchers believe that media users help determine the effects of media. Audiences are not passive.

Since the end of World War II, social scientists in many fields of study—especially psychoanalysts—have tended to rely upon a model of humanity that differs substantially from earlier models. The current approach refers to such human traits as the search for meaning, understanding, organization, growth, maturity, self-actualization, the need to know, to understand, and to be competent.[10] It may be significant that best-selling books on popular sociology of the 1950s included such titles as *The Lonely Crowd,*

A Nation of Sheep, The Seduction of the Innocent, and *The Hidden Persuaders.* By contrast, the dominant claim of pop psychology in the 1970s and 1980s has been that *I'm OK, You're OK,* and that there's nothing wrong with *Looking Out For Number One.* Granted, booksellers did unload a great many volumes of Wilson Bryan Key's *Subliminal Seduction, Media Sexploitation,* and *The Clam-plate Orgy,* but these attacks on media's supposed omnipotence have been in the minority and have not been well received in either academic or professional camps. Even advertisers seem to have gotten the message to "lighten up," and in the past decade or so commercials increasingly have tried to get us to laugh at ourselves and enjoy life rather than merely taking advantage of our collective guilts and insecurities. (We are still told we should worry about spots on our crystal, rings around our collars, and whether or not our breath is kissably fresh. However, more of our memorable—and effective—commercials and advertising campaigns find us sharing the outdoor fun of a rowdy bunch of beer-drinking ex-jocks, following the galactic battles between hamburger and soft drink sellers, and reaching out and touching someone.)

These new models and theories of human behavior developed fairly logically from the findings of the 1940s that interpersonal and group relationships intervened between us and the primary forces of influence upon us. Once scholars began to recognize the extent of these intervening variables, it was only a matter of time before some began to conclude that if the relationship between the media and the audience is all this complex, they had better delay making pronouncements about absolute cause-effect relationships. The hesitancy may never have been more poignantly stated than in Bernard Berelson's 1949 conclusion that *"Some kinds of communication on some kinds of issues, brought to the attention of some kinds of people under some kinds of conditions, have some kinds of effects."*[11] It was only natural for scientists to take a closer look at the subtleties. Significantly, few of them concluded that the media and messages made no difference in audience reactions. The problem became one of pinpointing all the variables in the equation to see if the cause-effect connections could be measured and controlled, or whether the impacts of media were due to a slow, imperceptible process of osmosis.

In short, the old hypodermic needle theories were being replaced by what might be called *stalagmite theories:* explanations showing how change occurs unnoticeably over a long period of time, as stalagmites form on cave floors after eons of steady dripping from above.

Numerous models and pretheories (essentially models that have not yet been found to be completely and theoretically sound) have emerged based on this stalagmite approach. Several of the most representative of these will be introduced; practical examples will show how they have helped us come to grips with the functions and impacts of mass communication in society.

Old theories are being replaced by "stalagmite theories," explaining the slow and subtle effects of mass communication.

EEK & MEEK ®by Howie Schneider

One of the most enduring theories of mass communication is referred to as *functional analysis,* or *structural functionalism.* It is based largely on older theories of sociology, economics, psychology, and political science. From the sociologist's point of view, functional analysis is concerned primarily with examining those consequences of social phenomena that affect the normal operation, adaptation, or adjustment of a given system—individuals, subgroups, and social and cultural systems.[12] Other disciplines view the process of individual and institutional checks and balances in much the same way. Obviously, mass media and their audiences can be analyzed along these lines.

The basic functions of the media were first proposed by Harold Lasswell in 1948 and expanded by Wright in 1960.[13] They postulated that media serve the functions of *surveillance* (surveying the environment and offering news reports about what is going on), *correlation* (interpreting information about the environment and editorializing or prescribing how people should react to those events), *transmission of culture* (binding time across generations, by educating people about information, values, and social norms), and *entertainment* (amusing people without necessarily offering any other functional values). These four basic functions were introduced in chapter 1 as being the information, persuasion, transmission of culture, and entertainment functions of both interpersonal and mass communication. Lasswell maintained that the media serve these functions for society as a whole, as well as for individuals and subgroups within that society.

In a recent work, Wright has outlined how each of these four functions can be seen to have negative effects or dysfunctions for both the individual and society.[14] For instance, learning about frightening aspects of the environment (*Death Wish, The Exterminator, Amerika, The Day After,*

Structures and Functions

Functional analysis explains the operations and consequences of mass communication.

Information, persuasion, transmission of culture, and entertainment can be both functional and dysfunctional.

and perhaps the evening news) can either help us prepare to cope or create fear. Being exposed to information about deviant social behavior ("Miami Vice" or any of a number of police and crime programs—including the evening news) can help individuals recognize the value of stability or can serve as a training ground for delinquency. Hearing numerous alternative persuasive political or commercial messages can bring about sophistication or render us unable to make decisions. Spending time on nonproductive entertaining diversions can bring pleasurable relaxation or detract from those things that need to be done. Naturally, the producers of the above media fare would insist that their purpose is strictly functional (even if only to entertain), and they have been known to go to court to prove they were not responsible for any dysfunctional effects.

Other observers have maintained that the four terms described here are inadequate explanations of the full range of media functions, that the functions—and dysfunctions—operate in a subtle chorus of harmony or disharmony, without either the sources or receivers understanding the full score. Daytime soap operas, in addition to the obvious function of selling soap, serve the function of relieving viewers of their personal burdens, and offering in their place numerous vicarious burdens and adventures of an extended "family" of television friends. Meanwhile, the shows also serve as a guide to personal behavior, establish fashion, cultural, and language trends for the hinterlands, and give viewers something to talk about. "Days of Our Lives" suggests a fascinating case study in media usage; one might find out which functions and dysfunctions it activates in its largely female and teenaged audience.

A rather broad typology or classification of structural functionalism has been proposed by McQuail. After merging functional analysis with newer types of research, he maintained that we should consider the fact that media serve audiences in at least the following ways:

Media serve audiences in a wide variety of ways.

1. *Diversion,* including escape from the constraints of routine by people whose job or family circumstances make them feel as though they are in a rut. Those who use the media for this purpose are also more likely to be less gregarious, so they seek diversion via media without leaving their homes;
2. *Escape from the burdens of problems,* sought by people whose jobs are taxing and tension-producing, or who are in difficult family situations, or who are troubled by personal and life-cycle worries such as ill health and old age, and those ill-equipped to cope with their problems, such as those with little education, in low status positions, or downwardly mobile;
3. *Emotional release,* sought by lonely people, introverts, or others whose inadequate personal relationships or cultural and psychological conditioning do not allow them normal emotional outlets;

4. *Substitute companionship,* for people with limited opportunities for social contacts (shut-ins, etc.), or those who have lost touch with former friends;
5. *Social utility,* for people who use media to "lubricate" social contacts, to aid conversation, or to gain information and opinions in case they find themselves serving as opinion leaders.[15]

Another effort to analyze a large variety of specific media functions maintained that mass communication is used by individuals to connect (or sometimes disconnect) themselves with (or from) themselves, their families, friends, nation, and so forth. The study attempted to determine what it means "to be connected" and to comprehend the whole range of individual gratifications associated with each means of being connected. It uncovered definite patterns showing which kinds of people, demonstrating which kinds of needs, utilized which kinds of media to serve as which kinds of connections.[16]

What has happened is that the traditional structural functionalist approach to mass communication research has been questioned by many contemporary researchers as being too simplistic. In some ways, older models have been discarded or modified to include newer concepts, such as the argument that greater importance be placed on analyzing what people do with the media, rather than merely what the media do to people. Regardless of its modifications, the functional view of mass media continues to focus on communication as though it were a systematic interaction between audience and media, a process in which balanced environments are sought between the opposing stress of audience needs and motivations on one end, and media functions and stimuli on the other.

Because the structures and functions approach represents insights from theories of psychology, sociology, political science, and economics, it remains a broad and disparate body of knowledge. Nevertheless, it is proving to be an especially useful framework for understanding the process of mass communication, what makes the media the way they are, and how audiences make use of them.

Uses and Gratifications

One of the most influential frameworks for media analysis in recent years is an offshoot of structural functionalism. Going by the name *uses and gratifications* research, it centers around a generalization stated earlier, that we should investigate how people use the mass media, rather than merely worry about how the media use people. The term "uses" implies that audiences are active rather than passive members of the communication process, and that they are willingly exposed to media; the term "gratifications" refers to the rewards and satisfactions experienced by audiences, and helps explain motivations behind and habits of media use. Pioneers in the field suggested that many research findings make more sense if communication is interpreted as a link between audiences and their environment, and if

"Uses and gratifications" research concentrates on how people use media rather than how media use people.

communication is explained in terms of the role it plays in enabling people to bring about more satisfying relationships between themselves and the world around them.

Three general groups of uses and gratifications studies have been conducted. One type looks at the satisfactions derived from mass communication, another looks at the social and environmental circumstances that lead people to turn to the media in the first place, and the third looks at the needs audience members are attempting to satisfy. Evidence from all three will be cited in the following.

"What reading does to people is not nearly so important as what people do to reading," Waples concluded as long ago as 1940. Maintaining that the effects of reading are determined by readers' predispositions, Waples posed two guiding questions to be utilized in the study of reading effects: "Who is the reader and what does he do and want and get?" and "What and how does the publication contribute to his wants?" The researchers ascribed the social effects of reading to the readers' search for prestige, for respite, for identification, for security and reassurance, and the enjoyment of artistic merit.[17]

Likewise, 1940s research on the impact of movies, drawing from recent insights into selective perception, concluded that "The motion picture is not a fixed pattern of meanings and ideas which are received by a passive mind. Rather, what the individual 'gets' is determined by his background and his needs. He takes from the picture what is useful for him or what will function in his life."[18]

A 1961 book about television and children put the argument in still clearer terms, when Schramm and his colleagues wrote: "In a sense the term 'effect' is misleading because it suggests that television 'does something' to children. . . . Nothing can be further from the fact. It is the children who are most active in this relationship. It is they who use television rather than television that uses them."[19]

The uses and gratifications research conducted in the 1940s and 1950s gave rise to a great many fascinating but unrelated studies on why people use the media. Surveys, case studies, and other audience analyses elicited information from media consumers on what functions they believed the media served for them. These early research projects demonstrated, among other things, the following:

Women fans of daytime radio serials obtained emotional release and stimulus, and some vicarious compensation for their own hardships, by finding scapegoats in story characters. Serials allowed them to identify with a more rewarding and exciting way of life, and provided some recipes for adjustment to listeners' problems. By praising and rewarding the wife and mother, serials helped reduce the woman's sense of futility and increased both her feelings of security and her acceptance of her position in society.

Research in the 1940s
and 1950s demonstrated
that comics serve three
functions for children: the
escape function, the
provision of an invincible
hero, and the provision
of information about the
real world.

Comics served three functions for young children: the "Alice-in-
Wonderland" escape function, the provision of an invincible hero
with whom they could identify, and the provision of information
about the real world. All three functions corresponded to and met
needs of three succeeding stages of children's mental and social
development. Youth who made excessive use of comics, however,
tended to have neurotic tendencies and/or physical disabilities.
Children seeking an escape from restrictive home backgrounds tended to
identify with aggressive heroes in comic books or on television
(they tended to use highly pictorial media). Children in general
used different kinds of media content in a way that closely
correlated with the extent of their attachment to their families or
peer groups, and middle-class children in particular who
experienced high levels of home-life frustration became heavy
television viewers.[20]

Research culminating in such generalizations gave us insightful descriptions of audiences and subgroups of audiences and their orientations
to various media content. People's attachment to media was carefully documented, in case after unrelated case. One limitation of these early studies
was their inability to determine whether the gratifications sought and the
gratifications received were one and the same. It was not easy to discriminate between the two when doing the kinds of research projects conducted
prior to the 1960s. Researchers managed to find out who the media junkies
were, but oftentimes—perhaps because the question was not always asked—
those junkies did not articulate precisely what they were getting out of the
communication experience.

By the 1960s scholars were
relating sociopsychological
characteristics and media
usage habits.

The 1960s saw the first attempts to carefully observe sociopsychological characteristics and to relate them to various patterns of media usage.
Extensive quantitative analyses were conducted, and definite patterns and
trends catalogued.

Among the most ambitious projects of this nature was the 1961 study
of American children and their television viewing behavior, cited earlier.
Schramm and his colleagues made a key distinction between the kinds of
rewards children received from "fantasy" content and "reality" content.
The former was associated with immediate gratifications and the latter with
deferred gratifications. The concept of *reward*—immediate and delayed—
has arisen from this and other studies of both entertainment and news media,
and has been applied in many studies that attempt to distinguish between
the ways in which media are used for escape or as coping mechanisms.
Sometimes the results are not as clear as researchers or subjects expect.
For instance, on more than one occasion people have told researchers they
watch television news to learn about current events in the world; however,
a simple comprehension test indicates they have actually learned very little
from watching the news. In other words, the needs remained ungratified,
even though audiences (and probably media gatekeepers) persisted in their
perception that the medium in question was the source of gratifying those
needs.

Among the studies focusing on audience rewards are an entire series
of newspaper readership projects, from which we have learned that:

Newspapers have their greatest following among people of higher
education and social status. Those who are habitual readers tend to
be mature people rooted by material self-interest and emotional
attachments to the community that the newspaper represents. They
generally are home owners, taxpayers, and parents of school-age
children; as shoppers and voters, they feel tied to their
community.[21]

News about public affairs is the backbone of the newspaper; it is the key
content which draws readers in the first place. Yet nonpolitical fare
is the content which separates the regular newspaper reader from
the occasional reader. Occasional readers typically use the
newspaper to get specific information, whereas daily readership has
more elements of "enjoyment" or "passing the time." Daily readers
devote more time to crime and sports news than do less-frequent
readers. Young people use newspapers less than older people. All
readers, however, learn how to use newspapers.

"Use of all media seems to be largely a habit learned early in life—like
party identification and brand loyalty," an American Newspaper
Publishers Association news research report concluded. The ANPA
has pursued uses and gratifications research as one of its priority
areas on the basis of preliminary insights it gained into why some
people develop into newspaper readers and others do not. It agreed
that diehard non-readers were unlikely to become newspaper

customers, but the occasional newspaper readers may be brought into the ranks of regular readers once more was learned about adapting the newspaper to their needs.[22] This most assuredly is not the type of conclusion adherents of the hypodermic needle theory of communication would have reached. To them, the task would have been to change the audience rather than adapt the medium to the audience's preexisting characteristics.

Early uses and gratifications research indicated that newspapers, along with radio and television, seem to connect individuals to society, whereas books and cinema appear to cater to more "selfish" needs such as those dealing with self-fulfillment and self-gratification. Further research, however, led to the argument that the same set of media materials is capable of serving a multiplicity of audience needs. A more contemporary view of the situation is that the relationship between the content of specific media and the needs of audiences is rather complex, and that one person's source of escape from the real world is a point of anchorage for another person. Thus it could be true that our reason for going to a violent war movie is a catharsis, diminishing our problems at school or work, while the person sitting next to us may be intently taking mental notes on how to become a soldier of fortune.

Media content that helps one person escape life's problems may help stabilize another person.

Meanwhile, studies over the past decade or two have shown a positive association in American adult television viewers between the level of stress they feel and the amount of "escape" rather than "reality" viewing they engage in. High anxiety has been found to lead to fantasy viewing primarily for those who are low in cultural participation and status. Similarly, those who are heavy consumers of popular fiction tend to have subjective feelings of depression and low scores on gregariousness.[23] Of course, there's a bit of a chicken-and-egg problem here: these studies do not necessarily tell us whether the people hooked on low brow television, checkout counter tabloid newspapers, and pulp novels got that way because they were already anxious, anti-social, and low status, or whether they got that way from their unbalanced media diet. That chicken-and-egg problem—the question of causality—remains one of the difficulties in uses and gratifications research.

Uses and gratifications research leaves us with some chicken-and-egg dilemmas.

As this research agenda continues, logical and rigorous methodologies are being applied to solving the causality dilemma. Substantive theory, rather than mere speculations, is the goal. In the words of one scholar, current researchers are concerned with (1) the social and psychological origins of (2) needs, which generate (3) expectations of (4) the mass media or other sources, which lead to (5) differential patterns of media exposure (or engagement in other activities), resulting in (6) need gratifications and (7) other consequences, perhaps mostly unintended ones.[24] Obviously, such research is a long-term proposition, and is not limited to a single methodology. Laboratory and field experiments, case studies and surveys have all proven of value in the drive to make theory out of uses and gratifications investigations.

© United Feature
Syndicate, Inc.

An entire book devoted to gratifications research concluded that whatever model and theory emerge from these investigations will have to take into consideration several points:

1. The audience is active and is using media to achieve certain goals;
2. Any links drawn between media and audience must consider the fact that the audience member can and does exercise choices;
3. Media compete with other sources that satisfy audience members' needs, so any conclusions about the meeting of needs must also consider the variety of ways in which media, in isolation or in conjunction with other sources, can satisfy a constantly shifting body of audience needs;
4. Much information about the goals of mass media use can be supplied by the audience members themselves; researchers should assume that audiences are sophisticated and rational enough to understand why they use media as they do, and devious or manipulative experiments should not be necessary when researchers seek this information.[25]

Research on audiences' needs has wide significance, and numerous scholars hope that a more rigorous methodology will convert the research studies into extremely useful theory. To date, that research still has not given definitive answers to the following basic questions. Why do we use the mass media? What individual needs lead us to use one mass medium more often than others or to choose some types of media content over others? How successful are the media in actually fulfilling these needs? It has not yet been determined precisely whether the media actually are fulfilling audience members' needs or whether the media are fulfilling needs that are only assumed to exist.[26]

Eventually, an understanding of exactly why we use the media the way we do should result in improvements in both the media and in our expectations of them. We as audience members will serve as a constant source of challenge to the media producers, who will cater to the multiplicity of requirements and roles we expect of our media.[27] As has been said elsewhere in this book, the consumers are the ultimate producers in a free market enterprise.

Consumers are the ultimate producers in a free market enterprise.

An interesting application of uses and gratifications perspectives, although developed independently, is William Stephenson's *play theory*. Stephenson, a British psychologist who did much of his research while serving as a distinguished research professor at the University of Missouri's School of Journalism, has proposed and rigorously and empirically supported a mass communication theory based around concepts of pain and pleasure, work and play.[28]

Like the researchers quoted in the preceding discussion on uses and gratifications, Stephenson maintains that audiences, whenever they are given the chance, will manipulate their media to serve their own needs. However, he goes one step further in pointing out that when pursuing media in their daily lives, audiences are engaged in pleasurable, ritualistic, and self-serving activities that are essentially play-like in nature. Enjoyment and contentment are inherent in activities that allow freedom of choice rather than social control. According to play theory and the psychological principles on which it is based, individuality is preferable to being forced to work and to conform to someone else's expectations.

Play theory research has explained some of the apparently strange things audiences do with media. We tend to read newspapers primarily for pleasure rather than strictly for information; we tend to read things we already know about (for instance, turning to the sports pages first thing Sunday morning to read about a game we attended Saturday evening); we read our magazines and newspapers in the same systematic and ritualistic pattern, day after day, week after week, because of the pleasure of the ritual; we tend to read the advice columns because they make us feel more confident about ourselves; we tend to enjoy many commercials and advertisements because we can put ourselves into the picture, and we know we have the freedom to decide whether or not to purchase the products being advertised; we organize our evenings and weekends around prime-time programs and movies that offer us vicarious and psychologically painless adventures, and so forth.

Not all communication is characterized by play and pleasure. Purposeful activities expected to elicit a specific reaction from us, according to Stephenson, have elements of work, pain, and social control. The distinction between play and pain rests not in the communication per se, nor in the motivations of the sources and gatekeepers, but rather in the minds and behaviors of the audiences. For instance:

Although reading a textbook might be some students' idea of pleasure—there are, of course, those for whom the acquisition of academic knowledge is a sort of "game"—for many others the experience is psychologically painful. This occurs because of the relationship between the medium and the audience. When the readings are assigned and students held accountable (by someone else's standards) for mastering the content, much of the pleasure has been removed. To the students experiencing communication pain, textbooks seem to be part of an academic system which the students have not created; it looks like a system replete with obstacles and rules imposed by bureaucracies. When looked at in this light, it is a prime example of social control.

Viewing news documentaries and educational programs merely because society has convinced us that we have an obligation to understand our environment is not a pleasurable experience freely entered into by the majority of us. For others who get turned on by knowledge for its own sake, or who view such programs while motivated in part by the desire to impress others with their knowledge, such viewing is a form of psychological play.

Viewing commercials or reading persuasive messages that play on our insecurities and senses of obligation, and which limit our alternatives ("If you don't use this product or vote for this candidate, you'll be a social outcast and responsible for the downfall of the republic.") can be painful to those who take them seriously. On the other hand, for those who fancy themselves as free thinkers, there is pleasure in ignoring the burdensome messages.

As these examples indicate, purposeful activities expected to elicit a specific reaction from us have elements of pain, work, and social control. In organized society, of course, such activities are essential for there to be any consensus or public opinion. Education and civics depend in part upon such activities. In totalitarian societies—which might include universities, churches, businesses, and even homes where choices are limited to either absorbing information or failing to meet institutionalized norms—communication is frequently characterized by pain and social control. Moralistic propaganda abounds in such conditions. However, even in the most totalitarian states, operating under the most rigid controls, leaders recognize the value of diversion and playful elements in their media. Mao Tsetung's wired villages, in which every Chinese citizen was continually reachable by radio tuned in to one central channel, heard as much music and drama as political polemic. If this were not the case, citizens would have sought other diversions from the overwhelming sense of responsibility imposed by the politicized media.

Stephenson justifies diversions in democratic states for the same reasons. Unless we have media content centered around superficialities such as fads, fancies, manners, tastes, and other escapes, we might become overly anxious and burdened. The research methodology employed by play theorists studies both the anxiety-producing conditions and those that diminish them, toward what Stephenson claimed to be the two basic purposes of mass communication: to suggest how best to maximize the communication pleasure in the world, and to show how far autonomy for the individual can be achieved in spite of the weight of social controls against him or her.[29]

Some communication is characterized by pain and social control. In other cases, diversions and superficialities provide communication play.

Play theory has been widely applied to research in advertising and political persuasion, as well as to the study of improving physical design of print media. Although the theory is demonstrable through the sophisticated research system (called "Q methodology") designed by Stephenson, there are many who take issue with the basic assumptions of the entire school of thought. Generally, Stephenson is seen as an apologist for the media status quo. His theory offends many who are calling for reform in the media and who wish to see the media raising instead of merely catering to popular tastes. Stephenson's response is that researchers and theorists have for too long been passing moralistic judgments on the media when, instead, what is required is a fresh glance at media audiences who are engaged in pleasurable self-interest.[30] Because most consumption of mass communication is done voluntarily, Stephenson wonders why the media shouldn't appeal to the basic psychological orientations of audiences. Otherwise, the media and audiences are wasting a lot of each other's time and energy.

In addition to the structures and functions, uses and gratifications, and play theories discussed so far, several other contemporary approaches to mass communication should be noted. Like the foregoing studies conducted since mid-century, these new theories hold that the media have important but difficult to measure long-term effects, and that information-seeking and discriminating audiences are active and important participants in the mass communication process. The theories to be singled out at this juncture have been called *political socialization* and *agenda setting* theories.

Socialization describes how people become functioning members of society, how they adapt to the values and norms of their communities and institutions. *Political socialization* is a special category in the process, explaining how political awareness is produced. Because political values begin at childhood, media researchers have studied children's mass media usage and have conducted extensive content analyses of media—particularly television—to determine the relationship between values conveyed by media and values adopted by children. How individuals learn about and participate in the political process—what they think of the use and misuse of power, law and order, economics, and the like—is subject to inquiry. Building upon the political studies of the 1930s and 1940s, cited earlier, researchers have combined insights into media usage with insights into children's peer and reference group behavior, their families, their schools, churches, and other socializing influences. The studies are necessarily of a long-term nature, for only a reckless scientist would claim a direct cause-effect relationship between a particular media message and specific subsequent political behavior. Nevertheless, evidence is accumulating from such studies that the media are an important, though subtle, influence in children's definitions of political reality and subsequent political behavior.

Play theory is faulted for giving support to the status quo.

Political Socialization and Agenda Setting

Political socialization theory explains how our political awareness comes about through social learning and media usage.

The process of political socialization is largely one of social learning, and depends upon a combination of direct experience and observational learning.[31] Much of that learning is based on imitation or modeling of successful (i.e., functional) behaviors depicted in media. From informative, persuasive, and entertaining media messages, audiences observe how power is achieved and used, how political and economic decisions are made, and how systems are maintained. (Sometimes, regrettably, they also learn dysfunctional messages—for instance, how to misuse power or tear down a viable system.)

Political information is an integral part of the mass media's contents, whether intended or accidentally incorporated into news, public affairs, or entertainment programs or even in commercials. An intriguing study of the 1972 presidential campaigns concluded that despite all the evidence of television's credibility and the heavy reliance voters placed on television news as their window to the world, they actually learned more solid information about campaign issues from watching political commercials than from the network newscasts.[32] The researchers maintained that due to its concentration on exciting visual images rather than substantive issues, the only effect of television news on the American voter is "to cheapen his conception of the campaign process and to stuff his head full of nonsense and trivia."[33] The study by Patterson and McClure, based on elaborate surveys and content analyses, shed new light on the political socialization process and caused a great deal of consternation among television news operations. It contributed to the field by indicating that the sources of political insights, values, and behaviors are not as clear-cut and predictable as intuition or traditional research might assume.

This relates directly to the concept of *agenda setting,* a theory built around the media's impact on the structure as well as content of audience perceptions. As we will see elsewhere in this book, particularly in discussing the effects of advertising and public relations (see Part IV), the media are powerful in their ability to create awareness of new ideas or topics, but somewhat limited in revolutionizing deeply held beliefs. Agenda setting theory maintains that the media may not be particularly successful in telling people *what to think,* but they are stunningly successful in telling people *what to think about.*[34] Merely by systematically allowing certain issues to appear in the news, gatekeepers give the public a pattern of things to think about.

A whole series of studies has found a direct and positive relationship between issues, events, and persons considered by editors to be newsworthy, and the issues, events, and persons considered by readers, viewers, and listeners to be newsworthy. Over a period of time, according to the theory, the very priorities for news utilized by professional gatekeepers become, almost by osmosis, the public's priorities as well. For evidence, look at the end-of-the-year lists of "Top News Stories of the Past Year" in your local paper, and see how many of the editors' choices are similar to readers' choices.

Agenda setting theory says the media are more successful in telling people what to think about than in telling them what to think.

BLOOM COUNTY by Berke Breathed

Fortunately, there is not always a one-to-one correlation between what gatekeepers choose to convey and what audiences think is important. Indeed, media critics can take some comfort in recognizing the checks and balances at work between gatekeepers' and consumers' priorities. Issues of growing importance to individuals and groups find their way into the media, just as the media give their audiences issues to consider. (This was referred to earlier as a multi-step flow of communication, in which audience feedback influences gatekeepers and vice versa.) Critics who fault the media for not giving the public what it says it needs would do well to consider this interdependent relationship, and look closely at who is influencing whom and how the influencing occurs. The fact that such research is difficult to conduct helps perpetuate the myth of powerful media.

Intriguing as they are, both the political socialization and agenda setting theories should be accepted only with a grain of salt. As noted, they generally depend upon longitudinal studies, and are subject to many contaminating variables (including, of course, all the interpersonal communication likely to either contradict or reinforce the mass-mediated messages). However, given the ubiquity of the information environment, it appears likely that in the long run, the nation's political self-identity and its sense of priorities are being subtly shaped as well as reflected by the mass media. The problem comes with not knowing what is meant by the long run, and being unable to pinpoint the many subtleties.

Political socialization and agenda setting research can be seen as separate components of an emerging emphasis on the mass media as "molders of meaning." That is, these theories attempt to assess the various manners in which the media create or modify the pictures of the world held by individuals and societies. Meaning can be purposefully or accidentally created. It is not easy to study, for it resides inside the heads of individuals, and cannot be accurately gauged by their overt behaviors. As media sociologist

"Meaning Theory" and Media Dependency

Media are "molders of
meaning" as they create or
modify our pictures of the
world. We depend upon media
to shape those images.

Melvin DeFleur has argued on several occasions, all the tools of the anthropologist, sociologist, psychologist, cultural historian, and communications researcher can be brought to bear in the understanding of how the mass media shape our cultural and "meaning-centered" environment.[35] When such tools are employed, they uncover evidence to support once again a stalagmite theory of mass communication, concluding that when media manipulate symbols, they establish and subsequently stabilize new meanings, change old meanings, and even subtly stimulate behavior. Fads, customs, clichés, and significant social values are involved, and the media are significant factors in their creation and continuance.

At this juncture, it is appropriate to propose what DeFleur and Ball-Rokeach have called an integrated *dependency theory* of mass communication.[36] Such a theory depends upon a recognition of various psychological and social factors that prevent the media from having arbitrary control over their audiences.[37] That means we should discard the bullet or hypodermic needle model of the mass communication process, because it assumes a homogenized, malleable "mass society." We have already pointed out that individual differences models show us how individual consumers psychologically cope with media messages; social categories models show us how groups of consumers actively respond to media messages; and social influences models show us how groups of consumers are connected with one another (via organizations, publics, demographic interest groups, etc.) in ways that are relatively stable and that help keep them from being too readily manipulated by external forces—including the media.

In the words of DeFleur and Ball-Rokeach:

Mass media not only lack arbitrary influence powers, but their personnel lack the freedom to engage in arbitrary communication behavior. Both the media and their audiences are integral parts of their society. The surrounding socio-cultural context provides controls and constraints not only on the nature of media messages but also on the nature of their effects on audiences.[38]

Nevertheless, the media—indeed, all our means of communication—have very important roles in holding society together. To understand this more fully, we should note what happens to societies as they change from traditional to modern systems of communication and cohesion.

In traditional societies, most information needs are derived from firsthand, direct contacts with "reality" or from listening to other people who have experienced that reality (tribal elders, hunters, or others who serve to bind time). Once that traditional structure breaks down, and life becomes more complex, individuals and institutions assume a wide variety of divisions of labor—serving different functions to connect them to one another. A breakdown in traditional, face-to-face oral communication culture brings about societies that have ever greater needs for second or thirdhand information, and ever greater reliance upon external sources to bind time and do the "reality testing." Modern societies expect their public officials to assemble and enforce their laws and codes of social conduct; they expect their scientists and engineers to establish and maintain mastery over their physical environment; and they expect their teachers and mass media to collect cultural data and to inform, educate, persuade, and entertain them. All these modern social systems, which are extensions of individuals, are interdependent; each serves and is in turn served by all the others.

Mass media obviously play a key role in establishing and maintaining the interdependent nature of modern societies. Just as individuals in a free society attempt to determine the levels of dependency they have upon political, economic, religious, and other systems, so do they attempt to exercise free will in deciding how dependent upon the media they will be. Nevertheless, simply by virtue of being part of a social system, today's citizens must put a great deal of reliance upon the media. Backtracking through our discussions of meaning theory, agenda setting, political socialization, play theory, uses and gratifications, and structures and functions, we will see how all this comes together.

The greater our needs to belong, to understand, and to cope, the greater our reliance upon the mass media. This gives media certain powers.

Media put our environment in perspective by giving its many parts various meanings. They help establish our agendas, by giving us things to think and talk about; they help us become socialized into our communities and political systems, and to participate in change when necessary; and they help us cope with or escape from life's realities in a wide variety of ways. In short, the greater our needs to belong, to understand, and to cope, the greater our reliance upon the mass media. From this, it follows that the media *must* have some pervasive influences upon our thoughts, beliefs, values, and even our behavior.

The dependency theory proposed by DeFleur and Ball-Rokeach emerges quite logically from other theorists' mid-1970s claims that the media are indeed powerful instruments for social change and maintenance. Several substantial research studies on persuasion, values clarification, information processing, and the effects of television violence on children gave tentative support to what one social scientist, Noelle-Neumann, called "A Return to the Concept of Powerful Mass Media."[39]

Basically, the claim for powerful media rests on arguments that the theories and models in vogue today have a great deal of logical validity, but cannot be adequately proven in traditional laboratory settings where only a few variables are manipulated and only immediate or short-range effects are studied. The only way to understand the long-range, accumulative, and indirect influence of media upon public opinion is to conduct cultural studies in the natural environment, according to this emerging school of thought.

Lab research is insufficient in proving long-term impacts of mass media.

Noelle-Neumann's argument is that the most basic and powerful characteristics of mass communication—the media's *cumulation, ubiquity,* and *consonance*—do not lend themselves to routine research. By cumulation she means the media's slow but pervasive influence—what we have been calling the stalagmite factor. By ubiquity she means the omnipresence of media—the simple fact that because media are everywhere, it is impossible to conduct classic experimentation in which we compare what happens to those exposed to media with those who are not exposed. (She also shares DeFleur and Ball-Rokeach's view that media and other social institutions are so interdependent that we cannot observe the influence of one without considering all of the others.) By consonance she means the unified picture of events and issues held by various media—an extreme case of agenda setting, in which public opinion cannot help but be shaped by media that rely upon the same sources of information and that share common definitions of what is going on, what is important, and what values should be presented.

"Powerful effects" arguments are difficult to refute.

The "powerful effects" argument, like the "meaning" and dependency theories of DeFleur and Ball-Rokeach cited earlier, is difficult to refute because it does not emerge from the empirical traditions which gave rise to so many of the other theories discussed in this chapter. For that very reason, Noelle-Neumann, DeFleur and Ball-Rokeach are being taken seriously in mass communication circles, where many scholars have grown tired of the limited findings uncovered by traditional research strategies. Their instincts tell them that the influence of the media is not trivial, but profound, and they are beginning a research agenda—with new tools—that will help them prove their case.

McLuhan's Media and Messages

McLuhan's "theories" are merely philosophical probings, but they are highly provocative.

The final theory introduced in this chapter is not really a theory per se, but a set of philosophical probings by Marshall McLuhan. He has been saved for last not because he offers the final word on mass communication theory, but because in some ways he demonstrates how far mass communication theory has evolved and how far it still has to go.

On one level, this Canadian philosopher, who died in early 1981, is a throwback to the turn of the century "grand theorists" whose single-causality perspectives explained away countless variables and leapfrogged over bothersome evidence in order to reach simple conclusions. But on another level, he typifies the latest in a long history of ever-widening perspectives about mass media in society. Indeed, his very breadth may explain why he has been the center of so much academic controversy. And, although many academicians have dismissed him and his work in the past several years, he seems deserving of special mention in this chapter because he reflects so many of the insights and problems faced by mass communication scholars throughout this century.

McLuhan saw himself more as a poet and artist than a social scientist, and said his explanations of media and society were less mass communication theory than they were mass communication probes or explorations.[40]

McLuhan's "the medium is the message" suggests that we are shaped by our communications tools and not just by their contents.

McLuhan gained his reputation talking about mass media and society in light of one principle: that human society has been and is being and will continue to be shaped by its *means* of communication, and not necessarily by the *content* of that communication. Some have labeled his approach "the principle of informational technological determinism." He preferred the simpler description: *"The medium is the message."* In either case, it means that the chief technology of communication in a society has a determining effect on everything important in that society—not only on politics and economics, but also on the ways in which people's minds organize their experiences.

Whenever we set out to shape our tools of communication, McLuhan argued, these tools end up shaping us. That is to say, a shift in informational media initiates significant and widespread social and psychological changes in any human system. McLuhan saw it as profound that today we communicate by electronic media, just as at one time we communicated with nonverbal languages, then with spoken words, and then with written languages.

Two of McLuhan's guiding principles should be noted:

1. Because we have only a limited number of senses, we constantly strive to keep those senses in balance. Any time those senses are thrown drastically off kilter, immediate and sometimes equally drastic adjustments are necessary.

2. Because media are extensions of our senses, they extend those senses in one of two ways—by being *hot* or *cool*. A hot medium extends one single sense in "high definition," by filling that sense with a great deal of data or information. This leaves us much less information to fill in or complete, meaning that that medium does not demand intense audience participation. For example, a high quality colored photograph is hot, because we know at a glance what it represents. At the opposite pole, however, would be an abstract drawing or a cartoon. It is a cool medium, extending senses in "low definition," providing little direct data or information. A cool medium requires us to use other senses and actively participate in completing the information we demand for a satisfactory understanding. (An abstract painting probably tells us more about what is in the head of the artist than what is "out there" in the real world, and it demands involvement on our part to infer the artist's intentions.)

Prior to the invention of the phonetic alphabet, our ancestors lived in close physical and psychological proximity with one another, in "tribal villages," to use McLuhan's term. Face-to-face speech was their basic means of communicating their perceptions of the world. Interpersonal contact extended all five basic senses—touching, tasting, smelling, seeing, and hearing, with a slight over-utilization of hearing. Hearing—auditory discrimination—tends to be less precise than other forms of discrimination, so this interpersonal environment saw all communication stimuli coming and being received at once. A simultaneous culture resulted. Communication was essentially cool, requiring the maximum amount of participation.

A second stage of communication history, as described by McLuhan, got underway with the widespread utilization of the phonetic alphabet, but really took hold with the invention of the printing press. McLuhan called

Tribal villages are typified by simultaneous "cool" communications.

the printed media the hottest ones there are, as they extend a single sense—sight—in high definition. Reading is a highly individualized and an artificially structured means of learning about or abstracting from reality. Print changed the communication environment from a simultaneous, indiscriminate perception to a limited, sequential, linear perception. This habit spread from reading to the entire way of looking at culture. Note reading's tendency to result in social alienation, detachment, individuality. McLuhan is one of several cultural historians who has argued that the invention of the printing press and subsequent mass literacy brought on a communications revolution known as the Renaissance and the Protestant Reformation. The ability to read and write became essential for anyone expecting to gain and hold power, because much essential knowledge was being preserved in books rather than being passed along by word of mouth by religious and political elites, as had been the case prior to the invention of the printing press. In essence, the print media detribalized us, McLuhan argued.

McLuhan says the "hot" print media brought about the Renaissance and Protestant Reformation. They detribalized us.

In his book, *The Gutenberg Galaxy,* McLuhan maintained that the detribalization became universal due to the linear, progressive way of thinking encouraged by print. Print encouraged people to be logical, to have "expertise" in narrow fields rather than to remain generalists. As this limited expertise grew, it had the effect of fragmenting society. McLuhan also described print as the lineal forebear of the assembly line, in which the mass production of things followed the same logical patterns as the mass production of meaning. From the assembly line, the industrial revolution, in combination with specialization, hastened technological advance. Technological acceleration led, in turn, to the communications or electronic revolution into which McLuhan saw society entering today, a revolution that has already altered humanity's relationship to its world as drastically as did the invention of the printing press five centuries earlier.[41]

This current stage got underway with the invention of electronic media, but did not begin to realize its full force until about a generation ago with the widespread adoption of television. Because it involves the senses of sight, sound, and even touch (McLuhan maintained that we psychologically reach out and fill in the mosaic of dots on the TV screen with our central nervous system), television is an extremely cool medium. Its use has had a widespread and unalterable impact on society and how we perceive ourselves, according to McLuhan. The Canadian philosopher noted the dominance of television on our lives, and particularly its sweeping influence on youth, whose time it commands from early childhood to puberty. The phrase "electronic babysitter" indicates some of the significance of this influence. McLuhan saw television as transferring more information haphazardly, even indiscriminately, to little brains during their most impressionable years than parental influence, peers, and schooling put together. TV means more than Transcendental Vegetation.

The coolness of television is reflected in the fashion in which it demands active participation from the viewers. They must mentally fill in the blank spaces in the coarse photographic screen, and, because the messages are continually interrupted by unrelated commercials, they must concentrate intensely and frequently change conscious and unconscious mental gears, according to McLuhan. (Note that many current theoreticians take issue with McLuhan's conclusions, maintaining instead that television is received by passive viewers who often appear to be in an alpha state or dreamlike trance.)

Electronic media, especially television, are cool. According to McLuhan, they have brought about a global village.

If the medium is the message, then the collage or mosaic of impressions in television has meant the restoration of multisensory perception, the return to a communal *global village* in which electronic audiences participate and act rather than withdraw. McLuhan has maintained that the electronic media substitute "all-at-onceness" for print media's "one-thing-at-a-timeness":

The movement of information at approximately the speed of light has become by far the largest industry in the world. The consumption of this information has become correspondingly the largest consumer function in the world. The globe has become on one hand a community of learning; at the same time, with regard to the tightness of its interrelationships, the globe has become a tiny village. Patterns of human association based on slower media have become overnight not only irrelevant and obsolete, but a threat to continued existence and sanity.[42]

McLuhan argued that television brings a whole world into a person's life. The young travel vicariously to the far corners of the world and universe. They observe situations known only vaguely, if at all, to their predecessors. They are participants in complex social situations. The depth of their experiences, knowledge, and ability to deal with the complexities they vicariously experience through television is a subject of considerable debate—a debate from which McLuhan largely excused himself because he maintained his job was to describe, not prescribe.

Critics maintain McLuhan's methodology was erratic and inconsistent, that many of his conclusions were not based upon accepted scientific research. He seldom cited journals or scholarly investigations because he considered such evidence to be biased by the same lineality and logic that pervades most print media. That old-fashioned scholarship is inconsistent with the thought processes of the electronic generations, he argued. On another front, McLuhan has been criticized for his loose borrowing of insights from anthropologist Edmund Carpenter, economic historian Harold Innis, and cybernetics scholar Norbert Weiner. All in all, the intuitive (and/or plagiarized) insights are dreadfully difficult to either refute or accept in whole. Nevertheless, because of the broad sweep of McLuhan's scholarly broom, and his impact on many media practitioners since the 1960s, he takes his place among those who have contributed to the literature of mass communication theory.

Critics fault McLuhan for his untraditional scholarship.

Summary

This chapter's basic purpose was to demonstrate the paths that have been taken en route to a more complete understanding of how the mass media operate in—and upon—society. A variety of assumptions, hypotheses, and theories about the media's impact on society and ways in which audiences utilize media have been presented. An overview of those perspectives discloses that our understanding of media appears to have gone through three basic cycles in the present century: early theories, generally single causation models, which assumed media had enormous and direct impact upon society; mid-century theories (or "pretheories"), lost in a minutia of data, which were hesitant to suggest any direct cause-effect relationships; and recent theories, of the stalagmite variety, cautiously suggesting that media are effective in influencing agendas, giving us a sense of self, and shaping our environment.

Hypodermic needle or bullet theories dominated the first third of the century. Powerful senders were seen as manipulating audiences at their whims. It was not until the 1940s, a time when new research tools were used, and important insights into the psychological and sociological natures of the audiences were taken into consideration, that researchers systematically broke down the study of mass communication into a logical series of questions about who, says what, in what channel, to whom, with what effect. The complexity of the relationships revealed by various research methodologies caused scholars to delay assuming causality. Laboratory and field studies, including complex experiments, surveys, and content analyses, all coupled with new statistical formulae, resulted in a vast quantity of interesting but largely unrelated findings.

Convinced that the media were indeed influential in society, but that the earlier methods of discovering the nature of the relationships were flawed, scholars since the 1960s have slowly and systematically accumulated data and formulated theories indicating that long, slow-term effects of media are both discoverable and significant. The bulk of the chapter is devoted to these more recent theories.

Today the emphasis is on the audience rather than the medium. Audiences are seen as obstinate and self-seeking, manipulating the media as much as, if not more than, the media manipulate them. Uses and gratifications theory, play theory, and some premises within structural functionalism assume that people consciously select media fare. Given the orientation of political socialization, agenda setting, DeFleur's "meaning theory" and media dependency theory, and even McLuhan's "medium is the message" arguments, we sense an increased value in having audiences understand their media environments so as not to be unwittingly controlled by them.

If there is validity to the emerging theories concerning powerful media whose influences are subtle and pervasive, it becomes increasingly important for us to withhold arbitrary judgment about the place of media among all the other forces impacting upon society. That is surely not to say we should ignore the potential effects of media; to do so would be a cop-out, and the stakes are too high. The social questions raised in the media scenarios at the beginning of the chapter—copycat crimes, antisocial messages in rock music, positive and negative repercussions of entertainment programming, political agenda setting—are important ones, and remain controversial regardless of the theories or research methods used to explain them. Just because there are differences of opinion concerning these issues is no reason to ignore the processes by which we are slowly gaining knowledge about media effects. We need to remember that in social science, as in any science, there is never the last word—just the latest. In this chapter we have tried to summarize some of the latest words.

The premise that knowledge of our media constitutes control over them permeates this entire textbook. It is introduced at this early stage and will be reintroduced at several points throughout, when individual media institutions are discussed. Audience interrelationships with each medium, and analysis of each medium's past, current, and potential impact on society, will be considered in each of the following sections. For now, a broad overview should suffice.

Notes

1. Harold D. Lasswell, "The Structure and Function of Communication in Society," in L. Bryson, *The Communication of Ideas* (New York: Harper, 1948).
2. Charles R. Wright, *Mass Communication: A Sociological Perspective,* 3d ed. (New York: Random House, 1986).
3. Neil Postman, *The Disappearance of Childhood* (New York: Delacorte Press, 1982), and *Amusing Ourselves to Death: Public Discourse in the Age of Show Business* (New York: Elisabeth Sifton Books, Viking, 1985).
4. Melvin L. DeFleur and Sandra Ball-Rokeach, *Theories of Mass Communication,* 4th ed. (New York: Longman, 1982), p. 187.
5. Ibid., p. 188.
6. Herbert Blumer, "Elementary Collective Groupings," in Robert E. Park, ed., *An Outline of the Principles of Sociology* (New York: Barnes and Noble, 1939), pp. 233–41.
7. Paul F. Lazarsfeld, Bernard Berelson, and Helen Gaudet, *The People's Choice* (New York: Duell, Sloan & Pearce, 1944).
8. Ibid.
9. Raymond A. Bauer and Alice H. Bauer, "America, 'Mass Society,' and Mass Media," *Journal of Social Issues,* Vol. 16, 1960, p. 15.
10. Milton Rokeach, "Images of the Consumer's Mind On and Off Madison Avenue," *ETC.,* September, 1964.
11. Bernard Berelson, "Communication and Public Opinion," in Wilbur Schramm, ed., *Mass Communication* (Urbana: University of Illinois Press, 1949), p. 500.
12. Charles R. Wright, "Functional Analysis and Mass Communication," *Public Opinion Quarterly,* 1960.
13. Ibid., and Harold Lasswell, "The Structure and Function of Communication in Society," *The Communication of Ideas.*

14. Charles R. Wright, *Mass Communication: A Sociological Perspective,* 3d ed.

15. Denis McQuail and Michael Gurevitch, "Explaining Audience Behavior: Three Approaches Considered," in Jay G. Blumer and Elihu Katz, eds., *The Uses of Mass Communication: Current Perspectives on Gratifications Research* (Beverly Hills: Sage Publications, 1974), pp. 290–91.

16. Elihu Katz, Michael Gurevitch, and H. Hass, "On the Use of Mass Media for Important Things," *American Sociological Review,* Vol. 24, 1973.

17. D. Waples, B. Berelson, and F. R. Bradshaw, *What Reading Does to People* (Chicago: University of Chicago Press, 1940).

18. F. Fearing, "Influence of the Movies on Attitudes and Behavior," *Annals of the American Academy of Political and Social Sciences,* 1947, pp. 70–80.

19. Wilbur Schramm, Jack Lyle and E. B. Parker, *Television in the Lives of Our Children* (Stanford: Stanford University Press, 1961).

20. Denis McQuail, *Towards a Sociology of Mass Communications* (London: Collier-Macmillan Limited, 1969), pp. 72–73.

21. Leo Bogart, *Press and Public: Who Reads What, When, Where, and Why in American Newspapers* (Hillsdale, N.J.: Lawrence Erlbaum Associates, Publishers, 1981), p. 54.

22. *American Newspaper Publishers Association News Research Report No. 13,* July 12, 1978.

23. Denis McQuail, *Towards a Sociology of Mass Communication,* pp. 73–74.

24. Elihu Katz, Jay G. Blumer, and Michael Gurevitch, "Utilization of Mass Communication by the Individual," in Blumer and Katz, eds., *The Uses of Mass Communications,* p. 20.

25. Ibid., pp. 21–22.

26. Sheree Josephson, "Uses and Gratifications," unpublished manuscript, April, 1984.

27. Katz, Blumer, and Gurevitch, *The Uses of Mass Communications,* p. 31.

28. William Stephenson, *The Play Theory of Mass Communication* (Chicago: University of Chicago Press, 1967).

29. Ibid., p. 205.

30. Ibid., p. 45.

31. Alexis Tan, *Mass Communication Theories and Research* (Columbus, Ohio: Grid Publishing Co., 1981), pp. 268–69.

32. Thomas E. Patterson and Robert D. McClure, *The Unseeing Eye: The Myth of Television Power in National Politics* (New York: G. P. Putnam's Sons, 1976).

33. Ibid., p. 144. See also John P. Robinson and Mark R. Levy, *The Main Source: Learning from Television News* (Beverly Hills: Sage Publications, Inc., 1986), for a strong criticism of the myth that the public gets—or thinks it gets—most of its news from television.

34. Bernard Cohen, *The Press and Foreign Policy* (Princeton, N.J.: Princeton University Press, 1963), p. 13; see also Maxwell E. McCombs and Donald L. Shaw, "The Agenda-Setting Function of Mass Media," *Public Opinion Quarterly,* 1972, pp. 176–87, and Jack McLeod, Lee Becker, and J. Byrnes, "Another Look at the Agenda-Setting Function of the Press," *Communication Research,* 1974, pp. 131–66.

35. Melvin DeFleur and Sandra Ball-Rokeach, *Theories of Mass Communication,* 4th ed. (New York: Longman, 1982), and Melvin DeFleur and Everett Dennis, *Understanding Mass Communication,* 2d ed. (Boston: Houghton Mifflin, 1985), and Shearon Lowery and Melvin DeFleur, *Milestones in Mass Communication Research: Media Effects* (New York: Longman, 1983), pp. 28–29 and 383–86.

36. DeFleur and Ball-Rokeach, *Theories of Mass Communication,* pp. 232–55.

37. Ibid., p. 234.

38. Ibid.

39. Elisabeth Noelle-Neumann, "Return to the Concept of Powerful Mass Media," in H. Eguchi and K. Sata, eds., *Studies of Broadcasting: An International Annual of Broadcasting Science* (Tokyo: The Nippon Hoso Kyokai, 1973), pp. 67–112. See also David L. Altheide, *Media Power* (Beverly Hills: Sage Publications Inc., 1985).

40. Marshall McLuhan, *Understanding Media* (New York: Signet Books, 1966) and *The Gutenberg Galaxy* (Toronto: University of Toronto Press, 1967).

41. McLuhan, *The Gutenberg Galaxy.*

42. Gerald E. Stern, ed., *McLuhan: Hot and Cool* (New York: New American Library, 1969), p. 151.

Part Two

The print media are the oldest vehicles for mass communication. They began essentially with movable type on Gutenberg's printing press in the fifteenth century. Gutenberg initiated a spiraling process as the availability of books prompted increased literacy, which created a demand for more books and more education. But the book was too slow to produce and to consume, which led to the development of other print forms. From the occasional tract evolved newspapers published on a regular basis and, later, magazines.

Society to date has been basically print oriented, and until very recently, the mass media network in the United States was exclusively print. All of that has changed since the electronic revolution, however. Newspapers infrequently publish "extras," and national general-interest magazines have come upon hard times. The book-publishing industry has turned toward paperbacks as a quicker and more inexpensive mode.

The print media are presently in flux. But as we will see in case after case, new media do not crowd out the old. Instead they stimulate innovation and a search for new markets. Paperbacks are, in effect, a marriage of the book and magazine. Specialty magazines have become commonplace in recent years. Newspapers have added

The Print Media

Sunday and weekday supplements, or small special-interest magazines. These and other innovations represent the strivings of the print media to seek their own level following the tremendous impact of the electronic media.

In chapter 3 we will look at newspapers, exploring the development of the medium during the second half of the nineteenth century and into the twentieth century, the control and distribution of newspapers, the nature of newspaper workers and audiences, and recent changes in the industry.

Chapter 4 is devoted to magazines, a hybrid print medium that allows for more in-depth analyses than is usually true of newspapers and is quicker to produce and disseminate than books. The chapter covers the history of magazines, the various types of current magazines, life cycles of magazines, their audiences and economic support systems, and their future.

The subject of chapter 5 is books, the medium that not only opened the world to mass communications but has remained the most durable. In addition to a history of the book medium, this chapter introduces the variety of books on today's market, how the business operates, who reads or does not read books, and the medium's future.

Newspapers

3

Introduction

The newspaper is a uniquely American phenomenon representing the highest achievement of the print medium.[1] Created in Europe, it was brought to North America where it was peculiarly suited to the rushed American temperament. The newspaper matured and flourished here, proving itself remarkably adaptable to changing times. Even though it has been superseded in some respects by the broadcast media, it nonetheless has provided the framework within which the broadcast media have grown. Local emphasis, advertising support, and audience appeal were all lessons that broadcast learned from the press. Further, the fundamental concept of press freedom was a newspaper doctrine disseminated by the founding fathers and subsequently applied to all media. The American newspaper is the traditional base of a growing mass communications network.

This chapter on the newspaper in American society begins with a definition of the medium, followed by a brief history of the newspaper in both European and American settings. Of particular importance is the rise of press freedoms. During the twentieth century the newspaper has had to compete with radio and, more recently, with television, which has prompted newspaper people to reconsider their medium's essential purposes and audience appeal. Contemporary newspapers are big businesses, typical of many industries in their mergers, consolidations, and potential for monopoly. Recent years have seen newspapers coping with technological revolution and economic retrenchment. We will discuss these changes, describe the basic characteristics of newspaper operations, the people who report and consume the news, and the types of newspapers currently available. Finally, we will consider the potential for this venerable medium to compete in an increasingly electronic communications environment.

The Medium Defined

Newspapers vary widely, but tend to mirror their own communities. There are several exceptions to the general rule.

Until very recently, it was appropriate to say that a newspaper is a regularly issued, geographically limited medium printed on unbound newsprint, which serves the general interests of a specific community with news, comment, features, photographs, and advertising. The definition has nothing whatever to do with frequency of publication, size, or format. Today the physical characteristics of the newspaper remain as described, but the recent satellite-aided distribution of *USA Today, New York Times, Christian Science Monitor* and *Wall Street Journal,* from scattered production centers to national audiences, causes us to reconsider the traditional "local nature" of the newspaper. The latter are significant exceptions to the general rule, unless we widen our definition of community to include a national collectivity of individuals sharing common interests and reading habits. Indeed, as the newspaper industry has found itself competing with national television for audience attention, the rules of the game have changed, as we shall see throughout this chapter.

Newspapers can vary as widely as the multiple editions of the economically plagued *New York Daily News,* whose 1.3 million copies per day are read by some four million people, to economically secure monthly newspapers serving only a few hundred residents of small American communities. There are also weekly newspapers and suburban dailies. Some of these carry extensive regional, national, and international news, but most do not. With rare exceptions, as noted above, most newspapers share a common preoccupation with their own locales.

By our definition, *Women's Wear Daily* and the *Wall Street Journal* are more like daily magazines than newspapers. They serve specific national-interest groups, the fashion and financial worlds, respectively. But in recent years the categories have blurred, and most observers list the *Wall Street Journal* as the nation's top circulation daily newspaper, with two million copies a day.

By and large, the American newspaper's allegiance is exclusively to the geographic area it serves. A geographically fragmented medium, each newspaper acts as a mirror of its community. This local emphasis is a twentieth-century heritage derived from an early nation of isolated and widely separated communities along the Atlantic seaboard, each self-reliant and each served by a self-reliant press. When regional and national newspapers survive, it is because they have managed to transcend local issues and to report, comment upon, and advertise more broadly based items.

Historical Development

Newspapers emerged from books, but they had predictable publication and circulation.

When movable type was invented in the 1450s, it was first used to print books. It was not until 150 years later that printers in England began distributing *tracts,* or pamphlets, containing topical news. By 1621, *corantos,* single-sheet tracts dealing with current foreign affairs, were common. Corantos were followed in 1641 by *diurnals,* four-page bulletins of local news.

Print was becoming available to wider audiences. Produced in days and sometimes in hours for only a few pennies, the print media's scope was expanding, bringing print to whole new audiences in terms of cost and interest. Tracts encouraged literacy and accelerated the dissemination of news.

These new print media offered a more immediate, although cursory, overview of contemporary events at a much cheaper cost than a book, though admittedly sacrificing the elegance, permanence, and depth of a book.

Each of these forerunners to the modern newspaper grew in popularity. Early printers began to publish them regularly, discovering that they could count on roughly the same number of customers for each issue. Thus, the concept of circulation was born—the formal, quantitative expression of audience numbers. Such a profitable concept, in turn, encouraged schemes to increase sales. One method was to step up the frequency of publication from monthly to fortnightly to weekly and eventually to daily issues as technology improved, doubling and quadrupling circulation.

The *Oxford Gazette* (later the *London Gazette*) became the first regularly published English newspaper in 1665; in 1702 the *Daily Courant* became the first daily in England. In America, Benjamin Harris tried to publish the first newspaper in 1690. But his *Publick Occurrences, Both Forreign and Domestic,* lasted exactly one issue. Apparently, Harris's version of the truth differed from that of the governing Massachusetts Bay Colony. He allegedly ran afoul of a 1662 ordinance that prohibited printing without governmental approval and consequently was shut down. The *Boston News-Letter,* a weekly started by John Campbell in 1704, became the first continuing American newspaper. Unlike Harris's *Publick Occurrences,* it had the full support of the governor and at one point was even subsidized by the state. During the next fifty years, some thirty other newspapers appeared in the larger American towns.

Although many of the first settlers supported education (establishing Harvard College in 1638, and at least a dozen other colleges before 1776), the majority of the population in pre-Revolutionary America was more concerned with farming or running small shops. Fewer than two hundred thousand of the colonies' three million people could read—hardly constituting the masses necessary for true mass media.[2] Still, the primary focus of media was political, economic, and literary, because that was what the small, sophisticated audience expected.

An illuminating study found that while small circulation newspapers in pre-Revolutionary days may have had high pass-on readership (or listenership, because many of the illiterates who gathered at coffeehouses and inns had newspapers read aloud to them), the primary subscribers were a kind of "Who's Who" of Colonial America.

In mid-1775 William Bradford's *Pennsylvania Journal* had only 220 subscribers in Philadelphia, the largest city on the North American continent, with a population of thirty-five thousand residents. Those subscribers were the rich and powerful leaders of the city; they were the heads of the political factions and, on the whole, were closely tied by business and religious connections, and sometimes by family bonds. In Philadelphia and in outlying areas, the *Journal's* readers were important political leaders, or important *information disseminators,* such as postmasters, tavern keepers, or newspaper publishers who tended to reprint significant news and opinions in their own local newspapers.[3]

By the time of the American Revolution, an unencumbered press actively participated in the dissemination of information and opinion. One might easily conclude that the press had always been able to print at will, but such was not the case. For more than two hundred years the governments of England and later the colonies had imposed restrictions on the press, and much of the early history of newspapers involved a struggle for the freedom to print.

Early newspapers catered to literate, sophisticated audiences.

Early newspapers had high pass-on readership.

From earliest times, authoritarian leaders have seen all too clearly the threat that the mass production of ideas imposes on established order, and most authoritarian leaders, at one time or another, have undertaken steps to control such threat. In the 1530s, the English monarch Henry VIII, recognizing the danger posed by the new medium of print, began requiring printers to be licensed. Only those individuals fortunate enough to obtain a license were allowed to print, and they retained the license only with the continued approval of the king. Needless to say, few printers were willing to challenge the crown by publishing material offensive to Henry. Because the strategy of licensing effectively prevented dangerous ideas from being printed, it came to be known as *prior restraint*. The practice continued in England in one form or another for the next three hundred years.

Control and Press Freedoms

English monarchs who followed Henry devised other means to control the press. Elizabeth I made use of the Star Chamber, a governmental body willing and able to punish severely those printers who offended the state. Selectively enforced taxation, another favored method, could put a printer out of business. And laws against seditious libel (criticism of the state) intimidated many printers.

The sixteenth and seventeenth centuries were years of restriction for printers in England. The medium that promised enrichment and promoted literacy was made to serve the interests of the state. No dissent would be tolerated, but on occasion there could be heard a voice arguing for change. In the late seventeenth and early eighteenth centuries those voices became louder and more insistent.

Freedom of the press is a relatively recent notion.

Press freedoms came slowly and were granted grudgingly. English poet John Milton, among others, spoke out for freedom of the press when, in 1644, he wrote *Areopagitica,* an impassioned appeal for the government to trust the people to be able to discern the truth. His eloquence had little immediate impact; press restrictions continued in England. But support for licensing of printers was eroding, and by the end of the century, the statute had been abolished.

In the colonies the practice of licensing continued until the 1720s when James Franklin, the older brother of Benjamin Franklin, started the *New England Courant.*

Franklin defied the Massachusetts Bay Colony by publishing his paper without a license. Trouble arose when he began printing articles that criticized both the colonial government and the established church. He angered the two dominant institutions in the colony, prompting a jail sentence on contempt charges. After his release, Franklin resumed his activities, to the chagrin of the government, which promptly forbade him from ever again printing the *Courant* without a license. Franklin's continued efforts eventually resulted in the grand jury refusing to endorse the charges of the government. Although the law proscribing unlicensed printers remained, it was no longer enforced.

Franklin and Zenger fought against authoritarian controls.

James Franklin's heroics gave printers freedom from prior restraint, but the colonial government still exerted control by using the laws against seditious libel. That block to press freedom began to break down in the case of John Peter Zenger, an immigrant New York printer. Zenger had been hired to publish the *New York Weekly Journal,* a newspaper started by James Alexander and other opponents of the incoming colonial governor, William Cosby. When the *Journal* challenged the governor's behavior, which was viewed by many of the local citizenry as greedy, self-serving, and tyrannical, Zenger was arrested for seditious libel.

According to the law, the task of the jury was to determine if the accused had, in fact, printed the offensive document; the judges were to determine if the material was criminally libelous. In the case of Zenger, the judges, appointed by Governor Cosby, were certain to uphold the charge. Imagine the impact on the audience in the courtroom that August day in 1735 when Peter Zenger's counsel, Andrew Hamilton, openly acknowledged that Zenger had printed the issues of the *New York Journal* on which the case was based. Hamilton, almost eighty years of age, was at the twilight of his career. Perhaps he felt he had nothing to lose, because before the attorney general could direct the jury to return a verdict of guilty, Hamilton launched Zenger's defense. He focused on the charge in the indictment that the newspaper contained "false, scandalous, and seditious libel." *False* was the key word, for as Hamilton admitted, if the words had been untruthful then "I'll own to them to be scandalous, seditious and a libel."[4] Hamilton's brilliant oratory, addressed to the jurors, strongly urged them to claim the right to determine if the printer's statements were true or false. When, at last, the jury retired to deliberate, they took less than ten minutes to reach a verdict of acquittal.

In retrospect, the Zenger case is perceived as important more as an indicator of a changing mood of the people than as a guarantor of press freedoms. At the time, the verdict was little noticed outside of New York and it set no legal precedent; seditious libel continued to be against the law. But it did serve notice on the colonial governors that the people were becoming less docile.

Forty years later the colonists rebelled. The press, which had been increasingly outspoken, was highly partisan in its support for the Revolution. The patriots used the press to instigate and sustain the fight. Samuel Adams agitated for rebellion in the *Boston Gazette*. John Dickinson, a supporter of business interests, presented logical arguments denouncing British restrictions on commercial activities. Isaiah Thomas of the *Massachusetts Spy,* James Otis, Thomas Paine, and others took up the cause. For the moment at least, freedom of the press was a fact used in the fight to secure other freedoms.

Within five years after the Revolutionary War, the Constitution was ratified. Three years later, in 1791, the Bill of Rights was approved. The First Amendment to the Constitution and the first of the Bill of Rights states in part, "Congress shall make no law . . . abridging the freedom of speech or of the press. . . . "

The statement is clear and direct, but as we will see in chapter 14, it is not absolute. Regulations exist for all the mass media, including the press. Many of the restrictions have been enacted by Congress and other governmental bodies; others are a result of controls imposed by the media themselves. The abundance of regulations notwithstanding, few would argue with the assertion that the press in the United States is among the freest in all the world.

Press freedom is guaranteed, but it is not absolute.

The American Press

America's first regularly published newspaper, the *Boston News-Letter,* came in 1704, published by the postmaster, John Campbell. The first daily papers came eighty years later in America. The history of newspapers on American shores can be split into eight eras of varying lengths, each making its own contribution to the press. They are: (1) the colonial press; (2) the Revolutionary press; (3) the political press; (4) the penny press; (5) the personal editors; (6) yellow journalism; (7) jazz journalism; and (8) the present age of consolidation. The first three periods occurred prior to 1833. From then until the early twentieth century we experienced the penny press, the era of personal editors, and the flamboyance of yellow journalism. Since 1920 we have seen the last two periods.

Before 1833 During the 130 years from the establishment of the *Boston News-Letter* in 1704 until the 1830s the press had limited circulation and tended to be written for the educated, if not the elite. Opinion and viewpoints were prevalent in published articles, especially after the 1740s when disenchantment with British rule began to spread.

The colonial press wasn't very impressive by present standards, but it served a need. From the rumor mills of the coffeehouses and taverns that had provided the original communications network in the colonies, the colonial press became established as a machine for institutionalizing gossip, a role that the newspaper still maintains. Its publishers were often undereducated printers. They dealt in rumor and in shipping news that was crucial to the mercantile concerns of the seaboard colonies.

Gradually, under pressures brought on by the rising mercantile class, specialization took effect, and by the Revolution, more educated and ideological editors were in charge. The Revolutionary press established a role of advocacy and reflected the political stirrings of a restless country. Thomas Paine's *Crisis Papers* and later Alexander Hamilton's *Federalist Papers* reflected the heavy doses of persuasion, propaganda, and public relations that from time to time have been important adjuncts to the press. Paine was an immigrant Englishman who arrived in Philadelphia two years before the colonies declared their independence. During the bitter winter of 1776–77, he published a series of pamphlets entitled *The American Crisis,* designed to stiffen the resolve of the rebellious colonists. Ten years later, the *Federalist Papers* were published. They consisted of a series of eighty-five essays written by Alexander Hamilton, James Madison, and John Jay, and they appeared in newspapers in 1787 and 1788 in support of the Constitution, which was then being ratified by the individual states.

After independence, the press quite naturally evolved into its role as political advocate. The successful part that the press had played in keeping the fires of revolution fanned led to extreme factionalism as it divided itself in vociferous support of the two warring political parties—the Federalists and the Anti-Federalists, or Republicans. Not only did the press support political parties, but political parties supported the press. The political press lasted until well into the nineteenth century, with subsidies paid newspapers by political factions to act as their mouthpieces. Since this was also an era of expansion, new publishers moving westward tended to take their political allegiances with them. Many were appointed postmasters, a kind of sinecure or political plum that permitted them to use the franking privilege and so distribute their papers through the mails at no cost. The official position further gave them unique access to news and gossip. From this era stemmed the American press's traditional political interest.

Throughout these colonial, revolutionary, and political press periods, the newspaper suffered from the same technological, geographic, and transportation barriers that hindered all communications at the time. From the dawn of mass communications until well into the nineteenth century, messages traveled no faster than people or horses could run, pigeons could fly, or boats could sail. There were a few exceptions—semaphors (signaling with flags), smoke signals, drums, or horns—but substantive messages entrusted to print media were limited by the terrain, weather, and durability of animals.

Carrier pigeons, used extensively in the early nineteenth century, could travel sixty miles an hour in good weather, and each pigeon could learn one basic route, not exceeding two hundred miles. Horses, the standard message bearers until the end of the nineteenth century, could average fifteen miles per hour on short runs. The much-heralded Pony Express of the mid-nineteenth century, for example, required 190 stations, 500 horses, and 80 riders to average five miles per hour from Missouri to California.

The difficulties of travel had particular impact on communications in America, where continental distances are great. Geography has shaped the social structure of America and has played a significant role in the development of the mass media. The hardships of wilderness life—the settling and subduing of a vast continent whose boundaries were unknown at the outset—left little time for recreational reading. Pragmatic settlers and villagers of the young nation sought information they could use as tools, much as they used other tools in coping with their geographic and technological constraints.

Early American settlements were isolated, with few interconnections and only a tenuous sea link to the Old World. Transportation systems grew slowly. This factor placed a premium on self-reliance, which complemented the independent spirit of the settlers. Circumstances, as well as temperament, fostered devotion to the concept of local autonomy, which became one of the guiding principles of the growing nation. Home rule, states' rights, parochial education, and the local emphasis of the press were to be manifestations of this focus.

Through the late eighteenth century, printing technology changed as slowly as did the methods of transportation. Printers of the fifteenth century would have been comfortable using the presses of America in the early years of the Republic, for the equipment was almost the same. It still took

Colonial and pioneer
printers operated
presses that had
changed little in three
hundred years.

an experienced printer nearly an hour to produce two hundred copies of a
single page. The pages, made of parchment or rags, had to be hung up to
dry, for the ink was still somewhat primitive and paper had to be wetted in
a trough to take a good impression. After handsetting each individual letter
for the opposite side of each page, the printer cranked down on the sturdy
hand-press, at the rate of two hundred impressions per hour. Each impression, or "token," required thirteen distinct operations.

Many colonial printers were true craftsmen, despite their crude
equipment. But 1800 spelled the beginning of the end of individual printers'
intense involvement with their newspapers' entire production, and in many
ways the beginning of the end of craftsmanship. The first year of the nineteenth century was the year Eli Whitney began mass producing muskets
for the U.S. Army, introducing the concept of interchangeable parts, which
encouraged other industries to innovate.

Technological developments during the century helped the press
emerge as a mass medium. From 1811 to 1813, Saxon journeyman printer
Frederick Koenig applied steam power and developed a printing press with
an impression cylinder, relegating to museums the converted wine presses
that had been the staple of the industry for centuries. The cylinder process
was improved, and by 1830 the New York firm of R. Hoe and Company
produced a flatbed and cylinder press that, within two years of its invention,
was capable of printing four thousand impressions an hour—compared with

the two hundred an hour of a few years past. Other improvements in printing presses followed, and by the end of the century presses were capable of printing seventy-two thousand thirty-two-page newspapers in an hour.

All the speed in the world wouldn't have resulted in faster production of print media if there were no paper to feed the presses. Due to the limited supply of linen and cotton rags and parchment made from animal skins, substitutes had to be found. In 1799, N. L. Robert patented a paper-making machine (the Fourdrinier) utilizing principles still central to paper production. In the 1840s (at a time when northern printing presses felt the threat of southern curtailment of cotton) technology was developed whereby wood pulp could be made into paper, and within two decades paper making was big business.

1833–1920 As we learned earlier in this chapter, the newspapers and journals of the eighteenth century had not been distributed in mass numbers. Geography and the limitations of technology had something to do with it. Literacy, or lack of it, also had an effect; people who could not read bought few newspapers. Another deterrent to mass circulation was the content of newspapers, which tended to stress opinion, shipping news, and literature.

When Benjamin Day started his penny press, the *New York Sun,* on 3 September 1833, the first true mass circulation newspaper was born. The cost per unit for newspapers had decreased with improved technology, and growing literacy and urbanization meant that primary circulation impediments—illiteracy, geography, and technology—had been overcome. The price of one penny was achieved because Day had shifted the costs of producing the paper to advertisers. This reasonable price contrasted with the five or six cents a copy for other papers and appealed to the public, which quickly made the *Sun* the most successful newspaper in the country.

Other papers copied Day's model. Like the *Sun,* they included the use of sensational stories about crime, violence, and mayhem, and their circulation also improved. This must have been an exciting time to be in the newspaper business. Almost overnight the form and format of newspapers began to change. In 1835 James Gordon Bennett established the *New York Herald,* challenging the *Sun* with a combination of lively stories and news about the affairs of government and the financial community. Like Day, Bennett emphasized the reporting of events rather than the analysis of ideas. He reached beyond the local community for news, setting up couriers to speed the news from distant points. When the telegraph proved viable, Bennett and other editors promptly tied into the system, all for the purpose of getting more news to the public as quickly as possible to sell more papers and, not coincidentally, to attract more advertisers whose contributions kept the price per unit within the ordinary citizen's pocketbook.

The penny press arrives: Ben Day's *Sun* ushers in the lively, mass-circulation newspaper.

Bennett, the first of the great personal editors, was followed by Horace Greeley (*New York Tribune*, 1841) and Henry J. Raymond (*New York Times*, 1851). These men established some of the basic tenets of journalism as a discipline. They improved upon the strategy of emphasizing stories about events and in the process created reporters who sought out news. In their newspapers they provided the first evidence of social concern. They established the newspaper in its role as public watchdog; in keeping a constant check on government, they brought about the adversary relationship between the press and government that has continued until today. They created new techniques for improving circulation, including the use of newspaper crusades—campaigns to comfort the afflicted (appealing for social welfare, reforms in big business, and changes in many other institutions) and campaigns to afflict the comfortable (exposing corruption wherever they could find it). They opted for specialized coverage of finance, religion, society, and the arts, thus beginning the departmentalization of newspapers that is known today. And most significantly, they discovered the potential of advertising. Each contributed to a golden age for newspapers, an era in which it all came together for the ink-stained journalists.

It was during this period, too, that newspapers seriously began to expand their coverage beyond the purely local and topical. Raymond, for example, organized relays of couriers stretching north to Newfoundland to meet incoming vessels from Europe so that his *New York Times* would be first with international news. He visited Europe to hire what were the first foreign correspondents to sniff out significant information exclusively for the *Times*. He personally accompanied presidential candidates on their campaign tours to give the *Times* firsthand political dispatches. His competitors soon followed suit, and news gathering became a specialized and, incidentally, an increasingly expensive function.

Samuel Morse's telegraph was prophetic in its first message—"What hath God wrought . . ."—as it ushered in the electric age and the communications revolution in 1844, connecting Baltimore and Washington. Newspapers adopted the telegraph immediately, and it brought about significant changes in them. With its many short, concise messages (longer messages were expensive, and correspondents feared mechanical failure and therefore wrote tighter stories, jamming the most important facts in the opening paragraphs), and its speed, the telegraph began to strengthen the connections between American cities and towns. Where the railroads went, there went the telegraph lines, and newspapers quickly followed, spreading into the outlying regions. From 1845 to 1856 in Illinois alone, some thirty daily newspapers sprang up with the appearance of the telegraph.[5]

The era of the personal editors, which lasted from about 1840 until 1870, was a bridge between the old press and the new. The personal editors who elaborated upon the principle of popular appeal also began a methodical organization of the press into a major social institution. Interlocking demands of expensive technology, advertising support, and popular appeal created a cause-and-effect relationship that characterized mass media from that time on. The press found it necessary to develop maximum circulation through sensational treatment of news events to attract advertisers who provided them revenue to competitively offer more popular appeal.

Inevitably, this process, which involved technological, sociological, and commercial factors, led to circulation wars in which the major newspaper publishers sometimes engaged in ruinous competition for audiences. The goal was to capture the growing volume of national advertising from businesses and industries prepared to sell material and services on a regional or national scale. This was the age of the great newspaper empires of Joseph Pulitzer, William Randolph Hearst, and Edward W. Scripps, who each owned several newspapers across the continent. Both Scripps and Hearst formed their own wire services—United Press (UP) in 1907 and International News Service (INS) in 1909—to consolidate this control. The great westward push following the Civil War opened new territory and new population centers; a growing urbanization was born of the Industrial Revolution; and an influx of foreign immigrants further concentrated populations in the cities where the newspapers could reach them. The telegraph also assisted in dispensing national advertising, which was attracted by the population concentrations and by the press empires to serve many different geographical areas.

Pulitzer had moved from the *St. Louis Post-Dispatch* in 1883 to take over the ailing *New York World*. Within four short years he increased the circulation to an astonishing 250,000 a day. Hearst, Pulitzer's arch competitor, came to New York from the West Coast where he had been the editor of his father's *San Francisco Examiner*. During Hearst's years at the *Examiner,* he had increased its circulation by applying the principles of sensationalism devised by the penny press fifty years earlier. In 1895 he bought the *New York Journal,* ready to take on Pulitzer and the *World*.

In his drive to add to the circulation of the *Journal,* Hearst hired a number of Pulitzer's best people, and Pulitzer, in turn, hired people away from Hearst. One of Hearst's catches was the artist who had created the cartoon strip called "The Yellow Kid," which featured a simplistic, folksy philosopher in a yellow nightshirt. The Kid became associated with the times and gave rise to the label *yellow journalism,* depicting the excessive sensationalism of that era.

This was flamboyant journalism in its most extreme form; an uninhibited time when editors invented incidents and headlines to go with them, and facts played a relatively small role in journalism. Indeed, Hearst was even accused of having started the Spanish-American War as a circulation device. It wasn't true; even Hearst wouldn't go that far. But inflammatory headlines proclaiming the sinking of the *Maine* to be the work of an enemy and an offer of $50,000 reward "for the detection of the perpetrator of the *Maine* outrage" could hardly be described as having a calming influence on the people who read the *Journal.*

Newspaper Czar William Randolph Hearst.

Pulitzer, Hearst, and Scripps engage in circulation wars as they build their newspaper empires.

Publisher Joseph Pulitzer, in competition with William Randolph Hearst, brought about yellow journalism.

The efforts of Hearst and Pulitzer paid off handsomely as circulation at both the *Journal* and the *World* increased to over four hundred thousand a day, making newspapers big business in a society that was becoming accustomed to big business. Remember that this was the age of industrial giants in steel, oil, and railroads, who made things happen just by flexing their muscles. The major publishers were simply the newspaper counterparts to those massive machines.

The 1890s and early 1900s also saw an explosion of feature and nonnews content in the newspapers. Comic strips and advice to the lovelorn, games and puzzles, and features and columnists made their entrance. Basically, yellow journalism was an appeal to the semi-educated urban populations swelled by foreign immigration who demanded a substantial measure of entertainment in their press.

A relatively high ratio of nonnews had been a part of newspapers for some time. Throughout the nineteenth century nearly a third of the content of major newspapers had been devoted to literary fare by such authors as Nathaniel Hawthorne, James Fenimore Cooper, and Mark Twain. In the 1880s the Bok and McClure syndicates furnished syndicated literary material to the press. But the period of yellow journalism redirected the material of appeal to the common citizen; and the comics, preprinted features, and popular columnists replaced serious authors as feature entries.

Gradually, readers reacted negatively to the press's excesses as they did to other flagrancies in society. By 1910 a notable decline in yellow journalism was evident. Even Joseph Pulitzer had come to shun sensationalism in favor of more serious and socially responsible coverage of national and international events. Taking a cue from the increasingly popular *New York Times,* with its accurate, documentary, and clean news for an increasingly literate readership, Pulitzer's *World* began to follow suit. Some readers, it seemed, were growing tired of the emphasis on crime, sex, and faked stories.

In a wave of social reform, the times were characterized by crusading writers—the *muckrakers*—who focused their pens and typewriters upon major problems caused by rapid industrialization and urbanization. Investigations of business, political, and social ills were conducted with regularity. Although the majority of muckraking efforts were undertaken by magazine journalists, as we will discuss in the next chapter, the metropolitan daily newspaper devoted column after column to fulfilling its important watchdog function. Like their counterparts of the 1960s and today, the muckrakers sought to interpret important new trends in society and effect change by arousing public opinion. One muckraker, Upton Sinclair, attacked the newspaper industry in a book he published privately entitled *The Brass Check*.

It is also interesting to note that it was during this period that the P.M. or afternoon paper made its appearance. Some say it was a direct result of the electric light; electricity extended the day and made it feasible to

Yellow journalism emerges from the Pulitzer, Hearst, and Scripps newspaper empires.

Muckrakers react to institutional excesses.

read an evening newspaper in the twilight hours. Perhaps more important, afternoon newspapers were better able to keep up with a day's activities that included news from Europe distributed by the newly-formed wire services, while new press technology permitted more rapid dissemination of spot or breaking news. Rapidly, P.M. papers captured a substantial portion of the market, and many of the newspaper barons rushed after the afternoon circulation, often issuing both A.M. and P.M. newspapers within the same city.

Since 1920 The last two eras of press history, jazz journalism, and the age of consolidation, bring us to the present. The early years of this century saw a more temperate press, epitomized by the resurrection of the *New York Times* by Adolph Ochs starting in 1896. Ochs set out to produce the "newspaper of record," and for many readers he succeeded in achieving his goal of providing the most complete and accurate coverage of local, national, and international news.

Although the *Times* became a powerful voice, Ochs did not speak for everyone. During the roaring twenties and into the depression of the 1930s, jazz journalism emerged, representing the renewed appeal of sensationalism. The *New York Daily News* was founded in 1919. Utilizing the smaller tabloid format (providing easier handling by subway riders) and extensive use of photographs, the *Daily News* quickly grew to be the newspaper with the largest circulation in America. Its audience was the working class and others in New York who were looking for excitement and entertainment in a newspaper rather than an in-depth analysis of current events.

Jazz journalism reflects the Roaring 20s.

These were the glamour years of journalism, with *Front Page* and a rash of other movies about the newspaper business during the 1930s that stereotyped the reporter as a tough, hard-drinking, fearless, sentimental man (seldom a woman), who spoke with a cigarette hanging from the corner of his mouth and never took off his hat with the press pass in it—a man who solved crimes, cleaned up city hall, and defeated dishonest politicians single-handedly, all in time to meet the everpresent deadline. This was a glorious age for reporters, arriving just before network radio began to usurp the market for both advertisers and audience.

The Great Depression of the 1930s took its toll of marginally profitable newspapers, as it did of other businesses. Newspapers found it difficult to adapt to the threats posed by radio, which had come into its own as a form of cheap mass entertainment as well as a viable news medium. Given its ability to bring instantaneous news to America, radio did much to eliminate the "Extra" editions of newspapers, which until the 1930s were issued whenever major news stories broke. It was only later that newspapers and radio reached a separate peace, each doing what it was physically suited to do: radio provided the up-to-date headline news, and newspapers gave the longer background stories.

Newspapers cope with threats posed by radio and television.

Bette Davis in *Front Page Woman,* one of many films of journalism's swashbuckling, sensationalistic times.

World War II forced the nation's press into a more responsible role than it had during its jazz days. Newspapers were joined by radio in bringing America the sounds and feelings of global conflict. Some two hundred U.S. correspondents were abroad in 1941, before Pearl Harbor. By the end of the war their ranks had been swelled to more than 1,600 accredited journalists. Perhaps the most famous were Scripps-Howard's Ernie Pyle and cartoonist Bill Mauldin.

The years following World War II saw newspapers having to adjust to the newest information, entertainment, and advertising medium: television. By the early 1950s, newspapers feared television as they had feared radio before it. In part, their fears were legitimate: TV was a far more attractive medium. So paranoid was the press about television that in this period many newspapers refused to mention television in their columns or even publish a television log. This was a far cry from today's pattern of special television editors on major metropolitan newspapers, and the publishing of elaborate magazine-type television supplements in the Sunday papers to rival the *TV Guide.*

In a sense television helped the printed press, just as radio had done a generation earlier. Because television could report the highlights of major events more quickly and graphically, newspapers were able to expand their interpretive content and provide a rational framework for understanding those events. Television's treatment of the news, typically incomplete fragments, stimulated the public appetite for more. In a familiar pattern, the electronic audience returned to print for explanation and elaboration.

Another change prompted by television was a move by newspapers to include more feature material, which network television has been reluctant to handle: food and fashion, travel, criticism, recreation and leisure, youth, and so on. Cable television, with its appeal to smaller audiences, has been quicker to handle feature material than the networks. This, of course, has posed a challenge to newspapers. Today, many newspapers rival magazines in the range and variety of coverage. In fact, many of the Sunday supplements to newspapers are weekly magazines, aimed at the same audiences as magazines and fully competitive with them.

Not all newspapers have thrived, of course. Those appealing to the lowest common denominator suffered the greatest circulation losses, which prompted many to go out of business. Generally, the more reflective newspapers have survived the competition from television, but even they have had to adjust. One of the most frequent forms of adjustment has been the consolidation of newspapers into chains. The costs of big business have forced this change and will be the subject discussed next.

The Structure of Contemporary Newspapers

At one time the local newspaper was a family-run operation. Many of them still are, of course, but more and more newspapers are now owned by chains and conglomerates. In this section we will look at the ownership patterns of American newspapers and then explore the operations of papers, including the roles of the reporter and other newspeople.

Ownership and Consolidation

In the late 1960s, press critic Ben Bagdikian wryly commented that the newspaper business is a great, clanking industry that buys paper at seven cents a pound and sells it at thirty-six.[6] He was speaking at a time when newsprint sold for $143 a ton, and the typical newspaper sold for 9.5 cents a copy on weekdays and 18.1 cents on Sundays.

By the mid-1980s, publishers were paying far higher wages to writers and printers utilizing incredibly more expensive equipment to implant ever more costly ink on a lighter weight newsprint costing more than $535 a ton. The product was an average 72-page daily costing twenty-five cents and a 265-page Sunday edition costing sixty-nine cents that reached a smaller proportion of the reading public. The newspaper business remained a great, clanking industry; with a workforce of 461,300, it ranked among the nation's largest manufacturing employers.[7]

As newspaper production costs have skyrocketed, economic pressures have forced all newspapers to change. The most noticeable trend has been toward consolidation and group ownership. Consolidation finds separate newspapers signing *joint operating agreements,* typically sharing printing plants and advertising departments and sometimes even editorial offices. More common is the movement toward *groups* or *chains,* in which two or more newspapers in different markets have common ownership. (The industry prefers to call them "groups," because of negative connotations associated with "chains.")

Economic pressures have forced newspapers to change; consolidation and group ownership are the most obvious trends. Head-to-head competition may be a thing of the past.

Daily newspapers were published in 1,528 different American cities in 1987, but only 125 of those cities had two or more dailies. Additionally, in all but forty-seven of those cities, the same firm owns or controls the operations of both papers. Head-to-head competition is practically a thing of the past, as is separate ownership of morning and afternoon dailies.

The current picture, so drastically different from those earlier phases of newspaper history during which a few dollars and a small printing press meant one could become a community's newspaper publisher, is a reflection of economic reality. A major reason communities today don't have competing newspapers is because the community and its advertisers cannot afford to support them. Just twenty-five years ago, the general formula used to place a value on a newspaper enterprise was established by multiplying its circulation by $100. Gannett's recent purchase of a nine-thousand-circulation daily in western Pennsylvania for $10 million is an indication that the going rate has multiplied tenfold in the past generation. Very few local capitalists can scrape together that kind of money, no matter how great their civic pride. Gannett's foray into a nationally distributed daily, *USA Today,* set the wealthy corporation back more than $100 million in 1983 alone. With 1984 losses running a mere $90 million, and 1985 losses perhaps half that, the media conglomerate was hoping to increase advertising sales and circulation enough to break even by 1987. No small potatoes, that.[8]

Individual, family-owned newspapers are becoming a dying breed for several reasons. Despite the expense of operation, they still tend to show a decent return on their investment, and are therefore attractive to chains whose larger capital base can better afford to update the plants and maintain operating standards. In addition, current tax laws encourage chains and conglomerates to reinvest pretax profits into the purchase of new papers. Because of the competition for their purchase, they have become increasingly expensive, and only chains and conglomerates can afford that escalated price. Some owners are happy to sell out while remaining aboard as local publishers, still reflecting the community, but with certain inherent management economies and securities. The newspaper gets the benefit of group administrative expertise and the advantage of some centralized functions, such as bulk newsprint purchase. Also, a more appealing package can be made available to national advertisers; syndicates, columns, and features are more or less standardized where appropriate.

Sometimes this arrangement appears to work without disturbing the local orientation of individual newspapers. Some can continue to provide the same kind of service both to their communities and their local advertisers. But if chains are interested primarily in profits and not editorial content, there may be more disadvantages than advantages to the move. Lord Roy Thomson, who owned seventy-seven American papers as part of his international empire, bought newspapers to make the money to buy more newspapers in order to make more money. To him, the editorial content served only to separate the ads. This same philosophy is found in concerns dealing with oil, automobiles, shoes, and deodorants. But in the case of newspapers, we are talking about the only institution specifically singled out by the Constitution, one that day in and day out helps shape and reflect the national consciousness. Little wonder, then, that media observers are concerned.

At the turn of the century, there were only eight chains in the United States. They controlled twenty-seven daily newspapers, which accounted for 10 percent of the nation's circulation. The average chain owned 3.4 newspapers, and was frequently a family concern that developed as children acquired newspapers in towns adjacent to the family stronghold. By 1986, there were 156 chains; a year later, after some of the bigger chains bought out some of the smaller ones, the number was down to 146. Those 146 chains control 73 percent of total United States daily newspapers, and 80 percent of daily circulation—a fact that cannot go unnoticed.

Seven chains control nearly one-third of the nation's total newspaper circulation. Listed in decreasing order, based on their daily circulations, they are the Gannett Company (with 92 daily newspapers, 5.8 million circulation); Knight-Ridder Newspapers, Inc. (33 papers, 3.8 million circulation); Newhouse Newspapers (27 papers, 3.0 million circulation); Times Mirror Company (29 papers, 2.7 million circulation); Tribune Company (9 papers, 2.6 million circulation); Dow Jones & Company, Incorporated (23 papers, 2.5 million circulation), and The New York Times Company (26 newspapers, 1.7 million circulation).

Other chains own greater numbers of newspapers than many of the "Big Seven" firms, but their total circulation is smaller. For instance, Thomson Newspapers Incorporated (U.S.) owns 99 papers with 1.5 million total circulation, and the DonRey Media Group owns 57 newspapers scattered throughout the Southern United States, with a combined circulation of only 751,000.

The case for decreased newspaper competition can be argued both ways. The absence of competition removes a balance of viewpoints, but actually the increased diversity of other media—in particular, local radio and regional television stations—compensates. It also removes pressure for timeliness, the "scoop," and in theory at least allows for more in-depth reflection. A newspaper without direct competition is generally healthier financially, more stable, and less prone to either advertiser or political pressure. Nevertheless, although there are advantages to reduced competition, the tendency is always to view this situation with alarm.

Much has been made of the decline in the number of newspapers, but as in most things statistical, figures can sometimes be misleading. Metropolitan newspapers have declined both in number and in overall circulation

Fewer family-owned papers and more chains concern many observers.

over the past couple of decades. However, this loss has been more than compensated by the growth of community or peripheral dailies, reflecting the overall social movement away from the central cities to suburbia. Let us not forget, however, that overall newspaper circulation, which has hovered between sixty and sixty-three million per day since 1965, has not kept pace with population growth. This means, of course, a net decline in readership and in the effect of this medium.

To put the situation in perspective, these figures may merely reflect the addition of other media—radio, television, and the proliferation of magazines—wherein all are simply achieving their respective shares of the changing market in a competitive situation. The addition of a still newer medium will undoubtedly dilute the total media market even further.

Newspaper Operations

Metropolitan papers have five divisions: editorial, advertising, production, circulation, and administration.

The news hole is determined by the advertising department.

The circularity of a newspaper's basic operation has already been noted: how it must attract readers as an inducement for the advertisers who pay the bills. The organizational structure of a major metropolitan newspaper reflects this by being divided into five major divisions: (1) editorial—to produce the copy and handle the news; (2) advertising—to solicit and coordinate the basic revenue-producing activities; (3) production—to physically print the newspaper, converting copy into editions; (4) circulation—to sell and distribute the finished product; and (5) administration (i.e., purchasing, promotion, and accounting)—to coordinate the activities of the other four.

A newspaper's deadlines for various editions are functions of several departments, but ultimately they depend upon circulation schedules, based upon the farthest home on the farthest route of the newspaper's distribution area. To meet those delivery schedules, production or printing schedules must be strictly followed. This places a burden on the editorial department to meet its news deadlines—the last second that new copy can be accepted. But the editorial department cannot begin filling its *news hole* (all the space devoted to non-advertising matter) until the advertising department has determined the number of ads sold and therefore the size of the paper to be printed on any one day.

At the top of the ladder in a major metropolitan newspaper is the publisher, who has overall control. Answering directly to the publisher are generally two people: a general manager in charge of administration, circulation, advertising, and production; and an editor in command of the paper's editorial content, who considers the significance of news.

Answering to the business manager are persons such as the comptroller, purchasing agent, advertising director, circulation director, production manager, promotion director, and industrial-relations manager, who is in charge of highly specialized negotiations with a half-dozen trade unions.

Under the editor is the managing editor, who is responsible for the day-to-day content of the newspaper. Under the managing editor are principally the city editor, charged with local news; and the wire editor or news editor, who is charged with other-than-local news—the regional, national, and international stories that come in from correspondents and the wire services. The photo editor runs a staff of local photographers and screens wire photos for suitability. The copy editor checks raw copy for accuracy and style, writes headlines, and sets approved stories into a *page dummy*—provided by the advertising department—on which the advertisements for the day have already been indicated.

The major departments outside the city room are more or less autonomous. Sports, society, business and financial, and entertainment sections often have their own editors and staffs and are generally allocated a certain number of pages daily to be filled with their specialties. The Sunday department has charge of special Sunday sections (i e., real estate, travel, television, home and garden, books, and opinion) and special editors are usually employed for these various sections. It is in these special sections that the majority of feature stories occur. The editorial pages have their own staff who answer to the editor or publisher or both.

Within these departments are specialists—sometimes called editors—on such topics as science, medicine, youth, education, religion, environment, military, farming, senior citizens, and politics. Generally, they are not allocated specific space but are required through their expertise to keep up with the developments in their fields and present timely articles.

At the lower level of the hierarchy are the reporters and correspondents who actually unearth and write the news, and who have most of the personal, on-the-scene contact. Some are general assignment reporters, delegated to stories at the direction of the city editor. They cover fires, murders, accidents, and Rotary Club luncheons, often taking a photographer with them. Special-assignment people stay on location at city hall; at state, federal, and county beats; at the courthouse; or at police headquarters. They, too, are specialists, assigned to the principal sources of the news, those locales where news is being made.

The News People

There are hundreds of reporters and correspondents and specialists on a metropolitan newspaper; they are supervised by dozens of editors. There are batteries of two or three dozen teletypes spewing out reams of copy daily, and all of this must be condensed by personnel into relatively few pages. In addition, there are scores of features on all conceivable topics, mail and phone calls from bureaus overseas and from Washington, New York, and the state capital. There are stringers—free-lancers paid so much per inch of copy—and correspondents in outlying communities, and there are scores of press agents advocating their clients' causes in numerous fields.

All of these personnel funnel their output into the newsroom, where a half-dozen or so harassed senior editors sift through it to decide what to print. They are the gatekeepers of a newspaper; on their judgments rest what visions of reality the public shall see—what events, in essence, take place. A rule of thumb is that only about one tenth of the available news

Gatekeepers can use only about 10 percent of available material.

Those at the lower level of the newspaper organization—reporters and correspondents—unearth and write the news.

Those at the lower level of the newspaper organization—reporters and correspondents—unearth and write the news.

gets into print, leaving about 90 percent (an overpowering figure) in the gatekeepers' wastebaskets, on the floor beneath the teletype machines, or as flickering images left on the gatekeepers' video display terminal.

Fortunately, because of the variety and pervasiveness of competing media all along the spectrum, the chances of running into a news item, if it is significant at all, are pretty good. But this does not detract in any way from the responsibility and, in a sense, the power of the gatekeepers or editors or the reporters who write according to their own perceptions and views of the world.

For years reporters were told that the average reader had at best an eighth grade education, and to write accordingly. Naturally, such reporters and editors thought of themselves as members of that amorphous middle class—or even lower middle class—based on journalists' relatively low incomes prior to the 1970s. The popular myth was that journalism was peopled by "everyman" producing a product for "everyman," inherently distrusting both authority and establishment figures.

Investigations of readers and reporters have disproven both stereotypes. Sociologist John Johnstone and his colleagues demonstrated very clearly in the 1970s that American journalists had higher levels of income, education, and social status than the average American citizen. The typical journalists were white, Anglo-Saxon, Protestant (although not highly active churchgoers) males, younger than the majority of the work force, and moderate to liberal in their political beliefs. In short, they were not representative of middle America; instead they were members of the same elite classes over which they thought themselves to be watchdogs and for whom they were unknowingly producing their daily products.[9]

The myth of the journalist as a normal citizen was further shattered in a 1981 study of some 240 journalists and broadcasters working for the most influential media outlets in America. Residing primarily in northern industrial states, 86 percent considered themselves capitalists. This is not surprising considering that nearly half of them earned more than $50,000 a year. Half rejected any religious affiliation; only 8 percent attended church or synagogue weekly and 86 percent seldom or never attended religious services. On political matters, 54 percent considered themselves left of center, while only 19 percent described themselves as being on the right. In presidential elections between 1964 and 1976, no more than 19 percent of those interviewed had ever voted for a Republican candidate. Nine out of ten supported abortion on demand, three-quarters supported gay rights, and nearly all supported sexual expression and the rights of women, environmentalists, consumerists, and journalists. When asked which group would be best suited to direct and remake American society, the media elite tended to choose themselves rather than government or business groups.[10]

The challenge for reporters and editors is to understand that there may be more distance between themselves and their readers than they realize; therefore they must take the necessary steps to become attuned to reader interests. It is never easy. Deciding what is important to readers and monitoring public preferences will continue to be vexing. In their recent book, Dennis and Ismach devote several chapters to the importance of journalists' understanding their own values and recognizing community needs. The authors conclude that journalists should stop thinking they can stay in touch with audiences through some mysterious process of osmosis. Rather, they should systematically utilize personal and recorded sources to track community trends and plan on never-ending professional development to intensify their own sensitivity.[11]

In the final analysis, newspeople and their newspapers mirror the community far more than they influence it. They must reflect the community in that the newspaper is a measure of public taste. It serves a basic economic purpose. If a paper does not ideologically pattern itself after the general mores of its community, readership will drop; this is the beginning of the end. Other media and other newspapers are anxious to steal readers and the additional revenue they represent, not only in subscriptions but, much more significantly, in added advertising volume and higher advertising rates.

Research reveals that neither journalists nor regular readers are "typical" Americans.

The Nature of News

Definitions of news abound; nowadays it means information that is useful or gratifying to audiences.

Although we have talked about the structure of newspapers and the people who gather the news, we have not described what it is that constitutes news. Perhaps that is because *news* is an elusive term to define. Walter Lippmann noted long ago that "something definite must occur that has unmistakable form" for it to be news. This is what happens at city hall and the courthouse and elsewhere when form and a tangible frame are given to human interactions that permit them to be isolated and written about. These are the events that comprise the news. Thus, while news is certainly the clearest expression of the information function, it is an artificially contrived form of information, leaving much unsaid, unreported, and unprinted.

Lippmann's definition of news reflects a concern with objectivity—the neutral, transmission-belt role of journalists. Dozens of observers have offered their own definitions of news that differ significantly from Lippmann's, some of which take into consideration contemporary thrusts for investigative and interpretive journalism. Typical definitions place value not only on the unusual and abnormal but, more recently, on materials deemed useful or gratifying to audiences. Additionally, cynics suggest that news is whatever editors and reporters (gatekeepers) allow to pass through institutional channels. Lack of consensus over what meets those gatekeepers' criteria may not necessarily be a flaw of the industry. That similar stories frequently appear from paper to paper is probably due to the influence of the wire services, whose daily news budgets suggest which stories "deserve" emphasis. Also, group norms within and among individual newspapers frequently result in the appearance of collective news judgments. But the pattern is more often (or more likely) a case of normal institutional behavior rather than one of conspiracy, as some critics suppose.

Contemporary editors sensitive to their audiences have been influenced strongly by pressures brought on by investigative and interpretative reporters in the past decade or so. In-depth and specialized journalism is not a new phenomenon, as our discussion of newspaper history has indicated. For several periods, the American newspaper stressed interpretation and opinion over straight news. During most of the twentieth century, however, news has become a value in itself. Only recently, with television's reign, has the journalist as interpreter returned to the fold. The return has not come about without a series of battles within and outside the newsroom.

"Purists" have objected to the recent tendency of newspapers to fill their columns with "soft" rather than "hard" news. By traditional definition, hard news is news that justifies its existence on the basis of: (1) immediacy, (2) impact on the greatest number of readers (minor impact on a great number is equated to major impact on a lesser number), (3) physical or psychological proximity to the newspapers' readers, (4) events surrounding prominent personalities, (5) uniqueness (first, last, biggest, smallest, oldest, youngest, etc.), and (6) conflict. Such news coverage relates to tangible events. Soft news—i.e., humor, offbeat incidents,

BLOOM COUNTY   by Berke Breathed

personal "life-style," and experiences—by contrast, may be of interest to people but may lack significance. In the past several years, newspapers have responded to critics' suggestions that news should not be defined strictly in terms of events, but rather in terms of trends or changes in subtle ways of thinking about and dealing with environments. As a result, more and more editors arc finding space for soft news stories calculated to help individuals cope with their circumstances.

There are no set categories for classifying newspapers, but we can easily see that differences between newspapers do exist. The *Los Angeles Times* and the *River Press* in Fort Benton, Montana, differ in more ways than in size. And yet, size is a logical category. So, too, is frequency of publication. For our purposes we will look at big-city newspapers, community newspapers, and the minority press.

Most of our discussions of the development and structure of newspapers have emphasized big-city newspapers and those papers that are, on occasion, in the news. Most of these newspapers still stress local interests, although the definition of *local* varies from community to community as does the population being served. The *New York Times,* for example, has an immediate market of fifteen million people living in the states of New York, New Jersey, and Connecticut; and the market for the *Chicago Tribune* stretches from the Wisconsin border to Northwest Indiana.

In order to maintain a local focus, many major metropolitan newspapers publish different suburban and/or regional inserts several days each week, carrying news and advertising for each suburban or outlying area.

In a few exceptional cases, newspapers that began as local media have expanded to a national audience through the miracles of satellite or electronic transmission of the paper to regional printing and distributing houses.

The Types of Newspapers

Big-City Dailies

Most metropolitan daily papers concentrate on the news of their own cities.

Big city dailies such as the *New York Times* and the *Washington Post* are able to reach a national audience by using satellite technology and regional printing and distribution.

Several newspapers now use satellites, electronic transmission, and regional printing plants to span the nation.

Most notable is the *New York Times,* long a paper of national prominence, which prior to 1980 was distributed by mail to any location and in bulk to many cities for newsstand sales. With the inauguration of satellite transmission to Chicago in 1980 and Lakeland, Florida in 1981, and later to the West Coast, the *Times* demonstrated marketing techniques to other ambitious publishers. The venerable *Washington Post* and fledgling *Washington Times* established what they called weekly "national editions" written and composed in the East and distributed by satellite for same day delivery to numerous major cities throughout the country. The *Christian Science Monitor,* founded in 1908 as a serious, analytical antidote to the then prevalent yellow journalism, is published in Boston but distributed nationally through regional printing facilities in New Jersey, Florida, Illinois, and California—and in England for the international edition. And, of course, the *Wall Street Journal,* flagship of the Dow Jones Corporation, managed its ranking as the top circulating daily earlier this decade after utilizing its satellite and regional distribution system to reach the nation's business-oriented readers.

Even the big-city dailies depend on the wire services for much of their national and international news, but their resources allow some to maintain bureaus in other cities, both within the United States and abroad. The larger newspapers (see table 3.1) also have the resources to support investigative teams that may work on an involved story for several months before their efforts appear in print. The daily papers in smaller towns rarely can afford to assign reporters to do investigative work.

In a recent assessment of America's "Ten Best" metroplitan daily newspapers, *Time* magazine noted that the 1970s and 1980s turned out to be both the worst and the best of times for the industry.[12] The magazine noted that dozens of papers had died, but circulation was being maintained by the national distribution of the *New York Times,* the *Wall Street Journal,* and *USA Today.* (A factor not pointed out by *Time* was the steady growth in smaller dailies—under twenty-five thousand circulation—which helped

Table 3.1 America's largest daily newspapers.

Rank	Newspaper	Daily circulation
1.	The Wall Street Journal	1,952,283
2.	New York Daily News	1,270,926
3.	USA Today	1,179,052
4.	Los Angeles Times	1,086,383
5.	The New York Times	1,001,695
6.	The Washington Post	748,091
7.	Chicago Tribune	744,969
8.	New York Post	731,668
9.	The Detroit News	680,800
10.	Detroit Free Press	656,477
11.	Chicago Sun-Times	612,686
12.	Newsday	603,172
13.	San Francisco Chronicle	551,545
14.	The Boston Globe	516,284
15.	The Philadelphia Enquirer	504,946
16.	The Miami Herald	462,887
17.	The Newark Star-Ledger	460,330
18.	The Cleveland Plain Dealer	452,343
19.	Houston Chronicle*	425,434
20.	The Baltimore Sun	397,584

Source: *Editor & Publisher,* Audit Bureau of Circulation ANPA from "Facts About Newspapers '87," American Newspaper Publishers Association. By permission.

*The *Houston Chronicle's* figures are from March, 1986; all others are 30 September 1986 figures supplied to ANPA by the Audit Bureau of Circulations.

The Ten Best U.S. Dailies, according to *Time* magazine.

In 1984 *Time* magazine selected these newspapers (listed in alphabetical order) as the top in the business, based upon the following criteria: imaginative staff coverage of regional, national, and foreign issues; liveliness in writing, layout, and graphics; national impact achieved through general enterprise, command of some particular field of coverage, or a track record of training top-rank younger journalists: the *Boston Globe;* the *Chicago Tribune;* the *Des Moines Register;* the *Los Angeles Times;* the *Miami Herald;* the *New York Times;* the *Philadelphia Inquirer;* the *St. Petersburg Times;* the *Wall Street Journal* and the *Washington Post.*

---

keep overall newspaper circulation near sixty-three million.) Television newscasts had cut into papers' influence, but print reporters' education, status, wages, and expertise had reached new heights, according to *Time.* During the 1970s and 1980s metros dropped their old-fashioned women's pages, replacing them with trend-conscious life-style reporting. There was also a sharp increase in the quality of stories about arts and popular culture. Special sections on television became commonplace. Daily papers were carefully selecting from a much broader range of syndicated news and features.

USA Today—
A Special Case

No discussion of contemporary daily newspapers would be complete without pointing out the unique role of *USA Today*—the nationally distributed Gannett Corporation enterprise. Even before its much publicized inaugural issue in 1982, the paper was stirring up more controversy within the industry than any other event in recent memory—controversy on a par with that seen when television had entered the marketplace some thirty-five years earlier.

The concerns were probably justifiable. The wealthy Gannett organization, which then owned eighty-six newspapers throughout the nation and in Guam (plus a hefty collection of broadcast and cable outlets), had earned a reputation in the 1960s and 1970s as a bottom-line oriented company which paid more attention to profit than to quality. Reputations die hard, and Gannett's recent dedication to high quality reporting and editing, and to local autonomy, had not yet won the hearts of its fourth estate brethren. Overall, the newspaper industry was not looking forward to being blitzed by a low-grade, mass-produced competitor. Many journalists were openly hostile to the interloper. Others took a wait-and-see attitude.

USA Today has been praised and condemned, but it has already had a major impact on established newspapers.

Although it is still too early to predict the ultimate fate of "The Nation's Newspaper," there is no question that it has had a significant impact on the industry. Even a casual comparison of the typical daily of the mid-1980s and its pre-*USA Today* version will show how drastically newspapers have changed in both physical appearance and marketing strategies. Most noticeable have been: the splashy (sometimes indiscriminate) use of colored weather maps, photographs, and information graphics; new experiments in design, such as more white space, larger photos, rectangular or horizontal layout, modular presentation of news stories and advertisements, news summaries, news indexes, wider columns, and clearly delineated sections intended to lure readers into topics or themes; shorter, more tightly written articles; greater emphasis on news about sports, leisure time activities, and "good news" at the expense of interpretative or analytical pieces; new methods of promotion and marketing, including the strikingly attractive street boxes (it is not by accident that *USA Today*'s vending machines look like freestanding television sets, designed to appeal to today's more visually oriented consumers); and efforts to give an overall modern, "user-friendly" image to a traditionally staid product.

In the sense of bringing about a new awakening in the industry, or making newspapers more consumer conscious, *USA Today* has undoubtedly provided a service. It has not come about without some concerns, however. Many journalists belittle what they say is the stress on form over function, on newspapers driven by dedication to appearance rather than substance. (One joke making the rounds was that the *USA Today* would be responsible for a new category in the Pulitzer prize competition: "Best investigative paragraph.") There has also been a running debate between traditionalists and Gannett's chairman Allen Neuharth over what Neuharth called *USA Today*'s "journalism of hope" as contrasted with the "journalism of despair" which he claimed dominated the news business. In an *Editor & Publisher* interview, Neuharth called the journalism of despair "a derisive technique of usually leaving readers discouraged, or mad, or indignant." In its place he called for efforts to cover all the news—the good and the bad—accurately, but without anguish; with detail, but without despair. Editorials, he said, should emphasize hope and ideas for the good life, advocate understanding and unity, rather than disdain and divisiveness.[13] Obviously, most journalists denied that they were in the despair business. Some, however, took a second look at their news policies, and began to more carefully balance negative and positive news and editorial items.

Gannett's impressive marketing strategies, designed to reach upscale readers rather than the broader general population, were not lost on the industry. (We should recall that *USA Today*'s initial appeal was to airplane travelers and others who tended to be cosmopolitan and transient.) To many, however, such strategies bordered on the unethical, as they tended to disenfranchise the broad general readership. By 1984 and 1985, however, Gannett's successful drives to have the national newspaper home-delivered, and readily available in even small communities, tempered the cries of "elitism." *USA Today*'s first official audit, by the Audit Bureau of Circulation, showed it to have two-thirds street sales, 29.8 percent home and office delivery, and 3.6 percent "blue chip" sales—bulk sales to hotels, airlines, and other businesses.[14]

By mid-decade *USA Today* was doing slightly better in the marketplace than even its founders had expected. Circulation was up to 1,352,897 in late 1985, despite a doubling in price to fifty cents for each copy. Advertisers were slowly coming aboard, lured by the newspaper's demographic readership and the opportunity to display their wares in highly attractive four-color formats. According to Gannett's five-year business plan, the newspaper would still continue operating in the red until it reached its target of 2,350,000 circulation (and five million readers) in 1987, underwritten in the meantime by the corporation's many other successful media holdings.

The jury is still out. The public may love the paper, but many leading newspaper journalists remain unimpressed. *Washington Post* editor Ben Bradlee, for one, has complained that "if it can qualify among the top newspapers in the country, I'm in the wrong business."[15] And press critic Ben Bagdikian has written:

USA Today will be no gain for the reading public, which gets a flawed picture of the world each day from the new paper, and (represents) a serious blow to American journalism, since the paper represents the primacy of packagers and market analysts in a realm where the news judgment of reporters and editors has traditionally prevailed.[16]

As our historical treatments of mass media will attest throughout this book, every time a new medium or a radical departure from traditional media has come about, the established media first react with territorial fear.

When *USA Today* was first published in 1982, its physical appearance and marketing strategies created quite a controversy in the newspaper industry.

Only later, if and when the upstarts become entrenched, do the criticisms die down and traditionalists give them the credit they probably deserve. If that cycle holds true in the case of this newest venture in newspapering, the *USA Today* experiment will prove of value to the business if only to force it to reconsider some of its longstanding general operating principles and assumptions about newspapers' place in the media-societal mix. In that sense alone, the grand experiment by Gannett is a positive one.[17]

*Community
Newspapers*

Community newspapers, generally small nondaily operations serving the provincial interests of isolated communities or discrete suburbs, have assumed unique roles in American media history. A look at their statistics throughout this century tells us much about their place in the media mix.

In 1910, there were 16,227 weekly and 672 semiweekly and triweekly newspapers published in the United States, most of them in small towns. Since that time the number of weeklies has been drastically reduced to some 7,711, or less than half the peak figure. The decline has been steady, particularly in rural areas, but appears to have leveled off since 1970. However, weeklies have been enjoying significant increases in both average circulation and total weekly circulation. With an average circulation of 6,497 in 1986, these 7,711 papers were selling 50 million copies a week—meaning they were being read by some 125–150 million Americans, nearly double their 1965 circulation figures.[18]

The biggest growth area in newspapers is in suburban or community journalism.

In the meantime, the number of semiweeklies and triweeklies bottomed out at 300 in 1944, reflecting the wartime economy and the move into the cities. Since then, however, their numbers have more than doubled, nearly reaching the 1910 peak. More often than not, those semiweekly and triweekly papers are published in suburban communities, serving clientele who have many of the same needs for local news and advertising as their parents and grandparents had when they lived on farms or in small towns.

These community papers are traditionally considered to be the real gossip sheets, concentrating on local news that is easy to acquire. Many have printing operations for independent jobs on the side and may be run by husband-and-wife teams. More typical, however, is the utilization of centralized printing plants that turn out perhaps a dozen or more nondailies in the same region. The practice has saved numerous existing weeklies and has allowed for the births of many others, papers that otherwise could ill afford purchase and upkeep of printing equipment they would use only a few hours a week. Rather than losing their autonomy, as some feared, the elimination of printing worries has allowed small operators greater amounts of time to devote to other concerns, such as newsgathering, advertising, and distribution.

The rationale for the increase in these smaller community newspapers in the outlying areas is directly related to American affluence. The major metropolitan press provides depth: national and international news,

features of all kinds, the comics and entertainment, broad fare. Historical tradition, however, emphasizes the local community; the suburbs are where many people live, where they identify, where they belong. It is the local city councils and civic organizations that most directly touch residents' lives and it is the local school districts that educate their children. Further, it is generally in the regional shopping centers and from the local merchants that community members buy most of their goods. Suburban dwellers need information about their own communities, and local merchants need a vehicle in which to advertise their products and sales. These two needs meet in the community newspaper because the gross national product is large enough to provide sufficient disposable income to support the whole broad spectrum of media.

Community newspapers fill a void left by the major metropolitan dailies. No matter how large and wealthy, metros lack the resources—time, money, and personnel—to adequately cover all the news of each community. There may be dozens of separate communities within their circulation area. They are not necessarily in competition with community newspapers, as each fulfills essentially different roles. Even in their metropolitan editions, major newspapers cannot cover all the news of a dozen city councils and a half-dozen school boards, and the speakers and activities of several hundred civic, women's, veterans', labor, cultural, or charitable organizations. Yet it is these very activities that are meaningful to suburban dwellers.

Similarly, local advertisers cannot afford the advertising rates of the metropolitan newspapers, which are based on overall circulations of several hundred thousand or more. They do not need all that exposure; they need simply to reach the 10 or 20 percent of that number who live in their trading area. The community newspaper provides local merchants an advertising showcase pinpointed toward the community they serve at prices proportionate to exposure.

There has been a trend toward the establishment of minigroups of several suburban newspapers under the same management, attempting to serve several adjacent communities. The major portion of the content of these newspapers is the same, little more than editorial filler to carry advertising. However, as a gesture to locality, they prepare a new front page, filling it with specific information about each separate community. Often they publish under different *flags* (the name bar at the top of the front page) to further the locality illusion. With rare exception, these minichains abdicate their basic responsibility to provide a full quota of local news and gossip and become little more than *shoppers,* advertising throwaways.

The typical newspaper maintains an average of 40 percent editorial content to 60 percent advertising, which means that if there are sixty pages of advertising sold, the newspaper will have one hundred total pages. Since the customer pays only the cost of the newsprint, advertising covers the balance: equipment, salaries, overhead, and profit, for example. By contrast, shoppers or throwaways carry 25 percent or less editorial content (everything other than advertising, which does not necessarily mean news). These newspapers are often supported entirely by advertisers and are distributed free throughout the communities. Almost everyone is happy with the arrangement: consumers get a free newspaper filled with advertising information and some trivial fluff; advertisers are guaranteed marketplace saturation because of free distribution; and publishers, with less effort than newspapers traditionally demand, profitably serve the consumers and advertisers.

Sometimes shoppers discover that their readers have more in common than spending their money at the same shopping mall, and the shoppers may evolve into "real" newspapers, with a bit of news, a television viewing guide, some wedding announcements, school lunch menus, and words of wisdom from a local philosopher; and *voila!,* the publisher finds it possible to charge a token subscription price and produces a real newspaper, even though it has a lopsided advertising-news ratio. But it is not all that different from the mercantile press of colonial days, if we stretch our imaginations a bit. As these new newspapers become profitable, they will doubtless undergo the same cycle of economics seen in their larger brethren: small, shoestring operations bought out by chains, which are in turn bought out by larger chains and conglomerates.

Where there are audiences and advertisers, newspapers are bound to appear.

The audiences for some newspapers are determined by characteristics other than geographic proximity. Foreign-language newspapers and newspapers published for specific ethnic groups have a history that stems from colonial days. From the mid-nineteenth century until well into the twentieth century (during the years when immigration was highest), newspapers written in German, Polish, Swedish, Yiddish, and other languages were common in the cities where the immigrants settled. As each immigrant group became assimilated, the number of readers dwindled until the audience was no longer large enough to support publication. Some foreign-language papers did survive, of course, especially in close-knit communities such as the Chinese settlements in New York and San Francisco. In recent years the number of Spanish-language papers has increased due to the wave of immigration from Puerto Rico, Mexico, Cuba, and other lands where Spanish is spoken.

Freedom's Journal, the first newspaper published for blacks in the United States, was started in New York on 16 March 1827. In the initial issue the editors, Samuel Cornish and John Russwurm, expressed the desire to speak up for their own cause and not rely upon others to do it for them. Slavery still existed in both the North and the South, and even free blacks in the North were being attacked by some newspapers. Said the editors of *Freedom's Journal,* "Too long has the publick been deceived by misrepresentation, in things which concern us dearly, though in the estimation of some mere trifles." The quest for freedom, dignity, and moral improvement filled the pages of the journal during the three years of its existence.

In 1847 Frederick Douglass, a former slave, founded *The North Star* in Rochester, New York, for the purpose of promoting the abolitionist cause. Douglass was a brilliant orator and a talented writer who used the podium and the pen to argue against slavery, helping to build public sentiment for abolition.

Numerous other black journals appeared prior to the Civil War, most intended for intellectuals, regardless of color. By 1905, when Robert S. Abbott started the *Chicago Defender,* the black community was hungry for a general-interest newspaper. Abbott's plan to include sensational news for a mass audience paid off. The *Defender* was a success, prompting other black newspapers to pursue the general reader.

Several of the black newspapers started at that time are still published today. Among them are the *Afro-American,* a Baltimore paper founded in 1892 by John H. Murphy, Sr.; the *Amsterdam News* in New York started in 1909; and Robert L. Vann's *Pittsburgh Courier,* which began publishing the next year.

The circulation of black newspapers reached an apex during the 1940s and has dropped since then. Only three survive as daily papers, one of which is the *Chicago Defender.* It has a circulation of about thirty-five thousand and is part of a chain that includes the *Pittsburgh Courier* and eight other newspapers.

Minority Newspapers

Foreign-language and ethnic newspapers have a long, but not always successful, heritage.

The circulation of black newspapers reached an apex during the 1940s, but has dropped since then. Today the *Chicago Defender* has a circulation of about 35,000.

The Changing Newspaper Business

There are fewer daily newspapers now than at any other period in this century.

There has been a shift from P.M. to A.M. newspapers.

Economic and other pressures have caused significant patterns of change in the newspaper industry throughout this century (see table 3.2). Fewer daily newspapers are published today than at the turn of the century; the number at first increased, peaking out at about 2,400 during World War I. Then there was a steady decline until World War II when the total was 1,750, a figure that has held relatively constant since. By 1987 there appeared to be a slight decrease in the totals: a drop to 1,657 daily papers. However, because thirty newspapers were being published "all-day," in several editions, another way of looking at the figures reveals 1,687 "different" newspapers on the streets of America each day. Even so, the apparent constancy is deceiving. Large metropolitan cities and small cities with less than twenty-five thousand population have suffered losses in the overall number of papers published, while mid-sized cities and metropolitan suburbs have shown contrasting growth.

The figures reflect more than a shifting of American population; they reflect changes in newspaper readership patterns. Since 1955, the number of morning newspapers has increased substantially in all but the smallest cities, while the number of evening papers has declined in both major metropolitan areas and smaller cities. Between 1960 and 1985, the number of evening papers decreased by 238, and circulation dropped by 8.4 million. In that same time period, the number of morning papers increased by 1,170, and circulation by 12.3 million. The shift from evening to morning publication was due not just to deaths of some P.M.s and births of A.M.s, but to a shift in publication schedule within many American dailies. Since the end

Table 3.2 Number and circulation of United States newspapers, 1950 to 1985

Year	Number of A.M.'s	A.M.'s circulation	Number of P.M.'s	P.M.'s circulation	Total A.M. & P.M.*	Total circulation
1950	322	21,266,126	1,450	32,562,946	1,772	53,829,072
1955	316	22,183,408	1,454	33,963,951	1,760	56,147,359
1960	312	24,028,788	1,459	34,852,958	1,763	58,881,746
1965	320	24,106,776	1,444	36,250,787	1,751	60,357,563
1970	334	25,933,783	1,429	36,173,744	1,748	62,107,527
1975	339	25,490,186	1,436	35,165,245	1,756	60,655,431
1980	387	29,414,036	1,388	32,787,804	1,745	62,201,840
1985	482	36,361,561	1,220	26,404,621	1,676	62,766,182

Source: *Editor & Publisher,* from "Facts About Newspapers '86," American Newspaper Publishers Association.

*Totals may differ from A.M. and P.M. numbers combined because of some "all-day" publications. There were twenty-six "all-day" newspapers in 1985; they are listed in both morning and evening columns but only once in the total.

of the 1970s, several dozen existing daily papers have rotated their publishing times from an evening to a morning cycle. The flurry of conversions, according to the American Newspaper Publishers Association, has been motivated by several factors: (1) to provide more timely information (including late-breaking news), final sports reports, and complete stock market tables; (2) to compete more effectively with television and other demands on readers' time; (3) to extend the publications' circulation areas; (4) to avoid traffic congestion during delivery; (5) to provide same-day mail delivery; (6) to save energy by distributing at night and in the mails and by not running printing presses during peak energy usage hours; (7) to provide all-day exposure and same-day response for advertisers; and (8) to increase advertising revenues.[19]

Publishers are reacting to perceived demographic changes when they shift their production from evening to morning. Typical newspaper readers are middle-aged with some college education. They can be described as middle-income and employed in professional or managerial positions. Frequently newspaper readers' jobs are in the suburbs, where they and their families are also likely to live, shop, pay their taxes, attend school, and engage in leisure-time activities. The growth of white-collar occupations means that a larger percentage of workers have more reading time in the morning and use their evening hours in other pursuits, particularly television viewing.[20]

Are typical newspaper readers representative of the larger population? Do most people still learn about the news by reading newspapers? The answer, it appears, depends on whom you believe. When asked where they get most of their news about what's going on in the world today, Roper Organization survey respondents since 1963 have listed television more frequently than newspapers, with the spread growing more pronounced in recent years. But the American Newspaper Publishers Association has found that every day more people get news from newspapers than from any other

Demographic Changes

Newspaper readers are in the "upscale" demographic groups.

medium (about 70 percent, compared to about 60 percent for television and 50 percent for radio). Other surveys indicate that somewhere between 73 and 80 percent of America's adults claim they read a newspaper daily, that nine out of ten read at least one newspaper every week, and that they spend more time reading it than they spend with any other news sources.

Surveys on TV versus newspaper consumption are contradictory.

Circulation Another way to answer the questions concerning newspapers' impact is to look at circulation. Overall, newspaper circulation has not been keeping pace with population growth. Indeed, daily circulation peaked in 1973 at 63.1 million, then dropped off because of the energy crisis, years of newsprint shortages, and spiraling production costs. Not until 1985, by which time population had increased substantially, did circulation move back up to the sixty-three million mark—and that achievement, according to some analysts, is attributed solely to the million-plus new readers of *USA Today.*

Circulation has not kept pace with population growth.

Current circulation trends contrast with the early years of the century, when newspaper readership increased steadily. Between 1900 and 1930 circulation expanded from 15.1 million per day to 39.6 million. The depression slowed growth for a decade, but between 1940 and 1950 newspapers added 12.7 million readers, bringing the total to 53.8 million. Since then, large segments of the population have found other sources of information and entertainment.

A different data base—penetration of newspapers into households—shows equally depressing news for newspapers. A 1966 study by the Newspaper Advertising Bureau indicated that 79 percent of American adults read a daily newspaper. By 1984 that percentage had dropped to 66 percent.[21] In 1970, for the first time in contemporary record-keeping, total United States daily newspaper circulation fell below the number of households. The gap between households and newspaper circulation has widened since 1970, and by late 1984 the penetration figure had dropped to .74 newspapers per household, indicating that fully one-quarter of American households seemed to be managing quite well without a newspaper.[22]

The lack of growth in circulation over the past generation does not mean that all newspapers are suffering financially. In spite of the impact of television, the press still generates very substantial revenues from advertising, and, as the president of one newspaper consulting firm has recently maintained, "Newspapers just talk poor. This is a healthy and outrageously profitable business."[23] The stock market concurs. Between 1980 and 1984, newspaper stocks recorded an increase of 115 percent, while the Standard and Poor 500 index averaged only an 80 percent increase.[24]

Newspapers are "a healthy and outrageously profitable business."

Advertising Advertising takes up nearly two-thirds of the space in the average American daily newspaper (63 percent, at last count). Not only is it the principal content of most newspapers, but in many instances it is their

Whereas display ads seek the buyer, the buyer seeks the classifieds.

entire reason for existence. Advertising volume in the nation's daily papers was up to $27 billion in 1986. That figure, 26.4 percent of America's $102 billion expended on advertising for the year, was nearly equal to the combined totals for television ($22.4 billion or 21.9 percent of the total) and radio ($7 billion, 6.8 percent of the total). Actually, 1984 was the first time in history that the newspaper industry had a smaller gross income from advertising than did the broadcast media. It remained the largest single advertising medium, but by a smaller percentage than in prior years.[25] By any yardstick, however, newspaper advertising is a significant function to both newspapers and the economy as a whole.

Newspaper advertising can be separated into two broad categories: *display* and *classified*. Displays are showcase advertisements occupying considerable space and distributed throughout the paper; they are generally measured and sold on a column inch or lineage basis. They sell goods and services, and sometimes ideas. Classifieds are notices—sometimes of sales, sometimes purely informational. They are concentrated in the classified sections of the newspaper, which generally carry no illustrations, and are often read as news. Whereas display seeks the buyer, the buyer seeks the classifieds.

Display advertising can be broken down further into two less concrete subdivisions: retail and national. Retail advertising is local advertising. It sells at a lower rate than national advertising and is aimed as a service to the local merchant. National advertising tends to be reinforcing or persuasive in nature. Retail advertising is a point-of-sale display. It is designed to

Newspapers still get the lion's share of the nation's advertising dollar.

Display and classified ads serve different purposes.

promptly move goods and services, often being read as news—pure information as in the case of the market ads, theater and entertainment announcements, and notices of sales and weekend specials. The audience is presold by necessity or disposition and the retail advertisements give advice concerning where, when, and how much.

Newspaper advertising is sold by contract over the course of the year or off a rate card, a published schedule of prices for space and position in the paper. The rates are based on circulation, with larger circulation naturally demanding a higher rate. Recognizing the advantage that community newspapers have in soliciting advertisements of purely local merchants, metropolitan papers have gone increasingly to zone advertising in an attempt to appeal to the merchants of local communities. Breaking their press runs down by different geographical areas of a metropolis, they sell advertising in the editions going to these areas alone at a far cheaper rate than the general metropolitan run. This technique does not prove too successful unless the metropolitan newspapers include the other elements of community appeal—local news, gossip, and civic information.

Reader Interests

Newspaper reporters and editors joined broadcasters in the 1960s in paying diminished attention to governmental news and intensified attention to social issues. It was a decade of social awareness of issues such as women's movements, minority concerns, youth cultures, and antiwar and antiestablishment movements. Quite obviously the public was increasingly interested in such issues. The following decade, dubbed the "Me Decade" by journalist Tom Wolfe, found media catering to and creating increased self-awareness issues. "How-to" pieces on entertainment, finance, health, sex, and similar ego-centered concerns filled the news columns. This stress on news of personal utility (news to meet individuals' personal goals of self-fulfillment, self-gratification, and self-expression) continued into the 1980s, egged on by the success of the *USA Today.* However, a wide-ranging 1984 study concluded that newspapers may have gone overboard in appealing to hedonism at some cost to their more traditional role of presenting and interpreting hard news.[26]

The American Newspaper Publishers Association's Newspaper Readership Council, in several dozen reports, has recently joined other researchers from the academic and research communities in telling news gatekeepers how to fulfill audience needs. Without such studies, journalists would continue to rely on their often flawed instincts and news values for clues to reader interests. With the studies, they at least had indications of what would be read, whether or not they followed the guidelines.

Researchers help newspapers understand the readers.

A particularly revealing study published in 1978 set off a flurry of research activities and breast-beatings among newspaper journalists. Pollster Louis Harris, after interviewing a national cross section of 1,533 adults, eighty-six top editors and news directors, and seventy-six major reporters and writers, concluded that journalists really have not understood their readers' tastes in news. On topic after topic, the survey revealed that journalists either seriously underestimated or overestimated reader interest in news and editorial topics.[27]

The Harris survey findings, which indicated that dedicated newspaper readers expect a great deal from their medium, were amplified in an important mid-1984 study by Ruth Clark. She concluded that readers appreciate the innovative 1970s feature and service columns such as calendars, consumer briefs, community notes, TV logs and the like, but that good old-fashioned hard news continues to be the primary reason readers are loyal to the print medium.[28]

Readers of large and small newspapers alike reaffirmed traditional readership patterns; they claimed they want "complete newspapers" that give them all the important news from the world, Washington, and their own hometowns. Readers of smaller papers faulted their newspapers for inadequately covering national and international news, for doing a minimum of investigative reporting, and for not offering enough news on health and consumer tips, family and childrearing issues, and changes in the increasingly computerized world. Readers of all types of newspapers expressed interest in new topics of concern to both men and women: business, consumerism, health and health care, the environment, family, children, and education.

Despite the large number of readership studies funded by such organizations as the American Newspaper Publishers Association, the American Society of Newspaper Editors, the Newspaper Advertising Bureau, and others, some editors are hesitant to change their product based on the data generated. They argue—with some justification—that there is a great difference between what the public tells pollsters it wants to read and what in actuality it spends its time reading.

There is, of course, the possibility that surveys may generate distorted data. Nevertheless, several unmistakable patterns have emerged in recent years. Study after study (some by newspapers themselves and others by academic and external agencies with no vested interest in the outcome) demonstrated through the 1970s and into the 1980s that many different kinds of newspaper audiences in America are interested in and give different responses to various kinds of news items. The findings, which parallel much of the discussion on media structures and functions and the uses and gratifications research quoted in chapter 2, indicate the following:

> Those readers with strong community affiliation use the newspaper for hard news, while others with weaker local ties use it principally for entertainment—which helps explain the success of *USA Today* as a newspaper marketed for either the transient reader or one who has more of a national and regional than a local affiliation.
> Young readers in the age group from twenty-one to thirty-four who are regular readers are more attuned to the aforementioned self-awareness stories than are older readers.

Some readership research seems contradictory. But several distinct patterns have emerged.

The largest percentage of readers rely on the newspaper for its current events and other hard information; a second group uses it especially for sports; a third, for personal entertainment; and a fourth, for a guide to daily life in their community and from role models.

Perhaps the best news, from the industry's point of view, is the evidence that almost everybody is a newspaper reader. Ninety percent of those surveyed in Clark's 1984 study said they had read a daily newspaper in the past week, and most have read a paper three or four times in the same period. Nearly two-thirds agreed that "There is no substitute for a newspaper every day," and 85 percent agreed that newspapers are "one of the biggest bargains there are these days."

The studies also revealed some depressing news for editors. A clear majority of readers indicated they feel the news is biased by the newspaper reporters and editors, and two-thirds claim newspapers ignore good, pro-social news in favor of sensationalistic and bad news. In the Clark survey, 58 percent said they did not believe the newspaper industry as a whole was fair, although a substantially lesser percentage—39 percent—described their own newspapers as biased. (On the plus side of the ledger, 88 percent said they think their own newspaper cares about the local community.)

Other national studies have indicated a decline in the credibility of newspapers—a decline mirrored by public reactions to other institutions. (See chapter 15 for a fuller discussion of media credibility.) Harris surveys showed that confidence in the press (newspapers and other media) plunged from 29 percent in 1966 to 14 percent in 1982. That decline in credibility parallels the decline in newspaper penetration—that is, the percentage of households receiving newspapers. Critics of the Harris survey, while agreeing that there has been somewhat of a decline in newspaper credibility, complain that the questionnaires used have been too simplistically worded, and that no follow-up questions have been used that might have given readers the opportunity to define their understanding of such terms as "having confidence in newspapers." The critics have also noted the enormous decline in public acceptance of all institutions since the mid-1960s. Whereas 61 percent of the public gave "colleges" a high credibility rating in 1966, only 36 percent did so in 1983; over the same period, the military dropped from a 61 percent to a 35 percent approval rating; Congress, from 42 to 20 percent; and major corporations, 55 to 18 percent. According to sociologist Amitai Etzioni, two-thirds of the historical decline of any institution can be attributed to the society-wide trend, and the other one-third to the performance of the institutions themselves.[29]

One 1980 study receiving wide circulation indicated that 37 percent of Americans thought curbs placed on the press were not strict enough; that total was up from 21 percent in 1958. Further, the poll indicated a sharp decline in the public's perception of newspaper accuracy. Most disturbing of all, according to pollster George Gallup, was the finding that 76 percent of all respondents could not answer the question, "Do you happen to know what the First Amendment to the Constitution is, or what it deals with?" Gallup concluded that the American press was operating in an environment of public opinion increasingly indifferent, and to some extent hostile, to the causes of a free press in America.[30] A series of studies since that time, including the Clark study, indicate that the decade of the 1980s has seen an improvement along these lines, that readers express stronger support of the concept of press freedom even if they cannot recall the contents of Constitutional amendments and even if they do have a series of complaints about newspaper performance.

For years, editors have been receiving hard statistical evidence that they are losing ground in the battle for circulation among the young, the minorities, the elderly, the poor, those who live in rural areas or apartments, transients, and those weaned on or subsequently addicted to television. Attempting to reverse this trend has posed a challenge to the American Newspaper Publishers Association. It has suggested that journalists create the newspaper that today's readers want, by letting the editor and reporter be more personal, by mixing good news with the bad, by telling people the news in terms of people they know, and by making the paper more attractive by giving it personality.

The young, poor, elderly, rural or apartment dwellers, and TV addicts tend to be nonreaders of newspapers.

Although not all journalists agree with the conclusions, many newspapers are making some changes to attract new readers. Some newspapers have attempted *demographic breakouts*—specialized sections delivered to different parts of the community. Others have experimented with magazine-style formats, "supermarketing" the newspaper with different themes or interest areas spotlighted on different days of the week. The addition of color on the front page and colored flags for special sections is giving newspapers a brighter appearance. Spanish language editions of papers in cities with large Spanish populations, such as the *Miami Herald,* are broadening the appeal of some newspapers. Most editors, however, have recognized that a newspaper and its audiences and advertisers exist in a delicate ecosystem, and that drastic adjustments to cater to one group probably means disaffection of another. For instance, hard-core loyal readers who utilize the newspaper for serious social, political, and economic matters are readily offended when editors water down the intellectual level of their newspapers. Such pandering to the nonreader, especially in terms of increased soft-core and self-awareness news, is best done inconspicuously.

In sum, newspapers and their readers exist in a delicate ecosystem.

Newspapers Today and Tomorrow

New electronic technology and segmented marketing techniques typify the newspaper of the 1980s.

The massive newspaper industry, with its workforce of 461,300, has not been highly regarded for its innovativeness. Contemporary metropolitan papers in particular have been cursed with an inertia born of ponderous organization and bureaucracy. Therefore, change has frequently come from the smaller community dailies, which are more flexible than their metropolitan cousins and generally operate on a closer profit margin. They were among the first to move forward toward *offset printing,* a cleaner operation that drastically cuts makeup time for the newspaper and reduces the amount of highly skilled, unioned *shop time* that is a part of production. Relatively unskilled personnel can *paste up* full pages, which are then photographically transferred to thin metal or plastic plates for the presses. Likewise the smaller papers were quick to employ the *electronic newsroom,* creatively marrying computers and *video display terminals* (VDTs) for writing and editing. This system cut down the size of the work force in the backshop and put editorial employees closer to a more rapidly produced finished product.

The larger metropolitan papers were shaken out of their lethargy by such innovations. For nearly a century they had plodded along, using archaic techniques that had served them well in the days before electronic media. But the offset press and VDTs offered economies they could not ignore, despite the enormous capital investment required to transform huge plants and retrain hundreds of people. The *Sacramento Union,* a former Copley newspaper in the medium-metropolitan range, was the first in the nation to go fully offset in 1967. By 1980, 75 percent of the nation's dailies had followed suit, and the same percentage had initiated VDTs and electronic newsrooms.

Still in their infancy are portable VDTs, which allow reporters on assignment to enter stories directly into the system. Another technological innovation will be inkjet printing coupled with the use of lasers, allowing editors to update news items and advertisers to modify ads while presses continue to roll at top speed. These changes and others are aimed at decreasing the time lag between a news event and its delivery to readers. But what about the problem of a decreasing market penetration?

To reach readers some publishers have experimented with the aforementioned segmented marketing by geographically zoning their papers, by redefining separate editions of the same paper to appeal to different groups within the same zones, or by publishing special types of papers on one or two days of the week. In the mid-1970s, the *New York Times* began publishing "Weekend," a leisure/entertainment guide, in its Friday editions. The *Los Angeles Times* put out "YOU," a consumer/leisure guide, on Tuesdays. The *Boston Globe* offered "Calendar," a cultural guide for families and youth, on Thursdays. And the *Louisville Times* came out with "Scene," a full-color Saturday magazine filled with advice on dating and astrology, tips on auto repair, how to spend the weekend, how to shop, and even a children's pullout section called "Jelly Bean Journal."[31]

To some this change from newspaper to "use-paper" is troublesome—a professional cop-out. To others the trend is a natural and necessary one, particularly to market researchers who argue that traditional editors have been producing newspapers for other journalists rather than for readers. "I think everybody is going to win on this. It will help us find a niche as an industry and give readers a better choice of product," asserts Robert G. Marbut, president of the innovative Harte-Hanks chain.[32]

Segmented marketing of newspapers became popular once it was realized that technology and audience desires had melded. Faced with soaring postal costs, many advertisers turned to newspapers for distributing pre-printed ads in *tabloid* or brochure form. Newspapers, they reasoned, had already established efficient distribution patterns, allowing them to reach the audiences most likely to be interested in the advertising. The *New York Times,* for instance, brags that it can guarantee an advertiser that a particular ad supplement will go only to readers on specified blocks of New York City, and smaller newspapers can insert or remove ads for separately zoned editions.

How far will this trend go? Theoretically, segmented marketing is limitless, with each individual receiving an individualized newspaper, unique to his or her own tastes, reading levels, interest levels, and purchasing behavior. How far in the future that theoretical day lies is anybody's guess, but it will never come about so long as present distribution patterns prevail. Marketers are stymied by the realization that the final link in the distribution channel is usually a twelve-year-old delivery boy or girl on a bicycle. They will admit that these youthful entrepreneurs are on their endangered-species list, once they figure out how to tie the newspaper to an in-home electronic delivery system.

Many signs point to the electronic newspaper. Astronomical price increases for and shortages of paper have already resulted in streamlined newspapers with smaller-sized pages. Despite apparently successful experiments with *kenaf,* a fast-growing plant from which newsprint is being made, society recognizes the benefits of finding alternatives to the problems of ravaging forests for wood pulp and polluting the atmosphere with the residue of waste paper.

It is only a matter of time before the newspaper industry, cognizant of readers' tendencies to read only what appeals to them, ceases distributing the same mass-oriented product to everyone. Futurists looking toward tomorrow's newspaper see homes equipped with receiving units (modified teletype machines similar to today's VDTs) that offer electronic scans of the news and hard-copy printouts. Users will exercise almost limitless control over content. By checking the daily news index, they will call up, in as much detail as it takes to satisfy them, those stories of greatest interest to them. In the process they will be free to overlook whatever disinterests them, thereby voluntarily missing out on entire chunks of important daily events and ideas. This disturbs many socially responsible editors

The newspaper of tomorrow may cause some social upheaval.

who know that readers in pursuing their self-interests will find it easier to overlook substantial information and in the process become less responsible citizens. We could, as Merrill and Lowenstein suggest, build political, social, and educational cocoons around ourselves, and our society could become divided into highly polarized, and probably unempathic, segments.[33] Considering how far Knight-Ridder and Dow Jones have already come in their experimental electronic newspapers, that future may already be upon us.

George Bernard Shaw used to refer to the daily newspaper as "the poor man's university."[34] When, not if, the electronic newspaper becomes fully operational, its first subscribers will be those customers who are wealthy enough to own the sophisticated electronic receivers. Families low in income and in educational and occupational achievement, who are already on the nonreader lists, will be even less likely to become readers.

It seems ironic, but the parent medium, the "poor man's university," may be at a turning point in its efforts to remain viable to serve the citizenry. As technological and economic factors coalesce, the newspaper is becoming more specialized. It has the option of either moving toward *internal specialization,* with each issue a potpourri of features, some of which will appeal to many groups, or moving toward *unit specialization,* in which each issue is tailored for each subscriber. It is uncertain which track the industry will follow, but from what we see, it appears likely the newspaper may opt for unit specialization because of economic realities. If it does, the newspaper will change its primary historical function and cause still another shifting in the delicately balanced media-society ecosystem.

Summary

A newspaper historically has been a geographically limited, regularly published print medium, serving the general interests of a specific community. However, some new satellite-distributed national papers—the *USA Today* in particular—are causing a revision of that definition.

Predecessors of newspapers, called corantos, appeared in England early in the seventeenth century, but the first daily paper did not appear until 1702.

The history of the press in America can be divided into eight eras: (1) the colonial press, (2) the Revolutionary press, (3) the political press, (4) the penny press, (5) the personal editors' press, (6) yellow journalism, (7) jazz journalism, and (8) the present age of consolidation.

During the eighteenth century, newspapers were essentially journals of opinion written for the elite. Benjamin Day changed that in 1833 with his penny press, the *New York Sun,* which appealed to the masses. Mid-century was the era of the personal editors who created many of the features that exist in newspapers today. Late in the century, circulation wars among New York papers led to excesses that have become known as yellow journalism.

Many newspapers today are owned by one of 146 chains, which account for two-thirds of the daily circulation in the country. Regardless of whether the paper is part of a chain or is independent, the person who runs the paper is the publisher. Most newspapers are divided into five departments: editorial, advertising, production, circulation, and administration. The newspeople—reporters and editors, wire services, stringers, and correspondents—feed material to newsrooms that print only about 10 percent of what is available.

In addition to big-city daily newspapers, the medium includes a recent surge of nationally distributed dailies, most influential (and controversial) of which is the Gannett corporation's *USA Today.* There are also smaller dailies, community papers, and minority papers. Each type of newspaper fills a need; but when needs change, some papers lose their audience and go out of business.

Newspapers are changing. Recently dozens of daily papers have changed from P.M. to A.M. editions in order to reach middle-income suburbanites. But that change alone has not yet increased circulation, which has been flat the past fifteen years. Fortunately, advertising has remained strong in spite of competition from other media.

Recent studies have shown that reporters and editors have not been very good at assessing what readers claim they want from a newspaper. Other studies reveal that what readers say they want and what they read are not always the same thing. However, one unmistakable trend of late, uncovered by polls, has been a decrease in the newspaper's credibility among the public.

In the final section we discussed some of the efforts of newspapers to rekindle reader interest and loyalty, and we proposed some changes (not all of them positive) that are likely for the newspaper of the future.

Notes

1. Edwin Emery and Michael Emery, *The Press and America: An Interpretative History,* 5th ed. (Englewood Cliffs, N.J.: Prentice-Hall, 1984).

2. Ben H. Bagdikian, *The Information Machines: Their Impact on Men and the Media* (New York: Harper & Row, 1971).

3. William E. Ames and Dwight L. Teeter, "Politics, Economics, and the Mass Media," in *Mass Media and the National Experience,* eds., Ronald T. Farrar and John D. Stevens (New York: Harper & Row, 1971), p. 46.

4. S. N. Katz, ed., *A Brief Narrative of the Case and Trial of John Peter Zenger* (Cambridge, Mass.: Harvard University Press, 1963).

5. Donald L. Shaw, "Technology: Freedom for What?" In *Mass Media and the National Experience,* eds., Farrar and Stevens, p. 70.

6. Ben H. Bagdikian, "The Press and Its Crisis of Identity," in *Mass Media in a Free Society,* ed., Warren K. Agee (Lawrence, Kansas: University Press of Kansas, 1969), p. 7.

7. American Newspaper Publishers Association, *Facts About Newspapers '86* (Washington, D.C.: ANPA, April, 1986); and *Facts About Newspapers '87* (Washington, D.C.: ANPA, April, 1987).

8. John Morton, "*USA Today*'s Ad-versity," *Washington Journalism Review;* September 1984, p. 18.

9. John W. C. Johnstone et al., *The News People: A Sociological Portrait of Journalists and Their Work* (Urbana, Ill.: University of Illinois Press, 1976).

10. Robert Lichter and Stanley Rothman, "The Media Elite," *Public Opinion,* October/November 1981.

11. Everette E. Dennis and Arnold Ismach, *Reporting Processes and Practices* (Belmont, California: Wadsworth Publishing Co., 1981).

12. William A. Henry III, "The Ten Best U.S. Dailies," *Time,* 30 April 1984, pp. 58–63.

13. Donna Ventimiglia, "Neuharth Defends *USA Today*'s 'Journalism of Hope,' " *Editor & Publisher,* 29 October 1983, p. 9.

14. "ABC Audits *USA Today,*" *Publishers Auxiliary,* 2 July 1984, pp. 1–2.

15. *A Look at* USA Today, *A Changing Newspaper Committee Report,* A Report to the Associated Press Managing Editors Association, November, 1983, p. 9.

16. Ben H. Bagdikian, "Fast-food news: a week's diet," *Columbia Journalism Review;* March/April 1983, p. 33.

17. Robert A. Logan, "*USA Today*'s Innovations and Their Impact on Journalism Ethics," *Journal of Mass Media Ethics,* Spring/Summer 1986, pp. 74-87.

18. ANPA, *Facts About Newspapers,* '86.

19. Kathleen Hunt Baird, "P.M. to A.M.: Is a Trend Building?" *Presstime,* December 1979, pp. 6-9, and Clark Newsom, "The Beat Goes on for P.M.s to A.M.s," *Presstime,* December 1980, pp. 46-49.

20. Christopher H. Sterling and Timothy R. Haight, "Characteristics of Newspaper Readers," in *The Mass Media: Aspen Institute Guide to Communication Industry Trends* (New York: Praeger Publishers, 1978), p. 338-39.

21. American Press Institute, *The Public Perception of Newspapers. Examining Credibility* (Reston, Va.: American Press Institute, 1984).

22. John A. Finley, "Special Report: Reading," *Presstime,* September 1984, pp. 22-23.

23. David Skylar, "Why Newspapers Die," *The Quill,* July/August 1984, p. 15.

24. ANPA, *Facts About Newspapers,* '86.

25. ANPA, *Facts About Newspapers* '87.

26. Ruth Clark, *Relating to Readers in the '80s* (Washington, D.C.: American Society of Newspaper Editors, 1984).

27. Louis Harris, "Public Prefers News to Pablum," *Deseret News,* 1 January 1978.

28. Clark, *Relating to Readers in the '80s.*

29. American Press Institute, *The Public Perception of Newspapers.*

30. Reported at the First Amendment Congress, Philadelphia, 16-17 January 1980.

31. David Shaw, "Newspapers Challenged As Never Before," *Los Angeles Times,* 26 November 1976.

32. Michael T. Malloy, "Newspapers May Some Day Let You Pick the News You Want," *National Observer,* 21 February 1976.

33. John C. Merrill and Ralph L. Lowenstein, *Media, Messages, and Men: New Perspectives in Communication,* 2d ed. (New York: Longman, 1979), p. 235.

34. Shaw, "Newspapers Challenged as Never Before."

4

Magazines

Introduction

The ability of a medium to survive and to thrive after having been superseded by a faster, more efficient form is nowhere better illustrated than in a magazine. Nor is the remarkable diversity of American life better demonstrated than through the extraordinary variety of magazines. There are almost twenty thousand magazines published in the United States today. Some look like newspapers and others look like books. Magazines are sold by subscription or on racks in over a hundred thousand retail outlets, and some are given away to preselected groups, and are called controlled circulation.

Magazines are published by magazine groups, by individual publishers, by newspapers, by small esoteric societies, by giant corporations, by trade associations and churches, by varying levels of government, and by all political parties. The roster of publishers is almost as varied as the titles themselves. Magazines for the most part are issued daily, weekly, semimonthly, and monthly; some are published bimonthly or quarterly, and a few come out once a year. They can be as general as the *Reader's Digest* or as specific as *Overtones,* which is published by the American Guild of English Handbell Ringers, for handbell choirs and directors. As varied as the medium is, all magazines do share two characteristics: they are published regularly, and each appeals to the interests of some specific segment of society.

Our discussion begins with a general review of the historical development of magazines as a hybrid medium. As in our discussion of newspapers, we will concentrate on the medium's adaptations to twentieth century pressures, particularly those brought on by television and by rapid increases in the costs of materials and postal rates. The rise and decline of some mass circulation magazines will be considered, and we will look more closely at a couple of attractive magazines that have succeeded in recent years. Marketing techniques, and trends in ownership and audiences, will be surveyed. We will conclude our discussion with speculations about the future of the magazine.

Historical Development

Historically, magazines developed as a hybrid form of print and appealed to audiences in the gap between newspapers and intellectually pinpointed books. They provided some entertainment and some culture in discrete doses, reaching neither populistic nor philosophical extremes that seemed to characterize the press and book-publishing businesses prior to and during most of the nineteenth century. For a period of time and to a limited extent, magazines seemed to provide the only national medium.

Magazines discovered early on publics other than geographical ones and began to appeal to these specific groups: farmers, women, professional people. They also discovered advertising and began to develop a specialized form of advertising geared to their specific publics, as was their content.

From European Roots to the Civil War

The history of magazines is somewhat obscure because print technology and accompanying literacy had to become fairly well developed before clear lines of demarcation could be drawn between newspapers and magazines. Generally, the first publications classified as magazines were the *Tatler* and the *Spectator,* published in England during the first quarter of the eighteenth century. Joseph Addison and Richard Steele contributed much to these periodicals by way of topical essays, satirical material, international news, and local gossip. Their formats included more opinion and entertainment than news.

The first magazines on American shores clearly defined as such were the *General Magazine and Historical Chronicle, for All the British Plantations in America* and the *American Magazine, or a Monthly View of the Political State of the British Colonies,* both originating in Philadelphia in 1741. In 1740 Benjamin Franklin announced his proposed *General Magazine,* but a competitor, Andrew Bradford, got his *American Magazine* off the press three days ahead of Franklin. Both magazines quickly folded— Bradford's after three issues and Franklin's after six issues. America was not quite ready for magazines. Despite imitators, no American magazine during the eighteenth century lasted for more than fourteen months. Some of the nation's outstanding writers and politicians, including George Washington, Alexander Hamilton, John Jay, John Hancock, Philip Freneau, and Thomas Paine, were involved either as editors or contributors, but lack of advertising and limited circulation spelled doom for most of their efforts.

By the turn of the century the situation had changed; *Port Folio* was followed by *North American Review* in 1815 and the venerable *Saturday Evening Post* in 1821 (which, contrary to popular legend, Benjamin Franklin did not found). The *Post*'s format of fiction, poems, and essays was typical

during most of that period. By 1830, a hundred or so magazines were being published in the United States, which proved the need for the new medium. Also, 1830 marked the founding of *Godey's Lady's Book*— the first magazine to cater specifically to women and, more significantly, the first medium to attempt to identify an audience of its own.

Additional entries into the field were made by mid-century, with literature, graphics, science, and travel being added to the bill of fare. *Harper's New Monthly Magazine,* the *Atlantic Monthly, Gleason's Pictorial,* and *Harper's Weekly* were typical of the titles that filled this newly created interest vacuum between newspapers and books. Some of the new magazines were commercially successful; however, many were not. The little *Nation,* which made its appearance at the close of the Civil War in 1865, is significant because in over one hundred years of continuous publication, it consistently lost money—every year.

The last part of the nineteenth century, from the end of the Civil War until 1900, has been called the Golden Age of the magazine industry, and for good reason. Several significant factors came together during that period. Channels of distribution were greatly facilitated when the transcontinental railroad network was completed in 1869. Then in 1879, Congress passed the Postal Act, allowing magazines to be distributed at less expensive second-class rates. Pulp paper made from cheaper wood and improved printing presses, coupled with the invention of the linotype (automatic typesetting)

The Golden Age: 1865–1900

The Golden Age evolved with improvements in distribution, printing, education, and urbanization.

machine by Ottmar Mergenthaler and photographic reproduction techniques, meant the average magazine could be produced less expensively, more rapidly, and more attractively than ever before. Also significant were the growth in secondary education and national urbanization, which created broader and more literate audiences for magazines.

Between 1865 and 1885 the number of American magazines increased from 700 to 3,300. Their circulations were larger than those of prewar days, but a magazine with a circulation of 100,000 copies was still considered a giant in the industry. Then, in 1893, publisher Frank A. Munsey reduced the annual subscription price of *Munsey's Magazine* from three dollars to one dollar, while dropping the price of a single issue from twenty-five cents to only a dime. His philosophy—not a new one, since it had been successfully applied by the newspaper industry sixty years earlier—was to sell the magazine for less than the cost to produce it, with advertisers carrying the financial burden. Sure enough, as circulation grew—from 40,000 in 1893 to 500,000 in 1895—magazines became increasingly attractive to advertisers, who recognized the value of a national marketplace.

Other magazines followed suit, particularly those magazines aimed primarily at women: *Ladies' Home Journal, Women's Home Companion,* and *McCall's.* Of these three, the *Journal* may be the most significant. Founded in 1883 by Cyrus H. K. Curtis and his wife, the magazine became a reflection of Edward Bok's editorial philosophy after he joined the staff in 1889. At that time the circulation was 400,000 per month, although the magazine was considered typically mundane in editorial content. Bok broadened the appeal of the *Journal* by including advice columns and articles about health, fashion, home decorating, cooking, and child rearing. National advertising expanded as the circulation grew; soon the magazine was reaching over one million readers each month.

Among the other magazines seeking to compete for the increasing amount of advertising by national manufacturers were *McClure's, Cosmopolitan, Collier's,* and *American,* to name a few. *Cosmopolitan,* for example, was published by William Randolph Hearst as an "insurance publication" during his efforts to establish a newspaper empire. Those older magazines that refused to lower their newsstand prices below twenty-five or thirty-five cents, and counted on subscriptions or street sales to give them the finances needed to continue running expensive literature, suffered in the increasingly competitive marketplace.

The Twentieth Century

Modern magazines have low cost, large circulation, advertising support, and diverse audiences.

As the twentieth century began, the features of the "modern magazine" had begun to take shape: low cost, large circulation, advertising supported, and service to diverse audiences. Over the next several decades, magazines offered numerous contributions to American culture. Historian Theodore Peterson credits the industry with providing social reform, putting issues in national perspective, fostering a sense of national community, providing low cost entertainment, serving as an inexpensive "instructor" in daily living, providing a cultural heritage for Americans, and offering a variety of entertainment, education, and ideas representing a wide range of tastes and interests.[1]

The Muckrakers Given their nationwide audience, low cost, and timeliness, it was perhaps inevitable that magazines became an instrument of social reform. In the first dozen years of this century many of them became "people's champions," investigating and attempting to correct political, social, and business ills. Theodore Roosevelt branded them muckrakers, after a famous painting of the *Man with the Muckrake,* in *Pilgrim's Progress,* who missed seeing the celestial crown because he was so busy raking the barnyard filth. Rather than accepting the derogatory term as intended, various writers and editors in the early 1900s took the title as a mark of distinction.

Political, social, and business ills were investigated by muckrakers.

Leaders among the muckraking magazines were *McClure's, Cosmopolitan, Munsey's, Ladies' Home Journal, Collier's, Everybody's,* and the *Saturday Evening Post.* Outstanding writers such as Ida Tarbell, Lincoln Steffens, Ray Stannard Baker, Finley Peter Dunne, and David Graham Phillips, fired by missionary zeal, attacked and sought changes from oil companies, meat-packing firms, patent medicine manufacturers and distributors, city governments, labor organizations, abusers of child labor, and even the United States Senate.

Spurred by the muckrakers' efforts, businesses and legislators enacted numerous remedies to the exposed inequities. The Pure Food and Drug Act of 1906 and legislation regarding finance and fair trade followed the public outcry. In 1911 the magazine *Printer's Ink* prepared the Printer's Ink model advertising statute that urged states to make fraudulent and misleading advertising a crime. Many states passed bills in support of the model statute. But the movement tapered off after Theodore Roosevelt's third party candidacy, Woodrow Wilson's election to the presidency in 1912, and America's entrance into World War I several years later. After that period, criticisms of American institutions were somehow less substantial and seemed more sensational and frivolous, and the muckraking era drew to a close—at least for the time being. Even though the majority of Americans during the early years of the century were reading escapist literature rather than social commentary, enough of them were aroused by magazine (and newspaper and book) writers to spark changes in the business community and to ensure social progress.[2]

1920s to the 1950s During the decade or so following World War I, several giants were established, including *Reader's Digest* (1922), Henry Luce's *Time* (1923), and the *New Yorker* (1925), as well as H. L. Mencken's spirited *American Mercury* and the *Saturday Review of Literature.* Improved production processes made possible the reproduction of photographs, which required a higher quality paper. Magazines were becoming more attractive, using bolder graphics to complement the articles and fiction. The mid-1930s

Striving for large circulation meant many changes for the magazine business.

saw the great photojournalism efforts of *Life* and *Look,* and the first of the slick men's magazines, *Esquire.*

The battle for increases in circulation first experienced by newspapers also afflicted magazines. This trend continued as radio came into its prime. The mass media, it seems, were committed to a "numbers game."

Some of the more familiar characteristics of magazines are an outgrowth of this trend. The cover girl is probably the best known of these phenomena. Brilliant color photography of slim, suggestive, young, and beautiful models adorned the front covers of American magazines in an effort to dress up the magazines' competitive appearance on the racks of 100,000 public outlets. This was the genesis of the "cheesecake syndrome," the attempt to use sometimes scantily clad but always alluring young females as bait for both male and female readers.

With scores of national magazines competing for attention on the newsracks of the nation, cover girls became big business indeed. At least two major modeling agencies (Powers and Conover) thrived on the aspirations of thousands of girls from every state who yearned for the recognition and prestige that being a cover girl or a Powers' model would bring. The depression years of the mid-1930s was a strange time, a time of dance marathons, crooners and swooners, gangster movies, cover girls, and magazine fiction.

Magazine marketing also became a reality as distributors fought for preferential position on the retail racks, making concessions and deals to outlets to place their magazines at top center on display racks and to bury their competitors' magazines. Magazines also outdid one another in seeking surefire material for public appeal. This generally took the form of the memoirs of famous people. Intense bidding among the national general-interest magazines in the post-World War II period resulted in magazine rights for Admiral William Halsey's memoirs going for $60,000, while General Eisenhower's (long before he became president) sold for an amount unheard of at that time, $175,000. Later the *New York Times,* a consortium of *Life* magazine, and a book publisher paid Winston Churchill $1,000,000 for his story, while *Life* ventured $600,000 for the magazine rights to Harry Truman's autobiography.

For twenty years, until the mid-1950s, magazine publishing flourished in the United States, and national circulations grew as the magazines provided a specialized visual quality that national network radio lacked. Then, beginning in 1956 under the impact of network television (which provided its own pictures), the days of reckoning came. *Collier's* was the first to go under in that year, followed in time by *Coronet, American, Look, Saturday Evening Post,* and *Life.* Some have reappeared in modified form, but they are unlikely to seek or achieve the mass-circulation status of their heydays.

The *Saturday Evening Post* and *Life* were two of the mid-century magazine giants.

Television and Advertising

The magazine, more than any other medium, was drastically affected by television.

Many observers assert that no other medium has been so drastically and permanently affected by television as the magazine. Most of the major national magazines were essentially entertainment oriented, and television performs this function much better. However, the arrival of television does not explain completely the troubles encountered by magazines during the 1950s and 1960s. Although television has played a role, the basic reasons for magazine failures are economic. Examples are the two graphic giants, *Look* and *Life*. Both folded with extremely healthy circulations, each numbering more than six million copies, even while they were losing money. You may ask then, why did they fail economically? The reason, of course, is because of advertising. With huge national circulations of six, eight, or ten million at their peak, the big magazines had grown fat off national advertising revenues. Remember that, as with newspapers, only a fraction of the cost of a magazine is borne by its subscription rate or sales price. The largest portion of its cost and all of its profit are carried by advertising.

When television appeared, it gradually acquired huge national audiences in previously unimagined numbers. The circulations of six million or so of the national magazines began to pall alongside the twenty million homes—of thirty-five million individuals, more or less—that a television network was able to reach on any weekday evening during prime time. Moreover, the advertising costs of television were comparable with those of magazines. A full-page advertisement in *Life* sold for around $60,000, color and production costs excluded. A network prime-time minute generally went for the same figure, production costs again excluded. In addition to having larger audiences, the presentation on television was dynamic, with action and complete control of sound, and, by the 1960s, color. Television advertising had real impact while a magazine display offered only a static presentation. Television was, all things considered, a more effective sales tool for many nationally advertised products. As national manufacturers and their agencies increased their use of television to advertise products and services, the sales results supported their choice.

Magazines found that they could not reduce the costs of their advertising. Production costs and postal rates were up, squeezing printing, paper, and editorial budgets. They were caught in a vise of rising costs and declining revenues that even six million subscribers could not offset, and the effects of this dilemma were spiraling. As advertising fell off, gradually at first, magazines sought to economize. They spent less on production and research. Consequently, the quality of the product suffered and this, in turn, had an effect on circulation as subscribers, disappointed in the "new look," began to cancel subscriptions. Further, since the size of a general-circulation magazine, like that of a newspaper, is based on the volume of advertising committed (generally, at least in the days before television, on a 60 to 40 percent basis of advertising to editorial content), the magazines grew thinner as advertising fell off. Customers began to feel that they were not getting as much for their money. Thus, as circulation dwindled, national advertisers began to see that they were getting even less for their advertising dollar in the medium, which prompted a more rapid shift of advertising to television.

The Circulation Wars

When the impact of television was first noticed, many of the nation's most popular magazines were engaged in a ruinous subscription war, which lasted from the early 1950s to the late 1960s. Caught up in a competitive struggle, they agreed to deliver eighteen, twenty-four, thirty, and even thirty-six monthly issues to new subscribers for a fraction of the listed cost, often offering thirty-six months for less than the original subscription price for a single year. At one point, *Life* was receiving only twelve cents per subscriber for a magazine that cost forty-one cents to edit and print.

While not apparent at the time, the cheap contracts would return later to haunt the magazines, often making it more expensive for them to go out of business than to continue publishing at a loss. Hundreds of thousands of three-year subscriptions representing little initial revenue were to become prohibitively expensive to buy up if a magazine wished to suspend publication. Since these subscriptions were indeed contracts to deliver, they were enforceable in court; hundreds of thousands of lawsuits would have been a ruinous undertaking.

The purpose of both the street sales and the subscription promotions was to build circulation, and no matter how that circulation was achieved, it represented solid figures in the Audit Bureau of Circulations' annual measurement. These audited figures served advertising agencies in making their selection of magazines. Furthermore, advertising rates were directly pegged to circulation size. The insanity of this particular course—sacrificing sales revenue in an era of rapidly mounting costs to try to attract additional advertisers in a declining market—became apparent after the fact. It is little wonder that so many of the giants folded willingly after such a long, hard struggle.

The next step undertaken by the large, general-interest magazines in their struggle for survival was exactly the reverse of the circulation war. They made an effort to curtail circulation: first, as a cost-saving device, and second, as an attempt to more specifically identify audiences in terms of their interests. To accomplish the latter goal, publishers turned to the computer to break down audiences or subscribers in terms of where they lived and their relative affluence, age, sex, and the like. These demographics of their audiences, the magazines reasoned with some cause, could categorize those persons to which an advertiser could beam a specific sales pitch, something television, by its massive generality, could not do. An additional reason for trying to curtail magazines' audiences was to remove a lot of the deadwood, those subscribers having only a marginal interest in the magazines who had bought long-term subscriptions at a fraction of the cost. If such subscribers would voluntarily remove themselves from circulation rolls, it would greatly assist in any subsequent cessation of publication.

Across America in the late 1950s and early 1960s, the public was confronted with the unexpected circumstance of magazines writing to subscribers, who only a year or so earlier had been actively solicited, telling them that they fell outside the parameters of the kind of people desired as subscribers. This was a technique scarcely calculated to win friends at a time when the magazine industry needed them so desperately.

This pruning technique was not successful to any great degree. At best it was a desperate measure. It did not sufficiently pinpoint audiences demographically for it to have been of any major appeal to specific advertisers, although it did reduce publication costs to the degree that it curtailed circulation and, to that extent, made subsequent liquidation less costly.

Regional Editions There were other attempts to use demographics to forestall the inevitable. Computer technology, for a while, seemed to be coming to the assistance of the national magazines. Computers permitted

Adjustment and Retrenchment

Finally, in desperation, the falling giants tried to prune circulation and seek new audiences.

magazines to break down their circulation geographically, a lesson learned from the *Wall Street Journal,* which, since the early 1950s, had been published in four regional editions as well as a national edition. The news and comment—the editorial content—were identical in each edition, but it did permit regional advertisers an opportunity to reach their spheres of interest at much less cost than if they were forced to buy for the entire national circulation of the *Wall Street Journal.* This was an advertising technique specifically designed for utilities, banks, and other financial institutions that were basically organized on a regional basis and comprised the natural interest market of the *Wall Street Journal.*

When this technique was adopted by *Life* it permitted the magazine to substitute purely local advertising for some of the national advertising that had been lost to television. Further, it permitted a Chattanooga utility or a Seattle bank the prestige of advertising in *Life.* At the peak of this experiment, *Look* was running seventy-five regional editions at fantastic cost. *Life,* more conservative, published in seven regions and eleven metropolitan areas.

Cost differentials are apparent. To buy *Life*'s full-run peak of eight million or so would have cost an advertiser $65,000, but *Life*'s 150,000 subscribers in Minnesota could be bought for a mere $2,500, and the advertiser could still have *Life*'s prestige. The theory behind publishing regional editions was that through an active sales effort the volume of advertising in each of the magazine's regions or districts would balance out. By and large it did, but meanwhile sales costs had increased considerably, constituting an additional drain on already dwindling reserves. At best, the use of multiple editions was a stopgap for the general-interest magazines.

However, such regional breakdowns in more specifically oriented journals, such as *Time* and *Newsweek* with their high news orientation, proved eminently successful. Regional editions in such instances permitted advertisers and manufacturers to reach customers whose demonstrated interest in news indicated an interest in the world, a relatively high educational level, and a degree of presumed affluence. These qualities are considered valuable to many advertisers.

The computers assisted briefly in yet another way. They were also capable of breaking down magazine subscribers by their relative affluence. Such affluence was established by the census tract on the theory that those who lived in wealthier neighborhoods were wealthier and those who lived in middle-income areas were less so. Thus, reasoned the magazines in their death throes, they could approach such national manufacturers as the Ford Motor Company with the suggestion that the company could reach the most affluent circulation of *Life,* for example, with advertisements for Lincoln Continentals; they could advertise Pintos in the less affluent areas, and Fairlane models filling in the middle ground. This was ingenious thinking, offering an element of product differentiation that television could not, but it did not work at the time.

As desperate as the situation appeared to be, not all magazines went out of business. By 1969 the *New York Times* said that magazines were thriving despite a few dramatic failures. Obviously, not all the effects of television were negative. Some magazines were born because of television, and others changed significantly—and successfully—due to the television environment.

The most significant change has been the move away from the national magazine of general interest, as represented by *Life* and *Look,* to more specialized magazines, many of which still have very broad appeal.

Television has also been largely responsible for a change in the editorial mix of many magazines. For instance, of those magazines that continue to publish fiction, the present ratio of nonfiction to fiction, about three to one, is almost exactly the reverse of what it was a half century ago.

Magazines may be categorized in several ways. Traditionally, they were broken down into consumer- or general-interest magazines, business magazines, and farm and company magazines. The Audit Bureau of Circulations has a more specific breakdown: automotive, brides/bridal, business/finance, campers/RV/mobile homes/trailers, computers, crafts/games/hobbies/models, epicurean, fishing/hunting, fraternal/clubs/associations, gardening, general editorial, health, home service/home, mechanics/science, men's, motorcycle, music, news weeklies, news biweeklies/news semimonthlies, photography, science, senior citizens, sports, travel, women's, women's fashions, beauty/grooming, and youth.

A third way of breaking magazines down is functionally, into three categories corresponding closely to the functions of communication. These are: (1) entertainment/escape; (2) news/information; and (3) advocacy/opinion. Within these categories fall some rather obvious examples. For instance, comic books, confessions, and science fiction all fall within entertainment/escape. The news/information category includes the well-known news weeklies of *Time, Newsweek,* and *U.S. News and World Report,* as well as a host of trade, professional, and scholarly publications, each specializing in some facet of the culture. Magazines with the category of advocacy/opinion that are oriented toward persuasion include not only those of the underground press but also most of the vast array of organizational or corporate publications, plus such watchdog publications as *The Progressive,* a periodical devoted to investigative reporting of governmental affairs. Most magazines represent some combination of these communication functions, with specific appeal being determined by the magazine's basic thrust or orientation.

Looking at magazines from another perspective, we can categorize them by the relationship between their content and their audience. The general-interest magazines of the 1950s had broad coverage of issues of importance in American society, and all of America was considered the audience. The adjustments made by many magazines since the 1960s redefined the content and the audience in more selective ways.

The Types of Magazines

Magazines may be categorized by audience appeal or function.

Today the majority of magazines are narrow either in subject, or audience, or both. An example of this is *The American West,* a magazine devoted to the history, geography, and arts of the West, having a circulation of 100,000 each month. To really appreciate the size of today's specialized magazine market, one need only look through *Ulrich's International Periodical Directory,* which lists 66,000 journals published around the world. (If we add the listings in *Ulrich's Irregular Series and Annuals* and *Ulrich's Quarterly,* the grand total of periodicals would come to 123,000, published by 63,000 publishers in 181 different countries!) In a tongue-in-cheek article about the joys of specialized magazines, author Murray L. Bob said the "little magazine" offers hope to people worried about growing media monopolies and conglomerates. Some of his favorites (with subtitles in parentheses) found in the Ulrich directories:

Negative Capability; Hard Pressed; Rabie's Magazine (Creative Writing); Damn You; Your Friendly Fascist (Pigs Arse Press); Dental Floss; Sadie's Chatter; Brussels Sprout Haiku Poetry; Nurse's Hipflask Poetry Broadsword; Mythprint; Urbane Guerrilla; Poisoned Pen; Grinning Idiot; Another Small Magazine; Some Other Magazine; Rag Bag; Fourteenth Century English Mystic Magazine; Overspill (For the Longer Poem); Margerine Maypole Orangutang Express; Pandemonium; This is Important; Truly Fine Press; Bedsitter (A Labour of Love); Hard Row to Hoe; Pundit; and *Difficulties.*[3]

Few magazines today can be described as being broad in both subject and audience. The *Reader's Digest* is one of the few, although it is narrowing somewhat as its audience becomes older and more conservative. The magazine prints more than thirty-one million copies each month, a third of which are published in seventeen different languages and sent overseas. (That thirty-one million copies makes the *Digest* the largest circulating journal in the world.) Originally supported by its subscription price alone, it began to take advertising in limited quantities late in its history. While this advertising is important to the magazine today, it is far from its principal means of support. The Reader's Digest Association, one of the largest privately held corporations in the United States, does not have to reveal its income. However, according to an article in the *Columbia Journalism Review,* the association took in some $1.25 billion in 1983 from its magazine, record collections, books, and audio equipment sales.[4] The *Reader's Digest* also goes to some pains to keep current with popular taste, running a random monthly sample of its readers to determine their continuing interests. Thus, to a degree, it has borrowed television's rating technique to ascertain the relative popularity of its offerings and to assess its gatekeepers' judgment. The technique has proven successful, as surveys since 1976 have indicated a steadily rising "enthusiasm quotient" among readers, who have been renewing their subscriptions at a fairly high 70 percent level.[5]

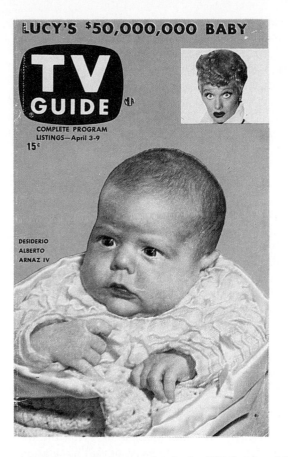

LUCY'S $50,000,000 BABY

TV GUIDE

COMPLETE PROGRAM
LISTINGS—April 3-9

15¢

DESIDERIO
ALBERTO
ARNAZ IV

Cover of the inaugural
issue of *TV Guide*
magazine, first published
in April, 1953.

Another atypical magazine is the *TV Guide,* which is narrow in subject but very broad in audience. The *TV Guide* and magazines representing other categories deserve a closer look.

    TV Guide's weekly circulation of some seventeen million ranks it nearly equal with *Reader's Digest* as one of the largest circulating domestic magazines (recall that much of the *Digest's* circulation is to foreign countries). *TV Guide* is the direct result of television's effect on the magazine field and is a classic example of one medium catering to another in terms of both interest and complementary functions. Television's ubiquitous popularity and ever-increasing numbers of programs demand a ready reference in more tangible form than its passing self-promotions can offer.

    On April 3, 1953, *TV Guide* first appeared with ten editions from coast to coast and a circulation of 1.5 million. Publisher Walter Annenberg, who had made his fortune putting out the *Daily Racing Form* and who went on to become United States ambassador to Great Britain—and head of the Triangle publishing empire that included *Seventeen* and other magazines

TV Guide

TV Guide, with seventeen
million circulation, is narrow in
subject but broad in audience.

and daily newspapers—correctly guessed that the public would spend money each week for a thorough guide to what was showing on television in their community. The magazine gathers together in one handy place all programming data for the week, including massive listings of cable programs (stored in its computers are summaries of more than 50,000 episodes of syndicated reruns and 8,000 movies), and adds a few articles and commentaries about the medium. The magazine's basic function as an index illustrates a curious admixture of communication functions: it is information about entertainment. On another level, it is persuasive, not only in "selling" the concept of TV viewing, but increasingly in the past several years in its hard-hitting albeit right-wing exposes about television news and the economics of the industry.[6]

Apart from its huge circulation and reference function, *TV Guide* is noteworthy from a technical standpoint: it is forced to publish in more than one hundred geographical editions to accommodate the various FCC (Federal Communications Commission) channel allocations on the TV band, and, since the early 1980s, the proliferating cable television channels available. In so doing, the magazine also allows the same number of options for local advertisers to ply their wares, which explains why your local television channel's news anchors smile out at you in full page ads each week.

The Newsmagazines

Henry Luce, founder of the Time-Life Magazine group.

Another combination of characteristics is represented by the weekly newsmagazines, of which *Time*— with its circulation of 4.7 million—is the largest. It is interesting that *Time*'s circulation, despite the magazine's popularity, is only about half that of some of the major magazines that have gone under in the past dozen years, including its stablemate, the original *Life*. This is fairly conclusive evidence that circulation alone is not the answer to a magazine's success or failure; the homogeneity of that circulation is the prime operant.

From its inception in 1923 by Henry Luce and Briton Hadden, two young Yale graduates who scraped together $86,000, *Time* magazine has attempted to couple entertainment with information and, according to many observers, has overlaid the entire package with a dose of persuasion. *Time* uses a "group journalism" concept of gatekeeping. This means that few articles in the magazine are the work of only one person. Editors assign stories to correspondents who gather data and prepare a file that is sent to New York where the material is rewritten for style and restrictions of space. The result is a weekly package of facts put in *Time*'s perspective, a perspective that for most of its existence was Henry Luce's peculiar set of economic, philosophic, and political values.[7]

For all their influence, newsmagazines represent only a tiny fraction of magazine publishing in terms of both number and audience. Often overlooked is the fact that newsmagazines publish weekly, or four times more often than most magazines, a factor that both provides additional exposure and establishes firm habit patterns in their readers.

Weekly publication also makes newsmagazines a cross between monthly magazines and daily newspapers, filling an information void by providing a limited perspective on current events. Newsmagazines are afforded a little more time than newspapers to digest the significance of contemporary developments. Also, their space limitations are acute in proportion to the scope of events demanding interpretation, and they must be highly selective in what they present. Another of their problems involves that of writing to make the quasi-historical appear contemporary—last week's news must seem current. This is a technique at which their team-writing staff has become particularly adept.

As much as any other magazine, _Time_ has profitably perfected geographical zone distribution, both for printing facility and as an additional source of advertising revenue. _Time_ publishes seven different geographical editions, permitting local advertising to be carried at a far more reasonable rate than national distribution would justify. In addition, _Time_ appears in five separate "demographic breakouts," with editorial and advertising content specifically aimed at: (1) "Big Time" or significant American markets where airline travel, wine consumption, and foreign car purchases are catered to; (2) "College Student" market, published during the school year; (3) "Exurban" markets for luxury products; (4) "T" or top business managers; and (5) "Doctors" or the medical profession market. The marketing strategy is working well: with only one quarter as many magazines circulated weekly as the giant _TV Guide,_ it nevertheless ranks ahead of _TV Guide_ in overall advertising revenues.[8]

Time was one of the first magazines to utilize communications satellites. In the spring of 1980 it bragged that it had made the jet age obsolete by its utilization of RCA's _SATCOM II_ and the _COMSAT Intelsat IV,_ satellites hovering some 20,000 miles above the earth's surface. The system allowed _Time_ to be the first publisher equipped to beam four-color finished

Weekly newsmagazines are crosses between newspapers and monthly magazines.

pages as well as black-and-white photos and text by satellite, almost instantaneously connecting New York and Hong Kong. Satellite technology was combined with other printing and distributing innovations; full page, press-ready pagination via computer, and use of the telephone to relay completed pages between computers in the United States and Europe.

City and Regional Magazines

A relatively new phenomenon in the magazine field is the so-called city and regional magazine, such as *Los Angeles Magazine, The Washingtonian Magazine, Chicago Magazine, Houston Living,* and some two hundred others listed in the annual *Writer's Market* under various categories: general interest regional publications, regional business and finance oriented publications, and recreational and travel publications.[9] They range in scope and influence, from small regional publications such as the *Bend of the River Magazine* published in Perrysburg, Ohio and circulating to fifteen hundred readers interested in Ohio history and antiques, to the venerable *New Yorker.*

Interestingly, the famous *New Yorker*—first published in 1923—was the first and for decades the only example of this kind of magazine. However, as the *New Yorker* progressively became a national magazine catering to the sophisticated across the nation, it left a void in New York City, which in the late 1960s began to be filled by *New York,* published by the staff of the defunct *New York Herald-Tribune. New York* concentrates on life in the city and the problems of New Yorkers, but ironically because these subjects are of such magnitude they are capturing a certain national audience, and *New York* is well on its way toward becoming another *New Yorker.* If that comes about, its place will be taken by a new publication or some of the other magazines already vying for Gotham's attention: *Focus* (a 250,000-circulation guide to New York City and to New York shops for hotel guests and New York residents), *Our Town* (a 135,000-circulation weekly tabloid covering neighborhood news of Manhattan), *Staten Island* (a 25,000-circulation quarterly covering locally-oriented, general interest topics), *New Brooklyn* (similar to *Staten Island,* covering a different borough), or *New York Affairs* (a 5,000-circulation quarterly emphasizing urban problems and catering to academics, public officials, corporation presidents and intellectuals).[10]

A brief look at *New Yorker* and *Los Angeles* magazines reveals why city and regional magazines have established themselves so firmly in the media mix, and why magazines serving smaller communities look to the future with enthusiasm.

Long recognized for its editorial quality, *New Yorker* is also recognized for its ability to deliver the "ideal" audience to advertisers. It boasts a paid circulation of 505,000 and a high pass-along readership estimated at 5.6 people per copy. The median income of its readers is $92,400; their net worth averages $622,300. Nearly all—94 percent—own their own homes, with a median value of $194,962. Given this market, it is little wonder that the magazine carries more than 3,000 pages of up-scale advertising annually, more than larger circulation magazines such as *People, Time,* and *Newsweek.* It can be choosey, and in so doing has led the nation's magazines in its ad pages devoted to wine, travel, hotels and resorts, luggage and leather goods, and has long been among the top ten print carriers of advertising for men's and women's clothing, liquor, and jewelry.[11]

Meanwhile, on the opposite corner of the nation, *Los Angeles* magazine has become one of the most financially secure regional magazines in America. Its nonadvertising pages are devoted to feature, photo and news coverage of the Southern California lifestyle—much of it superfluous and frivolous, but some of it substantial and incisive. Month in and month out, the magazine's cover photo is of an area celebrity—usually from the film or television industries. Glamour and "how-to" articles, coupled with reader surveys on life-style issues and humor columns, constitute the bulk of the magazine's editorial content, successfully appealing to the citizens of what is regarded as Los Angeles' relatively unstructured metropolitan environment. The advertising and editorial mix work well: the average reader household is college educated and earns $89,000 annually in the professional or executive work force, owns two automobiles, travels widely and dines out frequently.[12]

The original city magazines served major metropolitan areas. Some developed as the offspring of restaurant guides, television guides, chamber of commerce promotional pieces, and the like. Today, with circulations ranging from several hundred to hundreds of thousands, they have become significant media in their own right. They vary widely in appearance and editorial substance. While some still bear chamber-of-commerce overtones, others increasingly assume a watchdog role, probing local issues too delicate or complex for established media. Such magazines serve as a showcase

Originally local boosters, city and regional magazines have become substantial media, despite heavy doses of commercialism.

for local authors, poets, critics, photographers, and sometimes frustrated newspaper journalists seeking a less constrained writing outlet than their hometown newspaper allows.

As we have seen in the cases of the *New Yorker* and *Los Angeles* magazines, the real viability of this magazine form is in its commercial nature. The city and regional magazines are geographically and demographically selective—the merchandisers' dream. Audiences on the whole are decidedly "upscale," or well heeled and well educated, discriminating in taste and culture. A short browse through the descriptions in *Writer's Market* will prove to be an interesting affirmation of that generalization, as one can see the call for free-lanced articles and photos that would appeal to this generally elite audience.

In his national survey of city magazine publishers, Alan Fletcher found the medium to be flourishing. But he uncovered several problem areas that needed to be resolved if the business intended to become significant:

1. For city magazines to survive, they must establish strong local identities, taking advantage of voids that exist in other local and national media;
2. Circulations must be increased because of problems associated with ad revenue and production costs;
3. A continual supply of good editorial matter must be found;
4. Paper, printing personnel, and distribution costs must be dealt with;
5. Advertisers must be convinced it is worth their while to sell via the high quality graphics and upscale audiences available.[13]

Magazines without Ads

A few magazines survive with no ads but their subscription costs are high.

Magazines continue to cast about for survival formulae. One branch of the business avoids advertising completely, basing its entire financial operation on sales price alone. The best known is probably *Consumer Reports*. This type of magazine copies the formula under which the *Reader's Digest* was originally founded. Magazines that do not use advertisements come close to the book-publishing business; some are even in hardcover, such as *American Heritage*. They seek to combine the security of a subscription list with the relative freedom from advertising pressure that a book enjoys. The resurrected *Saturday Evening Post* was an experiment along these lines, which sought to capitalize on America's nostalgia as well as the six million subscribers it had owned before it went bankrupt. At a cost of $1.50 per copy, it served as an index of increasing American affluence.

Controlled Circulation Magazines

Controlled circulation magazines are delivered for free to predetermined target audiences.

Some magazines may pass along all the costs to the consumer. Others have been attracted to a new phenomenon in magazine publishing—the *controlled circulation* magazine. Controlled circulation generally is a euphemism for a free or marketing-directed magazine aimed at certain classes of people in terms of their presumed interests. Credit-card companies were the first in this field. They used magazines as vehicles for advertisers of luxury goods who were interested in the established credit ratings that cardholders had demonstrated. The facility of purchase was also appealing to the advertisers—they urged readers to "put it on your credit card."

The Business Press

One of the most rapidly expanding areas of the magazine industry consists of the publications produced for specific industrial, service, or professional business audiences. There are perhaps twenty-five hundred such publications. An obvious, although minute, portion of the magazines in this category are the nationally distributed general business or business news magazines such as *Barron's, Business Week, Dun's Business Month, Forbes, Fortune, Inc.,* and *Money.* Less well known are the dozens of regional business magazines, focusing on the financial climate of individual cities or regions of the country; examples include *Austin Magazine, Business News* from San Diego, *The Business Times* from Connecticut, *Kansas Business News, Mid-South Business, Northwest Investment Review, Regardies: The Magazine of Washington Business and Real Estate,* and the *Western New York Magazine.*

Of the two types of magazines described in the previous paragraph, the general business or business news magazines are frequently produced by large publishing houses, while the regional business magazines are usually independently produced, often by chambers of commerce or local publishing houses. The giants in the industry produce dozens of titles apiece. Harcourt Brace Jovanovich Publications, Inc. is the most prolific, with fifty-three titles, followed by McGraw-Hill with twenty-eight, American Broadcasting Companies with twenty-six, Penton/IPC and Reed Holdings, Inc. with twenty-five each, Williams & Wilkins Co. with twenty-one, Capital Cities Communications (Fairchild Publications, Inc.) and Dun & Bradstreet with nineteen each.

Company Magazines

A separate category of magazines, frequently confused with the business or trade press described above, is reserved for the nine thousand or so individual company publications. Almost as an afterthought as a management effort to keep employees in touch with one another, the company magazines, formerly known as "house organs" because they tended to be produced entirely "in house," have grown in quality and respectability. Such magazines rarely include advertising since they are basically informative or persuasive in content, and their cost is borne as a business expense by the organization or corporation involved. They continue to serve a management function, encouraging cooperation and dedication to the company's goals.

Major changes in company magazines over the past decade or so may be traced to management's recognition that their house organs should have journalistic integrity and should compete with general circulation magazines for space on the employees' coffee tables. Ever larger budgets have been allocated for such publications, and many companies seek out trained

There are nine thousand or so company publications, formerly called "house organs."

magazine writers and editors to join their public relations teams in producing the magazines. In some cases, such as the magazines produced by Alcoa Aluminum, lavishly illustrated articles of general interest are found among the companies' public relations pieces. Graduates fresh from journalism schools frequently find themselves among the fifteen thousand or so writers and editors working on such publications, which gives journalism deans reason to insist that their public relations and magazine students take a variety of courses in technical and skills areas.

Company publications comprise an integral part of the mass communications network, providing members, employees, or stockholders with data pertinent to their interests, jobs, or investments, which are obtainable in no other way. These organizational publications comprise a sort of media sub-network serving even smaller groups. Further, they often provide grist for the mills of the larger general media. For example, a corporate financial report may lead to extended articles in the general financial or business news press; or a technological development reported in an employee bulletin may spark an expanded piece in one of the trade magazines.

Life Cycles of Magazines

It has been said that in the magazine world, more than in any other industry, the success or failure of a general-interest magazine can be traced to a specific individual, one having a unique personality and singular vision. Indeed, magazines so frequently take on the ideosyncracies of their founders that when the founding editors move on or die, the magazines are guaranteed to change. Clay Felker has observed that American magazines follow the human life cycle: "a clamorous youth eager to be noticed; a vigorous, productive middle age marked by an easy-to-define editorial line; and a long, slow decline, in which efforts at revival are sporadic and tragically doomed."[14]

Magazines may be founded by individuals, but they tend to become institutionalized.

Perhaps because magazines historically have been the brainchildren of visionary individuals, today's chains, groups, and conglomerates still seem willing to let individuals experiment, but are happy to buy them out once the magazines appear to be making it. High prices are paid by companies recognizing the potential of a small magazine, so entrepreneurs are increasingly willing to sell out. There is a good reason why most magazines are published by multi-magazine groups: a single title, especially one of limited audience circulation, must carry too great a burden of overhead to make economic sense. Periodicals reach a saturation point, ad revenue becomes limited, and the natural advantages of group ownership come to the fore: bulk acquisition of paper, better printing contracts, more attractive subscription packages, utilization of corporate research, management, and circulation expertise, and a greater possibility of gaining favorable deals with national distributors.[15]

On the other hand, since the various magazines owned by a group will differ so greatly editorially, each group finds itself hiring a separate editorial staff and frequently a separate advertising staff. Of course the objective of the group is profit, which the magazine industry has realized by its steady profit margins, but on a small scale. Because group ownership is interested in the steadiness of its profits, it can tolerate the percentage. As a whole, the industry reports 3 to 6 percent pretax earnings, even though many profitable magazines make 15 percent pretax earnings.[16]

Starting a New Magazine

We still hear of the occasional magazine that was begun on a shoestring. Hugh Hefner's *Playboy* was started with $16,000, and the first issue was laid out on a card table in his apartment. Bob Anderson's *Runner's World* got underway for a measly $100 in the mid-1960s when Anderson was a seventeen-year-old high school cross-country runner. That $100, coupled with Anderson's production and distribution work, and his success in persuading other runners to contribute free articles to what was at first called *Distant Running News,* eventually mushroomed into an impressive empire. Anderson went on to develop a complex of profitable publishing enterprises comprised of biking, soccer, canoeing, and cross-country skiing magazines, and a successful mail-order operation. The 1960s and 1970s saw additional low-budget magazines become successes: *High Times* began with $25,000, *Rolling Stone* with $6,500, and *Mother Earth News* with only $1,500.[17]

When such publications become successful, it is through a combination of hard work, good luck, recognizing a readership pool ready to subscribe, and remaining flexible as time and circumstances dictate. An interesting case in point is *Rolling Stone.* Targeted at the baby boomers of the late 1960s, the magazine originally stressed rock and protest music and life-style questions. As the youthful audience changed, so did the magazine. In the early 1970s its median reader was twenty-one years old. Now he (two-thirds of its subscribers are male) is twenty-four years old, and does not represent the same socio-economic-political demographic mix of earlier readers. According to *Rolling Stone*'s associate publisher, Henry Marks, the magazine's reader of today is a child of the rich:

. . . an early achiever, someone with an income of $27,000. Of those who are not employed, the income profile of the homes they come from is exceedingly high. These are young people who come from privileged families. They have the advantages of seeing those privileges early. Of unemployed single males, the household income of their parents is in excess of $50,000. The most important thing about our readers is that they're youthful. We reach 6 percent of the universe of 18-to-24 (year-old) range, but they are much more privileged than the rest of that 18-to-24 range. It is a group blessed with a number of advantages.[18]

But for every *Rolling Stone, Playboy* and *Runner's World,* there are probably a dozen—or perhaps even a hundred—failures; nobody knows for certain. Few individuals can sustain a struggling magazine through the decade-long, $30–million-in-losses gestation period that Time, Inc., tolerated for *Sports Illustrated* before it became profitable, or the $13 million Bob Guccione poured into *Viva* before it took on life. The leading causes

We remember the success stories, but they are the exceptions.

for failure include underfinancing, inexpert management, insufficient advertising, and public indifference. Sometimes even the old pros—including Hugh Hefner—have failed miserably when trying to imitate their own successes. Hefner made it with *Oui,* but lost a bundle on *Trump, Show Business Illustrated,* and *VIP.*

Even the well-heeled Time, Inc., has proven its fallibility. In 1983 it poured $47 million into a weekly broadcast and cable TV guide that would compete directly with *TV Guide.* After conducting thousands of marketing interviews with TV families, Time, Inc., launched *TV-Cable Week* in April, 1983, with a promise to spend up to $100 million to assure the magazine's success. The uniqueness of *TV-Cable Week* was its aim of publishing not only personality and feature stories about television, along with broadcast listings, but cable listings specific to each community's cable TV system. There are over four thousand cable systems in the United States, so the task would only be manageable if the magazine were marketed by the individual cable companies.

Because Time, Inc., is the second largest operator of cable systems in America, and also owns Home Box Office, the most successful cable pay-TV service, the corporation figured it had a natural lead into the marketplace. What it had not counted on was the reluctance of other rival cable companies to share subscriber lists with *TV-Cable Week,* and it had also forgotten that many of those companies already had binding contracts to distribute competing cable listing guides. Thus, despite the efforts of a 250-person staff, working out of a brand new building that housed a sophisticated computer loaded with information about the industry, the $2.95 a month magazine went belly up after only five months of publication. It had managed to lure only nineteen cable systems into its fold—including five owned by Time, Inc. "We misread the tea leaves," Time, Inc.'s vice president for magazines Kelso F. Sutton explained. Not coincidentally, *TV Guide* over the same time period was engaged in a multimillion-dollar promotional campaign and was successfully changing the face and contents of its venerable product to include more substantial articles and full cable listings. All in all, it was another case of the marketplace forces prevailing in the complex and sometimes unpredictable media business.[19]

Magazine sales—for new and established publications alike—occur either through subscriptions or on a retail basis. Often the emphasis on one or the other is intentional; the publisher may want the security of the annual payments made in advance by subscribers or it may want to avoid postage costs. In some cases the subject matter dictates whether the sales will be by subscription or at newsstands. For example, 87.4 percent of the sales of *Muscle and Fitness* occur at newsstands. Almost all copies of *Woman's Day* and *Family Circle* are sold at newsstands and checkout counters, according to *ABC Magazine Trend Report.* Sex-oriented magazines also tend to be sold at newsstands: 91.1 percent in the case of *Oui,* and 94.3 percent in the case of *Penthouse.* At the other end of the spectrum, 99.9 percent of the sales of *Health, National Geographic,* and *Smithsonian Magazine* are by subscription. One hundred percent of the sales of *Modern Maturity, The American Hunter, The American Rifleman,* and *Boys' Life* are by subscription.[20]

Subscription sales provide considerable security once a magazine is established, but that takes time. Nicholas Charney, who started *Psychology Today,* folded a second magazine venture, *Careers Today,* because buying subscription lists proved too expensive—this despite the fact that his new venture was already showing more than $100,000 worth of advertising per issue. Other magazines have chosen the newsstand for launching publications, hoping to develop sufficient interest and to acquire subscribers. The difficulty with this approach is the uncertainty of being allotted space for display at supermarkets and newsstands. Even the larger newsstands make room for only 100 to 200 different magazine titles, and, of course, give preference to the big sellers. Since magazines are frequently considered to be impulse purchases, chances for survival under such conditions are minimal for new magazines. Careful field testing in metropolitan newsstand markets is often in order. *People* was tested in eleven markets prior to its regular weekly publication in 1974. At thirty-five cents per copy, it sold an amazing 86 percent of available copies. That figure is especially significant because approximately half of all copies of nationally distributed magazines remain unsold on the shelf; they are then returned—at the publishers' expense—to the respective companies. Considering that each

Magazine Sales

Each magazine seeks its own successful mix of subscription or retail sales.

copy of a magazine like *Us* costs the publisher twenty-five cents in paper, printing, and delivery, such inefficiency in distribution makes it prohibitive for all but the wealthiest publishers to gamble on attracting customers at the nation's newsstands.[21]

In addition to the field testing and mass mailing of promotional materials to identified target audiences, marketing of new magazines is sometimes handled by buying space in existing magazines, such as by inserting a sampler of the proposed magazine's editorial content. This technique gives the existing magazine some additional editorial content assumed to be of interest to its readers, along with a few dollars in profit, while providing the new magazine exposure to potential readers and advertisers.

Contemporary Successes

In the United States today four magazines have a circulation greater than ten million copies per issue. These four—*Reader's Digest, TV Guide, Modern Maturity,* and *National Geographic*—represent the pinnacle of the medium in terms of number of copies printed and sold. All four magazines reach a much larger percentage of the population than did *McClure's* when it was selling one million copies each month at the turn of the century. Table 4.1 reveals that at present twenty-eight magazines have circulations larger than two million copies, and the top fifty magazines all sell more than one million copies per issue.

How have each of these magazines become so successful? As we have seen, for the most part, a key element has been the ability of the publisher to identify a market segment and then tailor editorial content to attract that market. *Cosmopolitan, Psychology Today, Esquire,* and *World Press Review* illustrate this point.

Cosmopolitan effectively serves a market—the unmarried working woman—designed to replenish itself. Advertisers love it.

Cosmopolitan In the early 1960s *Cosmopolitan,* one of the fourteen magazines published by the Hearst group, found itself a poor competitor in the national general-interest field, suffering from dwindling advertising revenues and a dwindling readership as public taste departed from the essentially fiction format so successfully utilized by the magazine.

Cosmopolitan felt the winds of change and decided to leave the general-interest magazine business, seeking a specific audience, chosen by interest rather than by demographics. Casting about for an audience not efficiently served by any medium, it hit upon that of the unmarried working woman. This was a happy choice in that it was a market designed to replenish itself. In addition, there was no other magazine catering to the interests of this group—to their tastes, habits, and life-style. To accomplish its purpose, *Cosmopolitan* hired Helen Gurley Brown (author of *Sex and the Single Girl*) as editor and began to offer a titillating choice of articles designed to help the reader become sexy and alluring, to snare a mate or shed a date. In some ways the magazine resembles soft-core pornography, especially in its daring covers, which are primarily responsible for 90.5 percent of its 2.9 million circulation occurring at newsstands and grocery checkouts.

Table 4.1 United States magazines with 2 million-plus circulation, for six months ending 31 December 1986.

Rank	Magazine	Total average paid circulation
1.	TV Guide	16,800,441
2.	Reader's Digest	16,609,847
3.	Modern Maturity	14,973,019
4.	National Geographic	10,764,998
5.	Better Homes and Gardens	8,091,751
6.	Family Circle	6,261,519
7.	Woman's Day	5,743,842
8.	Good Housekeeping	5,221,575
9.	McCall's	5,186,393
10	Ladies' Home Journal	5,020,551
11	Time	4,720,159
12	Redbook	4,409,450
13	National Enquirer	4,381,242
14	Guideposts	4,260,697
15	The Star	3,706,131
16	Playboy	3,477,324
17	Newsweek	3,101,152
18	People	3,038,363
19	Sports Illustrated	2,895,116
20	Cosmopolitan	2,873,071
21	Prevention	2,820,748
22	American Legion	2,648,627
23	Glamour	2,386,150
24	Penthouse	2,379,333
25	Smithsonian	2,310,970
26	U.S. News & World Report	2,287,016
27	Southern Living	2,263,992
28.	Field & Stream	2,007,479

Source: Audit Bureau of Circulations FAS-FAX Report, 31 December 1986. By permission.

When *Cosmopolitan* did these things, it did something else too. It attracted the advertising of manufacturers of perfumes, uplift brassieres, cosmetics, and other glamour products. Although circulation became far smaller, advertising revenues were assured and they grew as the magazine continually expanded its reach. *Cosmopolitan* made a successful adjustment during difficult years and is now financially healthy. Most likely the change could not have been made without the close correlation of the three factors we have mentioned: (1) the interest level of the magazine's editorial content; (2) the makeup of its audience; and (3) the kinds of products it advertises.

Psychology Today A magazine success story of a different sort is that of *Psychology Today*. *PT* was the $10,000 brainchild of Nicholas Charney, who reasoned that sufficient interest in human behavior existed to justify a popular magazine, rather than a scholarly journal, on the subject. He was right. Five years after starting *PT* in 1966, Charney sold the magazine to Boise-Cascade for $20 million.

Psychology Today's success formula has been to give young, affluent, and educated readers insights into themselves.

Psychology Today's success stems from knowing its educated, curious, and affluent audience.

Psychology Today was born at a time when the postwar emphasis on education, stimulated by the GI Bill, and an accelerating public affluence had begun to merge. With less time required to produce the necessities of life, Americans turned inward, in a sort of mass introspection that manifested itself in curiosity about behavior and what motivates us to think as we think, act as we act. *PT* was an instant success. It attained a circulation of 600,000 within four years, with advertising revenues to match. By the mid-1980s it ranked in the top fifty general magazines, with a circulation of 862,193. Advertisers were secure in the knowledge that 90.5 percent of that circulation was being mailed to the homes of subscribers, a stark contrast to the nine out of ten *Cosmopolitan* issues picked up at newsstands.

Consider the profile of *PT*'s subscribers. Subscribers are relatively young, well educated, nearly equally divided between males and females, and affluent. What kind of advertiser seeks this sort of "new generation" marketing mix? It is not the manufacturer of esoteric psychological paraphernalia. Rather, it is the manufacturer of sports cars, modern clothing, modern furniture, liquor, and other amenities of fine living. Thus, *PT* differs from the *Cosmopolitan* approach in that its interest focus is not specific to its advertisers' products, but rather productive of a real demographic mix. It is important to note that this demographic mix is achieved naturally through self-selection on the subscribers' part and not artificially by the computer. There is a certain arrogance to demographically predetermining readership affluence by census tract and then arbitrarily directing advertisements for sleek or sleazy product lines. However, an audience may preselect itself along these same lines, and in the end the magazine and the advertisers may be justifiably self-satisfied.

Esquire In its half-century of existence, *Esquire* magazine has had its ups and downs, but an analysis of its recent history justifies including it in the list of contemporary success stories. When founded in 1933, it concentrated on men's fashions, with the lines between its ads and its editorial content somewhat blurred. Within a few years, despite the depression, it was selling a half-million copies monthly, based in part on its prestige factor, but also due to its quality fiction by such authors as F. Scott Fitzgerald and Ernest Hemingway. During the 1940s it changed its focus, dishing up cheesecake in such large portions that it became embroiled with the United States Post Office over questions of pornography. In a landmark Supreme Court decision *Esquire* was cleared for postal distribution, and the titillation continued into the 1950s.

The 1960s saw the magazine returning to its "quality" days as an outlet for mainstream fiction and non-fiction writers—Norman Mailer, William Buckley, and Kurt Vonnegut, among the more noteworthy. Even though the 1970s saw it publishing seminal works by such essayists and journalists as Tom Wolfe and Harrison Salisbury, the magazine fell upon hard times. Its circulation peaked out at 1.25 million, placing it among the largest general circulation magazines (although it was still recognized as a "men's magazine"). It suffered an identity crisis in the eyes of advertisers, contributors, and readers, and when it was sold in 1979 to Phillip Moffitt and Christopher Whittle—a pair of Tennessee businessmen in their thirties—it was losing $25,000 a day. *Esquire*'s main problem seemed to be that it slipped between the cracks; it was not entirely a men's magazine, nor a general interest publication. Like *Playboy* and *Penthouse,* it was suffering from reader defections. Its imminent demise was forecast in numerous trade journals.

However, within a matter of months, Moffitt and Whittle turned the magazine around. By 1984 the magazine had regained some 100,000 of its lost circulation, and by 1986 it was circulating 710,604 copies. More importantly, it had managed to convince advertisers that it was a vehicle well worth spending money in. In 1981 *Esquire* carried only 535 pages of advertising; in 1983 the total was up to 1,312. Reason for the rapid turnabout? According to the new owners (Phillip Moffit is also the editor), the magazine successfully appeals to increasingly affluent members of the postwar baby boom generation by combining service-oriented, "how-to" features with pertinent advertising. Advice on trendy places to live, items to collect, hobbies to pursue, and wisdom on how to cope in today's troublesome world are nestled up against glossy ads. One recent issue's sixty-eight-page section on bars and drink recipes included twenty-three pages of full-color liquor advertisements. Many manifestations of popular culture are seen in the pages of the magazine. As *Writer's Market* informs prospective contributors, articles for *Esquire* should be slanted for sophisticated, intelligent readers, but should not be highbrow in the restrictive sense.

Esquire changed its orientation and appeal numerous times over the past half-century.

Marketing surveys indicate the typical reader of the magazine is a male college graduate between the ages of twenty-five and forty-four who earns $35,000 in a managerial or professional job.[22] All in all, *Esquire* has successfully gone full circle, once again delivering an audience to eager advertisers.

World Press Review What public affairs magazine crammed full of articles lifted directly from the news media of Red China, Soviet Russia, Latin America, Africa, and other hotbeds of international intrigue and revolution—but devoid of articles written in the United States—is subscribed to and read religiously by chief executive officers of America's Fortune 1000 firms, the presidents of top United States banks, members of the United States Congress, and network anchorpersons?

World Press Review, which provides "News and Views from the Foreign Press," is composed entirely of material from foreign journalists and translated into English: news items, commentary, and cartoons on politics, business, science, the arts, travel, and, of perhaps the greatest interest to its readers, items on America as seen from foreign eyes.

In a decade noted for the demise of numerous public affairs magazines, *World Press Review* reaches more than 60,000 subscribers because, in the words of its former editor, Alfred Balk, it serves a previously unfilled need:

That need is a global perspective—a picture of events, issues, and trends as seen from outside the American cultural prism. This provides an indispensable opportunity to see ourselves and the world as others see us and it. No other publication does that—and decisionmakers in all fields, along with ordinary conscientious citizens—increasingly require it.[23]

The magazine began publication in May, 1974, as *Atlas World Press Review,* and dropped *Atlas* from its masthead in 1980. (It succeeded a defunct monthly, *Atlas,* a semi-scholarly journal published from 1961 to 1972.) Unlike most magazines, it is produced by a foundation rather than a profit-oriented corporation. The magazine is part of the Stanley Foundation out of Muscatine, Iowa; the organization sponsors conferences, research, the syndicated radio discussion program "Common Ground," and other international policy-related activities.

World Press Review monitors scores of foreign publications, including forty of the world's top-ranked "elite" newspapers and many prestige magazines. Because foreign reporters have access to sources and information denied to United States journalists, the magazine offers scoops and insights not available through normal channels. "The American press is very good, but it is American. We need more than one voice," said former editor Balk. "Americans have the habit of believing they have access to it all, to everything we need to know. That just isn't true anymore."[24] Apparently the forumla adapted by *World Press Review* is working to add that additional voice.

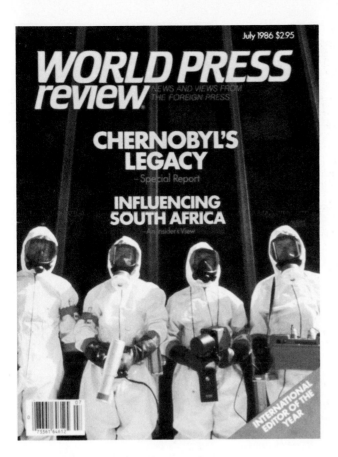

World Press Review is unique in that it is composed of material from foreign journalists and translated into English.

Cosmopolitan, Psychology Today, Esquire, and *World Press Review* perceived a need and acted to fill it. For an example of magazines that have filled different needs, consider Meridian Publishing, Inc., of Ogden, Utah. With only three editors and two artists in the national editorial department, the company produces nearly half a million copies of more than a thousand different titles each month. Incredible? Not really, when you consider that Meridian actually creates the basic ingredients of only seven magazines, each of which is sent tailor-made to hundreds of different sponsors. In each case, the organization sponsoring the magazine buys the entire issue, 80 percent of which is preprinted for all customers. It contracts to have its own logo on the cover as well as local advertising and local editorial copy to fill the remaining 20 percent.

The oldest of Meridian's basic magazine formats is *Your Home,* founded in 1947. It is used as the house organ of 250 firms, such as home improvement outlets, decorators, realtors, and builders. *Accent* is a travel-oriented magazine used by resorts, airlines, and travel agents interested in articles about how and where to travel. *People in Action* is an upbeat magazine filled with interesting personality profiles and success stories. Among

Filling the Gaps

Meridian Publishing uses a small staff to produce seven generic magazines that end up carrying one thousand different, customized titles.

Meridian Publishing Company makes it possible for a thousand small companies to have their own "personalized" magazines.

its sponsors are civic organizations, community colleges, and credit unions. *Sports Parade* features overviews of specific sports and profiles of past and present sports figures. Meridian's three newest magazines are *Bridal Fair,* in which practical articles for the new bride are featured; *Business Today,* which covers tried-and-true business principles as well as personalities; and *Your Health,* which highlights interesting people, technological advances, and helpful advice having to do with personal health care.

Free-lance writers contribute 65 percent of the editorial matter in Meridian's magazines; thousands of different writers submit articles annually for the publishing house's editors to pick over. About twenty-five free-lance articles are used each month in the seven outlets; the authors are paid fifteen cents per word. Fifteen artists are needed to produce the customized advertising and sponsor copy for the various purchasers, who expect each magazine to look uniquely its own.

In short, it is a perfect business arrangement, with the publisher, writers, institutions, and readers all happily mass producing and consuming the "personalized" media.

From a production standpoint magazines fall somewhere between their print cousins, books and newspapers. They are not on the relentless daily schedule of a newspaper, nor do they experience the painstaking leisure of book publishing. Most magazines work further ahead of schedule than one might be led to believe. The average monthly general-interest magazine is prepared four to six months in advance of publication. That is, the staff will start "dummying up," or composing, a July issue in January, committing some articles and commissioning others with the thought that, for all intents and purposes, the issue will be closed by April except for some minor loose ends of material.

Newspapers are almost entirely written by staff members, and books are almost entirely written by outside authors. Magazines fall between these two poles. While some are completely staff written and others are completely contracted, most employ some balance of the two techniques, with an edge toward staff writing because, by and large, it is less expensive and more reliable than the work of outside authors. Good authors are hard to come by, and authors who have successfully published and have established some reputation generally command considerable fees for their material. The names of some of these authors also constitute an additional drawing card for the magazine. All this notwithstanding, there are surprisingly few full-time free-lance authors in the United States. Because of this, it is reasonable to ask where magazine articles come from. For the most part, they come from experts in a given field who hold jobs in the field and who write as a sideline. For example, *Psychology Today,* which is about half staff-written and half contract-written, commissions its outside articles from people such as practicing psychologists and university professors. A considerable number of newspaper journalists augment their earnings with outside articles and feature material, as do book authors. The latter, of course, bring considerable "name attraction" to a magazine.

As in book publishing, magazine editors have either a stable of acceptable authors whose work and expertise they know, or they rely on literary agents whom they trust. Agents and authors also constantly suggest story ideas to the magazines in the form of queries or outlines in hope of a contract. Each magazine editor zealously tries to maintain a constant tone or flavor to the publication—a proven format of exactly the right material mix, both written and graphic—that will continue to appeal to the magazine's specific readership. The editor's gatekeeping function is dictated less by the surplus of material than by the rigid requirements of the particular

Magazine articles are written by staff members or free-lancers; balancing content to meet reader interests demands sensitive gatekeeping.

audience. The editor must ruthlessly prune everything, regardless of intrinsic merit, that does not exactly fit both the overall format of the magazine and the particular mix of a given issue. The task is, in part, a balancing act in which judgmental factors enter more sensitively than in the case of either newspaper editors, whose principal concern is the news, or book publishers, whose concern is with appeal in the singular sense.

Magazines and Advertising

As we have seen, advertising is of critical importance to the magazine industry. Without substantial advertising, most magazines would fold, as did *Collier's,* the original *Life, Look,* and *Saturday Evening Post* during the 1960s and early 1970s. In this section we will explore the relation between magazines and advertising—the use of reader surveys, the ways in which ads become part of the content of a magazine, and the potential for marketing directly to the readers of a magazine through the use of the subscription list.

Readership Surveys

Magazine readership is far wider than simple circulation figures indicate.

Magazine circulation is deceptive. There is an exposure factor that is not reflected in raw circulation figures. In other words, the number of people who read or are at least exposed to magazines is far greater than the number who buy or subscribe to them (as happens, for example, in doctors' offices). This exposure factor will vary widely, of course, with the nature of the magazine, running as high as six times for the circulation of national general-interest magazines, to little more than single exposure for some of the smaller, more esoteric business or scholarly publications. In any event, this exposure factor is far higher for magazines than for either newspapers or books among the print media, or for all broadcast media.

Readership surveys help to determine exposure. Most magazines subscribe to some rating service such as the Starch Report, which helps advertisers determine their true costs in reaching audiences. Magazine advertising, while sold on the basis of circulation, is calculated on estimated exposure. This is to say, a kind of double standard exists whereby advertisers purchase space off the rate card, which is a reflection of audited circulation, but calculate their *cost per thousand*— the cost to reach each one thousand members of the audience—on presumed exposure as indicated by the Starch Report.

From various readership surveys, magazine publishers have learned that their audience is changing, and those changes are being mirrored in new types of content for established magazines and a proliferation of new publications. Heavy magazine readers have for years tended to have college educations, incomes above the norm, and professional/managerial occupations. Between 1981 and 1984, the average adult magazine reader increased his or her magazine reading from eight issues to nine, and increased the time devoted to each issue from fifty-six minutes up to sixty-two minutes.[25] The average consumer is exposed to any given magazine on 3.2 different days, opens 53.4 percent of the magazine's pages each reading day, and is exposed to the average page 1.7 times—all statistics of great interest to advertisers.[26]

Newsweek

444 Madison Avenue, New York, NY 10022

Gerard Smith, Publisher

How are we doing?

Dear Newsweek Subscriber:

One of the reasons Newsweek has continued to grow and maintain its high level of journalistic excellence is that we at Newsweek try never to lose sight of who the magazine is for: you, the subscriber. In an attempt to keep in touch and find out more about our subscribers we have periodically asked them what they like about us. Their answers have helped us to keep Newsweek both timely and relevant. Won't you help us now?

Please take a moment and complete the enclosed questionnaire. All your answers will be totally confidential and will be used only by Newsweek.

For your completed questionnaire, we will make a donation to one of the four charities indicated on the survey. The choice is yours. So far we have donated over a quarter of a million dollars as a direct result of subscribers, like you, participating in our surveys.

I look forward to your opinions.

Sincerely,

Gerard Smith
Publisher

GS/ct

P.S. Please mail your completed questionnaire in the enclosed postage paid envelope provided.

26008

continued

Newsweek
SUBSCRIBER STUDY
Newsweek, 444 Madison Avenue, New York, NY 10022

Gerard Smith, Publisher

Dear Reader
How are we doing?
please let us know.
G.S.

July 24, 1987

7308277431

The following questions can be answered quickly and easily.
Please use the postage-paid envelope provided for your convenience in returning the questionnaire to us.
Your answers will help us make Newsweek more valuable to you.

1. Which sections of the magazine most interest you?

1. National Affairs 2. International 3. Business 4. Lifestyle 5. Society 6. The Arts

2. What is the most important reason you read Newsweek?

1. Summarizes week's events 2. Presents the news without bias 3. Provides in-depth analysis
4. Covers more than just national and international news 5. Provides different points of view

3. What is your *primary* business or industry? (choose one)

A	Agriculture/Forestry/Construction/Mining	E	Real Estate
B	Communications/Publishing/Advertising	R	Religion
N	Data Processing/Computers	I	Services, Personal (Lawyer, CPA, etc.)
C	Education	J	Services to Business (Consultant, CPA, Lawyer, Architect, etc.)
D	Finance/Banking/Insurance	G	Trade, Retail
S	Fine Arts/Crafts/Photography	P	Trade, Wholesale
F	Government/Public Administration/Military	K	Transportation/Public Utilities
Q	Health Care/Social Services	L	Travel/Entertainment/Recreation
H	Manufacturing	M	Other (please specify):_____

4. What is your *primary* title or position? (choose one)

E	Chairman of the Board	H	Scientist/Engineer
B	Company President/CEO/COO	J	Technical Specialist
A	Owner/Partner	M	Teacher
K	Director of the Board	Q	Retired
C	Other Company Officer (Vice Pres., Treas., etc.)	N	Skilled Craftsman/Laborer
D	Manager/Department Head	L	Clerical
F	Supervisor/Foreman	O	Homemaker
G	Other Administrative Position	P	Student
T	Professional (Doctor, Lawyer, Architect, etc.)	R	Military (Present Rank:_____)
U	Professional Non-Business (Artist, Reporter, etc.)	S	Other (please specify):_____
I	Sales		

5. Which one of the following categories most clearly approximates your individual total annual income from all sources, including salary, dividends and so on, before taxes?

M	$40,000 and over	J	$25,000-$29,999
L	$35,000-$39,999	I	$20,000-$24,999
K	$30,000-$34,999	P	Under $20,000

6. Please indicate the charity of your choice (choose one):

A American Cancer Society B Red Cross C American Heart Association D UNICEF

YOUR ANSWERS WILL BE HELD IN STRICT CONFIDENCE. THANK YOU.

The close ties between magazines and advertising frequently extend beyond the economic support provided by advertisers. In fact, advertising sometimes becomes a very real part of the editorial content itself, and may be one of the prime reasons for reading the magazine. Evidence of this fact stems from World War II when editions of national magazines with the advertising deleted were sent to troops overseas on the naïve theory that soldiers were interested only in the reading material and were in no position to purchase the goods advertised anyway. This brought immediate complaints from the troops, who felt that they were being shortchanged. They missed the ads and the memories of home evoked by the ads. A part of the pleasure experienced by the troops was their contemplation of the things to which they would return. This speaks much for the commercial base of American society, even for a serviceman lingering in a foxhole. It also demonstrates the integrated nature of mass communications, which act as a carrier of culture, reflecting quite accurately the life-style of a society.

In fact, in many glossy-paged magazines, advertising has developed into an art form of colorful and aesthetic and sensuous appeal, which can be described as "word, camera, and brush pictures of new products."[27] The smashing illustrations in *Vogue* and *Harper's Bazaar* come to mind immediately. In such magazines, the editorial content plays either a secondary or a supporting role. In many of the specialized magazines dealing with topics such as fashion or travel, the editorial matter serves purely as supportive material for the advertising, with substantial editorial preference and even "trade-out" material being awarded to major advertisers.[28] Trade-out material is that editorial matter promised to an advertiser in exchange for a substantial volume of advertising. Some magazines match advertising and editorial content for major advertisers on a page-for-page basis.

In *Psychology Today* and other such publications, the audience is far more than a source of subscriber revenue and a target for advertising. The audience is a self-identified market for a variety of goods and services made available by the publishers; in *PT*'s case, this includes posters, games, videotapes, and a whole series of books. The magazine, in essence, provides the means for establishing a credible roster of potential customers for certain kinds of products. *PT* founder Nicholas Charney once said, "The magazine is simply a tool to identify an audience with a particular interest . . . and that special list of subscribers is a prime target for a whole range of other services."[29]

Other publications have followed *PT*'s lead. *Family Health,* a relatively new magazine, had little over a million subscribers in 1971 and immediately moved into the business of marketing medical encyclopedia sets. *Southern Living* has used its subscribers for marketing quality Southern cookbooks, while *New York* envisages its readership as a base for selling a how-to-do-it series on survival in New York.

Advertising as Content

Advertising is more than mere persuasion; it has an important informational component.

Marketing Audience Interest

Once identified, magazine subscribers become targets for all sorts of related goods, services, and persuasion efforts.

Magazines' complementary services reach beyond the magazines themselves. It did not take publishers long, for instance, to realize that a painstakingly developed, self-identified list of potential customers was in itself a valuable asset. Some magazines have taken to selling—or, rather, renting—all or portions of their subscription lists to other organizations as sales tools. If a test mailing is satisfactory, an organization or manufacturer may want to buy the entire list at so much per thousand names. Magazines also make a healthy profit selling their subscription lists to other magazines at three to five cents a name, and more than one entrepreneur has made a go of it by merely selling different sets of lists to interested buyers. A "good" mailing list is likely to bring as high a response as 15 percent, though 10 percent is considered very satisfactory.

The Future of Magazines

The magazine, more than any other medium, has seen the future: an age of specialization.

In many respects, the future of the magazine industry is already here. The prevailing trend in media is the logical progression from elitism, to the pursuit of the largest possible audiences, and on into the age of specialization. The latter stage finds media seeking out, and catering to, individuals and groups identified by social, economic, educational, professional, cultural, recreational, religious, and many other interests. Surely the magazine industry has already gone that route; indeed, some of these interests were being catered to over a century ago, and we see no likelihood of a complete reversion to mass-circulation magazines that seek the broadest national common denominator.

Technologically and economically, the magazine of tomorrow will differ from that of today. As we have already seen in our forecasts about newspapers, the marriage of print to electronics will radically alter the production and distribution of print media. Magazines are no exception. In some ways, perhaps most significantly in terms of distribution techniques, where the bulky magazine has been hard hit by exponential increases in postal rates (nearly 500 percent in the 1970s alone), the magazine industry is ready for change. Smaller issues on lighter paper sent through the mail or distributed by individual carriers contracted by publishers have constituted one change. Experimentation with electronic, satellite, and laser technology for production, as seen in *Time* and *TV Guide,* is another. Publishing houses of the future may find themselves contracting with electronic information-dispensing companies, or they will start up their own, to assure rapid and widespread dissemination of magazine matter that appeals to a few or to many consumers, each having his or her own in-home computer with page-printing capacity.

As discussed in chapter 3, individual consumers will be given freedom of selection in the privacy of their own homes, using their home information systems—video-display terminals with printout capacities. Consumers will select news, information, entertainment, and advertising to suit their own purposes, much as they do now but with wider choices. More than in any other media form, tomorrow's magazine will be able to pander to the most individualistic characteristics delineated by demographic and psychographic research. This raises, of course, the issue of self-indulgent masses of heterogeneous individuals who lack a sense of community, a serious concern facing not only tomorrow's magazine publishers but all media producers.

Summary

Magazines have developed historically to occupy a middle position between the time-consuming and self-selective book and the hurried, geographically circumscribed newspaper. All magazines share two characteristics: they are published regularly, and each appeals to some specific segment of society.

The first magazines were the *Tatler* and the *Spectator,* published in England in the first quarter of the eighteenth century. By the early nineteenth century magazines were available in the United States, the first market-specific magazine being *Godey's Lady's Book,* first published in 1830. The last third of the century is known as the Golden Age of magazines. The industry flourished because of technological developments, the existence of the transcontinental railroad, and the Postal Act which created less expensive second-class mail.

Muckrakers characterized the early part of this century, but their influence waned after the failure of Theodore Roosevelt's third party candidacy in 1912. By the 1950s, television began to usurp the national audience, and several general-interest magazines eventually ceased publication after costly circulation wars. The advertising had gone to television.

There are several ways to categorize magazines. One way is to categorize them functionally, for example, as entertainment/escape, news/information, or advocacy/opinion. We can also consider the relation between content and audience. Most magazines today are narrow in subject and narrow in audience. *TV Guide,* however, is narrow in subject but broad in audience. There are also newsmagazines and city or regional magazines. Some magazines carry no advertisements and others, called controlled circulation, are given away free. Finally, there are the business press and company magazines, which represent a burgeoning area of magazine publishing.

Most magazines are owned by multi-magazine groups to share costs and expertise and to protect against possible failure. While some magazines are started on a shoestring, the costs of publishing magazines today make that method rare; few companies can afford the kind of $47 million gamble Time, Inc., lost with its *TV Cable Week.*

Four quite different magazines that represent contemporary success stories are *Cosmopolitan, Psychology Today, Esquire,* and *World Press Review.* The four differ in scope and appeal, attracting different audiences and different advertisers.

Some magazines rely on staff writers for material while others buy the work of free-lancers; many use a mix of the two sources for their articles.

Advertising is crucial to most magazines. Reader surveys help to determine the pass-along readership, which is a measure of exposure for advertisers. In some cases, the advertising may be more important than the content to the reader, as is often the case with glossy fashion magazines. In recent years subscription lists have provided advertisers with an additional channel to markets, and have provided additional revenue to their owners; magazines may use these lists to sell their own products or they may rent them to organizations and manufacturers.

Technology will soon make it possible for readers to select only that material they want to see. This is of concern to media producers who fear that it may isolate people, leading to a lack of a sense of community.

Notes

1. Theodore Peterson, *Magazines in the Twentieth Century* (Urbana: University of Illinois Press, 1956).
2. Edwin Emery and Michael Emery, *The Press and America: An Interpretative History of the Mass Media,* 4th ed. (Englewood Cliffs, N.J.: Prentice-Hall, 1978), p. 274.
3. Murray L. Bob, "Long Live the Little Magazine," *St. Louis Journalism Review,* September 1985, p. 2.
4. Miriam Jacob, "*Reader's Digest:* Who's in charge?" *Columbia Journalism Review,* July/August 1984, p. 42.
5. Ibid, p. 43.
6. Charlene Canape, "The Chase Is On: Can *TV-Cable Week* Catch *TV Guide?*" *Washington Journalism Review,* June 1983, pp. 25–28, and Eric Nadler, "Guiding TV to the Right," *Mother Jones,* April 1984, pp. 17–20.
7. David Halberstam, *The Powers That Be* (New York: Alfred A. Knopf, 1979), and Herbert J. Gans, *Deciding What's News: A Study of CBS Evening News, NBC Nightly News, Newsweek and Time* (New York: Pantheon Books, 1979).
8. Publishers Information Bureau, "Revenues & Pages Set Records in 1983" (New York: Magazine Center, 1984).
9. P. J. Schemenaur, ed. *Writer's Market* (Cincinnati: Writer's Digest Books), published annually.
10. P. J. Schemenaur, ed. *1983 Writer's Market* (Cincinnati: Writer's Digest Books, 1983), pp.414–15.
11. Karen Heller, "Hot On the Press: Magazines That Are Making A Mint," *Washington Journalism Review,* April 1984, p. 29.

12. Bruce Cook, "*Los Angeles:* The Monthly That Wants Respect," *Washington Journalism Review,* April 1984, pp. 31-34.

13. Alan D. Fletcher, "City Magazines Find a Niche in the Media Marketplace," *Journalism Quarterly,* Winter 1977, pp. 742-43, 749.

14. Benjamin M. Compaine, "Magazines," In *Anatomy of the Communications Industry: Who Owns the Media,* Benjamin Compaine, ed. (White Plains, N.Y.: Knowledge Industry Publications, Inc., 1982), p. 172.

15. Ibid, p. 176.

16. Ibid.

17. Ibid, p. 173.

18. Karen Heller, "Hot On The Press," pp. 27, 29.

19. Charles Kaiser and Neal Karlen, "Time Inc.'s $47 Million Mistake," *Newsweek,* 26 September 1983, p. 94, and Charlene Canape, "The Chase Is On: Can *TV Cable Weekly* Catch *TV Guide*?" *Washington Journalism Review.*

20. *1985 ABC Magazine Trend Report.* (Chicago: Audit Bureau of Circulations, July, 1985).

21. A. Kent MacDougall, "Magazines: Fighting for a Place in the Market," *Los Angeles Times,* 9 April 1978; reprinted in *Readings in Mass Communication: Concepts and Issues in the Mass Media,* 4th ed., eds. Michael Emery and Ted Curtis Smythe (Dubuque, Iowa: William C. Brown Company Publishers, 1980), pp. 284-93.

22. William A. Henry III and Richard Bruns, "*Esquire* at Mid-Century," *Time,* 21 November 1983, p. 54, and Bill Thomas, "What Does A Man Want?" *Washington Journalism Review,* December 1984, p. 21.

23. From a news release produced by *World Press Review,* Spring 1984.

24. Michael Wantzel, "*World Press Review* helps cure the 'all-American' outlook," *Baltimore Evening Sun,* 30 September 1983.

25. John A. Finley, "Special Report: Reading," *Presstime,* September 1984, p. 22.

26. Marvin M. Gropp, "New You See It! Now You Don't!" *Magazine Newsletter of Research No. 44,* March 1984.

27. Roland E. Wolseley, *The Changing Magazine: Trends in Readership and Management* (New York: Hastings House, 1973).

28. John Hohenberg, *The News Media: A Journalist Looks at His Profession* (New York: Holt, Rinehart and Winston, 1968).

29. Lynne Williams, *Medium or Message?* (Woodbury, N.Y.: Barron's Educational Series, 24 May 1971).

5

Books

Books serve a historical role in the development of mass communications and qualify as a mass medium on the basis of their aggregate audience, which is huge, rather than on the basis of the number of persons who might read any given title. Even blockbuster best-sellers rarely sell more than two million copies each year. Given pass-along readership, they still don't reach as many people as the *New York Daily News* reaches each day in New York City alone, or as many as *Time* or *Newsweek* reach in a week. But size is not the only measure of a medium. The hardback book is considered by many to be the most credible print form, followed by the paperback book, the magazine, and finally newspapers.[1] Books are credible because of their association with formal education, and because they are durable.

Books last; some books printed several hundred years ago are still in existence. And the ideas contained in some books last even longer, becoming transmitters of the culture from one generation to the next. They represent the consciences of times and people gone by, and they speak to each reader individually. They are private in a harassed and public time; according to Marshall McLuhan, they enhance individuality, orderliness, and logic.[2]

In our consideration of the book as a mass medium in society, we begin by taking a look into history and then concentrate on significant twentieth-century factors relevant to the publishing industry. Of special interest is the manner in which the oldest mass medium has adapted to contemporary media—with which it has come to coexist. We conclude the chapter with a discussion of the relative freedom in the content of books and the future of books in an increasingly electronic media world.

Introduction

Books are durable and credible; they enhance individuality, orderliness, and logic.

Historical
Development

Book publication and
readership date back about
fifty-five hundred years.

Books have a history that dates back nearly fifty-five hundred years. The oldest known alphabet, on which the Phoenician alphabet is based, dates from about 1400 B.C. Using this alphabet, the Phoenicians created a writing system that they derived from speech rather than from unwieldly pictographs or abstract symbols, such as wedge-shaped cuneiforms. At the great library at Alexandria, in Egypt, scholars seeking to codify the knowledge of the ages accumulated more than half a million scrolls. Through the Greek and Roman periods writing became more standardized. Later during the Dark Ages, the Christian influence made itself felt as monks did their part in keeping literacy alive by transcribing ancient literary and religious works onto parchment.

As the writing process developed, materials used to write on became more portable. They progressed from clay tablets to animal skins to expensive but durable parchment, to linen and other ragstock, and finally—within the past two centuries—to wood-pulp paper, which arrived in time to accommodate the rotary printing press.

The physical appearance of books, particularly their binding materials, also changed considerably. Papyrus or parchment scrolls, which were frustrating to handle, were replaced in the fourth century A.D. by the Roman invention of the codex, a system by which sheets of papyrus or parchment were tied by cords between wooden boards. This binding led to greater ease of reading and made possible the preparation of indexes, which enhanced the book as an information medium.

Gutenberg's Bible and the Spread of Literacy

During the 1430s Johann Gutenberg, a little-known German printer, began experimenting with ways to improve the primitive methods of printing in use at that time. Although the Chinese had invented movable type several hundred years earlier, Europeans were unaware of this Asian achievement. Consequently, Gutenberg was credited with having invented it. In 1456 Gutenberg printed the Mazarin Bible, reputed to be the first book printed with movable type. And so began the era of mass communications.

At that time most of the population was illiterate. With rare exceptions, priests and monks were the only ones who had access to hand-printed books and the advantage of some schooling. Nonetheless, only a few years had passed before the practice of printing had spread across Europe. In 1476 William Caxton set up a press in England where he printed *Dictes* or *Sayengis of the Philosophes*. By 1500 some thirty thousand different books had been produced, and several million volumes were in the hands of the clergy and the secular elite and had started to become available to the developing middle classes.

Gutenberg's printing press ushered in the age of mass communications, with enormous social and political ramifications.

With the growth of the printer's craft the opportunity to become literate was enormously enhanced. It became possible for persons to sequester themselves with the Bible, read it, and develop their own interpretations of the Gospel—interpretations that were sometimes at variance with church dogma. As more people formulated new opinions, they won converts and

More than five hundred years ago, Johann Gutenberg printed beautiful Bibles using movable type.

then began writing their own interpretations of the Bible. Disputes occurred within the Roman Catholic Church, which caused a fragmentation into different denominations, forming the social matrix of that period known as the Protestant Reformation.

Even in its earliest history, the book demonstrated its capacity to bind traditional societies together and to give rise to new and sometimes revolutionary thought patterns. The book's power is reflected in a quote by educator and social commentator Jonathan Kozol, author of *Death at an Early Age* and *Illiterate America:*

Whether within the biblical tradition of the Hebrew people, or those of Martin Luther and John Milton and the Puritans who came to Massachusetts in the 1600s, precisely out of the determination to be able to empower every citizen to have uncensored and unmediated access to a sacred book, literacy has always flourished in a context of respect for history, reverence for truth, and adulation of the beauty which has been passed down through generations by no other force so potent, yet so fragile as the bound and printed book.[3]

Book Publishing in America

Colonial book publishers served an elite clientele.

Given a newly literate audience, textbooks and paperbacks were mass produced in the 1800s.

McGuffey's *Eclectic Readers* taught nineteenth century Americans to read.

The first printing press in America was established at Harvard College in 1638, less than twenty years after the Pilgrims landed at Plymouth Rock. However, the orientation of the colonists and their successors was less toward the book than toward shorter, more quickly digested forms of print. Little leisure time was available to the frontier settlers, and they took their information on the run.

Different reading patterns existed among the affluent, who were the primary owners and readers of books. Aristocracy, even in colonial America, was frequently measured by the number of leather-bound books one possessed. Paperbacks of a sort didn't make their appearance until 1777, when Americans began buying some of the first 190 volumes of John Bell's "British Poets" series.

Technologically, the production of books did not vary much between 1450 and 1800. Hand-operated presses continued to crank out handset type a page at a time. Literacy rates were low and the cost of books was high, so there was no great pressure on printers to develop a more efficient process. It is estimated that only about 10 percent of Americans could read and write in 1800, and the average book cost about one dollar, or a week's wages for the typical worker. But when compulsory public education was instituted later in the nineteenth century, technology was also improved, which resulted in a lower per unit cost. Paper in a continuous roll and the durable iron press were two European inventions that improved the printing process. Steam-powered presses and ultimately mechanical typesetting machines coalesced during the nineteenth century, resulting in large-scale book production.

Despite the production of some fifty thousand different titles (books, magazines, and newspapers) rolling off American presses between 1640 and 1820, most Americans were still relying on European printers to satisfy their literary appetites. Once mechanical, literary, and economic bases permitted it, however, home-grown publications flourished. By the 1840s, inexpensive editions of mass-produced books were available for ten cents a copy, spawning the age of books as a popular mass medium in America. E. F. Beadle, a New York publisher who started by publishing dime songbooks, discovered a ready market for entertaining prose as well. The first big era of the paperback was born with *Beadle's Dime Dialogues,* devoted to pioneer adventures.

Textbooks also got their start in the 1840s. William Holmes McGuffey, an educator in Ohio, may be called the father of the schoolbook for his work in compiling the McGuffey *Eclectic Readers.* Between 1836 and 1857, McGuffey wrote six primers (graded for difficulty) that became the standard reading textbooks for more than half a century. Total sales of the McGuffey *Readers* are estimated at over 122 million copies.

During the Civil War, soldiers seeking ready diversion constituted a market for escapist literature in inexpensive, portable packages. Horatio Alger, before his death in 1899, cranked out 120 different titles of such fiction, which sold to the tune of thirty million copies. America loved Alger's rags-to-riches sagas, which have helped to form America's social and cultural legacy.

Paperback publishing expanded for a time during the 1870s and 1880s. When the cost of newsprint dropped, the *New York Tribune* began publishing paperbacks in the form of newspaper extras, which sold for five to fifteen cents a copy. By 1877, fifteen different firms were competing for the paperback trade, and by 1885 about a third of the nearly five thousand titles published annually were produced in paperback and distributed through established bookshops and by subscription.

The international copyright law of 1891 spelled the end of the second paperback generation in America, by granting royalties to overseas writers whose works had previously been pirated. By 1900 the paperback business went into a coma, after years of heavy competition, price-cutting among competing companies, soaring costs, large numbers of unsold books, and a dwindling supply of new writers.[4]

The hardback book business, on the other hand, had expanded steadily. Such authors as Henry James, Mark Twain, Edgar Allan Poe, and William Dean Howells contributed immeasurably to the growth of book publishing and distribution in the United States before the end of the nineteenth century. They were followed by the naturalists and the muckrakers at the turn of the century, who also found a ready market for their wares—gripping narratives and descriptions of institutional corruption and greed, the nature of power, and the realities of life in an increasingly urban society. Social historians need look no further than these books to recapture America.

Hardback books, reflecting society's changes, were eagerly read by a newly literate audience.

Publishing houses before the turn of the century had developed into significant forces, and book publishing had gained a reputation as a gentleman's profession. Publishing houses that were family owned and operated respected each other's writers and products, and went so far as to acknowledge colleagues' rights to republish imported books on the basis of whoever gained first access to them. Decisions about what works would or would not be published rested as much on aesthetic as on economic grounds.[5]

By 1900 literacy rates had reached 90 percent, quite a jump from the 10 percent a mere century before. Free public libraries were giving even the penniless access to books. Works of history, fiction, science, and classical literature, as well as newspapers, magazines, and inexpensive dime novels, were being read by an increasingly literate population whose work status and economic conditions found them better able to utilize their time and finances.

Most of the significant changes in the publishing industry during the twentieth century have occurred since World War II. The years prior to the war were characterized by the growth of major publishing houses catering to a mass demand for works of fiction, a decrease in the number of family-owned companies specializing in a single type of book, and cutthroat competition

The Twentieth Century

among literary agents that altered the gentlemanly understandings formerly prevalent among publishers. The time might be characterized as the *commercialization of literature,* when printing and marketing practices became streamlined in conjunction with other big businesses. While some segments of the industry seemed to change very little—plodding along as usual with only marginal profits—the paperback industry was gearing up for a renaissance of unequaled proportions.

In 1939 the scene was set. Taking a cue from Britain's Penguin Books, which were successfully sold through newsstands and chain stores, Robert F. deGraff introduced Pocket Books to the American consumer. Within two years they were sweeping the country. Their rapid popularity came about largely because of the twenty-five-cent price and because deGraff marketed them through independent magazine wholesalers, who could distribute them to newsstands in very busy train stations where Americans were passing to and from work each day. Avon Books followed in 1941, Popular Library in 1942, Dell Books in 1943, and Bantam Books in 1945. All of them published numerous war books—fiction and nonfiction—to qualify for paper allotments. This in turn made possible the publication of vast numbers of general-interest paperbacks, which were being consumed as rapidly as they could be produced and shelved.

The period since 1945 has been referred to by Charles Madison as the *publishing goes public* era. By the end of World War II publishing had become a risky business. Vast financial commitments for presses and production equipment forced some publishers to become timid about gambling on untried authors or untapped markets. Modernization was expensive, and many independent, family-owned companies realized the only way to survive was to go public, or to sell stock in themselves to individual investors. Many mergers occurred as publishing houses pooled resources for economic survival. Conglomerates also emerged as a result of rapid growth during the 1950s and 1960s—realized by the sales in paperbacks and textbooks—which began when the large number of postwar babies reached school age and an equally large number of returning servicemen enrolled in college on the GI Bill. Ties between Hollywood and the book industry also strengthened after World War II, providing publishers with new opportunities for marketing books and, of course, with additional revenue. All of this painted a rosy picture for the large corporations. From 1952 to 1970, the industry as a whole grew at a rate of more than 10 percent annually, although it slowed down to around 7 percent shortly thereafter.

Given that growth rate, it was only natural for larger corporations to look at book publishing as a good investment. Instead of being controlled by family patriarchs, the finances of publishing houses are more likely today to be in the hands of bookkeepers and financial experts of massive corporations. Electronics corporations such as Xerox, ITT, IBM, Litton, and others who entered the publishing marketplace recognize that the future of the book, and particularly the textbook, might well be closely aligned to electronic multimedia information storage and retrieval. Media conglomerates also became interested in publishing. Gulf & Western bought Simon & Schuster and Pocket Books; MCA now owns G. P. Putnam's Sons and Berkley/Jove.

CBS, as another example, entered publishing in 1967 with the purchase of Holt, Rinehart & Winston. By the early 1980s it had added Popular Library, W. B. Saunders (a professional publishing unit specializing in medical texts and journals), Dryden, and Praeger Special Studies to its book holdings. CBS also publishes a host of special interest magazines. Taking advantage of its overall corporate knowledge and connections, the CBS Publishing Group has begun to develop educational telecourses, integrated learning systems that employ broadcast, cable, or closed-circuit television in conjunction with various print materials to provide complete courses to students in the convenience of their own homes. Overall, CBS has become one of the nation's largest publishing corporations.[6]

Other print-oriented companies, such as Reader's Digest, Time, Incorporated, McGraw-Hill, and the Times-Mirror Company of Los Angeles, have carved out places of their own in the book industry. Time, Incorporated, for instance, owns the Book of the Month Club, a profitable venture that might give rise to some interesting questions about conflict of interest: what happens when a company owns an influential weekly news magazine that reviews books, while at the same time another arm of the company is selecting books for a book club and assuring those books best-seller status?

Even though the entire mass communications network is becoming more interwoven, no more than a dozen major corporations seem to dominate the book industry. Given the worldwide trends toward consolidation and economic efficiency, any change in that overall picture appears unlikely during this decade. On the other hand, the past few years have witnessed the birth of a significant number of new small publishing ventures, many of them specialist houses. One source estimates that there are currently nearly seventeen thousand individual book publishers with works in print in the United States, and that something like two hundred new ones are being added to the rolls every month.[7] Although the death rate of new, special interest publishing houses is high, and the survivors may not make a sizeable dent in the overall book sales figures, such ventures are a sign of health in the industry; they demonstrate that it is still possible for many individuals or small groups who have ideas they consider worth sharing to enter most sectors of book publishing with relatively limited funds.[8]

Book publishers are likely to be in other communications businesses, such as electronics.

The Types of Books

The book industry is made up of fairly distinct publishing divisions, determined primarily by the market the publisher wants to reach. To the layperson, the most familiar type of book is probably the *trade book,* the industry's term to describe books sold through bookstores to the general consumer. *Paperbacks* sold in bookstores and at newsstands form another category of books, although some paperbacks are considered as either trade books or textbooks. *Textbooks* are the third major category of books. They are published for elementary, high school (el–hi), and college-level students. Books published for post-college specialists are categorized as *professional* or *scholarly* publishing. Most university press books are for this market.

Americans are spending some $10 billion a year to purchase books. Despite other claims on their time—especially the lure of television—they are devoting larger proportions of their available dollars and leisure time to a wider selection of increasingly expensive books. Although the number of titles published each year (including new books and new editions) increased from 38,000 in 1970 to some 50,000 by the mid-1980s, most of the increase in revenue book publishers enjoyed during that period can be attributed to rising prices. Reflecting the inflationary spiral, the *average* hardcover volume sold for $31 in 1985. Disregarding the special books in the over-$80 price range, the average volume cost $26.57, enough to put a dent in the book lover's budget. College students will not be surprised to learn that much of this expensive market was comprised of textbooks; science books, at an average of $40.06 per copy, led the list. Fiction books, at an average of $15.24 per copy, were the least expensive hardback books being marketed for the adult audience. Some comfort could be found in the proliferation of paperback texts, but the prices in the paperback industry were also inflating. In 1985 the average trade paperback cost $13.98, and the average "mass market paperback"—primarily fiction—cost $3.63.[9]

Trade Books

Typically, trade books are hardcover editions, written for adults and children, although increasingly in recent years paperback versions of substantial trade books have established a stronghold in the market. As a nation, we are spending about two dollars on hardbacks for every dollar we spend on paperback trade books. These books deal with subjects in a multitude of categories: fiction, current nonfiction, biography, literary classics, cookbooks, hobby books, popular science books, computer books, travel books, art books, and books on self-improvement, sports, music, poetry, and drama. They are distributed through approximately 12,000 retail outlets, 200 book clubs, and 30,000 libraries, primarily by corps of *travelers,* the trade name for salespeople.

Each year thousands of new trade titles are published. Fiction books account for only about 5,000 new titles annually, or a mere 10 percent of the total of new books published.[10] Eleven book publishers, led by Random House and Simon & Schuster, dominate the trade business with 71 percent of the market. Most publishing is concentrated on the East Coast, primarily in New York, Boston, and Philadelphia, its antecedents dating back to the Revolution, but many eastern publishers are branching out, with West Coast divisions becoming more common. Small trade publishers are located in all

Books are categorized as *trade books* for general audiences, *paperbacks, textbooks,* and *professional books.* It's a $10 billion-a-year business, with 50,000 different titles published annually.

Trade books are produced for mass or general readership, and are usually sold through bookstores.

Trade books are intended for the general audience, and constitute a publishing risk.

parts of the country. Some of them do a thriving business publishing books with a regional emphasis. Others focus on books that will appeal to special-interest groups.

Book publishing entails an enormous gamble, which probably accounts for its relatively small size in the sprawling communications field. For instance, fewer than 5 percent of the new trade titles published each year sell as many as five thousand copies, and sales of at least that amount are necessary for publishers to break even on their investment. This figure is particularly significant because it demonstrates that only a tiny fraction of the books published are money-makers; furthermore, the profits derived from this tiny fraction must compensate for 95 percent of the speculative titles that don't make it at all. This is one reason why books are so high priced and why the industry has had to economize somewhat and use a number of imaginative merchandising techniques. Hardcover trade books typically lose money for publishers. Where profit is made, it is most likely to come from selling rights to publishers who will reprint the trade book in paperback form or by developing a paperback version in-house to supplement the higher status but unprofitable hardback editions.

Paperback Books

The postwar paperback boom was the publishing counterpart of a general communications explosion in the United States that was spawned by affluence, attendant leisure, and new technology. High-speed presses and offset printing from rubber mats made the printing process faster and more economical. The introduction of *perfect binding*—the practice of using adhesive to bind books instead of sewing them—produced additional savings. As the business grew, it gradually evolved along two lines: mass-market books and quality paperbacks. Mass-market books are compact volumes designed to fit paperback racks in bookstores and at newsstands. Quality paperbacks have a larger trim size and a higher price than the mass-market variety. They resemble other trade books except for their soft cover.

There are about twenty major paperback publishers in the United States, although 75 percent of the mass-market business is controlled by eight publishers. Like their hardcover counterparts, most tend to specialize, some in entertainment, some in reprinting classics. Most try to achieve some sort of a balanced inventory of offerings, in an attempt to reach all segments of the potential market from several directions. Publishing is one of the few fields where an operator can be in competition with itself by bringing out a number of competitive and even contradictory titles. Its audience, more than in any other segment of publishing, is made up of individuals whose purchases need not be single and whose interests are not mutually exclusive.

Paperbacks try to be all things to all people. Classics and trash abound.

When one thinks of paperbacks, the cheaper mass-market editions of former hardcover best-sellers generally come to mind. While this is certainly one practice, it does not tell the entire story. For example, there is a good deal of classical publishing in paperback for the simple reason that as copyrights expire, no royalties are required to be paid. Thus, with cheaper paperback publishing costs, profit can be shown on smaller sales. In this way paperback publishers seek to tap the college and university market in the same fashion as hardcover publishers who produce texts.

There is also a substantial amount of original publishing in paperback. Authors' royalties per copy are generally less, hovering between 2 and 4 percent (contrasted with the 10 to 15 percent for hardcover editions), but the greater projected sales more than compensate. Generally, these originals are concentrated in the entertainment field where their lack of staying power is not significant. Occasionally, however, a paperback original of considerable merit appears. Western novels and romances are also published originally in paperback. In recent years these areas have become "hot," with some publishers producing ten or more titles every month. Harlequin romances is an example. The company has a stable of writers who follow a standard formula, cranking out books that often read alike and sound alike.

Sometimes paperback publishers will pay more for a reprint title or an original than they can hope to make in profit. They do this with the thought of enhancing their position with their distributors and their retail

outlets, as well as gaining certain prestige in the upper echelons of the limited paperback business. If they were not to acquire the rights, a competitor would, and the monetary loss suffered is sometimes worth the less tangible advantages of having a best-seller.

Increasingly, as paperback publishing has developed in stature and status in the book world, some authors have found it advantageous to market their wares directly to paperback companies, who in turn sell reprint rights to the hardcover industry and perhaps even to Hollywood.

In the publishing industry, the $2.8 billion textbook market is an appealing alternative to trade books. Instead of passing quickly from the scene, a successful text keeps on making money for both the author and the publisher, year after year, as new classes, and even new generations, of students enter schools and universities. Because a successful textbook becomes dated three to five years after publication, the author must update the material in light of new knowledge. The publisher can then bring out a new edition, profiting from the text's already established reputation. It is not unusual for a successful text to last for six to eight editions.

Textbooks differ from trade books in other ways. The number of potential authors is smaller, since texts are generally written by professors; and the market per title is smaller, since it is restricted to those persons enrolled in schools and colleges who are taking the pertinent subjects. However, the relative permanency of a long-term audience and the mandatory nature of sales more than compensate for these shortcomings.

It should be apparent that professors are not only a primary source of text material, but the means of its distribution. It is their responsibility to order the texts for the classes they teach; the bookstores merely stock the books. *College travelers*—salespeople from the publishing houses—are

Textbooks

Textbook publishing is less of a gamble than trade book publishing.

familiar with academia and the professors within the respective institutions and disciplines. In addition to selling textbooks, they are manuscript scouts for their publishers, dealing with potential authors and distributors in one.

Somewhat different procedures are used in the development and sale of elementary and secondary school (el-hi) texts. In this billion dollar market, which represents the most profitable division of many large publishing houses, the textbooks are typically written by a team of in-house editors. This is especially true of basal series for elementary programs in reading, mathematics, social studies, and science. The reasons for the team approach are to make manageable the task of putting together a complex package, which may include more than two hundred pieces, and to give the publisher more control over the final product.

Control is necessary because of the substantial investment in development, which may exceed $20 million, and because the publisher has to tailor the series to appeal to the broadest possible market. A textbook accepted for use in Texas high schools, for instance, is guaranteed sales in the hundreds of thousands of dollars. The total California textbook market is worth more than $100 million. Given markets such as these, which economists refer to as "inelastic demand," publishers tend to bow to the dictates of public opinion when producing texts.[11]

Twenty-two states have statewide el-hi text adoption codes that are weighted with readability formulae and taboos. The result is often a text that is criticized by educators as bland and inoffensive, a case of not only "dumbing down" a book to be understandable by what are perceived as less and less literate children each year, but also a case of removing all references to controversy. (Tom Sawyer no longer says "Honest injun," but merely "honest." Richard Nixon supposedly became involved in Watergate because "he tried to help his friends.")[12] Two recent criticisms of textbooks indicate that publishers' self-censorship—self-imposed blandness—can backfire. A National Institute of Education study has accused social studies and history textbooks of blatantly overlooking religious aspects of American development, and the California Board of Education rejected every textbook submitted for the state's seventh- and eighth-grade science classes for shallow treatment of controversial subjects such as nuclear war, the environment, and human sexual reproduction.[13]

From the publisher's point of view, however, blandness and inoffensiveness remain safe and positive attributes in el-hi texts. The textbooks are frequently purchased by committees made up of parents and school-board members, so the series that is the least offensive often stands the best chance of being selected.

Textbook sales obviously reflect the fluctuations in school populations. In 1945—at the end of World War II—textbooks accounted for one-fifth of the total gross sales of books in America. Once children from the postwar baby boom reached school age that total increased dramatically.

The Print Media

Today the total exceeds 400 million books (hardbacks, paperbacks, assorted workbooks, reference books, manuals, etc.), which account for one fourth of all books sold. This shows a fairly steady increase in spite of a decline in the school-age population during the past ten years.

Works written by specialists to be read by other specialists are generally referred to as professional or scholarly books. A monograph explaining the symbolism in James Joyce's novels qualifies; so too does a review of research on color perception in monkeys. For years publishers have been active in the fields of law and medicine because attorneys and physicians must keep informed about their fields. Other disciplines have not been as attractive, but in recent years growth in the behavioral sciences has prompted substantial interest by publishers.

Print runs on some books may be as low as one thousand copies, resulting in an elevated per unit cost and a high list price. Consequently, a book of only two hundred pages may be priced at more than thirty dollars.

Professional books in law and medicine are usually sold by salespeople who are assigned territories. This is possible because the price of each book is relatively high, often greater than fifty dollars per volume, and because the professionals may purchase several books at one time. Professional books in other disciplines are sold by direct mail to clinicians, practitioners, and college professors.

The process of book publishing is basically the same for each type of book publisher. Editorial people are needed to acquire projects and to shape manuscripts; production people, including designers, produce the books; and the marketing and sales staff promotes the books. In addition, a company has an operations group that is responsible for processing orders and billings, and for keeping straight all financial matters for the company.

The overwhelming number of published manuscripts are solicited or commissioned. The idea for a book may arise from either an author or a publisher. When it originates with the author, he or she generally writes a relatively brief prospectus describing the idea—the approach, theme, and development. A couple of sample chapters generally accompany it to indicate style. It then goes through a successive and brutal screening process. In the trade industry the first screening is done by recent literature graduates, who can reject the manuscript, and often do. Most authors have impressive files of rejection slips, the number of which forms a sort of badge of professionalism.

If a manuscript passes the first screening, it goes to an editor for consideration. Should the editor like the idea, the manuscript will probably be sent out for evaluation by recognized authorities in the field. If the manuscript makes it this far, it may then go to an editorial board for a final

Professional Books

Professional books are by and for specialists, meaning high per unit costs.

The Business of Book Publishing

Editorial

The gatekeeping process in book publishing is often ruthless, because each product, or book, must try to make it on its own in the marketplace.

decision. However, the manuscript may be rejected at any of these stages, or the author may be asked to further elaborate his or her idea. But it should be obvious that only a tiny percentage of the abstracts or summaries submitted survive to reach an editorial board decision. The competition for attention at the channel level is intense, and the gatekeeper's role in publishing is even more ruthless than at a newspaper. The numbers involved are frightening: less than 2 percent of the prospectuses submitted will ever see print, and less than 5 percent of that number will sell well enough to pay expenses.

The process can be exceedingly discouraging, but rejection is not necessarily a measure of the quality of an author's work. As an example, John Toole, author of *Confederacy of Dunces,* became so despondent by his inability to attract a publisher that he committed suicide. Toole's mother then took up the cause, approaching publisher after publisher until at last the Louisiana State University Press agreed to publish the book. Recognition followed. The book was praised by critics and *Confederacy of Dunces* received the 1980 Pulitzer prize for literature.

The other route a manuscript may take occurs when an idea originates with the editor, whose business it is to keep in touch with the public pulse. The editor has to know what people are reading, what kinds of books are selling, and where the public interest lies. The process is somewhat simplified by the fact that publishers tend to specialize, first, in certain types of books—whether they be text or trade—and then in certain concentrations within these fields: such as fiction or nonfiction, mysteries or westerns, or international affairs or politics. Consequently, most publishers only have to keep track of their specialized fragment within the market. If a publisher has an idea, the search begins for the right person to write it. The publisher does this from personal knowledge of an author's prior work or from gleanings from magazines and newspapers. Or the publisher may go to a literary agent to find the proper talent.

Agents play a significant role in trade book publishing. Each has a stable of significant authors whose work is marketed. Agents maintain close contact with editors and publishers. They know their inventories and the kinds of material that editors and publishers are looking for and they know the market as well as the publishers do. Most successful agents have been in the business for years and have earned the confidence of the editors with

Manuscript ideas originate with either a writer or editor; the latter stands a better chance of being published.

Agents earn their percentage by serving the needs of authors and publishers alike.

whom they deal. Agents generally take 10 percent of a book, both advance and royalties, and generally earn it by winning more sales and more generous advances for their authors than the authors could win for themselves. In addition, agents have the time and contacts many authors lack.

Regardless of how a book gets started, the next step is the contract, which specifies the author's advance—a flexible figure depending on the author's reputation for dependability, performance, and popularity. An advance is a sum of money paid to an author for writing the book; it is chargeable, for the most part, against the royalties when and if they develop. Successful authors obviously can command larger advances than beginning ones. Generally, royalties are 15 percent of the list price of the book for every copy sold, less the author's advances. If the book does not sell enough copies to return its investment, the author keeps the advance anyway. Thus, the larger the advance, the better the author is protected regardless of subsequent sales. Successful authors gamble little; beginning authors gamble with their time and their publisher's investment of money.

Production

Several months before the final manuscript is ready for composition, the production team begins work. Production is the process of transforming a manuscript into a bound book.

Production teams put it all together.

Every book has to be designed. Someone has to determine the trim size of the book, the typeface to be used, the length of each line, how illustrations will be used, and whether footnotes will be placed at the bottom of the page or grouped as notes at the end of the chapter. These and a hundred other questions must be resolved, often at a meeting of the editor, the project editor who will shepherd the book through production, the production manager, the marketing manager, and the designer. Each person brings certain skills and experiences that will affect the appearance and quality of the book.

As we mentioned earlier, the risks involved in publishing increased dramatically following World War II. As a result, few book publishers today own the equipment for composing and printing books. Most will hire the services of a compositor to set the book according to the design specifications, and a printer to print and bind the book. Although a few compositors still use linotype machines—known as hot metal—the vast majority use computers, which are a faster and cleaner method of type preparation. By making only a small change in the program a compositor can alter the design without having to reset large chunks of manuscript; likewise, editorial changes and proofreading corrections can be made with relative ease.

The liaison between the publisher and the compositor is handled by a production person. He or she hires the suppliers and arranges the schedule. When galley proof (the first printed version) is available, the project editor, proofreaders, and the author read for typos, dropped lines, and other problems. Galleys are usually printed on long sheets of paper without pagination or illustrations. Errors located at this stage can be corrected at less expense than later in the production process, when the book has been prepared in pages.

Page proof is also read by the author and project editor. Were the corrections made as requested? Do the pages look right? Is everything in place? When everyone is satisfied, the compositor prepares film for the printer, whose plant may even be in another state. The entire process from final manuscript to bound book may take as long as twelve months for a heavily illustrated technical book. That is why books in some rapidly changing fields will be dated even before they are published. For most trade books (especially fiction, biography, cookbooks, etc.) this does not present a problem.

As production nears completion, the marketing department gets more actively involved. The promotion of school and college textbooks will begin at least six months before the book is published, while trade promotion may occur a little closer to publication. All the work that went into developing and producing a book will mean very little if the book is inadequately or improperly promoted.

Marketing

Techniques of marketing and promotion have been refined over the years, but at base they still include complimentary copies of new titles to critics and book editors in the hope of gaining a review or a recommendation, plus a wide range of public relations activities that take advantage of the relationships among various media of information and entertainment.

The most desired review forum for books is the *New York Times Book Review* section, the bible of the publishing business. Here a review, even a bad one, is a treasured commodity, demonstrating credibility and assuring at least some sales. Competition for review by the *New York Times,* however, is intense, and publishers spend a lot of time and effort seeking it.

Making the *New York Times Book Review* section is nice, but not nearly as nice as making the *Times* best-seller list. That is nirvana, a lofty goal that publishers strain to reach.

In an intriguing article that takes book promotion to task, *Los Angeles Times* media critic David Shaw seriously questioned the legitimacy of most best-seller lists.[14] While sparing the *New York Times* list from some of his criticism, he described a variety of ploys used by publishers and authors to reach best-seller status. The compilation of newspaper and magazine best-seller lists was accused of being so haphazard, so slipshod, so imprecise, and, at times, so dishonest that most publishers would gladly do without it, were it not such a necessary evil.

The three major national best-seller lists, appearing in the *New York Times, Time Magazine,* and *Publishers Weekly,* are usually similar, but rarely identical, because they use three different means of gathering information. According to an article in *USA Today,*[15] the *New York Times* uses the most elaborate—and controversial—system. Insiders say the paper sends a printed list of thirty-six titles to some 2,000 bookstores around the country, and asks booksellers to fill in the number of each title sold for that week. There is space at the bottom of the list to include others that might be selling well locally. The major chains, B. Dalton Bookseller, with 680

Book reviews and best-seller lists are keys to marketing books, but they aren't always what they appear to be.

stores, and Waldenbooks, with over 800 stores, make up their own best-seller lists based on computerized sales records, and those totals are factored into the *New York Times*'s listings each week

Time Magazine and *Publishers Weekly* determine best-sellers by somewhat different means. *Time* does not contact independent bookstores, relying instead on several major chains to give it exact sales figures of mass-appeal books—ignoring those that independent stores might be selling well—over a prolonged period of time. *Publishers Weekly,* meanwhile, contacts more than sixteen hundred bookstores, including independents in major cities, and asks for the top five to fifteen best-selling books on order. Stores are given a point value, meaning that chain stores' sales count more heavily than those of independents.[16]

Inertia is seen as a major factor in book sales. "Once you get on the list, the combination of inertia, self-generated momentum, and self-fulfilling prophecy keeps you on for months," one publisher told *Los Angeles Times*'s David Shaw. For this reason authors and publishers stretch the limits of propriety to get on the lists. They use advertising and promotion, puffery and tricks, to get there. If it is possible to find out which bookstores will be called by newspapers and magazines, some authors will buy—or have their friends buy—their own books in quantity from those stores. The late Jacqueline Susann and her producer-husband Irving Mansfield had been known to introduce themselves to the bookstore clerks, then purchase and autograph a copy of their book for every employee in the store.[17]

A unique marketing technique adapted by the William Morrow publishing house pushed the mystery novel *Who Killed the Robins Family?* to eleventh place among 1983's hardback bestsellers. In an "unprecedented publishing event" the publisher offered a cash reward of $10,000 to the reader who came up with the best solution to the mystery posed by the novel. A year later, after a nurse from Denver solved the crime, the solution was incorporated into a final chapter of the paperback version of the book. That publication coincided with the release of the hardback sequel, *Revenge of the Robins Family,* which offered a $10,001 prize for the most ingenious reader-sleuth. Knowing a sure thing when it saw one, Pocket Books printed a first-run release of 600,000 copies of *Prize Meets Murder,* offering $15,000 to the reader who unraveled the crime.[18]

Such ploys are probably less frequent, and less useful, than the technique of promoting one's book on radio and television talk shows, where an author can reach millions of potential readers at a shot. Scientist Carl Sagan sold nearly a million copies of his book on evolution, *The Dragons of Eden,* largely because of his frequent appearances on the Johnny Carson "Tonight" show. An appearance on the "Today" show is said to generate sales of 3,000 copies of a particular book on the day of the appearance, and a guest spot on the "Donahue" show can move 50,000 copies. Couple that with a mention in Ann Landers's advice column, and the author and publisher can smile all the way to the bank.[19]

Peddling books on TV and radio talk shows is a perfect example of media feeding off each other.

A follow-up to the best-selling mystery novel, *Who Killed the Robins Family?*, is this reader-involving package, *The Picture Perfect Murders,* in which the clues to the mystery are hidden in the photos as well as in the text. Note that the package has both a creator and an author!

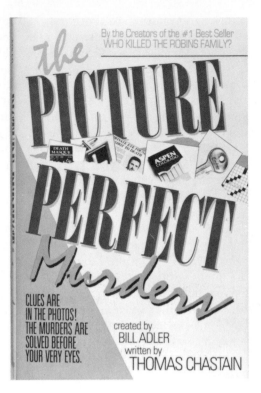

Some writers balk at having to peddle their books, and themselves, since writing is, by nature, an individualistic and introverted profession. Novelist John Cheever says he fears the artificiality of promoting books. "Writing is a highly intimate exchange of communication, and merchandising puts it in a different light," he laments.[20] But purely from a marketing standpoint, charisma is more important than how well the author writes, according to a publicist for one publishing house. A case in point is the success of Erma Bombeck's *The Grass Is Always Greener Over The Septic Tank,* which became a best-seller after the author's publicity tour on radio and TV shows. "People loved her as well as her book," her McGraw-Hill publicist said. "A wonderful human being comes across as a wonderful human being, and that can make people want to read the book."[21] Obviously, authors who are shy and who don't project well are likely to lose out on the broadcast circuit, which causes one to wonder whether Nathaniel Hawthorne would have scored with *The Scarlet Letter* had he bombed on the "Tonight" show.

The recent proliferation of radio and television talk shows, including all the opportunities offered by cable television, may be resulting in an overexposure of authors and some diminishing of the impact of authors' tours. According to one critic, supply is now in excess of demand; there are too many similar books fighting for reader's attention; and there is increased emphasis on authors and books of local or regional appeal. Authors and publishers have to be more clever and selective in their "media blitzes" when promoting their books from city to city, talk show to talk show:

Although author tours are by no means dead, there is no longer a standard formula. They are still one of the most cost-efficient and effective means yet developed to sell books, if used prudently and creatively—which means analyzing the possibilities on a book-by-book basis, exercising selectivity, localizing the pitch, choosing more targeted media and, most of all, creating alternatives to the tour, when necessary, for reaching the public. The author tour is alive—but it needs careful nursing to bring it back to its former health.[22]

The reputation of the author continues to be an important ingredient in book sales. Some, such as the late Truman Capote, were credited with best-selling books almost as soon as they had signed a contract to begin writing. Lavish promotional parties also help as part of publishers' sales campaigns. One such party was a $15,000 soiree in Beverly Hills promoting Judith Krantz's novel *Scruples*. Although the novel was seen by critics as "fun trash," or worse, it quickly climbed onto best-seller lists.

Utilizing demographics, book publishers have focused their promotions on seemingly appropriate audiences with great success. Joseph Wambaugh's *The Black Marble* deals with a dognapping, so promotion was aimed at California dog lovers. The author attended all the most important dog shows, presenting several of the best-in-show awards. The twenty-eight hundred dog-breed clubs with their own publications were sent news releases saying dogs were featured in an upcoming major novel. As a result, *The Black Marble* was a number one West Coast best-seller *before* publication and quickly hit the national lists.[23]

Authors' reputations, appeal to demographics, and advance publicity help guarantee success.

How can a book be a best-seller before it is published? This disturbs media critic Shaw, who pointed to several examples where there was so much advance publicity about a book that bookstores were reporting booming sales before copies of the book were even on the shelves.

Lately a total packaging concept has come into the fore, and the author is only a small cog in the machinery. The aforementioned *Robins Family* mysteries, for example, were noted for their publishers' and agents' creativity as much as for the authors' texts. Agent Bill Adler, who coaxed mystery writer Thomas Chastain in the production and marketing of the *Robins Family* stories, is quick to admit that he first hits upon a marketable idea, then seeks an author to flesh it out. Watching the "Donahue" show on television one day, he saw syndicated TV cook Jeff Smith bill himself as "The Frugal Gourmet," and knew the concept was marketable. Adler signed Smith to a contract with William Morrow & Company, and the resultant book quickly hit the best-seller lists.[24] Adler is credited with having initiated the book "packaging" concept in the 1960s—taking salable ideas and finding the right authors to make them work. Today, the American Book Producers Association estimates that some 1,000 book titles a year are produced by 250 different packagers.

If potential authors begin to suspect that their reputations no longer play a part in the process, they should consider the plight of British author Doris Lessing. Twenty-five of her books had received critical and commercial success; her *The Golden Notebook* sold 900,000 copies. However, when she decided to test the system, and submit manuscripts under the pseudonym Jane Somers, even her own publisher turned down the books. When they were eventually published, they received almost no reviews, and none of the reviewers recognized the familiar writing style. The books sold only 1,500 copies in Britain and 3,000 in the United States.

"I wanted to highlight that whole dreadful process in book publishing that 'nothing succeeds like success,' " the author said. "If the books had come out in my name, they would have sold a lot of copies and reviewers would have said, 'Oh, Doris Lessing, how wonderful!' " She admitted that she devised the ruse as a monument to the thousands of unknown writers who get bombarded by rejection slips.[25]

In the meantime, steady sellers—the Bible, Dr. Spock's books on child care, the *Weight Watchers Program Cookbook,* plus dictionaries and some others—do not appear on the weekly best-seller lists. Even cookbooks, which are staples in many publishing houses, rarely make the top ten. The capriciousness of booksellers can also keep other books off the lists. Although *The Joy of Sex* sold 700,000 copies in hardback in 1973, several stores refused to list it in their weekly totals because they thought it was a dirty book. *Jonathan Livingston Seagull,* a thin, little book, sold 750,000 copies before finally making the *New York Times* list, while some "bigger" books made the list with only 30,000 sales. And *The Total Woman,* which sold 369,000 copies, never made any best-seller list.[26]

Book Clubs Book clubs are a tempting marketing device. They offer, through membership rolls, an opportunity that publishers usually lack, the capability of predicting the approximate size of sales in the same way that magazines and newspapers can. A book club selection is a valuable publishing-house asset, one that can be additionally merchandised in advertising and publicity. Publishers are eager to have their books chosen. However, there are certain shortcomings to this particular system. Recognizing their leverage on publishers, the book clubs often beat down the price on titles they accept, to the point where there is little profit to the publisher beyond the exposure that a book club selection offers. This, in turn, has led many publishers to organize their own book clubs, with varying results. The odds are against their success because of the temptation to restrict themselves to their own inventory, and the lack of broad selection that this implies.

An essential ingredient of the book club operation is the so-called *negative option.* Book club operators doubt that the system would work without it. Under negative option, club members receive their monthly selection unless they specifically decline it. Each month they are advised of

Book clubs generate predictable sales for a given title, even though the "negative option" is a questionable marketing ploy.

what the selection will be. They receive it (and are charged for it) unless they immediately return a notice declining purchase. Since people have a tendency to be negligent in such matters, club operators rely upon their members' "laziness" not to decline the selection.

During the early 1970s, the negative option came under fire by consumer groups and was tested in the courts. The Federal Trade Commission, by a four-to-one vote, allowed the practice to continue despite complaints that consumers were being taken advantage of.

Book clubs have survived and proliferated. Today there are 143 book clubs in America, peddling some $661 million worth of books annually.[27] The original Book-of-the-Month Club (BOMC), founded in 1926, has attracted some 1.5 million members, making it about one and a half times as large as its nearest competitors, the Literary Guild and Reader's Digest Condensed Books. As an indication of the specialization that is possible in print media, consider the Literary Guild's nearly two dozen subsidiary clubs, which deal with topics such as gardening, cooking, science fiction, and mystery stories. While earlier critics maintained that book clubs would hurt bookstore sales, it appears that the clubs have tapped new markets in areas where bookstores are scarce. They have provided opportunities for readers to fulfill their specialized interests, including such interests as pornography, without shopping for them openly in bookstores.

Serialization A technique that has been used sparingly but with considerable success is the prepublication serialization of a book in magazines or newspapers. Truman Capote's successful *In Cold Blood,* William Manchester's *The Death of a President,* and Richard Nixon's and Henry Kissinger's memoirs are among the major works serialized in this fashion, having appeared, respectively, in *The New Yorker, Look,* and America's daily newspapers. The case for prepublication serializing can be argued both ways. While the additional revenue that comes to the publisher for the magazine and newspaper rights offsets a considerable amount of publication costs, serialization in a popular magazine or in many newspapers may detract from the eventual audience. For example, the sophisticated readers of *The New Yorker* had to be subtracted from the eventual sales of *In Cold Blood.* However, the massive exposure accorded a title by serialization can precondition a responsive public to buy the book when it does go on sale. The pros and cons of serialization have not been resolved; like everything else in book publishing, serialization is a bit of a gamble.

Prepublication serialization is not the only form of serialization on the market. Beginning in late 1984, numerous Sunday newspapers began carrying *The Sunday Novel,* republications of major twentieth-century American novels originally printed prior to 1970. The novels are being offered in magazine format, surrounded by expensive four-color ads costing $36,000 per page, to run as monthly inserts initially in major metropolitan

Prepublication serialization in magazines and newspapers may help or hinder eventual book sales.

Sunday papers in New York, Los Angeles, Chicago, Dallas, and San Francisco. Producers of *The Sunday Novel* expect to capture more than a million upscale demographic audiences with their republication of such works as John Steinbeck's *The Pearl,* Thornton Wilder's *Bridge of San Luis Rey,* Virginia Woolf's *Flush,* and Ray Bradbury's *The Martian Chronicles.*

Paperback Marketing Paperback distribution accounts for nearly all the basic differences between softcover and hardcover publishing, including the price differences between the two forms. Popular opinion has it that paperbacks are only a fraction of the price of hardbacks because of the less expensive binding, but such opinion is incorrect. In reality, differences in binding account for only a miniscule part of the price differential. More significant is the lower initial investment by publishers; if the book has already been printed in hardcover, it will need little if any additional editing. Also to be considered are the lower royalties generally paid to authors for paperbacks. Most significant, however, are the sheer numbers involved and the means of distribution utilized.

Paperbacks are likely to be printed in runs up to, and sometimes exceeding, a million at a time, drastically reducing the cost per unit. Given the quantities involved, efficient channels of distribution must be maintained. Taking their cues from England's Penguin books, American publishers quickly learned the value of saturating the shelves in the retail establishments where their smaller, pocket-sized books could compete for the premium spaces and be sold alongside incidental merchandise.

Distribution of paperbacks is essentially magazine distribution. It utilizes basically the same distributors, some 800 wholesalers across the country, or the American News Company with its 350 distributors. Together, they serve around 100,000 retail outlets (i.e., drugstores, supermarkets, shopping centers, and airports), always locations where the casual traffic is high. Like magazines, paperbacks are displayed in racks that generally have about 100 pockets holding three to five copies each. The racks occupy valuable floor space. However, there is a tremendous turnover in the paperbacks on display, with premium space going to the fast movers. Books that do not sell rapidly are either returned or destroyed, often after no more than a week. Paperbacks are fully returnable by the retailer for either cash or credit, and one of the biggest problems faced by the industry is to keep the percentage of returned books to a minimum. Through computerized merchandising, the most efficient companies are able to hold returns to 25 percent, but at some companies returns are more than 50 percent. This is one industry where a unique channel gatekeeper plays a role: the truck driver faced with finding shelf space for a new delivery sometimes determines what books to remove from the channel.

On the retail floor, paperbacks are in competition with magazines for the customer's purchase, and the paperback proliferation has further hurt the magazine industry by cutting into its casual or street sales. But more

The paperback is an omnibus medium— something for everyone.

often the book is the medium that is overlooked. People tend to associate a new issue of a magazine with previous issues. Even with highly visible authors, paperbacks are lacking such a conditioned acceptance and need a longer display period than do transient magazines. Since they are generally not permitted it under competitive conditions, only the best-selling titles remain on display.

While this system does not detract from paperback sales in the aggregate, it does tend to remove some of the "heavier" titles from circulation in favor of flamboyant fare represented by sex, western, mystery, romance, and science fiction offerings. Experience has shown that classical reprints will move well if given a long enough display time. As with hardcover marketing, a constant gamble is involved. Even if publishers do have to take back half of their books, a few fantastic successes can often balance the judgmental errors. As Clarence Peterson says, Fawcett's *Peanuts,* New American Library's *Fear of Flying,* or one blockbuster by Harold Robbins or James Michener could support "a kennelful of dogs."[28]

Stores specializing in paperbacks have become popular in recent years. Paperbacks have also found their way into traditional bookstores and the book areas of department stores where previously they were unacceptable. It was once believed their cheaper cost would detract from the more profitable hardcover book sales, but this has not proven to be the case. Rather, they have served as traffic builders for these outlets. Publishers who have released both hardcover and paper editions of the same title have found that less than 10 percent of their readerships overlap. Surprisingly, though their contents are the same, their audiences differ.

New Strategies in Book Publishing

The "traditional" book business has begun to employ creative marketing techniques.

Cross-Media Ties

Books used to become films, but today it is a chicken-or-the-egg situation.

Book and movie tie-ins saturate the media market.

The variety of books available should convince us not to stereotype books as a monolithic form among the media. Compared with the other mass media, however, books have always been bound more by tradition and have therefore experienced less innovation. Although innovation is not central to the world of books, neither is it unheard of. One fairly recent development is the growing relationship between Hollywood and the trade industry. Book publishers are finding it attractive to work with movie producers, selling film rights for very substantial sums of money and earning more money from increased book sales when the film is released.

Lately, one sure way to make the best-seller lists is to produce a book in conjunction with a highly promoted movie or TV mini-series. For years Hollywood has capitalized on best-selling novels as the basis for films, figuring the exposure already achieved would pay off at the box office. Recently, a reversal of this technique has shown some promise. Yale professor Eric Segal's *Love Story* was written first as a screenplay, then as a novel. However, the book appeared shortly before the film version, and the two media fed off each other's popularity. There was no way, of course, of predicting the fantastic popularity of *Love Story,* but at least the promotional costs were minimized; and by virtue of the dual reinforcement, both media probably got better promotion than either could have afforded individually.

This publicity and the accompanying advertising is aimed not only at the critics, but also at the general public in order to create a demand, or at least a little curiosity—enough to make people seek bookstores (which is not always easy) and spend their money (which is always hard). Once a fire is ignited, word of mouth does the rest.

Such was the story of the *Jaws* phenomenon, in which author Peter Benchley and his publisher, Doubleday, auctioned off the paperback rights to *Jaws* for $575,000, picked up an additional $150,000 for the movie rights, and gained $85,000 more from book clubs. Doubleday spent $50,000 advertising the hardcover edition, and Benchley appeared on talk shows across the country as an expert on sharks. So much interest in the project was generated that a spin-off of the spin-off—*The Jaws Log,* a quickie describing the making of the book and the film—rested comfortably on the best-seller lists for some time while America enjoyed shark mania.

The big dollars reflected in Benchley's success with *Jaws* are not all that unusual in the industry. In 1984 and 1985 alone, publishers issued more than a million copies of each of the following paperback tie-ins: *The Thorn Birds* (10,640,000), *A Woman of Substance* (4,097,478), *Hollywood Wives* (3,300,000), *Gremlins* (2,800,000), *Dune* (1,700,000), *Indiana Jones and the Temple of Doom* (1,500,000), *Star Trek III: The Search for Spock* (1,116,000), and *North and South* (1,075,000).[29]

Given all the demands on their leisure time—by network, cable, and other forms of television, home computers and video games, radio and recordings, plus traditional forms of recreation—Americans today can be forgiven if they read fewer books than they used to. For thirty years, since television entered the home, readership has been on the decline, according to popular wisdom. This comprises the present argument about the place of book reading in America.

There is just one problem with the argument. It is too simplistic of an explanation, and does not account for what is happening in the book selling and library lending business. One cannot say unequivocally that book reading is on the rise across all segments of society, but there is renewed evidence that in several demographic groups there is an upswing in the reading habit, despite the competition for media consumers' attention.

Is book reading on the decline? Research is inconclusive.

The reason we cannot say unequivocally that reading is increasing is that various surveys offer contradictory conclusions. We do know that over the past decade or so library usage has increased significantly, the total number of books bought and apparently read has increased, that "heavy" book readers are devouring books at a record rate, and that overall, reading consistently dominates other leisure time activities. On the other hand, different surveys show that Americans are spending record numbers of hours per day watching television, a majority maintain they are getting more of their news from television than from print media, and the numbers of American adults who are either illiterate (can't read) or aliterate (don't read) are rising at an alarming rate.

A glance at a couple of recent studies should be instructive. "In the age of electronic entertainment and personal computers, books are thriving," according to a survey of American reading habits conducted by the Book Industry Study Group (BISG) in late 1983. Between its 1978 and 1983 national surveys, BISG found eight million new readers in the 21 to 60 age group; a total of 56 percent of Americans aged 16 and over were classified as "book readers" because they had read at least part of a book in the previous six months. Average Americans were spending 11.7 hours a week reading books, magazines, and newspapers, 16.3 hours watching television, and 16.4 hours listening to the radio, according to BISG's hour-long interviews with 1,961 subjects nationwide.[30]

However, the generally upbeat BISG report had some disturbing news for book publishers. Critics of the report noted that the increased number of readers over a five-year period could naturally be attributed to the "coming of age" of the baby boom generation, now in the 30 to 39 age group. There has actually been a decline in the percentage of book readers in the under-21 age group. Whereas 75 percent of this group could be classified as "readers" in 1978, the figure was down substantially, to 63 percent, in 1983. There was also a marked drop in book reading among senior citizens. Despite a 4.7 percent increase in the number of persons in this age

Table 5.1 Book reading trends, 1955–84*

Gallup Polls in March of 1955 and 1984 asked a nationally representative sample of adults if they had happened to do any book reading the previous day. Those who did were asked the title of the book and how many pages they had read. In 1955, one respondent in five (20%) reported having read a book; if reading the Bible is excluded, the total dropped to 14%. In 1984, one in four reported having read a book the previous day (25%), and 21% if reading the Bible is excluded.

In 1955, 25% of those who had either attended or graduated from college had read books other than the Bible the previous day; in 1984, that total was up to 35%. The only group for which book reading declined was for those with only elementary school educations, from 11% in 1955 to 6% in 1984.

These figures do not coincide very closely with the Book Industry Study Group data, reported in the text, primarily because of the different wording of the questions and definitions of what constitutes a "book reader."

The Question: *Did you do any book reading yesterday?*

	All books			All books excluding the Bible		
			Change in % points			Change in % points
	1955	1984	1955–1984	1955	1984	1955–1984
	%	%		%	%	
Sex						
Male	16	23	+7	11	20	+9
Female	23	26	+3	15	22	+7
Age						
18–34	20	23	+3	17	21	+4
35–49	17	27	+10	11	24	+13
50 and older	24	25	+1	12	20	+8
Education						
College	30	37	+7	25	35	+10
High school/High school incomplete	17	20	+3	12	16	+4
Less than high school	20	10	−10	11	6	−5
Region						
East	20	24	+4	16	22	+6
Midwest	18	21	+3	10	17	+7
South	21	24	+3	11	20	+9
West	23	31	+8	14	28	+14
Total	20	25	+5	14	21	+7

*Source: Leonard A. Wood, "Book Reading, 1955–84: The Trend is Up," *Publishers Weekly,* 25 May, 1984. By permission.

NOTE: The 1955 survey results are based on 1,630 personal interviews with a national sample of adults 21 years of age or older, conducted between 25 March and 3 April, 1955.

The 1984 survey results are based on 1,514 personal interviews with a national sample of adults 18 years of age or older, conducted between 16 March and 25 March 1984.

group over the five-year period studied, there was a 21 percent drop in the numbers of people who read. Many said they "used to read books" but no longer do. If both these trends continue—decreases in readership among the young and the elderly—the overall, long-term trend for book reading would appear to be downward. (See Table 5.1.)

Growth in book sales is not an across-the-board phenomena, but can be attributed almost solely to a doubling, from 18 to 35 percent, of those reading 26 or more books in a six-month period. They typically spend 21.3 hours a week reading, 14 of which are devoted to books. This group reads three-fourths of all the books that are read. While those "heavy readers" keep the presses rolling, some two-thirds of Americans seem to be reading less than before. The typical book reader is a white woman under fifty, college-educated, single, and affluent in a white-collar job. The average non-book reader—who reads no books, but has read newspapers and magazines in the previous six months—is a black man over age sixty, with only an elementary education, retired or unemployed from a low-paying blue-collar job. Some 37 percent of the non-book readers said they found books boring; 22 percent said books took too much time; 15 percent preferred other leisure activities; 14 percent cited physical problems; and 9 percent said they couldn't read books because of other demands on their time, such as housework or raising children.[31]

According to *Publishers Weekly*'s assessment of the data, "if there appears to be a heavy dropout among the elderly as the baby boom generation matures, and if the age group following the baby boom generation is reading less, the trend could turn any current elation over a thriving industry into a thing of the past."[32]

The problem appears to be a dual one of both illiteracy and aliteracy. The former category, of those who can read simple passages but cannot read or write well enough to function effectively in contemporary society, includes somewhere between 17 and 73 million American adults—depending upon whose definitions and surveys we accept.[33] This large group, representing as many as one-third of the adult population, is of special concern to the nation's educators. The latter category, of those who have learned to read but choose not to do so, constitutes an even more bothersome problem for the book, magazine, and newspaper businesses. "What really worries me," states Art Seidenbaum, book editor of the *Los Angeles Times*, "is that in our system, I don't believe it's healthy for one-third to be exquisitely educated and informed while 65 percent are not. I think that the very society we live in requires a large, informed audience to work at all. Our political machinery requires readers."[34]

As many as one-third of adult Americans may be illiterate.

Book Publishing Today and Tomorrow

Traditionally, books have been characterized by a certain freedom of content, a freedom resulting from their smaller audiences. They are free to explore dimensions of radical politics or erotic interest that are not open to the broader-based family media. Further, books are free to experiment with new language and techniques—new expression. They are not tied to the "tried and true" because they can exist on much smaller audiences and they need not rely on the lowest common denominator of general appeal.

Charles Steinberg has made the point that the proportionately larger audiences that paperbacks attract show signs of eroding a part of this freedom. The paperback industry demands a certain level of proved acceptance to bring a title into existence. He notes that *Catcher in the Rye* aroused no criticism for its questionable content while it was in hardcover, but a rash of public indignation burst forth as soon as it appeared in the more widely distributed paperback edition.[35]

Merchandising techniques affect the freedoms and social roles of books.

A part of the traditional freedom of book publishers may be on unfirm ground as a result of what might be called vulgarization in paperbacks. As outlets multiply and press runs grow, there may be increasing pressure to emphasize formula writing, to avoid controversial themes, and to appeal to the largest possible audiences.

As the book becomes more a corporate product and less the handiwork of family-owned houses, the classic role of the book as a medium designed to enlighten is being altered. Under the impact of the communications explosion, book publishing is becoming a corporate activity promoted with techniques reminiscent of P. T. Barnum. Marketing becomes the dominant concern of publishers who sacrifice editorial considerations in their pursuit of profits. Book publishing still remains an essentially individualized mass medium, but that essential characteristic is gradually becoming eroded.

What can we expect in the future? Some change is almost inevitable due to the spread of electronic technology. No one knows for sure, of course, just what will happen in the years to come, but making predictions about an institution as diversified as book publishing is always interesting.

Publishers Weekly, the authoritative trade journal of the publishing industry, asked several of the best-known futurists what they foresaw as the most important changes likely to occur by the turn of the century that would have a major impact upon books. Responses varied, but a pattern seemed to emerge.[36]

Like other media, books will change forms and roles in the new electronic environment.

Alvin Toffler, author of *Future Shock* and *The Third Wave,* said that special-interest media, including diversity-producing cable and cassette television, will bring each of us more varied images through many more channels, and that less and less of the culture will be shared. Toffler's term for this process is *demassification.* Increasingly, Toffler said, we will live in a "blip culture" that bombards us with unrelated chips or blips of data. Forced individually to fabricate our own images of reality from these blips, we will cry out as a culture for synthesis. The answer to that cry, he maintained, is twofold: computers and "that powerful information technology called the 'book'."[37]

The book enhances individuality; it is private in a harassed and public time. But it, like all media, must change as our world changes.

Edward T. Hall—author of many books, including *Beyond Culture*—told *Publishers Weekly* he expected the reading public of the year 2000 to be far more sophisticated and particular than today's readers, and less captivated by the cliché. Publishers will find additional demands to be both more discriminating and more venturesome, ultimately surviving only if they attract true talent and manage to distinguish between genius and hack writers. "The market will favor the real pro over those basing decisions simply on bottom-line considerations," Hall warned. "Good book design techniques will be increasingly important," he added.[38]

Ernest Callenbach, author of *Ecotopia* and editor of the University of California Press, said that publishers should be able to make America the first culture to enjoy universally accessible books:

[They] could enter manuscripts [for a fee] on a centralized computer. The texts could then be printed upon demand by dispensing machines [rather like juke boxes] in libraries, bookstores, schools, supermarkets, post offices, banks, etc.— even in remote towns and in neighborhoods where books are virtually absent today. The would-be buyer could locate a title by punching Standard Book Numbers guided by a simplified title and author index, check reviews from an electronic *Book Review Digest,* and browse through sample pages electronically. The deposit of an indicated number of coins would produce carry-away copy. After such initial publication, some titles would then prove popular, and be bought up and printed by mass market publishers; illustrated books and fine printing would continue to be normally printed. In this Ecotopian future a bookstore's stock would consist of a few thousand popular titles, older publications not yet entered on the computer, imports and recycled copies of electronically produced titles.[39]

Meanwhile, in the shorter run, the marriage of electronics and book publishing portends some immediate changes in the reading business. Already some universities, led by Carnegie-Mellon University, are placing personal computers in every dormitory room, giving students access to the library resources, allowing them to interact with any other computers on campus, and, within a few years, to call up library book texts on their individual computer screens.

As the above example indicates, and as our earlier discussion of economic trends in publishing mentioned, many aspects of book publishing are continuing to merge with the electronics industry. A *Business Week* cover story reflects the pace at which the mergers are underway; the full title of its 11 June 1984 article is: "Publishers Go Electronic; An Industry Races to Relearn the Information Business." Martin Greenberger, professor of public policy and analysis at the UCLA Graduate School of Management, may have summed it up best with this advice to book publishers attending a forum on electronic publishing:

It may not be that books will become obsolete, but that the business of producing only books will become obsolete. So don't throw away the printing presses, but make sure you've got some computers humming along, too.[40]

As public libraries around the country eagerly enter the electronics age, a new question arises. Whereas in past times libraries have offered everyone who could read equal access to the world's stored knowledge, doesn't the increasing reliance by libraries upon electronic retrieval and storage threaten to destroy that equality?[41]

Another question is raised by new trends in publishing: for the past several years, best-sellers have increasingly dealt with special interest non-fictional, relevant, informational topics (including the many shelves of computer books), and there does not seem to be a concurrent growth in readership of enduring classics. Do these two trends indicate that book publishers are capitalizing on trendy, timely issues—basically the kinds of materials more traditionally published in magazines and newspapers—at the expense of their traditional role in transferring culture from generation to generation?[42] And, as readership studies indicate, does the increase in aliteracy threaten democracy? Only time will answer these important questions.

Summary

The industrial revolution, which was responsible for most of the technological innovations of the mass media, was the direct result of technical knowledge recorded and transmitted through print. The industrial revolution could not have occurred without the technical knowledge made possible by print, through its ability to record, transmit, and augment what went on before.

Book publishing in America has gone through several phases, the most recent of which has been the purchasing of independent publishing companies by conglomerates, larger shares of the market being devoted to paperback and textbook publishing, and the development of stronger ties between book publishers, film, and electronics industries.

The categories of books include trade books, paperback books, textbooks, and professional books. Trade books are general-interest books sold in bookstores. They present a tremendous gamble; only 5 percent sell well enough to recover the investment made by the publisher. Paperbacks are primarily reprints of books that were originally published in hardcover, but as the paperback industry develops in stature, some authors have found it advantageous to sell directly to a paperback company.

Unlike most trade books, textbooks can sell for many years. The markets are smaller but easier for the publisher to identify and reach. Revisions are published every three to five years, and some textbooks continue to sell for six to eight editions. The last category of books, called professional books, are written by specialists for other specialists to read.

Book publishing has three primary functions: editorial, production, and marketing. The marketing of books has been refined in recent years, but it still includes distribution of complimentary copies to reviewers and various other strategies for getting books to the best-seller lists. One of the best ways to insure substantial sales is for an author who has a lively personality to be a guest on one of the popular network television shows. Book clubs and serialization are also utilized by publishers.

The major difference between mass-market paperbacks and hardcover books is their distribution. Paperbacks are treated essentially as magazines and are distributed by wholesalers and the American News Company, while hardbacks are sold primarily through bookstores and book clubs.

Readership studies offer contradictory insights into the health of the book business. On the one hand, there are increases in the total number of books being sold and apparently read; on the other, apparent decreases in the numbers of young and elderly readers may pose a threat to the long-term health of book publishing.

Books, traditionally the freest of the media, in terms of content, retain that freedom when published in hardcover and given limited distribution, but they face increasing public hostility when questionable content is widely circulated in paperback form. As book publishing becomes more and more market conscious, the essential libertarianism of the product may diminish.

Despite widespread arguments concerning whether the book is a dying medium in this age of electronics, we garner from expert testimony and current policies that the medium is adjusting to changes in technology and will continue, although in one of several modified forms, to serve the needs of consumers who will expect individualized information, consumable upon demand.

Notes

1. Edward Jay Whetmore, *Mediamerica: Form, Content, and Consequence of Mass Communication,* 3d ed. (Belmont, Calif: Wadsworth Publishing Co., 1985), p. 18.

2. Marshall McLuhan, *Understanding Media: The Extensions of Man* (New York: McGraw-Hill, 1965); and *The Gutenberg Galaxy* (Toronto: The University of Toronto Press, 1967).

3. Jonathan Kozol, "A Nation's Wealth," *Publishers Weekly,* 24 May 1985, p. 30.

4. Clarence Peterson, *The Bantam Story: Thirty Years of Paperback Publishing,* 2d ed. (New York: Bantam Books, Inc., 1975), p. 5.

5. Charles A. Madison, *Book Publishing in America* (New York: McGraw-Hill, 1966).

6. Janice Castro, "Selling Off A Magazine Empire," *Time,* 3 December 1984, pp. 62-63; and "Publishers Go Electronic," *Business Week,* 11 June 1984, pp. 84-97; and *CBS/83 Annual Report to the Shareholders.*

7. John F. Baker, "1984: The Year in Review," *Publishers Weekly,* 15 March 1985, p. 30.

8. J. Kendrick Noble, Jr., "Book Publishing," in Benjamin M. Compaine, Christopher H. Sterling, Thomas Guback, and J. Kendrick Noble, Jr., *Anatomy of the Communications Industry: Who Owns the Media?* (White Plains, N.Y.: Knowledge Industry Publications, Inc., 1982), p. 119.

9. Chandler B. Grannis, "U.S. Book Title Output and Average Prices, 1983-1985," *Publishers Weekly,* 3 October 1986, pp. 89-92.

10. Ibid., p. 89.

11. Richard Martin and Carol Innerst, "Textbooks Besieged for Ducking 2Rs—Religion and Relevance," *Washington Times Insight,* 23 October 1985, pp. 60-61.

12. Ezra Bowen, "A Debate Over 'Dumbing Down,'" *Time,* 3 December 1984, p. 68.

13. Martin and Innerst, "Textbooks Besieged for Ducking 2Rs—Religion and Relevance," p. 60.

14. David Shaw, "Book Biz Best-Sellers—Are They Really? Laziness and Chicanery Play Major Roles," *Los Angeles Times,* 24 October 1976.

15. "How the best-seller lists are made," *USA Today,* 13 September 1983.

16. Ibid.

17. Shaw, "Book Biz Best-Sellers—Are They Really?"

18. Daisy Maryles, "The Year's Bestselling Books," *Publishers Weekly,* 16 March 1984, p. 29; and Alan D. Hass, "Some Businesses are Sheer Murder," *Family Weekly,* 18 November 1984, p. 18.

19. Laurence Bergreen, "Just Don't Get Booked After the Animal Act," *TV Guide,* 17 March 1979, pp. 33-36.

20. Maria Lenhard, "The Author as Peddler," *Deseret News,* 4 August 1979.

21. Ibid.

22. Eileen Prescott, "Author Tours: The Bloom is Off the Rose," *Publishers Weekly,* 3 August 1984, pp. 22-24.

23. Linda Deutsch, "Publishers Finance Lavish Book Promotions," *Salt Lake Tribune,* 28 April 1978.

24. Robert Garfield, "New book publishers look for concept, then author to write it," *USA Today,* 2 October 1984.

25. Ellen Goodman, syndicated column, "Author's hoax makes important point," *Logan* (Ut.) *Herald Journal,* 2 October 1984.

26. Shaw, "Book Biz Best-Sellers—Are They Really?"

27. Colleen O'Connor, "The giant book club turns 60," *USA Today,* 17 April 1986.

28. Peterson, *The Bantam Story,* p. 5.

29. John Mutter, "Paperback Top Sellers," *Publishers Weekly,* 14 March 1986, p. 33.

30. Howard Fields, "Survey Finds Eight Million New Readers in Five Years," *Publishers Weekly,* 27 April 1984, p. 14.

31. Ibid., p. 19; and John A. Finley, "Special Report: Reading," *Presstime,* September 1984, p. 19.

32. Fields, "Survey Finds Eight Million New Readers in Five Years," p. 14.

33. Kozol, "A Nation's Wealth," Howard Fields, "The View From Washington," *Publishers Weekly,* 24 May 1985, p. 31.

34. Finey, "Special Report: Reading."

35. Charles A. Steinberg, *The Communicative Arts* (New York: Hastings House Publishers, 1970).

36. "Predictions: Change to Conjure With," *Publishers Weekly,* 6 August 1979, pp. 26-28.

37. Ibid.

38. Ibid.

39. Ibid.

40. Howard Fields, "Forum Predicts Future of Books, Software," *Publishers Weekly,* 8 June 1984, p. 23.

41. Ibid., p. 22.

42. Marianne Yen, "Panel Inquires: Will the Book Cease to Exist?" *Publishers Weekly,* 27 June 1986, p. 19.

Part Three

In the next five chapters we discuss the media that depend on electronics: radio, recordings, television, film, and television's offspring. Together they demonstrate the influence, potential, and problems inherent in instantaneous communication.

Electricity is significant for the speed with which it transports messages across distances. Instantaneous information transmission has radically altered our senses of time and space, and of knowing and being known. It has introduced a new factor into the information business. Whereas print media rely on the mass production of separate units of books, newspapers, or magazines for individual consumers, the electronic revolution has made it possible for a single electronically produced message to serve masses of individuals separated physically (and perhaps culturally, intellectually, and emotionally).

Radio is no longer a national medium; it has fragmented to serve specific publics, often in terms of their special interests. In the process radio has applied electricity to the principle of specialization by which mass media have traditionally survived, making it a sort of electric magazine. Radio fought, won, and solidified the battle of commercial and advertising support for electronic media. It achieved a conditioned audience, the basis of programming, network and affiliate structures, and federal governmental control. Then it turned around and gave the game plan away to television. Afterwards, radio had to find a new content for new audiences and new prime time. Strangely, radio has not only survived under these conditions; it has actually thrived.

Recordings, the dominant content of radio, constitute a mass medium in their own right. Popular music has always been a diversion, yet it has never been too far from controversy. Today's popular music binds together some segments of our culture while ostracizing others, and there is little sign that such a situation will change in the near future.

The Electronic Media

In terms of its massive audiences and its total effect, television is a medium more speculated about than understood. Politically, socially, and economically, its ramifications are enormous. More particularly, it has radically affected all other media—driving some magazines out of business, encouraging others, and altering the role of the newspaper. It has reduced radio to a supplemental medium closely aligned with the recording industry. On the one hand, television has offered movies freedom from the tyranny of the masses; on the other hand, it has taken this freedom away as it offers movies vaster audiences in a single night than the average film could expect in its lifetime.

Significant competitors have arisen to threaten network television. The emerging media, still in their developmental phases, offer hundreds of channels, and hence service to special interests. In some cases (interactive cable in its latest forms) the new electronics offer built-in, immediate feedback. Coupled to home computers and satellites, we have the world at our fingertips.

What we do with that new access to information remains uncertain. Computerized information delivered by media whose natures we scarcely understand has become available to us more rapidly than we can consume it. The producers of such information have responded to our requests by providing massive quantities of the same materials (movies, sports, music, and escapist entertainment) that Hollywood and the commercial networks have so effectively dished up for generations.

Only recently have the "neovideo"—cable, home computers, satellite-delivered information, and the like—begun capitalizing on their own unique potential for reaching specialized audiences with specialized messages. Our new tools lend themselves to novel ways of dealing with our postindustrial, informational society; schooling, banking, shopping, interacting with our neighbors, and participating in the day-to-day running of our democracy lie within our electronically enhanced grasp.

201

Radio

6

Introduction

By 1987 there were 10,046 radio stations in the United States, including 4,856 commercial AM stations, 3,936 commercial FM stations, and 1,254 noncommercial FM stations. The numbers are growing; the federal government has allocated frequencies for 1,000 new FM stations, which went on the air starting in 1986. With 507 million radio sets in operation, 99 percent of the nation's homes and 95 percent of its automobiles are equipped with radio. That means there are more than two radios for each man, woman, and child in the country.

Each of us uses our two radios, too—an average of twenty-five hours a week for all Americans, which is five hours a week more than in 1980. Women aged 65 and older and black Americans are the heaviest listeners at thirty hours a week. Among men, 18-to-34-year-olds listen the most: twenty-seven hours.

As a nation, we spend several billion dollars each year to purchase some fifty million additional radios for our cars, purses, nightstands, family rooms, offices, or to attach to our bodies while we work and play. At mid-decade advertisers were spending $6.6 billion annually to hawk their wares over radio—double the dollar figures of only five years earlier. That flow of money brought owners of larger radio stations annual pre-tax profits in the neighborhood of 37.6 percent, although many smaller stations, which often charge advertisers less than a dollar to air a commercial, insist they are struggling to stay in business.

Radio today is a bigger business than ever before in terms of stations, programs, listeners, and dollars.

Radio today is a bigger business than ever before in terms of stations, programs, listeners, and dollars. However, radio has become a supplemental medium. Furthermore, it has become a necessary adjunct to the recording industry.

It was not always this way. During the 1930s and 1940s the radio was America's prime mass medium; it was a time when most people got a good deal of their news and most of their entertainment from radio. Ironically, there were fewer stations, fewer sets, and fewer listeners then than there are now. Many of those stations were affiliated with a major network, so for most intents and purposes radio was a national medium. Today, by and large, radio is a specialized or local medium, although new types of national networks have begun to emerge.

It is a medium of paradoxes. While larger in scope than ever before, it is the most readily overlooked mass medium. It ranks behind television and newspaper as a source of general news and in terms of news credibility. Yet as the most portable and instantaneous medium—although television is rapidly catching up on both fronts—it remains the one to which most people automatically turn in times of emergencies. Due to its portability it is also the number one source of news in the morning, when America gets dressed, eats breakfast, and drives to work. At the same time, many listeners are using it as a background medium, and don't even realize they have heard a newscast during the hours the set has been within earshot!

Radio set the stage for television. It developed over-the-air advertising—the commercial base of broadcasting; it established major networks to serve local stations with a quality of programming that none could afford by itself; it prompted the creation of the Federal Communications Commission and the doctrines of fairness and public interest, convenience, and necessity—and then became the first broadcast medium to be deregulated by the federal government, which came to believe that the medium was so pervasive and variegated that marketplace forces rather than legal controls would suffice; it pioneered ratings as a feedback device; it adopted the star system from the movies, developing its own personalities; it designed the formula of 90 percent entertainment and commercials, and 10 percent information, which has carried over into all of United States broadcasting; and it borrowed from print the idea of using content to deliver audiences—broadcasting's real "product"—to the advertisers.

On nearly every front, radio paved the way for television, and then in its moment of glory, radio did something daring: it turned the nation's theater over to television and walked right off the stage and into the audience. It learned how to speak to the audience one on one, one person at a time, achieving an intimacy which, to this day, no other medium possesses.[1] This was no noble gesture; it happened in spite of efforts of people in the industry. Radio was hard hit for a while under television's dynamic impact, but it recovered and went on to a healthier life than it had ever before experienced.

Radio set the stage for television; then it used new formats to reach new audiences.

Radio developed new prime times during the driving hours; it became portable—a constant companion; and it served special interests. It developed a demography of its own and began to discover automation and cheaper ways of broadcasting, even as television grew more expensive. And then, as we shall explore more fully in chapter 7, radio became the right arm of the record industry, the aural means of marketing aural products—records, then albums, then tapes, and, most recently, compact disc recordings.

Samuel Morse's telegraph (1844), Alexander Graham Bell's telephone (1876), and Thomas Edison's light bulb (1879) constituted the first technology to escape the physical net of printing. The miracle of electricity grew quickly, heralding a revolution in electronics that continues today.

Historical Development

During the 1890s, Guglielmo Marconi's incessant tinkering with electromagnets paid off. He had spent several years testing out the theories first proposed by physicists James Clerk Maxwell and Heinrich Hertz. Maxwell had laid the groundwork by investigating electromagnetic fields, and Hertz produced electromagnetic waves in his laboratory in 1888. The theories holding that electromagnetic waves could be detected and put to practical use for communications intrigued many engineers, including the young Italian, Marconi. With financial help from his father he invented a way to transmit sound—the dots and dashes of Morse code—without using wires.

The Early Days

Maxwell, Hertz, and Marconi were significant in nineteenth century radio.

This is the fifteen watt spark transmitter in the Herrold College of Engineering and Wireless, San Jose, California, which Charles D. Herrold (left) and Ray Newby (right) used to transmit "voice and music" in 1909.

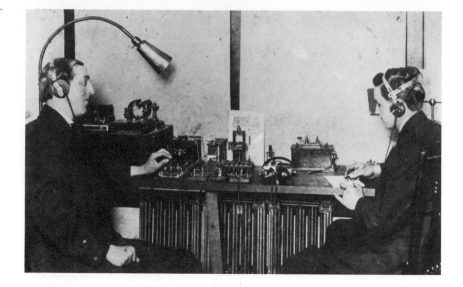

Almost immediately Marconi's wireless became a functional way to communicate with ships at sea. Marconi, however, had more ambitious plans, including the creation of a communications link between Europe and North America. By 1901 he had succeeded, which spurred the growth of his wireless companies.

Other inventors, such as John Ambrose Fleming and Lee De Forest, were at work on other elements crucial to radio. In 1906, De Forest perfected the audion, which became the vacuum tube, making possible the clear transmission of voice and music.

The earliest radio broadcasts were aired before the second decade of the twentieth century. On Christmas Eve, 1906, Reginald Fessenden made what is generally regarded as the first broadcast. Lacking access to De Forest's new audion tube, he coupled a telephone microphone to his transmitter and put together a short program of music and talk. The primitive broadcast was picked up by many wireless operators aboard ships within a radius of several hundred miles of Fessenden's Brant Rock, Massachusetts laboratory. We can only imagine the surprise of those operators, who heretofore had heard only the staccato clicking of dots and dashes coming through their earphones! The experiment was successful, and within a couple of years Fessenden's primitive telephone microphone had been replaced with De Forest's audion tube, which offered far greater fidelity of sound.

Charles Herrold dispensed a little music and a little chatter from San Jose, California, starting in 1909. In 1910 De Forest helped to broadcast an Enrico Caruso performance from the Metropolitan Opera House in New York. The equipment was crude and the reception poor, but it worked. Under the sponsorship of the New York *American* newspaper, De Forest reported the results of the tight Wilson-Hughes election in 1916. Ironically, radio's first newscast was in error: "Charles Evans Hughes will be the next president of the United States," it announced.

Fleming, De Forest, Fessenden, and Herrold were early twentieth century radio pioneers.

World War I intervened, postponing the development of commercial broadcasting for several years. It was not until 1919 that the government stepped back and private inventors, developers, and operators returned to the scene. In that year General Electric, Westinghouse, and American Telephone and Telegraph pooled their various patent rights for broadcasting equipment and receiving hardware, and they formed the Radio Corporation of America (RCA). The new company sought to control the machinery of radio and prevent competition from entering the medium. They were especially concerned about protecting their rights to radio receivers, which RCA believed would be the primary source of profit to be gained from radio. The whole idea of wanting to capture as much of the business as possible through the exercise of patents was not unique to radio. It happened in the film industry just a few years earlier and then again when television came along. In each case, the effect was to retard development of the medium.

Entertainment, News, & Ads Much credit for the development of radio as a national entertainment, news, and commercial medium must go to David Sarnoff. Sarnoff, a Russian immigrant who started as an office boy with the American Marconi Company, became a wireless telegraph operator when he was seventeen years old. Four years later, in 1912, he was on duty in New York City when he heard the faint signal "S. S. Titanic ran into iceberg. Sinking fast." For the next seventy-two hours, he was the key link between the Titanic disaster and the rest of the world. Due partly to his fame and partly to his ability and perseverance, he rose rapidly within the industry, becoming commercial manager at Marconi's American operations and then at RCA when it bought out Marconi in 1919. In 1921, he became general manager for RCA and was largely responsible for its formation of the National Broadcasting Company in 1926. In 1930 he became the president of RCA, and from 1947 until his retirement in 1969 he was chairman of the board. During his long association with RCA he fostered the emergence of monochrome (black-and-white) and color television from experimental stages to market saturation. He is generally credited with being the most influential person in the growth of American broadcasting.

Sarnoff deserves credit for developing radio as a national entertainment, news, and commercial medium.

While Sarnoff was assistant traffic manager of the Marconi company in 1915 and 1916, he sensed that one of radiotelephony's major liabilities, its lack of privacy, could be turned into its major asset. In a memo to Edward J. Nally, vice-president and general manager of American Marconi, Sarnoff wrote:

I have in mind a plan of development which would make radio a "household utility" in the same sense as the piano or phonograph. The idea is to bring music into the home by wireless.

While this has been tried in the past by wires, it has been a failure because wires do not lend themselves to this scheme. With radio, however, it would be entirely feasible. . . . The receiver can be designed in the form of a simple

"Radio Music Box" and arranged for several different wave lengths, which
should be changeable with throwing of a single switch or pressing of a single
button. . . .

The manufacture of the "Radio Music Box" including antenna, in large
quantities, would make possible their sale at a moderate figure of perhaps $75.00
per outfit. The main revenue to be derived will be from the sale of "Radio Music
Boxes" which if manufactured in quantities of one hundred thousand or so could
yield a handsome profit when sold at the price mentioned above. . . .

Aside from the profit to be derived from this proposition the possibilities
for advertising for the Company are tremendous; for its name would ultimately
be brought into the household and wireless would receive national and universal
attention.[2]

David Sarnoff's vision was incredibly accurate, except for his conviction that sales of receivers rather than ad sales would furnish the major
revenue for the fledgling industry. Within a decade radio had become the
household utility he had envisioned.

Westinghouse engineer Frank Conrad had been broadcasting phonograph music over a transmitter in his Pittsburgh garage in 1920 as part
of his experiments in radiotelephony. Once Westinghouse realized that many
area residents were tuning their homemade receivers to Conrad's signal
and were even calling or writing him with special requests for music, Westinghouse stimulated a demand for its receivers by formalizing the programming over KDKA. Conrad's Westinghouse station was inaugurated
on 2 November 1920 when it gave the results of the Harding-Cox presidential election, using telegraphed reports from its newspaper sources. The
impact of the broadcast was so great that the public flocked to buy primitive crystal receivers from local department stores for as little as $10, not
the $75 minimum Sarnoff had estimated.

There is some controversy as to whether KDKA in Pittsburgh, WWJ in Detroit, or WHA in Madison, Wisconsin, was the first true broadcasting station in the country; however, the first station to receive a regular broadcasting license was WBZ in Springfield, Massachusetts, on 15 September 1921. Those earliest commercial stations were owned by companies interested in promoting their primary business, such as the efforts of Westinghouse at KDKA. Several newspapers started radio stations in order to sell newspapers, the first being WWJ in Detroit. The *Detroit News* began broadcasting news in August 1920 and became a licensed commercial station the following year.

Prior to 1927 about sixty-nine different newspaper or other publishers built or bought radio stations. Some saw radio as a threat to their own newspapers and wished to keep control over local news dissemination; some saw it as a prestigious venture into new technology; and some, as a means of expanding their community service.[3] Other radio stations were owned by educational institutions, department stores, car and motorcycle dealers, music and jewelry stores, and hardware stores. According to Christopher Sterling, none of them were in the business of broadcasting for broadcasting's sake alone; they were selling their own image, their service, or their own name as a precursor to commercial advertising.[4]

Commercial advertising on radio probably began in 1922 when the American Telephone and Telegraph Company's New York station, WEAF, accepted $100 from a local real-estate firm for the broadcasting of a fifteen-minute message. The idea of such sponsors taking over the airwaves was not immediately accepted throughout the industry, however. Secretary of Commerce Herbert Hoover, for years in the thick of radio's struggles for identity and self-control, reacted to radio advertising with the statement, "It is inconceivable that we should allow so great a possibility for service, for news, for entertainment, and for vital commercial purposes to be drowned in advertising chatter."[5] But his voice, and the voices of other industry and civic leaders, was quickly drowned in the sea of chatter that was to become the lifeline of the industry. Quality programming cost money, prompting broadcasters to seek sources of capital provided by advertising. Audiences discovered early on that it was more convenient to put up with advertising in order to get good programming. The options, direct subsidies to each station or governmental funding of the medium, seemed less attractive and probably less realistic.

Establishing the Networks The networks and advertising are inseparable. In 1924 Eveready Batteries sponsored the "Eveready Hour" over a network of twelve stations that were linked together by telephone lines, which allowed for simultaneous broadcast in a dozen cities. This leasing of phone lines was the principal interest in ATT in the RCA consortium. By 1925 ATT had linked together twenty-six stations, stretching as far west as Kansas City, from its own master station, WEAF, in New York. The thought of

Early broadcasters sold their image, their service, or their name. Advertising was quick to follow.

Advertising and networks grew rapidly. NBC and CBS dominated the airwaves.

reaching an evergrowing number of listeners in population centers across the nation proved irresistible to national advertisers, and the character of radio was no longer experimental; it had become highly organized. Catchy jingles were developed to promote product familiarity. Radio began to develop its own stars and its own drama; names and serials were guaranteed to attract listeners night after night, week after week.

Also in 1925, RCA (now including only Westinghouse and General Electric) set up its own flagship station, WJZ, in New York, and began to tie together stations to compete with ATT.

The network concept was perfected during the following year, 1926, when ATT sold WEAF to RCA for $1 million, an unheard of sum in the 1920s. Charges of antitrust by the Federal Trade Commission prompted ATT to get out of the broadcasting business and devote its energies to leasing lines to the broadcasters for network purposes. RCA created the National Broadcasting Company and suddenly had two fledgling networks to manage. WJZ became the flagship of NBC's Blue network, and the newly acquired WEAF operated as the key station in the Red network. Before the end of the year the NBC networks included about fifty stations operating coast to coast.

In 1927 the Columbia Phonograph Record Company invested some money in a struggling network and renamed it the Columbia Broadcasting System (CBS), indicating even then close relationships among radio, the music industry, and the record business. CBS was spearheaded by William S. Paley, who continued to dominate it throughout its formative years and into the 1980s.

Local stations are paid to carry network programming and, of course, commercials.

A large number of stations became affiliated with networks, whereby they would carry network programming in return for a fee from the network. This financial arrangement surprises many people who think local stations should probably pay the networks for the privilege of receiving programming, stars, and newscasts that they could not possibly afford on their own. But the networks realize—as do the stations—that the programming assets attract large local audiences that add to the national audience. This permits the network to charge more for national advertising. The local stations, in turn, can charge higher rates for their purely local advertising. Network affiliation is a lucrative opportunity. By 1934 CBS had ninety-seven affiliates, NBC Red had sixty-five, and NBC Blue had sixty-two. About 40 percent of the stations broadcasting were network affiliates; the rest were independents.

The history of radio, and subsequently television and newer forms of electronic communication, has been a story of corporate struggles. The communications companies that had been so instrumental in establishing the medium squabbled among themselves, especially after they became aware of the commercial potential in radio. Other companies had entered the industry, starting stations across the country. There were more than five hundred radio stations broadcasting by 1924 and more than three million receivers in operation. At the end of 1925 one in five hundred American homes had a radio receiver; only a year later, the ratio had changed to one in twenty. By 1927 the number of stations had increased to more than seven hundred. Without a doubt, the industry was booming.

Radio required no physical distribution and it crossed municipal boundaries and state lines indiscriminately, particularly on the crowded East Coast where it was born. But technological proliferation, unchecked by any systematic internal or external regulation, resulted in so many stations operating so close together on the band that none could be received clearly. Radio had become "a tower in Babel," according to broadcast historian Erik Barnouw.[6]

When the radio industry finally sought regulation, the federal government was the only logical agency to turn to, because cities and states could regulate only within their own jurisdictions, and radio signals carried across political boundaries.

Governmental Regulation Government's first direct involvement with radio came in 1912, when Congress passed the Radio Act. The law had modest provisions: radio operators were expected to pass an exam before being allowed to transmit; and the Secretary of Commerce was responsible for assigning wavelengths or radio frequencies to whoever applied for a license. These regulations seemed reasonable enough, considering that so few stations or individuals were seeking to use the airwaves at that time.

But the 1920s saw a boom in radio growth, particularly in commercial stations. With the boom came chaos, since all commercial licensees were on the same frequency (remember, most of the earlier licensees were amateurs, hams, or naval operators) and had to work around each other's broadcast schedules if they hoped to be heard. In addition, the 1912 Radio Act gave government no authority to regulate the content of programming; consequently, the airwaves filled with commercial and noncommercial propaganda, and promotion of everything from snake oil to salvation.

A National Radio Conference, called to solve some of broadcasting's problems, began in 1922. The conference met again in 1923, 1924, and 1925. Secretary of Commerce Herbert Hoover, at the request of some broadcasters and other interest groups, began reinterpreting the 1912 Radio Act. He expanded the broadcast spectrum so that more broadcasters could be heard. He imposed schedules on stations and organization thus followed. Some station applicants were denied licenses on technical grounds, generally because of limited frequencies. When broadcasters appealed what they said were Hoover's illegal actions, chaos returned to the airwaves and Congress again stepped in.

The solution proposed by Congress was known as the Radio Act of 1927. It established the Federal Radio Commission (FRC) and a comprehensive set of regulations that affected every facet of the industry. The Radio Act included provisions for the FRC to assign frequencies to stations, but of broader significance was the Act's coupling of broadcasting to the public welfare. There were those both inside and outside of government who felt

In the 1920s, radio, left to its own devices, became "a tower in Babel." It then asked for governmental regulation.

The federal government has been involved from radio's inception, and it happily stepped up regulating once invited to do so.

that broadcasting had a public significance surmounting commercial considerations. They argued for public broadcasting such as England had with the governmentally controlled British Broadcasting Corporation (BBC). However, the commercial precedent for radio was already several years old and the battle for dominance of the mass media by commercial interests had already been fought and won in the newspapers. The clear statement of the First Amendment to the U.S. Constitution, that Congress shall make no law abridging freedom of the press, was difficult to circumvent, if radio was a form of the press.

As a partial salve to the proponents of public radio, the Radio Act of 1927 made the requirement that the individual stations operate in the "public interest, convenience, and necessity" of their own communities. Congress was saying that the airwaves belonged to the public and could be leased to stations so long as they were behaving responsibly. That concept was retained by the successor to the Federal Radio Commission when the Communications Act of 1934 established the Federal Communications Commission (FCC) to oversee radio, telephone, and telegraph (i.e., all wire and radio communication). Defining the public interest, convenience, and necessity has constituted untold thousands of hours of bureaucratic time and energy in the meantime, as broadcasters attempt to meet the spirit of the law while turning a profit.

America's Prime Mass Medium

The Great Depression gave rise to radio's Golden Age. News, entertainment, and advertising flourished.

The Great Depression of the 1930s gave an unforeseen boost to radio. In a time of poverty, radio reception cost nothing more than the price of a receiver and a few pennies for electricity. Radio became America's home entertainment, establishing and solidifying a pattern that would be inherited intact by television. President Franklin D. Roosevelt also clearly saw that the broadcast media entered America's homes directly and without the interpretation and gatekeeping of the printed page. He made radio his own political instrument as he reached into America's living rooms to reassure the people through his fireside chats, a device that he parlayed into four terms of office and that set the pace for the political use of broadcast media in the future.

By the early 1930s the networks were in full competition with one another, vying for the attention of listeners. Recognizing the demography and differing sizes of audiences, time classifications developed: *AA* for prime time, seven to eleven P.M.; *A* for the hours adjacent to prime time; *B* for most of the rest of the day; and *C* for the late, late hours that belong to stations broadcasting for insomniacs, students, night owls, and people with unusual work schedules.

Of course, *AA* time was the most costly for sponsorship and commercials, and *C* was the least expensive. These classifications, based on presumed audience size, helped to formulate a rate card for determining *costs per thousand* (cpm)—that is, cost to the advertiser per thousand listeners.

FDR used radio to reassure an anxious nation during his fireside chats in the 1930s and 1940s.

A measuring device was needed to accurately determine, and prove, this presumed audience size, so an embryonic form of rating was developed, based largely on telephone polls. Public opinion polling was in its infancy and the yardstick was crude, but telephone polling was the first form of feedback, born of commercial necessity, which grew to the present-day ratings system.

Originally, advertising on radio had taken the form of sponsorships. National commercial advertisers would sponsor and often pay for a program that would carry their name ("The Eveready Hour," for example) and would permit them to insert commercials during the body of the program. As hour-long time slots became more expensive, half-hour sponsorships developed. Subsequently, some of the independent stations discovered that their survival lay in catering to the local community, and while few local merchants could afford a sponsorship, they could afford a minute or two, the price of which, when added up, netted the same amount of, if not more, advertising revenue. Network affiliates followed suit, selling advertising for their station breaks, or for the minute or so they were permitted to identify themselves each hour. Thus, the commercial came into being, which, by and large, has supplanted sponsorship.

Amos 'n' Andy entertained a nation but perpetuated racial stereotypes.

Comedy, thrillers, westerns, and music filled the airwaves.

Entertainment A pattern of general prime-time entertainment catering to the largest possible audience developed in the 1930s with a packet of comedy shows, thrillers, and westerns. The 1930s were also a musical era, the time of the big bands: Benny Goodman, the Dorsey Brothers, and Guy Lombardo. These bands appeared periodically on radio specials and their records were daytime stock. One of the most popular prime-time programs of the 1930s was "Fred Waring and His Pennsylvanians," a weekly musical potpourri that even included baseball scores in season. "The Bell Telephone Hour" brought symphonic music into America's homes; the networks and some of the major stations even had their own house orchestras in which musicians of considerable stature played. Together, radio stations offered a spectrum of classical, pop, and "swing" for America.

"Amos 'n' Andy," starring a blackface comedy team, perpetuated the Negro stereotype in the early evening. The program was so popular, even in the late 1920s, that President Calvin Coolidge supposedly refused to miss an episode, and many theaters delayed their evening programs so people could listen to the radio show before coming to the theater.[7]

Jack Benny for Lucky Strike cigarettes and Fred Allen with "Allen's Alley" on Sunday evenings were comedy stars of the era. Benny was among the first to inject humor into advertising. Other comedians and entertainers of the 1930s continued their careers on television in later years, including George Burns and Gracie Allen, Edgar Bergen and Charlie McCarthy, Ed Wynn, and Kate Smith. Burns and Allen had the first continuing situation-comedy series.

"The Whistler" was the first of the Gothic thrillers to send chills up and down spines in darkened rooms each Sunday evening. Once each week "The Shadow" opened with a sinister voice asking, "Who knows what evil lurks in the hearts of men?" Mysteries, in particular, activated the imagination of the listener, and broadcasters heightened the impact with sound effects—solitary steps in a lonely alley, creaking stairs, and a clap of thunder when least expected.

Then there were game shows, including the "64 Dollar Question," where the maximum a contestant could win was $64, arrived at in a geometric progression (a sign of depression values and subsequent inflation). There were westerns, of which "The Lone Ranger" was the model, with his super horse, Silver, and faithful Indian companion, Tonto—yet another racial stereotype. "Little Orphan Annie" and "Jack Armstrong" were for the children but were avidly followed by adults, too.

During the daytime hours, radio discovered that there was a demography present. The audiences were no longer mass and inclusive but were composed of different kinds of people at different hours. The late afternoon was kiddie time, served by a variety of breakfast food sponsors. The early mornings were news times for arising breadwinners, upstaging the morning newspaper. During the bulk of the day radio found itself appealing to a largely female audience of housewives, and the soap opera was born—so named because it was almost universally sponsored by soap manufacturers such as Lever Brothers, Procter and Gamble, Colgate, and Palmolive-Peet. Soap operas were depression-era escape, offering either poignant sorrow that listeners could sob over or a kind of saccharine wisdom of sweetness-and-light, or both. The soaps engendered incredible loyalty, with listeners sending gifts to the newborn babies, letters of sympathy to victims of series troubles, or letters of thanks to characters whose actions had helped listeners cope with their own lives. "One Man's Family," "The Romance of Helen Trent," "Our Gal Sunday," and "Portia Faces Life" were leaders of this genre.

The importance of radio during the depression cannot be underestimated. Sociologist and media theorist Melvin DeFleur has said that radio actually thrived on the depression. Advertising revenues and the number of sets in operation grew exponentially. Poverty-stricken families would scrape together enough money to repair their radio sets even if they had to let the furniture go back to the finance company or stall on rent payments.

And on a summer night Americans could walk down a city street and hear their favorite program uninterrupted through the open windows of every house they passed.[8]

By 1938 more American homes were equipped with radios than with telephones, automobiles, indoor plumbing, or electricity. Of the 32 million households in the nation, 27 million had radios, a higher household penetration figure than enjoyed by either newspapers or magazines. Radio had become accepted, not only as a means of escape from depression-era burdens, but as a connection to an increasingly complex and troubled world. If there remained any doubts as to the popularity and power of the aural medium, they were laid to rest on the night of 30 October 1938.

On that Halloween night CBS carried to six million American listeners a memorable broadcast of "Mercury Theater on the Air." The program, intended as a Halloween spoof, was a realistic dramatization of novelist H. G. Wells's science fiction masterpiece about an invasion from Mars, *War of the Worlds.* Young Orson Welles, who starred in and wrote, directed, and produced many of the Mercury Theater's weekly shows, worked with John Houseman and writer Howard Koch in presenting the invasion as a series of news bulletins that consistently interrupted a routine program of dance music. At the outset of the program, and three other times during it, listeners were told it was only a dramatization. Those who listened carefully would have recognized that the events presented in the hour-long show would have taken days or weeks to occur in real time. However, many tuned in late, having switched over from listening to the more popular Charlie McCarthy and Edgar Bergen show on a rival network. By then the invasion was in full force, and was being presented so effectively that the latecomers and even those who had heard the advisory notices were being swept along by the biggest media-induced panic in history.

The CBS switchboard was flooded with calls from concerned listeners. Streets were crowded with carloads of families fleeing to escape the eastern seaboard where the fictional invasion was occuring. Bus terminals were packed. Priests and religious leaders were called upon to take final confessions and offer comfort. Police all across the nation, but particularly in New Jersey and New York, reported levels of mass panic never before seen. Little mind that it was only a Halloween spoof, as Welles called it. America, which had learned to trust radio, had been tricked.

For many, the episode helped support the argument that mass media had enormous direct influence on public opinion. (See chapter 2.) Researchers, taking advantage of the natural laboratory experiment that had fallen into their lap, conducted extensive investigations of the phenomenon, trying to find out how many and what kinds of people had been panicked. The studies revealed that a variety of forces had been at work: the public's general confidence in radio; the historical timing of the show (following the depths of the depression and coming at a time when Europe had erupted into war); the technical brilliance of the show, with its reliance upon news bulletins and interviews with observers and "experts" from government and

Radio's power was never as obvious as it was in 1938, when Orson Welles scared the nation with *War of the Worlds.*

science; and the problems that occurred when people tuned in late, and had missed the opening announcement about the play. A closer look at why some panicked while others did not revealed a variety of psychological and sociological variables at work, and helped pave the way toward more sophisticated research and theories concerning the impact of media.[9]

In the aftermath of the show, the Federal Communications Commission adopted a resolution putting a stop to broadcasters' use of "on the spot" news bulletins in dramatic shows, CBS issued an apology, the now-popular Mercury Theater got a longterm sponsorship from Campbell Soup, and America turned its attention to the serious matters of World War II.

Early Radio News As early as 1922 the newspaper-dominated Associated Press refused to serve radio stations with news. Radio stations threatened to turn to the United Press (UP) and the International News Service (INS), an argument of some weight, since many of the station owners were newspapers that already subscribed to AP.

However, when the depression came, competition between radio stations and newspapers grew more intense. Newspaper advertising lineage was dwindling and radio was absorbing much of it. Furthermore, radio had gone heavily into the news business. For example, in 1930, KMPC, an independent in Los Angeles, had ten reporters on the newsbeat. Newspapers in duress just could not tolerate the direct competition with radio for both advertising and news.

Newspapers sensed a threat to their news dominance, and years passed before the media learned to cooperate.

The battle reached a climax with the 1932 election results. The contest between incumbent Republican Herbert Hoover (the depression president) and Democrat Franklin Roosevelt attracted wide attention, and a good part of America learned of the results over the air. The American Newspaper Publishers Association (ANPA) had had enough, so in 1933 all news services stopped furnishing news to radio. Of course, this simply spurred the networks to organize their own news bureaus. The affiliated stations provided a natural national network of sources akin to the wire service bureaus and member papers, but abroad, where no such facility existed, they were more or less on their own. The irony of this is that the ban by the ANPA strengthened the position of the networks in a news-hungry America and probably would not have come about had it not been for the association's ill-advised pressure. Further, the loss of revenue to the wire services, particularly during the depression, was damaging.

An attempt to compromise resulted in the December 1933 "Biltmore Agreement" among publishers, networks, and press associations. Radio networks were to receive only enough news items from the press associations each day to present two five-minute news summaries. Morning news was not permitted until 9:30 A.M., and evening news not until 9:00 P.M., so as to avoid competition with the newspapers.

The Biltmore Agreement was short-lived, because Trans Radio Press, a new organization, started selling news to stations. The publishers then had to direct their guns at Trans Radio Press and they found they could

whip the upstart by also selling news to the stations, just as in the good old
days. By 1935 both UP and INS began selling full service to the radios
again, while the more reactionary and press-controlled AP held out until
1940.

Throughout the 1930s and later, Walter Winchell bridged news and
entertainment with a breezy, nightly report of gossip, opinion, and innu-
endo, plus a smattering of headlines, becoming the first radio columnist.

The forceful impact of dramatic international events on radio was
demonstrated on 11 December 1936 by Edward VIII's poignant worldwide
radio announcement that he was abdicating the throne of England for Wallis
Simpson, a divorced American commoner. Meanwhile, the Spanish Civil
War, which had begun six months earlier, provided a testing ground not
only for Nazi weapons, but also for radio reporting techniques, which would
be put to use in the inevitable, approaching holocaust of World War II. In
1937 CBS sent H. V. Kaltenborn overseas, where he broadcast live from
battlefields in Spain, on one occasion getting dramatic effects by hiding a
CBS microphone and transmitter in a haystack right in the middle of the
battle. In the late 1930s the cerebral Kaltenborn translated Adolf Hitler's
dramatic speeches into English for an increasingly worried American public.
Later, working out of New York, he coordinated and interpreted the war
reports from a large CBS crew of correspondents, becoming history's first
anchorman.

Following is a verbatim sample of Edward R. Murrow's wartime radio journalism. It is from "Buchenwald," filed April 15, 1945 after Murrow traveled with American soldiers who were liberating a German concentration camp.

Permit me to tell you what you would have seen, and heard, had you been with me on Thursday. It will not be pleasant listening. If you are at lunch, or if you have no appetite to hear what Germans have done, now is a good time to switch off the radio, for I propose to tell you of Buchenwald. It is on a small hill about four miles outside Weimar, and it was one of the largest concentration camps in Germany, and it was built to last. As we approached it, we saw about a hundred men in civilian clothes with rifles advancing in open order across the fields. There were a few shops; we stopped to inquire. We were told that some of the prisoners had a couple of SS men cornered in there. We drove on, reached the main gate. The prisoners crowded up behind the wire. We entered.

And now, let me tell this in the first person, for I was the least important person there, as you shall hear. There surged around me an evil-smelling horde. Men and boys reached out to touch me; they were in rags and the remnants of uniform. Death had already marked many of them, but they were smiling with their eyes. I looked out over that mass of men to the green fields beyond where well-fed Germans were ploughing.

Source: Lawrence W. Lichty and Malachi C. Topping, *American Broadcasting: A Source Book on the History of Radio and Television* (New York: Hastings House, 1975), 387.

The impact of World War II on radio news was enormous. The war itself engendered interest far beyond usual human curiosity. Real drama, written on an international stage with real actors and real stakes, was rendered the more piquant because survival was at stake. Radio brought the war home, and responding to public demand, it upped its coverage. For example, in 1937 NBC broadcast a bare half-hour or so of news each day; by the war's end in 1945, a quarter of its daily broadcast was news. For a twenty-day period during September 1938 (the Munich Crisis), NBC's Red and Blue networks carried 443 separate news broadcasts, including "flash bulletins" of the war, taking up 58 hours and 13 minutes of airtime, while CBS was sending out 54.5 hours of broadcasts.[10]

Radio provided an international forum for the unmatched rhetoric of Winston Churchill, whose immortal speeches held a badly battered Britain together by the sheer power of words. He also consolidated the Allied cause in a time of doubt. With radio reporters, America witnessed the invasion of Poland and the surrender of France.

Edward R. Murrow, who had been sent to Europe in 1937 as CBS's "Director of Talks" with the intention of sending back cultural and feature programs, became the most famous radio war correspondent. He broadcast the battle of Britain from London rooftops amid falling bombs, which provided a dramatic backdrop for his remarks. Later he broadcast memorable reports about his flights in bombers over Germany, bringing the war ever closer to his audiences.

The impact of World War II on radio news was enormous. Edward R. Murrow brought the war home to American listeners.

Most of the world learned of Pearl Harbor on 7 December 1941 by radio, and a day later a record audience listened soberly to Franklin Roosevelt's measured words to Congress for a declared war on the Axis powers. News developed its own star system, and commentators, including the feisty Fulton Lewis, Jr., became major figures, interpreting the events of the world into bite-sized pieces for America to consume. At times, the commentators didn't allow the facts to interfere with their own perceptions of reality. Consider, for example, the words Fulton Lewis, Jr., zealously mouthed to an anxious America only five hours after the attack on Pearl Harbor:

The attack on the ships in Pearl Harbor [was] a very very foolish thing, as a matter of fact, suicidal fool-hardiness as a matter of fact, because the Japanese must know, as all the rest of the world knows, and all the rest of the navies and military men of the world know, that Pearl Harbor is the one invincible, absolutely invulnerable base in the world. It's stronger even than Gibraltar itself, and as far as an attack or siege of it is concerned, there could have been no possible sane intention on the part of the Japanese to such an end.[11]

Costs of Broadcasting As radio grew, it became increasingly expensive. Studio staffs on major stations included news bureaus with up to a dozen reporters, a full orchestra with a musical director, program people, and a corps of engineers to balance sound and splice in both network programming and commercials, as well as people to tend the transmitters. Advertising sales forces, and continuity folk to develop the daily log, added to costs. Radio's hardware was far from cheap as transmitters became increasingly more powerful—reaching hundreds of miles—and remote facilities traveled to on-site broadcasts. Top announcers and star commentators drew huge salaries, and subscription to the radio wires added to costs. All this had to be paid for by commercials, and consequently split-second timing was the earmark of radio's peak when it was the preeminent mass communications network in America. As technology improved and radio reached larger audiences, its commercial rates went up and up until none but the wealthiest of advertisers could afford network exposure. This limited radio's potential to national advertisers of basic consumer goods, as they were the only ones who could use such large, undifferentiated audiences. More local advertisers found the newspaper their best buy, since it reached out into entire communities and metropolitan areas; and more specialized products sought advertising in the increasing number of specialized magazines on the market. Some, of course, found it possible to reach the desired local audience during network station breaks or on local radio programs.

Eyeing the successes of CBS and NBC, two large independents, WOR in New York and the *Chicago Tribune's* WGN (for World's Greatest Newspaper), organized the Mutual Broadcasting System in 1934, but they found themselves unable to compete on equal footing with the solidly entrenched networks. Subsequently, Mutual filed a complaint with the FCC against the networks for restraint of trade. Adjudication of the suit brought about the sale of NBC's Blue network in 1943, which became the American

The networks' costs began to prove prohibitive to many advertisers.

Broadcasting Company in 1945. The suit filed by Mutual did not particularly help that fourth-place network, a loose affiliation of weaker stations. In addition ABC, divorced from NBC's experience and in direct competition with its parent, also found the sledding rough.

Paramount Theaters later bought ABC in 1953 after the motion picture industry had been forced to divest itself of its theaters. It was a wise move to diversify interests while still remaining in the mass communications business, and to further exploit the threat of television, which, in 1953, was beginning to make serious inroads into the box office of the motion-picture industry. But Paramount's management was oriented toward motion pictures and Hollywood rather than toward broadcasting, and for nearly two decades ABC remained a distinctly third-place network, never really gaining on CBS and NBC, who were until the late 1970s the first and second place networks in television.

By the end of World War II, radio had set the stage for television, and America grew accustomed to listening. The principle of public interest had been established and its parameters defined. A strong national trade association (National Association of Broadcasters, NAB) had been established and a rating system had been pioneered. Advertiser and agency loyalties were developing, and the methods of doing business were understood.

The Rivalry with Television

It took television only a short while to catch up with radio, which was largely the result of network influence. The broadcast system was inherited ready-made: major networks rapidly acquired wholly owned stations in major cities across the country, and they found it relatively simple to make affiliate agreements with other licensees. With their wealth and experience, the nets were able to develop national programming far superior to anything a local independent could afford, just as they had in radio. This expansion happened at the expense of their radio interests, despite the fact that radio income from the networks and local stations supported television for many of TV's years. However, radio gradually became supplemental and downgraded, and then was often supported by the increasing profits from television.

Just when it got it all together, radio lost it to television.

The radio industry was scared; it viewed television as radio with pictures, and the death of radio was predicted. Throughout most of the 1950s this was the prevailing attitude in the industry.

While radio lacked the visual appeal of television, it had certain inherent strengths: it demanded less attention than television, it could be perceived almost unconsciously while the listener went about doing other things, and it was also superior to television as a music medium; in general it offered less distraction. Finally, the airwaves allowed for more channels in radio than in television, which permitted far greater diversification. This potential was never fully explored while radio was under domination of a limited number of networks and while the costs of radio production remained high.

TV and the expressway
helped radio find a new
prime time and new
audiences.

Radio had to find new
techniques, new sponsors,
and new audiences.

As television gradually took over the evening prime-time hours, radio discovered that it had large audiences all to itself in the early morning when America arose and dressed. Radio was further assisted during the transitional decade of the 1950s by a number of technological improvements. Miniaturization of parts and certain technical innovations made car radios practical and cheap. Radio found it could extend its prime time to driving hours when America went to work and returned home again in the evening. Radio concentrated on a new split prime time from roughly 6:00 to 9:00 A.M. and again from 3:30 to 6:00 P.M. In fact, it increased its prime time from four to five and a half hours and changed its programming.

Instead of the costly dramas and comedies that had been staples of the 1930s and 1940s, radio experimented with music and new formats presided over by disc jockeys. This strategy kept the medium profitable enough to keep it alive while it was trying to adjust to TV, which was taking its advertising and audiences.

Miniaturization was helped by the development of the transistor, which did away with unwieldy and perishable vacuum tubes, and by printed circuits that reduced both the bulk and assembly expense of complicated wiring. The addition of miniature batteries freed radio from the necessity of an external power source and made it fully portable. Radio could take its instantaneousness anywhere that people could go. It became a constant companion, and as its parts became simpler and assembly less complex, it became cheap—almost disposable. It was possible for everyone to have a radio, perhaps several.

Such portability assured that there was an audience regardless of the time of day. Radio could be where people were; it no longer had to wait for them to gather around it. That was television's problem. Radio became a personal rather than a family medium, which naturally extended both its sales and reach.

Radio received another boost from the final development of FM, or frequency modulation. FM radio had two characteristics different from AM (amplitude modulation) radio. Its signal, like that of television, was a line-of-sight signal, which meant it had a relatively short range because of the earth's curvature. It also had remarkable clarity of tone, which meant that it was ideal for broadcasting music. This characteristic was further enhanced by the fact that FM could be readily adapted to stereophonic sound, adding an orchestral dimension to the previously crude single-source sound of traditional radio.

FM

The FM radio was originally developed in 1933 by a Columbia University professor, Edwin H. Armstrong, who successfully demonstrated its static-free, high-fidelity sound. Shortly before his death in 1954, Armstrong also patented multiplexing on the FM channel, which permitted stereo broadcasting. Experimental broadcasts in FM were held during 1939, and the FCC authorized regular FM broadcasting over thirty-five commercial, and five noncommercial, educational channels starting 1 January 1941.

FM was developed in 1933, but did not take off until the end of World War II.

Operation of FM had been delayed first by the depression and later by World War II. Then, in 1945, the FCC, seeking to encourage VHF television broadcasting, moved the FM frequencies to a higher and less vulnerable position on the radio spectrum. The FCC argued that the move on the dial was based on fear that sunspot activity would interfere with the lower frequencies, but Armstrong and other engineers charged that the FCC was trying to protect AM radio and VHF television by weakening the FM signals. This fight with the FCC, and later fights over royalties he felt were due him because of his original inventions, made Armstrong so despondent that he committed suicide. He never saw the proliferation of FM radio, nor the application of its multiplex capabilities.

A series of FCC freezes imposed upon AM radio during and after World War II (for both technological and economic reasons) served as a boon to FM. The AM market was saturated, and broadcasters who wanted to expand their share of airwave space, plus newcomers to the industry, found FM a logical place to invest. That fact, along with pressure prompted from television plus the simple matter that FM broadcast a much better sound than AM, encouraged rapid perfection of FM. It took a while to develop FM's technology and to lure people into purchasing FM receivers to augment their AM sets, but by mid-century the new broadcast band was starting to be recognized as the coming thing.

Broadcasting hardware became miniaturized along with receivers, and magnetic tape came into use. Just as 78 RPM records yielded to the polyvinyl chloride 45s, and to the long-playing 33s, the record and radio industries discovered the values of magnetic tape. Cassettes and eight-track tapes provided greater clarity; there was no needle noise and there were no grooves to wear out (the same factors that were to delight audiophiles in the mid-80s when compact disc recordings were developed). Used with automated equipment, tape greatly reduced the operating costs of broadcasting. Because large antennae and powerful wattage were not needed for local use, broadcasters discovered that they could build and operate FM stations for purely local consumption with minimal investment.

FM stations began to pop up in isolated and suburban communities by the early 1960s. Financially, FM stations had a rough time. Many of the FM stations that survived those early years were operated by owners of more profitable AM stations. The FM stations served the same function in broadcast that the weekly or peripheral daily newspaper did in the press. They made it possible to carry a quotient of purely local news that the larger metropolitan stations had neither the time nor the resources to cover. Further, they permitted the local advertiser an outlet to a purely local audience at an affordable price.

By the 1970s FM had begun to demonstrate its audience-delivering potential with a vengeance. In 1973, when AM held 65 percent of the radio audience, industry forecasts held that, given current growth rates, FM would overtake AM by 1985. Their estimates were way off, as FM grew much faster than expected. By 1978 FM had taken a larger share of the audience than AM, and by 1980 FM, despite having fewer stations on the air than AM, had captured 55 percent of the audience—outstripping AM listenership during all hours of the day except during the weekday morning drive time. By the end of 1985, when the number of commercial and noncommercial FM stations was nearly equal to the number of commercial AMs, the FM band had won the battle, drawing 71 percent of the listeners. And, because the one thousand new FM station frequencies that the FCC had approved back in 1983 were beginning to go on the air in 1986, there was no question about FM's viability.

By the late 1970s FM had a larger overall audience than AM.

FM was a key factor in the massive changes radio underwent after 1960. Portability, availability to audiences, developing musical clarity, and local emphasis began to develop a new kind of radio based on economical specialization in either music or local news, or both. In addition the FCC began to pay attention to FM. It authorized stations to broadcast stereo and required that FM stations originate their broadcasting at least half of the time rather than merely carry the AM programming where both AM and FM stations were under single ownership. As Table 6.1 shows, FM has increased exponentially since 1960.

Table 6.1 Growth of radio, 1922–1985

Year	Number of stations (commercial only) AM	FM	FM non-commercial	Total	Industry revenues (millions)	Estimated % U.S. homes with radios
1922	30	0	0	30	N.A.	0.2
1925	571	0	0	571	N.A.	10
1930	612	0	0	612	N.A.	46
1935	616	0	0	616	$ 79.6	67
1940	765	50*	0	815	147.1	81
1945	933	3*	0	936	299.3	89
1950	2086	733	48	2867	444.5	95
1955	2669	552	122	3343	453.4	96
1960	3456	678	162	4296	597.7	96
1965	4012	1270	255	5537	776.8	97
1970	4319	2184	396	6899	1077.4	98
1975	4432	2636	711	7779	1479.7	99
1980	4559	3155	1038	8752	3200.0	99
1985	4754	3716	1172	9642	6600.0	99

Source: *Broadcasting/Cablecasting Yearbook*, with permission.
*Experimental stations

Over the past three decades radio has branched out. Prime-time radio of the 1930s and 1940s featured dramas, comedies, and soap operas, which disappeared in the 1950s. For obvious reasons music came to dominate radio after television entered the marketplace, but not all radio programming became musical. A number of successful new formats have been developed, some of which have all the trappings of programming from radio's Golden Age. But they are different this time around. They are packaged and delivered in new ways, and serve new audiences.

Radio today offers a diversity of content matched only by that found in America's wide variety of magazines. Just as with magazines, radio specialized because of the pressure of television, pursuing smaller, more loyal audiences and the advertisers who needed to reach those audiences. Local stations tend to be supported by local merchants, but in a day of new technology and satellite-delivered programming, there has come a resurgence of regional and even national advertising in radio to blend with the regional and national programming. In a very real sense, radio has become an electric magazine, just as film has emerged as an electric book.

Radio drama has made a comeback in recent years, partly due to the nostalgia trend sweeping America. CBS resurrected "Mystery Theater" in 1974, complete with name stars and high quality sound effects. Its success caused other networks—especially the National Public Radio web—and local stations to dig into their vaults for mysteries and soaps from the Golden Age of radio, and to create some new offerings.

Contemporary Radio

Radio's content today is as diverse as that found in magazines.

News

As radio sought new ways to reach and hold audiences in the era of television, some purely local community news stations came into being. They have met with mixed responses from their audiences; only those in major metropolitan markets seem capable of attracting sufficient ratings and revenues to stay in business. Meanwhile, network radio news and major independents continue to serve the least populated areas of the country where newspapers are distributed only by mail, and where it is not economically feasible for television to call the audiences its own.

Some radio operators find it profitable to broadcast pure news on a thirty- or forty-minute cycle. Traditional, established AM stations are maintaining, if not increasing, their emphasis on news, talk, and information, but for FMs it is a different story.

The majority of radio executives contacted in a 1985 *Broadcasting* magazine survey said they were programming about the same amount of news and information that they did in 1981, before the FCC deregulated radio stations and permitted them to make their own decisions about how much public affairs coverage they wanted to carry. However, another study, by the National Radio Broadcasters Association, gives a different perspective. It found that between 1982 and 1984 the average length of newscasts for FM stations had decreased by 26 seconds (down to 4 minutes and 5 seconds), while AM newscasts increased by 18 seconds (up to 5 minutes and 48 seconds).[12]

Radio news may be important, but stations are giving it a decreased amount of emphasis.

Many AM stations say they consider local news reporting the cornerstone of their programming, and recognize the value of such programming to their older audiences. To some extent, however, they are covering less "hard" news of a substantive nature than in prior years, and offering more feature coverage. In fact, the majority of stations—AM and FM alike—are guilty of "*rip 'n read journalism*"—wherein a disc jockey or talk show personality tears the headline news from the station's wire, and, if time permits, gives it a quick preview before reading it aloud on the air. Occasionally the staff localize stories. More often, they go with whatever scripts and sound bites their news service provides them.

The trend is particularly noticeable among FM outlets, with their concentration on music and entertainment programming, and their assumptions that their audiences do not want their hard news brought to them by radio. (Surveys typically show that media audiences believe radio gives them the headlines, while television gives them news in depth. Newspaper executives are not particularly pleased with such survey findings.)

Today's musically oriented FM stations all but ignore hard news, offering their listeners "life-style" coverage and purely utilitarian insights into finances, recreation, personal health, and other hedonistic pursuits. For many FM stations, particularly those with contemporary hit or album-rock music formats, news—soft news—is an integral part of their morning drive-time shows, but they air little if any news the rest of the day. Major stories are treated in one or two sentences, with feature-oriented stories given thirty seconds or more.

Station owners, program directors, and news editors have not hit upon this formula by accident. Many stations have "topic-tested" their newscasts, conducting research on what kinds of news and information their listeners are most interested in hearing. As a result, the news meshes with the music formats, and if the most popular radio stations are attracting the 18- to 34-year-old audience, the newscasts and informational tidbits—plus, of course, commercials—are demographically and psychographically tested to mirror that audience's interests. More than ever before, the medium with which the average American adult spends twenty-five hours a week is being driven by marketplace considerations.[13]

This is not to say that one approach to news is right and the other wrong, although we cannot help but be concerned about a generation whose myopic media habits result in almost a conscious avoidance of hard news and an incessant pandering to ego and id. (As noted, newspaper people are particularly disturbed that radio audiences consider television, not newspapers, to be their prime source of substantive news.) However, even some radio executives are worried, and publicly hope that the relatively young audiences weaned on FM will eventually move over to the AM band where they will be exposed—even accidentally—to more news, talk, and information. To capture that audience, AM is starting to adapt the same technology that has made FM so attractive, programming in stereo and using compact disks and digital audio fidelity. Some AMs are experimenting with new formats that include comedy, game shows, children's programming, ethnic shows, and other shows particularly geared to mesh with their traditional information and music formats. It is an interesting reversal when the old borrows from the new, but so far the jury is still out.[14]

The future of radio news is anyone's guess, but a couple of predictions seem safe. Regardless of what it does in terms of trimming back its overall news coverage, radio can be expected to continue emphasizing news and information in the early morning and early evening drive-time hours. Surveys reveal it has a stranglehold on audiences from six to ten A.M., despite the increased attention television is paying to early morning news. A recent survey of 1,009 adult Americans revealed that 52 percent of them called radio their major source of news each morning, whereas only 30.9 percent listed television, and 12.8 percent newspapers.[15] Radio's strength in this time block is due, of course, to its portability and accessibility—it is still very difficult to drive to work while watching TV or reading a paper.

More and more radio stations are using computers in their newsrooms and are supplying their reporters with cellular telephones for field reporting. That should result in a greater amount of coverage of breaking and hard news. If FM stations continue to segment the market, they should begin to offer more newscasts. However, that will not necessarily mean an increase in (or development of) in-house news staffs, because much of that news will be coming from outside program suppliers.[16] Networks and external news sources are expected to continue feeding local stations news

Radio news is tending to be softer and softer, despite new technology.

and information segments via satellite, and the packaging is so slick and so well targeted to the audiences of a given station's format that in-house news productions will sound shoddy in comparison.

Talk and Information

Emerging from the social consciousness of the sixties and early seventies was the use of radio as a community forum. What started out as call-in shows airing debates over racism, sexism, and the military-industrial complex has turned into a cacophony of dialogues and diatribes, community pulses and ego trips—something for everyone. Psychologists, physicians, ministers, politicians, anarchists, and T. C. Pits (The Celebrated Person In The Street) have all had access to the microphone, and have not hesitated to use it.

Radio psychology has swept the country. Ratings are soaring for call-in shows of all sorts—sex and marriage counseling, family counseling, and religious counseling. Radio personalities, who may or may not be certified counselors or clinicians, offer listeners a chance to expose their inner lives and opinions, their fears and insecurities.

Radio offers an open mike for views from all sides, including those crying out for advice.

Critics maintain the programs are questionably ethical and can be misleading, because even the most qualified talk show hosts are forced to offer rushed and offhanded advice while operating under radio's time and commercial constraints. Complex questions are given simplistic answers, and counselors make broad generalizations without recognizing the uniqueness of each caller's problem.

On the other hand, supporters of radio's healers insist that the short interaction offers preventative mental health care by informing listeners that there are many social support systems available. One research study indicated that callers to the top-rated Dr. Judith Kuriansky show on WABC in New York took a personal liking to the hostess, and especially enjoyed listening to other people's problems. In general, they felt less depressed and less lonely after talking on the air with Dr. Kuriansky.[17]

From one perspective—probably an elitist one—there seems to be a bit of the "eavesdropper" in all this, as people expose themselves, their inner lives, and their opinions to one another electronically and anonymously. On the other hand, however, many find such programs a badly needed outlet and sympathetic ear in an increasingly depersonalized world. For example, consider the successful experiment conducted recently by KPFA-FM in Berkeley, California. For a half hour each Monday and Thursday, KPFA set up a microphone on a downtown street corner to give people who walked by free, unedited air time to discuss whatever happened to be on their minds. The open access show, initiated prior to the 1984 elections on a trial basis, drew a predictably varied and sometimes zany collection of ideas, proposals, and diatribes. However, no real wackos or foul-mouthed pranksters used the microphone, according to the station's managers, who said they were proud to offer the first United States anarchist radio show. "You don't have to be famous any more to get on the radio," one delighted participant told reporters.[18]

At the opposite end of the spectrum from anarchist radio has been the carefully orchestrated use of radio as a political tool by President Ronald Reagan. With his five-minute Saturday afternoon broadcasts, Reagan resurrected radio as a major communications instrument whose like had not been heard since the fireside chats of Franklin Delano Roosevelt. The broadcasts, begun in April of 1982, afford the President an opportunity to reach millions of Americans directly, without having to subject himself or his remarks to reporters' nagging questions or heavy editing. Audiences for the presidential radio addresses may be relatively small, but because Saturday is such a slow news day, the President's messages get major play in most news media all weekend. It was through radio that Reagan launched his professional career in the 1930s—as a sports announcer on WHO-AM in Des Moines, Iowa.[19]

There are parameters to radio's talk format that are forever being stretched. In New York City one radio station has catered to homosexuals; in Long Beach, California, and elsewhere, other hosts developed a large following by discussing their intimate sex lives with their callers. At WKQK in Chicago disc jockey Robert Murphy helped pump up his station's ratings by entertaining listeners with off-color takeoffs like his spoof of television soap operas called "The Young and the Impotent."[20] More widely known, perhaps, are the frank radio sex therapists such as Dr. Ruth Westheimer in New York and Phyllis Levy in Chicago.

The increasingly candid, risqué talk shows of the early 1970s were put to a stop by the combined efforts of the FCC and NAB. Such programming, when found today, has been toned down, and as often as not is found away from the mainstream stations, on peripheral FM outlets. Meanwhile, the highly popular nationally syndicated nightly talk show hosted by Larry King on Mutual Broadcasting reminds us that radio can be an ideal medium for exploring the broadcast range of public issues.

Pacifica The best-known and most controversial of the "peripheral FM outlets" is the Pacifica Foundation, a nonprofit network of five stations across the country. Founded in 1946, the foundation has been committed to unpopular ideas and minority issues. Its stations have gotten into difficulty for dramatizing a nuclear attack on the United States (shades of "War of the Worlds"), offering sex counseling over the air, and constantly criticizing conservative causes while openly advocating ultra-left, feminist, and minority views. The Pacifica stations can afford to be anti-establishment because most of their revenue comes directly from their listeners, who voluntarily pay thirty dollars a year to "belong" to one of the affiliates and receive a monthly program guide. Some funding, however, does come from federal grants for programming, administered by the Corporation for Public Broadcasting, which gives rise to criticisms about the stations' political orientations.

Pacifica radio stations provide a forum for controversy.

Blacks are more avid radio listeners than non-blacks, but their loyalty has not paid off in increased advertising revenues.

Pacifica officials insist that their programming demonstrates a dedication to free speech. "If you think it's off the wall or I think it's off the wall, that doesn't mean it shouldn't have a forum. We believe absolutely in the First Amendment," argues Sharon Maeda, executive director of Pacifica.[21]

The radio band seems broad enough to accommodate such programming in an era of increasing permissiveness and one which finds the federal government relaxing its control over the public airwaves. An argument heard frequently of late is that the radio spectrum is broad enough to permit whatever diversity of programming—short of that which is totally obscene, libelous, or which endangers national security—the marketplace will bear. As that marketplace reaches its own natural level of saturation, the types of programming heard over the Pacifica stations are settling in amidst mainstream musical formats and innocuous talk.

Black Radio Minority interests have been addressed by many radio stations and networks, but not without some difficulties. The National Black Network, founded in 1975 with seventy-five affiliates, successfully provided ethnic news and entertainment programming to millions of Americans. More recently the Sheridan network was renting transponder space on the *Telstar 301* satellite to reach more than 100 black-oriented stations. Indeed, between 300 and 400 stations that program specifically for black audiences have gained strongholds throughout the nation, particularly in metropolitan centers. Such stations are consistently at or near the top of the ratings in New York, Philadelphia, Atlanta, and Houston.

Research indicates that blacks tend to listen to radio much more than the general public—about thirty hours per week for blacks versus twenty-five hours for all audiences.[22] Unfortunately, those ratings have not necessarily equated to commercial success. Advertisers have been somewhat reluctant to buy time on black stations, particularly to advertise expensive items, fearing that the potential audience is likely to consist of many inner-city unemployed and underemployed who may be less apt to support the merchants. To counteract this economic crisis, and to reach affluent blacks as well as whites, many black stations have adopted an "urban contemporary" format, playing a mix of mainstream black and non-black music.[23]

Black radio stations may not be profitable, but they have attracted avid audiences.

Radio's uniqueness has been best explored by National Public Radio (NPR), which first went on the air in May, 1971. By the mid-1980s NPR was a well-respected network made up of 304 noncommercial stations, overseen by the Corporation for Public Broadcasting (CPB). It brings award-winning, thorough newscasts and cultural programming to some ten million listeners each week. Though still miniscule when compared with the reach of commercial stations, public radio has made its mark in America's media mix.

Public Radio

NPR stations can secure grants for operating expenses and can broadcast according to the guidelines established by the CPB. However, the level of federal support has been erratic, contingent upon budgetary moods of Congress and the White House. Under the Reagan administration, with its policies advocating private and business support of broadcasting, CPB and NPR have not fared well. In late 1984 President Reagan vetoed the 1987–89 public broadcasting bill, calling the funding excessive and "incompatible with the clear and urgent need to reduce federal spending."[24] A year later, he and Congress were again at loggerheads, with Reagan seeking the eventual elimination altogether of federal support for public broadcasting by the year 1992.[25]

Public radio may be radio at its finest, but it is receiving very little taxpayer support.

One result of Washington's irregularities in supporting public broadcasting has been an increased flurry of local fund-raising activities, including on-air appeals directly to listeners and a great deal of scrounging around to local corporations who can get a tax write-off for supporting a non-profit institution. This means that stations in isolated university towns are less likely than ones in metropolitan areas to feel financially secure from year to year. However, the financial game can be won, as evidenced by the Minnesota Public Radio (MPR) group of nine stations serving 400,000 listeners in Minnesota and parts of North and South Dakota, Iowa, Wisconsin, and Michigan. MPR has become the nation's most successful public radio system by securing major funding from large corporations and setting up its own communications conglomerate in St. Paul, renting out studio space to commercial clients, producing advertising and printing, mail-order sales, and other entrepreneurial enterprises.[26] Operating from this base, MPR has been able to produce and distribute to a national audience the expensive and high-quality programs that public broadcasting aficionados love—especially the variety show "Prairie Home Companion" and the musical "St. Paul Sunday Morning."

The Minnesota Public Radio group is one of several that serve NPR stations. MPR is a member of the largest programming services group, American Public Radio Associates (APR). APR originated within NPR in 1982, but developed into somewhat of a rival to the parent organization. APR was formed by five stations that were among the largest and most aggressive of the stations that "own" NPR (WNYC of New York, WGUC of Cincinnati, KQED of San Francisco, KUSC of Los Angeles, and Minnesota Public Radio). It produces no programs of its own, but sells NPR member stations numerous new programs and series each month, primarily performances or studies of classical music, but also some news and public affairs shows. At last check, its 261 affiliates were paying a fee as high as $1,500 annually for access to APR's forty-two hours a week of programming.[27] Its most popular show has been "Prairie Home Companion," produced by the Minnesota Public Radio group but financed by APR's membership.

More than two million Americans became accustomed to staying home each Saturday night to listen to the whimsical "Prairie Home Companion," brought to them from mythical Lake Wobegon, Minnesota, "where the women are strong, the men are good-looking, and all the children are above average." Lake Wobegon, "the town that time forgot, and decades cannot improve," originated in July, 1974 from the fertile mind of Garrison Keillor. Interspersed with Keillor's down-home wit was bluegrass, folk (including Norwegian and Russian), gospel, and country fiddle music, plus "news" from the Sidetrack Tap and Ralph's Pretty Good Grocery. The commercials were tongue-in-cheek, often "brought to you by Powdermilk Biscuits, baked by bachelor Norwegian farmers, to give shy persons the strength to go out and do what needs to be done."[28] Keillor's homilies and wit brought him many admirers, enough to put his book on *Lake Wobegon Days* atop the best-seller lists in 1985. Unfortunately, by mid-1987 Keillor announced he had grown tired of producing the weekly miracle, and PBS resorted to the tried-and-true reruns.

"Prairie Home Companion," according to some observers, showed what radio can do best. Others point out the fact that "PHC" was only the second most popular program carried by the NPR web of 304 stations. In first place, reaching five million discriminating listeners each week, is the news and public affairs programming, "All Things Considered" and "Morning Edition." These daily programs come from NPR's Washington, D.C. headquarters, and are carried live by satellite to affiliates throughout

The producers, editors, technical directors, and administrative staff of National Public Radio's "All Things Considered."

the country. By organizing a news staff in the national capital and offering worldwide news coverage and analysis, NPR has gained a reputation as "*the* national radio news operation," dedicated to imaginative, enterprise reporting.[29]

Public radio seems to have established itself with its listeners, but its future remains questionable. The big problem remains one of funding. As APR President William Kling argues:

This country has never supported public broadcasting to the extent most other countries fund their systems. There has never been a real commitment to public radio here. Our culture needs the civilizing influence of non-commercial broadcasting just as it needs the civilizing influence of public libraries and art museums. Of course, there are taxpayers who never take advantage of the facilities available, but what kind of a country would we be without them?[30]

One solution to the funding problem has been proposed by Colorado Congressman Timothy Wirth, who wants to charge commercial broadcasters a "usage fee" or tax for using the public airwaves, and to turn that tax over to non-commercial broadcasters. Putting such funds into a tamperproof public trust would relieve public broadcasters of the stresses caused by federal administrative vagaries, and allow noncommercial radio the chance to concentrate on producing and distributing quality programming. Needless to say, the proposal has not been particularly well received by all commercial broadcasters, and the battles over support for public radio are likely to continue for some time.

New Formats

Radio's search for audiences in recent years has led to a variety of innovations, an example of which is the experiment of ABC radio. In the 1960s ABC trailed the other two major networks, NBC and CBS, always seeming to come in a poor third. While NBC and CBS could afford to operate their radio networks out of television profits, profits from television at ABC were not correspondingly high. Consequently, ABC was forced to reconsider its entire radio network operation. Whereas previously it had affiliation agreements with a single radio station in most of the nation's major markets, it decided to scrap this arrangement in the mid-1960s and go to multiple affiliates. The truth was that ABC's affiliates, as well as those of NBC and CBS, were rated well behind the major independents in almost every market. The independents had found the new sound, or they had moved to the new format (time, news, and temperature), or to a variety of other highly specialized forms of programming.

Instead of offering a potpourri of national programming to a single station in a major market, ABC organized four subnetworks specializing only in the kinds of news, sports, or music that affiliates would seek: The American Contemporary Radio Network, The American Information Radio Network, The American Entertainment Radio Network, and The American FM Radio Network. It then made affiliation agreements with the principal independents for each of these specialties; in effect, there were four ABC affiliate opportunities instead of just one. This was a profitable plan all the way around. ABC had outlets for a wide range of national advertising by catering to specific audiences, which made it an easier and more economical package for advertisers of hi-fi equipment, for example, or aftershave. ABC managed to quadruple its advertising billings between 1968 and 1975 because of this arrangement.

In the 1980s, ABC once again saw new opportunities to specialize, and offered three new networks to fill the void: ABC Talkradio, ABC Direction Network, and ABC Rock Radio Network. Taking a cue from ABC, other networks have also begun to split their offerings, using satellites to reach their affiliates. CBS and NBC have both branched out, the former offering RadioRadio, the latter The Source and Talknet. Among the many other networking examples are the Associated Press and United Press International radio networks, CNN Radio Network, Music Country Network, Mutual Broadcasting, National Public Radio, RKO Radio Networks I and II, Satellite Music Network, Sheridan Broadcasting, Transtar Radio Networks, and the Wall Street Journal Report. As can be inferred from the networks' names, some are offering specialized musical formats, while others are concentrating on news and information programming.

Once the networks began using satellites to deliver programming to affiliated stations, national networking for radio reemerged. It was not exactly like the Golden Ages of radio, in the 1930s and 1940s, when popular shows reached an enormous cross section of Americans. Now, audiences

ABC's successful experiment with "subnetworks" gave other programmers some good ideas.

are targeted according to life-style and psychographic variables. Simultaneous multi-channel transmissions are being offered to stations, giving them greater flexibility than ever before in choosing what to air. The casual observer may conclude that the only thing being aired is music, but the packaging is much more subtle than that. NBC Radio Entertainment is exploring a variety of program areas outside of the contemporary rock-oriented programming now offered through its "The Source." Mutual Broadcasting has added more live country programming. ABC has been distributing a wide variety of popular weekly shows: "American Top 40," "American Country Countdown," "Silver Eagle Cross-Country Music Show," "Supergroups," and "The King Biscuit Flower Hour." CBS's youth-oriented network, RadioRadio, delivers music and entertainment news magazine programs ("Entertainment Coast to Coast") to contemporary music stations. The United Stations network offers "The American Music Magazine" and "Solid Gold Country." And, for the first time in thirty years, sports fans can once again get a major league baseball game of the week.[31]

Syndication and Automation Today approximately one half of America's commercial radio stations have turned to consultant companies that will help them package their programming to assure financial success. The consultants design and carry out survey research to discover local tastes, and broad marketing research to find out what works elsewhere across the nation. Then, armed with reams of data, they propose complete formats for their client stations—everything from suggesting what music to play and when to play it, what type of talk or chatter to air, what advertisers to bring on board, to what types of promotional campaigns to use to "sell" the new format in the competitive marketplace.

Once a station decides upon a format, it turns to one of the many syndication services for a complete package, centered usually around a given musical "sound." As seen earlier, these sounds are often delivered and picked up live via satellite. However, they are also available over audio tape and cartridges, delivered through the mails or by courier. At automated stations, the tapes and cartridges are pre-programmed by computer (with, of course, human guidance) to play specific music, talk, and commercials at specific times of the day. Ideally, the automated stations retain a unique identity—if nothing else, they have their own station identification and jingles, plus a hometown announcer doing news and weather.[32]

Without a doubt, radio has become more sophisticated and more reliant upon the computer. An example of the extremes to which new communications technology is being put is the sophisticated computer software now available to disc jockeys. At the push of a button, it tells them which of twenty-five hundred song titles should be played to fit into any one of several broad musical formats and to satisfy dozens of specific audience characteristics.[33] If the current trend to automation continues at its present

Radio stations frequently rely upon consultants and syndication services to give them a winning sound.

pace, much of the gamble and uniqueness of radio—and many of the personnel—may be lost forever. From the station owners' point of view, that may not be such a bad thing. From another perspective, however—that of the audiences and those who are looking for careers in radio—it may be something else altogether.

Pay Radio So far we have been discussing radio as a vehicle for *broadcasting* to the general public. Another aspect of this business, however, is rapidly coming to the fore. Known as *narrowcasting,* it utilizes the airwaves to deliver specialized programming to highly homogeneous and self-selective audiences. In its most advanced form, narrowcasting involves the use of *sidebands*— signals on a frequency that every FM broadcast creates but that cannot be received by a regular radio. The only people able to receive those signals are those who have bought, leased, or have been given special receivers tuned to those sidebands. The entire system is called pay radio, because audiences are charged a fee by the programming service. Either they purchase the receiver, or they pay a monthly fee to use it, or both.

Pay radio has become an integral part of the communications system for many professionals such as stockbrokers or executives who want up-to-the-second reports on stock or commodity markets, for physicians, or for others with specialized interests and the money to cater to them. It has also been used somewhat like cable television. In Austin, Texas, for instance, listeners pay $1.50 a month to receive classical music station WFMT, which originates in Chicago but is "boosted" to Texas.

Some narrowcasting services, such as Dow Alert, a stream of specialized financial news, are customer programmable. Subscribers can punch certain codes into their receivers and get only the stock market reports they want. In fact, at least one sophisticated system permits the subscribers to receive data in written as well as oral form—once they hear a report they are especially interested in, they can call for a hard copy of it on their computer printer.

Perhaps the most specialized narrowcasting now in effect is the Physicians Radio Network. It reaches eighty-thousand doctors with the latest in medical information and advice. Unlike Dow Alert and others, it does not rely upon sales of receivers or monthly fees for its support, because so many drug companies and other advertisers want to reach this highly select audience that it is advertiser supported. It is not exactly "pay radio," but it shows the logical extension of the pay radio concept.

The FCC recently loosened its control over sideband communication, allowing broadcasters to use narrowcasting for practically any commercial purpose they desire. Pay radio is expected to boom; not just sound, but anything that can be carried over the phone lines will be narrowcast. That includes point-multipoint digital data transmission, which intrigues computer software developers, financial institutions, legislators, and many others.

Pay radio, utilizing the process of *narrowcasting,* is growing in popularity.

What until recently has been the province of a self-selected few wealthy individuals with an insatiable thirst for instant information and diversion may well become a routine, portable, and inexpensive way for any of us to pursue our special interests.[34]

Because it is so commercially oriented, the radio industry has always tried to keep in touch with its audiences. Its primary feedback mechanism is the Arbitron rating service, which continually samples members of the radio audience. Diaries filled out by Arbitron sample audiences indicate how much time they are spending with radio, what stations they listen to, and where— at home, work, play, or in the car—they are listening. By sampling a representative group of listeners, Arbitron can compile a reasonably accurate profile of both local and national audience trends. Advertisers and program directors are especially interested in the Arbitron results, and advertising rates and programming formats fluctuate widely depending upon the ratings.

Radio ratings services tell who is listening to what. The motive, of course, is economic.

A recent edition of Arbitron's *Radio Today* offers an interesting profile of the American radio audience.[35] It shows that the aural medium is being used more today than at any other time in recent memory. The average listener at mid-decade was tuned in for 25 hours a week, compared with 20 hours a week in 1979. The demographic groups who listen the most are females over age 65 and black Americans, who average 30 hours weekly. Among men, the 18- to 34-year-olds are the heaviest users of radio, at 27 hours a week.

One fourth of all radio audiences spread their listening to all areas— at home, in the car, and at "some other place." Forty-one percent listen both at home and in the car, seven percent listen only in their cars, and 18 percent listen only at home.

The radio audience is largest in the morning hours, from 7 to 10 A.M., when 26 percent of all potential listeners are tuned in. In second place is the Saturday 10 A.M. to 3 P.M. period, which draws 25.6 percent of the available audience. The youngest listeners—median age 29—tune in on Saturday nights, from 7 P.M. until midnight. Weekday morning audiences are older, with 35 as the median age. The oldest audiences listen on weekend mornings: on Saturdays the median age is 39, and on Sunday it is 40.

Even during odd hours radio reaches a good-sized audience. From midnight to 4 A.M., for instance, radio is still heard by 25 percent of the population. The overnight audience is young, with a median age of 30 on weekends.

Arbitron reports that almost all full-time working women listen to radio. Those aged 18 to 34 listen 26 hours a week, compared to 22 hours for their counterparts who are not employed outside the home. Another faithful group is the black radio audience. Radio reaches slightly more of them (97 percent in any given week) than the total listening audience (96 percent).

The Disc Jockey and Music

Music has been an integral part of radio from the very beginning. Today, it's the prime ingredient.

The term *disc jockey* gained notoriety in the 1950s, but the concept of a broadcaster playing recorded music is as old as radio itself. Reginald Fessenden played Christmas carols in radio's first broadcast, back in 1906. To promote his audion tubes, Lee De Forest broadcast music from the Eiffel Tower in 1908, and Frank Conrad established his reputation around Pittsburgh in 1920 when audiences responded to his classical music selections.

Record playing did not dominate the airwaves during the 1920s and 1930s, however. Live music was the mainstay of radio programming during this period, partly because the playing of recorded music was fraught with difficulties. Until 1940, musicians, who feared that playing their records on radio would hurt record sales and violate their exclusive network contracts, intimidated stations into avoiding much recorded music. But a 1940 Court of Appeals ruling held that broadcasters could do what they wished with records, and the floodgates were opened wide.

Al Jarvis of KFWB in Los Angeles probably initiated the contemporary concept of the record-spinning personality with his "World's Largest Make Believe Ballroom" back in 1932. Shortly thereafter Martin Block of New York's WNEW parlayed Jarvis's idea into a highly marketable package of music, advertising, and chatter. Block began by playing records during lulls in the famous broadcasts of the Bruno Hauptmann trial (the Lindbergh baby kidnapping trial) in 1935, and soon polished his routine so sponsors were waiting in line to reach the four million listeners Block was delivering.

After World War II, when drive time became prime time, the DJ's personality became a crucial ingredient in radio programming. As popular music fragmented into well-defined formats, DJs were needed who could program their own shows, choosing records in keeping with audience expectations.

It was at this point that radio started to become an extension of the recording industry. In the 1950s record manufacturers found that radio provided ready-made exposure for their wares. Production companies began sending thousands of copies of their new singles and albums to America's radio stations, hoping for the all-important air play that would give them exposure to the buying public. The practice continues today. However, because the typical radio station receives some two hundred or so unsolicited records a week, chances are slim that a given record will even be listened to by the radio personnel, let alone granted air play. With all the pressures for conformity brought on by syndication and automated programming, the radio system is biased in favor of established recording artists and recurring themes. This means that mainstream radio, like most big business, is inherently conservative. In fact, we might say that the prime objective of the radio industry is not necessarily to choose programming that might succeed, but to choose programming that cannot fail.

If, by some lucky chance, a sympathetic program director or DJ and a few enthusiastic fans cause a record to be heard a few times, the record's popularity will be reflected in *charts* posted each week in *Billboard* magazine. The bible of the recording industry, *Billboard's* weekly charts indicate where a record's sales place it in comparison with all other popular hits, whether it is moving up or down in the standings, and how long it has appeared on the list. At the local level, individual radio stations publish their own *playlists,* reflecting what is hot in their particular market area. (You will find your local radio stations' playlists posted near the record racks at major distributors, which is still another feedback and marketing force in the media mix.)

Radio stations that hope to keep an audience and make a profit follow the charts and play the hits. Over and over. Period. So success breeds success, and the odds are stacked in favor of the tried and true musical formulas and the popular artists. Of course, there is a flip side to this promotional equation. Once the listeners have gone out and bought the hit

Radio stations follow the charts and play the hits. Over and over.

tunes, it does not matter how often radio replays them—nobody is going out to buy another copy of the record or tape because of hearing it again and again. In a way, conservative radio play breeds a sort of economic stagnation, especially when there are a limited number of musical formats and pop artists being aired.

Occasionally a radio station struggling near the bottom of the ratings will experiment with new musical sounds. Sometimes it works. In 1979, debt-ridden KROQ in Pasadena, California, deciding it had nothing to lose, broke tradition and started playing some of the tunes its personnel liked but which were not on anyone else's playlists. The move was fortuitous. For a couple of years the popular market had been wallowing in repetitive and unexciting sounds, so, throwing caution to the wind, KROQ played what it thought was exciting: unknown Duran Duran, the rockabilly Stray Cats, and the like. Very soon the change got noticed. Records played on KROQ started selling, and the station earned a number one rating in the large and brutally competitive Los Angeles market. KROQ's happy program director said he had correctly sensed that "there was a big audience out there looking for something of their own. Music just didn't have the excitement of the early '60s."[36] As we survey the entire industry, however, we note that the instances of such successful deviance from the norm are rare, and that most economic and cultural forces militate against individualism.

We have pointed out that once record manufacturers found that radio provided ready-made exposure for their wares, they began providing the nation's DJs with records at no cost. This reduced the expenses of broadcasting, but created a system ripe for abuse. After all, in the 1950s, disc jockeys had an unusual gatekeeping capacity: they programmed their own shows, deciding how to fill their slots with a mix of music and talk. Recognizing the influence DJs had on a record's success, record distributors in many instances paid popular jocks to air the records. *Payola* became a household word in the late 1950s once federal investigations revealed that many DJs were earning more from their record payments than from their broadcasting salaries. In addition to outright cash bribes, payola included free trips and other special favors.

The FCC investigated the payola scandals, and temporarily curbed the practice. Today payola is no longer the key factor in the multi-billion dollar recording industry, although investigations in the late 1970s and again in the mid-1980s revealed that some radio personnel were receiving drugs and other freebies as incentives to air records. Broadcast staffs are now required by the FCC to sign a pledge swearing that they have not accepted any improper bribes—payola, drug-ola, what have you—but insiders insist that segments of the recording industry still count on such things as part of their promotional and marketing efforts.

An important offshoot of the 1950s payola scandal was the routinization of radio and a more sophisticated marriage of it to recorded music. Program directors, musical directors, entire radio staffs, outside consultants, and syndicated and/or automated program suppliers have by and

Payola has influenced the record-playing gatekeepers.

large taken the musical gatekeeping decisions out of the hands of the DJs. The competitive market, particularly in the "Top 40" or in "contemporary music" stations, has meant less likelihood of individuality and innovation. Playlists and charts from *Billboard* magazine tell stations what is working elsewhere and, by extension, what they should be playing.

In today's competitive radio marketplace, the search for audiences allows little room for experimentation with marginal music. The crucial factor in the radio-music alliance is audience ratings, because a drop in a metropolitan station's ratings of only one point (or about 15,000 listeners in a city the size of Washington, D.C.) would cost the station anywhere from half a million to a million dollars in ad revenues in a given year.[37] For that simple reason alone, the promotion and marketing efforts of the recording companies continue to have an enormous impact upon the radio industry.

The Future of Radio

Of all the media institutions we treat in this book, the broadcast media are the most risky to predict. We have already noted that the history of radio has been one of randomness—random inventions coalescing with random social and economic trends, and seemingly random governmental regulations. We will learn that the same is true of the recording, television, and film industries. And given that background, it is especially difficult to forecast the results of new technologies, changing demographics, economics, and federal controls. Nevertheless, the past few years have shown us some trends that will no doubt affect the future of radio.

Radio's obvious trend toward localization and service to specialized interests is reflected in the actions of the FCC and existing programmers. Recent congressional moves to deregulate radio, allowing it to reach its own level in the competitive marketplace, indicate some acceptance of the idea that radio is a localized medium. But such an idea must be questioned, for two basic reasons. First, most local stations attempt to be clones of the most successful parent station of their genre, whether they are all-news, all-talk, Top 40, or what have you. As long as playlists are restricted to those sounds selling elsewhere, and as long as advertisers seek the largest audience with appropriate demographics, the tendency toward imitation and sameness will continue, no matter how much we claim to admire radio's sense of individuality and specialization.

A second reason to question radio's localization is the phenomenal growth of networking and syndicated or automated services. In a sense, radio has become an infant industry again, discovering a world founded on satellite technology's multi-channel capacity. Centrally located networks have begun to provide advertising, news, talk, and special features to hundreds of affiliates and their well-defined demographic targets across the nation. At the same time, radio stations' reliance upon consultants who tell them how to capture and hold their target audiences bodes ill for individuality in programming formats. And, much as they would like to claim that automated radio is individualized radio, the ear tells us differently; there is much of a muchness across the nation.

Is noncommercial—or perhaps even pay—radio the answer? Some feel that the solution to the problem is for radio to be supported by taxes or subsidized by its listeners or by an institution. They argue that by divorcing such stations from commercial interests, diverse audience needs will be met. As an example, they point to the National Public Radio and its variety of programs, but they ignore the surveys showing that only a small percentage of Americans have been exposed to NPR. NPR, of course, has little intention of being a mass medium. Appeal to the mass, it argues, means ignoring the needs of audience subgroups. So long as it relies upon its member stations for program materials, it will maintain its unique appeal. But once again, we note that reliance upon syndicated services and satellite transmission of a fairly full programming day points toward a certain homogenization of appeal.

Technologically, the future of radio may lie not in the airwaves and their limited frequencies but in broad-band cable with its multifaceted capacity for two-way electronic transmission. We have already alluded to this in our consideration of print media's future. The place radio will have in such a system is open to some question, because the visual dimension of cable is the one most frequently considered. However, we can already see enormous growth potential for "narrowcasting" of programming and information of a financial or technical nature. Speculations are that such programming will come to us over several channels at once. Quadraphonic, or multiphonic, reproduction of sound will be linked to variable lighting, to change the "mood" of the room as the tempo of music changes.

Stereophonic broadcasting over the AM band has finally been refined, and AM broadcasters are finding themselves competing for the musically sophisticated audiences that were taken away from them over the past decade by FM. How this will change programming over the two bands is open to speculation.

Summary

Radio had its start before the turn of the century, but public broadcasting did not begin until 1916. The pioneers in broadcasting expected the primary profit from radio to come from the sale of receivers. However, during the 1920s companies began buying air time and sponsoring programs. It was also during this period that the first of the major networks were created, when RCA bought WEAF from ATT and organized two NBC networks. Soon after, William Paley started CBS.

In the 1930s and 1940s radio was the prime mass medium in America. Network programming included comedies, mysteries, westerns, soap operas, live big-band music, and children's programs. After a protracted fight among broadcasters, publishers, and wire services, radio news became established and expanded dramatically during World War II.

During this last third of the century, television has become the preeminent mass medium, while radio has become a supplemental medium in league to a large degree with the record industry. In adapting to a changing environment, radio has grown to several times its previous size in the number of stations it airs and the number of sets it reaches. Its profitability has been greatly increased.

In an attempt to appeal to audience interests, radio has become specialized; each station specializes in a "sound," in news or sports, in appeal to ethnic minorities, or in talk shows—but especially in music. The medium has found new audiences for different kinds of music in highly fragmented sections of society. The growth of FM radio attests to this segmentation. Magnetic tape and automation have permitted FM stations to be operated at minimal cost, and overall, more Americans listen to FM stations than they do to AM stations. In a real sense, then, radio is an electric magazine, just as film is an electric book.

Notes

1. "Jankowski foresees shakeout in radio," *Broadcasting,* 5 November 1984, pp. 36–37.
2. Frank L. Kahn, ed., *Documents of American Broadcasting,* 4th ed. (Englewood Cliffs, N.J.: Prentice-Hall, 1984), pp. 24–25.
3. Christopher Sterling, "Television and Radio Broadcasting," in Benjamin M. Compaine, ed., *Who Owns the Media? Concentration of Ownership in the Mass Communications Industry* (New York: Harmony Books, 1979), p. 63.
4. Ibid.
5. Alfred N. Goldsmith and Austin C. Lescarboura, *This Thing Called Broadcasting* (New York: Henry Holt, 1930), p. 279.
6. Erik Barnouw, *A Tower in Babel: A History of Broadcasting in the United States to 1933* (New York: Oxford University Press, 1966).
7. Eugene S. Foster, *Understanding Broadcasting* (Reading, Mass.: Addison-Wesley, 1978), p. 61.
8. Melvin L. De Fleur and Sandra Ball-Rokeach, *Theories of Mass Communication,* 4th ed. (New York: Longman, 1982), p. 90.
9. Hadley Cantril, *The Invasion From Mars: A Study in the Psychology of Panic* (Princeton, N.J.: Princeton University Press, 1940), and Shearon Lowery and Melvin L. DeFleur, *Milestones in Mass Communication Research* (New York: Longman, 1983), pp. 58–84.
10. Sammy R. Danna, "The Press-Radio War," *Freedom of Information Center Report No. 213* (Columbia, Mo.: University of Missouri School of Journalism, December, 1968).
11. Ernest D. Rose, "How the U.S. Heard About Pearl Harbor," *Journal of Broadcasting,* vol. 5, no. 4 (Fall 1961).

12. "News mixed for radio," *Broadcasting,* 16 December 1985, pp. 72, 76.
13. Ibid, p. 76.
14. Bill Stakelin, "Radio at 60: 'It's still a fun business,' " *Variety,* 8 January 1986, pp. 169, 196, and "Beating the bushes for AM resurgence," *Broadcasting,* 13 January 1986, p. 63.
15. "News mixed for radio," p. 76.
16. "Radio," *Broadcasting,* 31 December 1984, p. 91.
17. "Radio psychologists: Misleading or helpful?" Logan (Utah) *Herald-Journal,* 3 January 1984, p. 11.
18. "First U.S. anarchist radio show," Logan (Utah) *Herald-Journal,* 31 October 1984, p. 16.
19. "Reagan bringing radio back as political tool," *Broadcasting,* 29 October 1984, p. 62.
20. Kim Foltz, "Radio's Wacky Road to Profit," *Newsweek,* 25 March 1985, pp. 70-71.
21. Marc Gunther, "Pacifica: Radio's Outlet for the Outrageous," *Washington Journalism Review,* May 1983, p. 44.
22. "How the U.S. Uses Radio Today," *Beyond The Ratings,* vol. 7, no. 5, May 1984.
23. Marc Gunther and Noel Gunther, "Black Radio: Playing for the Big Time," *Washington Journalism Review,* October 1983, pp. 36-37.
24. Robert Rothman, "Public broadcasting bill shot down twice," *Washington Times,* 2 November 1984.
25. "White House, Congress at odds over public broadcasting funding," *Broadcasting,* 23 December 1985, p. 57.
26. Dennis Holder, "Mixing Public Radio With Private Enterprise," *Washington Journalism Review,* June 1984, pp. 43-47.
27. Ibid., and Eric Scigliano, "Public Radio Airs a Feud," *Washington Journalism Review,* December 1982, pp. 47-49.
28. Scigliano, "Public Radio Airs a Feud" and Richard Stengel, "The Sound of Quality," *Time,* 5 November 1984, p. 85.
29. John Wicklein, "National news: public TV and the NPR example," *Columbia Journalism Review,* January/February, 1986, p. 32.
30. Holder, "Mixing Public Radio With Private Enterprise," p. 47.
31. "Radio," *Broadcasting,* 31 December 1984, p. 92.
32. Linda Busby and Donald Parker, *The Art and Science of Radio* (Boston: Allyn and Bacon, Inc.), 1984, pp. 24-25, 148-150.
33. Harriet C. Johnson, "Radio refuses to roll over and die," *USA Today,* 10 October 1984, pp. 1-2.
34. Marc Gunther and Noel Gunther, "And now a word from pay radio," *Washington Journalism Review,* July/August 1983, pp. 64-65.
35. "How the U.S. uses radio today," *Beyond the Ratings,* vol. 7, no. 1, May 1984.
36. J. D. Reed, "New rock on a red-hot roll," *Time,* 18 July 1983, pp. 60-61.
37. "Recording industry: A 4-billion dollar hit," *U.S. News & World Report,* 30 April 1979, pp. 68-70.

Recordings

7

Introduction

When we use the term "recording," we mean it in its broad, generic sense. It includes not only the earliest cylinders and discs, but modern plastic records, reel-to-reel tapes, cartridges, cassettes, laser audiodiscs, and even videotaped music productions. We will introduce all of these terms in this chapter, and will have more to say about some of the most recent technology later, in chapter 9.

Today the recording industry is a multi-billion dollar a year enterprise, and it is in a constant state of flux. Music media have married. Popular music, as we have already seen, is a necessary ingredient in contemporary radio. Record companies have learned the necessity of producing a video version of their record if they expect to reach today's buying public. Live concerts augment record sales for many groups. Promotion of music has reached new levels of hype. There are 75 million or so recording systems in America's households, and millions of stereo cassette units in the nation's automobiles. And all of them are playing the widest variety of music in recent memory, as the music and recording industries attempt to be all things to all people.

In this chapter we will trace the history of the recording business from its beginning a century ago. We will consider the sociological and economic natures of the business, and discuss trends in popular music since the turn of the century. As we have done with other media, we will take a look into the future, venturing a guess or two about tomorrow's recording industry.

The Recording Industry

The recording industry has antecedents in electronics, as do radio, television, and the telephone, but its birth can be traced most specifically to the 1870s.

Early History

Edison's 1877 phonograph captured sound on tinfoil for later playback.

The scientific genius of Thomas Alva Edison was focused more than a century ago on capturing the sound of the human voice. In 1877 Edison and his machinist, John Kruesi, built a functional sound record-playback device Edison called a *phonograph,* from the Greek words for "sound writer." The original phonograph used a hand-cranked metal cylinder wrapped in tinfoil. A stylus attached to a diaphragm, vibrating from sound waves, cut an indentation in the tin. The sound was reproduced by reversing the process: a needle ran across the indendations, and the sound was amplified mechanically.

Edison's first experimental recording, "Mary Had a Little Lamb," amazed the scientific community. Although at first he did not exploit his invention commercially, Edison did envision its use for a wide variety of purposes: recording school lessons, books for the blind, business communication, music, and important speeches. However, the machine became little more than a sideshow curiosity item, until Edison was faced with some competitors who were intent on making money with their versions of recording devices.

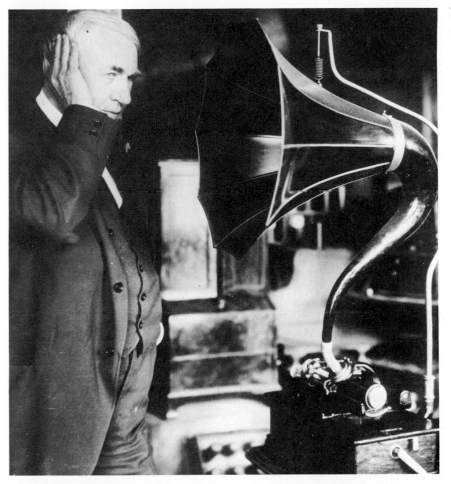

Chinchester Bell, who was Alexander Graham Bell's cousin, cooperated with Charles Summer Tainter in developing what proved to be an improvement over Edison's phonograph. Instead of using tinfoil, which was very fragile and could be recorded on only once, they created a wax cylinder in the late 1870s. Their system, the *graphophone,* was patented about the same time Edison was designing a solid-wax reusable cylinder to improve his original phonograph. The ensuing patent debates between Edison and Bell-Tainter were settled when Jesse H. Lippencott bought the business rights to both systems, and began leasing the machines as dictating machines for business firms. (Actually, the novel inventions made more money when used as diversions in saloons and bawdy houses during the 1880s than they did in established businesses, but that's another story.)

Bell's *graphophone* used a wax cylinder.

Berliner invented the flat disc *gramophone* system in the 1880s.

In the 1880s Emile Berliner improved upon the wax cylinder by inventing a flat disc system. His *gramophone* system made a lot of sense because the hard shellac flat discs were easier to store than the round cylinders, and had better sound. They also could be mass produced, unlike the earlier systems in which each recording had to be an original. Berliner's technique proved to be successful, and was profitably adapted by the Victor Talking Machine Company at the turn of the century. Millions of discs were sold, and audiences expanded the reach and impact of classical and popular music alike as the costs of gaining access to music dropped.

Not everyone could travel to a concert hall to hear famous performers such as Enrico Caruso, but most people could afford to buy the operatic tenor's gramophone records. In fact, Caruso earned more than two million dollars between 1902 and his death in 1921 from recordings alone, a clear indication that there was money to be made in the newest entertainment business. While doing a favor for the recording companies, Caruso did one for recording artists as well. Until he came along, the voices heard on records were generally unknown entities. Customers bought records to hear the songs, not to hear the singers, and the recording artists were paid a flat fee for each recording session. Since Caruso's success, however, performers have been paid royalties on the sale of each recording, so their income is commensurate with their popularity. To this day, royalty rights have been an important component of the business; two firms, BMI (Broadcast Music Incorporated) and ASCAP (American Society of Composers, Authors, and Publishers), keep track of recordings aired over radio and television or played commercially, and collect royalty fees that go back to the recording companies and artists. The system is now built into the United States Copyright Law, which protects the commercial interests of performers.

The gramophone may have brought popular and classical music to the masses, but at some cost to quality. Prior to the 1920s recording systems were mechanical, and sounds were somewhat rough and imprecise. An electronic recording system invented in the 1920s brought about great improvements in the phonograph industry. Increased quality and availability of record players brought on a boom in the business; in 1927 some 104 million records were sold. However, just when things were looking brightest for recordings, radio stole the phonograph's audiences by giving away the music. Reacting much like other successful media that have been threatened by an upstart, the recording industry at first stubbornly refused to join forces with radio, perceiving it as an enemy in the battle for audiences and profits. Eventually, however, it became obvious that the two should enjoy a symbiotic relationship and benefit from each other's strengths. Detente took quite some time—the Great Depression hurt the recording business, so dependent upon unit sales, while it helped the radio industry, as noted in chapter 6. Eventually, however, the cooperation between radio and recording led to increased sales of records and to more frequent playing of

During the 1930s radio hurt the recording business by "giving away" music to audiences.

records over radio. An important economic sidelight to all this was the popularity of jukeboxes during the 1930s and war years; individuals who hesitated to buy single discs were apparently quite willing to spend a nickel to liven up dinner out or a social gathering.

The years following World War II saw the recording business changed once again, this time by new technology and by radio's insatiable appetite for music, once television—the newest kid on the block—stole radio's traditional programming and audiences. Prior to the late 1940s, records had been cut directly from live performances in "real time," coughs, rattling paper and all. Electromagnetic tape dramatically improved the quality of muscial recordings, and permitted engineers and studio technicians to manipulate and enhance sounds on the master tapes that were to be dubbed onto discs.

Another important development of the 1940s and 1950s came in the speed and length of records. For a half century records had been played at 78 revolutions per minute, which meant that the average record offered only three to five minutes of playing time per side. A near disastrous war between two giants of the industry changed the system forever. Columbia developed a high quality, long-playing 33⅓ RPM record, and shortly thereafter RCA Victor began promoting its 45 RPMs. The new systems were exciting breakthroughs, but created problems because everyone involved— particularly home users and radio stations—didn't know what system they needed to buy to be able to play whichever speed would eventually win out. The debate was finally resolved when recording companies decided to record symphonies, operas, and musicals on the 33⅓ LP albums, and popular artists on the inexpensive 45s. By the mid-1950s, the 78 RPM records had become collectors' items.

The 1950s also witnessed the invention and refinement of hi-fi and stereo recording and playback systems, which rendered the monaurals obsolete and created an economic (and physical!) boom of sorts that very well may have saved the recording business from being ravaged directly and indirectly by television. TV, we will recall, was drawing new audiences with old fare—radio's. But that fare also included Dick Clark's "American Bandstand" and other record shows, which indirectly improved the lot of the recording business by motivating America's teenagers to buy their own discs and organize their own sock hops.

The past two decades have seen the ultimate marriage of musical media. Radio, television, and film have joined forces with live concerts in bringing music to the millions. Component stereo systems of all sizes, shapes, and costs have become *de rigueur*—practically essential in the home. At last check there were more than 75 million phonographs in America—nearly one per household. And they are no longer just record players. Millions of them include laser audiodisc players, and millions more include stereo or stereo-hi-fi videotape recorders and playback units. High quality tape decks

TV and new technology for recording music changed the industry after World War II.

Dick Clark, surrounded by teenaged fans, reads fan mail during an early 1950s TV broadcast of "American Bandstand."

Musical media have merged in the 1970s and 1980s.

The Industry of the 80s

and AM–FM radios are as necessary as wheels in today's automobiles. (Some wise shoppers insist the best way to tell what kind of driving a used car has undergone is to check to see which FM music stations the radio tuning buttons are set on. The assumption appears to be that hard rock aficionados abuse their cars, while classical or easy listening buffs pamper theirs. We have not seen any empirical studies on the matter.)

In the late 1970s the recording industry was on a roll. Record and tape sales hit the $4.1 billion mark in 1978, a phenomenal 18 percent increase over the previous year and an all-time high. Rock albums that cost a half million dollars to produce, and another million dollars to promote, were retailing briskly at $8.95. The soundtrack to *Saturday Night Fever* had sold 15 million copies—it was what the industry calls a "gorilla." Superstar Paul McCartney and his group, Wings, signed a $20 million recording deal with Columbia. Record companies were offering huge contracts and perks to relatively obscure new groups who had the right "sound."

Then, just a year later, and for the next three years, the bottom dropped out of the market. The youthful audience for albums had started a quiet rebellion over the high prices. There were no gorillas or monster hits that could salvage the industry, not until 1983, when Soul Rocker Michael Jackson's *Thriller* sold nine million copies. Overall, the late 70s and early 80s saw far fewer gold or platinum records than in any other recent period. (The Recording Industry Association of America calls a recording "gold" if the album sells 500,000 copies and the single sells one million, and "platinum" if sales are twice that level.)

Soul-rocker Michael Jackson.

Attendance at rock concerts was down, reflecting both the general economy and audiences that were getting more choosy about how to spend their time and money. Competition for audience attention was higher than ever. Some economists noted that at least a billion of the dollars that would have been spent on records and tapes was being poured into video arcade games each year in the late 1970s and early 1980s.

Just because Americans were not spending as much money to listen to music at the end of the 1970s did not mean the music was not being heard. With the growth of FM stations came an increase in broadcast music. Some 80 percent of the total programming on radio continued to be musical, most of it popular music. And a great deal of that free broadcast music was being tape recorded on home stereo systems for later playback. It is not technically illegal to tape a song from the radio, but the practice represents a billion dollars a year in lost revenues to the recording industry, because home tapers pay no royalties. Neither do organized rings of musical pirates. An enormous chunk of the potential sales revenues for any given recording are lost through pirates who duplicate the original and then sell the copies at discounted prices. The practice is particularly widespread in Latin America and Asia. All in all, insiders estimate that the recording industry would have twice its current income if home recording and pirating were kept under control.

Recordings boomed in the late 1970s, then bottomed out. Only recently has there been a resurgence in the industry.

As the 1980s got underway, television started bringing free music into America's homes at a pace never before seen or heard. Local and network musical shows were nothing new—"American Bandstand" has been around since the mid-1950s—but the extraordinary popularity of music videos was something else. By mid-decade the brokered marriage of video and music, coupled with the increased musical appetite of FM radio and availability of some exciting new sounds, outran the revenues lost to home tapers and pirates and helped bring the recording industry back to its bullish pre-1979 economic state. According to some observers, it was the best of times for the industry.

Even in the best of times, the economics of recording are uncertain. Three-fourths of all pop record albums released never make a profit. Things are even worse in other types of music; nineteen out of twenty classical albums are not commercially successful. The break-even point for success is somewhere between 100,000 and 150,000 sales. This explains why the thirteen hundred or so independent record companies are not nearly as healthy on the whole as the half dozen major producers. The big six—CBS, Capitol/EMI, RCA, MCA, Polygram, and Warner—can afford to lose money on the majority of their productions so long as they have an occasional chart buster. Those half dozen companies record, manufacture, and market their own albums and singles, controlling the entire process. They frequently capitalize on a particular musical trend and push it as far as the public appetite will tolerate.

The independents, or indies, are an important ingredient in the record industry mix, however, because they take the risks with unknown performers and new styles of music. Like hardback books, records produced by indies have a subtle but enduring influence in the marketplace, because they are the first to be consumed by music's opinion leaders. The indies generally do not market their own productions in the manner the majors do, but rather rely upon a network of wholesalers and retailers. They have a tougher time getting played on the radio, and count on word of mouth, specialized publications, concert tours, and the like to generate sales. Obscure forms of music, some of which will eventually become mainstream, are recorded by indies—jazz, reggae, new wave, rapping, folk, blues, and some classical releases. In all, the indies generate some 10 to 15 percent of the recordings sold annually.

Promotion and Marketing Numerous gatekeepers are involved in the success or failure of recordings. Some are directly employed by the record companies, but many others are peripheral decision-makers.

For instance, "rack jobbers" are gatekeepers of a sort. They work for the producers and wholesalers, going into the 60,000 or so retail outlets that sell records along with thousands of other items, and arrange the record rack displays in ways to most effectively show off the albums they want pushed. They essentially rent the space from retailers, because retailers in grocery stores, discount stores, drug stores, and high class department stores are not expected to know all the latest details about what is popular and what is not. The logical argument of rack jobbers is that someone who understands the record business and current public taste needs to be involved in the retailing efforts. What it means, of course, is that what's hot gets pushed, and what's not gets buried in the back of the rack, behind the hits. The vicious commercial cycle, already observed in our discussion of paperback book marketing, is again apparent in the ephemeral medium of music.

Record clubs are also important in the sales mix, accounting for about 15 percent of the business. In most cases, the record clubs are run by the major producers, who can afford to offer discounted prices on records they can reproduce inexpensively from master tapes they own, or can unload records that have been returned by retail outlets and have no other sales potential.

Until the mid-1980s, the most effective means of promoting records was to have them receive radio air time. Recently, however, radio play and music videos have gotten to be almost indistinguishable prime promotional tools. Since the early 1980s, no musical group or production company hoping to have a chartbuster dares put out an album without a music video to accompany it and to generate fan and radio interest. (The music video phenomenon is discussed later in this chapter.) But the tried and true route to success—at least for lower budget operations—is still to get a DJ to play the song and say something nice about it.

Because the average radio station receives two hundred records or albums a week, it is not easy to get a DJ to even notice a given release, let alone say something nice about it. This dilemma gives rise to some of the most incredible hype or promotional campaigns modern society has yet witnessed.

Modern recording industry marketing techniques are highly sophisticated, even though at times their execution appears banal and primitive— payola surely ranks among the more primitive promotional strategies. Record companies utilize pervasive and systematically researched sales campaigns whose elements of hype include visits by artists and promoters to radio stations, discos, and record stores (accompanied, of course, by T-shirts, mugs, posters, and very frequently under-the-counter trade-offs such as lower wholesale prices to retailers who will offer larger window-display space).

Promotions, including the production of sophisticated and often suggestive jacket covers, are more and more tied to demographic and psychographic studies of potential music audiences. The rock group, Kiss, was as aware of this as anyone, drawing media and public attention to itself with its bizarre costuming and promotional paraphernalia that included a comic book printed in "real (?) Kiss blood." Primitive? Perhaps, but the teenaged and younger audiences were turned on.

Popular music goes through a commercial cycle very similar to that of other media.

Radio play and music videos are key means of promoting recordings. Hype is part of the system.

Hype has impact, but hype alone will not guarantee extensive air play or commercial sales. If the music does not attract the public, the record will not be played. As described in chapter 6, the competitive radio marketplace has little tolerance for marginal music, so no matter how much hype is involved, the typical commercial radio station will play only the music that fits safely into its particular format or sound. Obviously, this practically guarantees that mainstream music will remain somewhat bland and innocuous. Were it not for the splinter FM stations and specialized discos, innovations in popular music would probably be few and far between.

Trends in Popular Music

With a few notable exceptions (primarily some types of ethnic music and that which has been deemed too risqué for airplay), trends in popular music and radio's musical formats have been inseparable throughout most of this century. A brief look at those trends should prove insightful. First, however, some background is in order.

Music Before 1900

Eighteenth century *broadsides* may have been the first mass-mediated music.

Religious and folk music have been a cultural given for centuries, passed along from generation to generation and community to community by— what else?—word of mouth. The first truly mass mediated music may have appeared in the eighteenth century, when musical *broadsides*—large printed pages with notes, words, and commentary—were published. Between the Revolutionary and Civil War years, patriotic sheet music for piano and voice gained enormous popularity.

From the late 1800s and into the early years of the present century, prior to the popular acceptance of the record player, America's music marketplace consisted of a random mix of classical, religious, folk, ballads, blues, ragtimes, and show music. All were available by attending live performances, by buying sheet music (as often as not, put out by "Tin Pan Alley" publishers), or by winding one of those terribly clever rolls with holes onto the player piano in the parlor.

Early Recorded Music

Recorded music has gone through cycles, many of which have challenged the status quo.

Among the first types of music to be recorded, as noted earlier, were shortened versions of classical and operatic works, popularized by the likes of Enrico Caruso. Then came Dixieland Jazz, a wildly frenetic mixture of New Orleans percussion, woodwinds, and horns that swept the nation in the teens, around the time of World War I. The roaring twenties brought with them a new and vigorous appetite for live and recorded dance music, and such crazes as the fox trot, turkey trot, the shimmy, and, of course, the infamous Charleston. Popular jazz music of the period (recall Louis Armstrong?) reflected youthful challenges of the status quo, as has popular music of every period since. To conservative adults, almost all such trends have indicated the fall of civilization as they believed they knew it. After all, what stronger signs of immorality could there be than the drinking, smoking, cussing, and cuddling that seemed to accompany the jazz age (and, a generation later, the age of rock 'n' roll)?

Music, like so much else in popular culture, moves in cycles, and the cycle following the roaring twenties and its age of jazz was the softer, moodier, more gentle, and very danceable "swing" music played by the big bands and sung by crooners. It was not an accident that caused the musical pendulum to swing so widely. After all, with the end of the twenties came the Great Depression, and hard times tend to bring about reflective music, music that brings people together. The bands of Benny Goodman, Glen Miller, Jimmy and Tommy Dorsey, Harry James, Duke Ellington, Count Basie, Woody Herman, and Guy Lombardo, plus the voices of Rudy Vallee, Bing Crosby, Frank Sinatra, Perry Como, and Ella Fitzgerald, carried the slow-dancing nation cheek to cheek through the 1930s and into the war years of the 1940s. Later on, in the post-war years, some of this type of music remained. Sinatra was probably the most popular male recording star with the bobby soxers, and Dinah Shore, Peggy Lee, Patti Page, and Doris Day the most popular females with all age groups. They "sang innocent," their songs mushy and sentimental, and sometimes just plain cute.

War causes institutional upheavals on more than one front, and World War II was no exception. The period ushered in some new types of music that hearkened back to World War I and the roaring twenties. Boogie woogie, bebop, and black rhythm and blues demonstrated energy, creativity, and abandonment of conservatism. Not all of these musical forms entered the mainstream in any meaningful way, but in sum they both reflected and affected a nation's popular taste. Nowhere was this to be made more clear than in the rock 'n' roll that began in the mid-1950s and which, despite a few detours along the way, has retained its central location in the media marketplace ever since.

The causes and effects of rock 'n' roll are still debatable. Some wonder why anti-establishment music would gain popularity during the tranquil Eisenhower years, years in which a healthy postwar economy brought about a national sense of security and optimism. In searching for answers, perhaps we could argue that early rock 'n' roll music was not as much "anti-establishment" as it was "nonestablishment." It didn't reject the dominant culture so much as it reflected an obsession with the here and now, the transient, the chaotic. In essence, it glorified the values of youth, which is pretty much what popular music has been doing all along.

Some have traced the rise of rock music to the youth culture's financial independence and acquisition of the automobile. High school parking lots were jammed with souped up hot rods—recall *Rebel without a Cause?* A car radio was mandatory, but somehow it was not cool to be cruising along listening to Patti Page ask "How Much is that Doggie in the Window?" or Rosemary Clooney lamenting the structural problems of "This Old House." Frank Sinatra, Dean Martin, Perry Como, Eddie Fisher and the other crooners gave teenagers in the early 1950s some nifty makeout music in the back seat of the '48 Hudson or a good excuse to dance cheek to cheek at a sock hop, but did not speak to the need for independence. Something else was needed to fill that void.

Rock 'n' Roll

Since the mid-1950s, rock 'n' roll has retained its central location in the media marketplace.

Bobby-soxers of the
1950s hung out at the
local drugstore listening
to the jukebox playing
early rock 'n' roll.

What filled the void was a fascinating combination of black rhythm and blues (R&B), jazz, country and western, and mainstream (i.e., white) popular music. How this all came about remains contentious, but much credit goes to Sam Phillips, founder of Sun Records in Memphis, and Alan Freed, a Cleveland DJ. Both of these white middle-class mainstream Americans had developed a love for black R&B music, but both were worried that the sound would not sell commercially. From today's perspective it must seem strange that black recording artists would not go over with white audiences, but the truth was that much of the earlier black music had a reputation for being particularly earthy and bawdy, and could not pass FCC or NAB muster for radio play.

Starting in 1951, Freed began taking a chance on playing some R&B records to his predominantly white Cleveland audience. His experiment was a tremendous success locally; his teenaged audience was turned on by the beat and mood of the music, and not particularly offended by the lyrics. On the national scene, major recording studios sensed there was a buck to be made, so they sanitized the lyrics and re-recorded the R&B music, this time sung by white groups. The songs, heavy with percussion, bass guitar, and piano, were dubbed "rock 'n' roll." Freed went on to New York City and fame and fortune as the nation's leading DJ. He promoted rock via radio, television, films, and concerts, but became a tragic key figure in the payola scandals in the late 1950s and was blackballed from the industry. His career was traced in the 1978 film *American Hot Wax.*

Meanwhile, in Memphis, Sam Phillips had founded a recording studio that offered outlets for many black performers. B. B. King, Chester Burnett (Howlin' Wolf), Ike Turner, The Prisonaires (a quartet of convicts who cut

Sam Phillips and Alan Freed
deserve much credit for
creating rock.

the original "Just Walkin' in the Rain," a song later mainstreamed by Johnnie Ray), and other bluesmen recorded there, and their discs earned modest incomes for them and for Sun Records. What Phillips kept looking for, however, was a white man who could sing black, a voice that could transcend the color line and bring the sensitivity and depth of black music to the white record buyers.

Phillips found what he was looking for in 1954, when a young truck driver for the Crown Electric Company stopped in at Sun Studios to cut a record for his mother's birthday.[1] The rest, as they say, is history. Elvis Presley, an introverted young man who played his guitar and sang country, R&B, soul, and gospel music like no one before him—black or white— went on to sell a half billion records and change the nature of popular music.

The speed with which rock 'n' roll swept through the national consciousness was astounding. About the same time Elvis was breaking into the business at Sun Records, young Americans who went to see the film *Blackboard Jungle* were taken in by the pulsating beat of a white group, Bill Haley and the Comets, playing "Rock Around the Clock." The song was released as a single, pushed by Alan Freed, and sold 15 million copies in 1955 to become the first rock song to top the charts. The following year Elvis, who had signed with RCA, had five of the nation's top sixteen songs, including "Don't Be Cruel" and "Heartbreak Hotel," and legitimized rock by appearing on the "Ed Sullivan Show" (filmed from the waist up).

Elvis Presley changed the nature of popular music.

Between 1956 and 1958 Elvis owned the industry, cutting fourteen straight singles that sold more than a million copies each. He was a cultural A-bomb.

Sam Phillips, who had sold Elvis's contract for $40,000, continued to scout around for marketable talent. He found it in the persons of Jerry Lee Lewis, Johnny Cash, Carl Perkins, Roy Orbison, Gene Vincent, and Charlie Pride, performers who refined the sound of *rockabilly* and ensured a place for Phillips on the rock 'n' roll honor roll—even if they all did leave him for bigger contracts with major record companies.

In the late 1950s rock took on new sounds, reflecting the creative genius of both black and white performers. Three decades later, in 1986, when the newly formed Rock and Roll Hall of Fame awarded Oscar-like statuettes to eleven rock pioneers, nobody made a big deal out of the fact that six of the honorees were black men, and five were white. As *Rolling Stone* put it, these artists remade the world, providing "a salutary tongue in the ear to the overstuffed status quo," and eventually displacing it.[2]

Rock and Roll Hall of Fame awards have gone to artists who have "remade the world," according to Rolling Stone.

The Beatles, etc. Strange as it may sound to some, popular music did not begin that Sunday evening in 1964 when Ed Sullivan introduced to America a group of four moppets from Liverpool. Rock had been around for at least a decade, as previously pointed out. It had manifested itself in several forms, but at the time none of them seemed capable of lasting. The record buying public was fickle, always ready to pounce on a new sound and a new way to dance. From the West Coast came beach blanket and surfing music of Jay and the Americans, Jan and Dean, and The Beach Boys. From Detroit—Motown—came very danceable black soul music of The Supremes, The Jackson Five, Stevie Wonder, The Temptations, Marvin Gaye, Smokey Robinson and the Miracles, and others. From Philadelphia came Dick Clark's "American Bandstand" and its influence on the Top 40 radio formats, particularly its devotion to a whole slew of clean-cut caucasions like Frankie Avalon, Paul Anka, The Four Seasons, Bobby Vinton, Fabian, and the late Ricky Nelson. All in all, however, it was a period dominated by conservative recording studios and equally conservative radio stations, both of whom had been burned by the payola scandals and were playing it close to the vest. If the public expected something *really* innovative, it was not about to find it in the mainstream musical business.

Mainstream music was fairly conservative until the Beatles came along.

Record buyers found the excitement coming out of the guitars, drums, and mouths of the Beatles. Even in those early days when the skiffle band members from Liverpool could not read music, they seemed to exude a certain magnetism reminiscent of Elvis Presley when he had debuted a decade earlier. Within a few months of their emergence on the American scene, they had conquered the nation. In April, 1964, they achieved the impossible, by holding down all five of the top spots on the hit parade with "Twist and Shout," "Can't Buy Me Love," "Please Please Me," "She Loves You," and "I Want to Hold Your Hand." By the end of the year, they had dominated the charts with seven separate number-one records, and over the next two years they were to have six more.

Every album the Beatles produced became a million seller, and as musicians and talents they improved with age. Not only did they learn to read music, they mastered literally dozens of instruments—ranging from the esoteric Indian sitar to the contemporary American computerized moog synthesizer. With the release of *Sergeant Pepper's Lonely Hearts Club Band* in 1967 and the cartoon film *Yellow Submarine* they proved themselves not only to be versatile, but to be creative and financial geniuses. By the time they broke up in 1970, to the chagrin of much of the civilized world, they could lay claim to being the most widely recognized group of individuals in history.

So long as American teenagers were adopting the Beatles, the time seemed ripe for other Anglos to venture across the Atlantic. The Dave Clark Five, Herman's Hermits, and Freddie and the Dreamers brought their up-beat, cheerful rock music, and became short-lived successes here. On the other hand, England also exported some not so upbeat, cheerful music, in the persons of Mick Jagger's Rolling Stones, the Yardbirds, and the Animals.

The Rolling Stones' American career has been long but checkered. They reflected and probably expanded the drug culture of the sixties and seventies; they gained notoriety when they hired the Hells Angels motorcycle gang as security for a concert in Altamont, California, and the Angels killed a rowdy fan—on camera, no less; they—or at least Jagger—survived the turbulent sixties and seventies, and even into the eighties continued exhorting fans to challenge authority. Like the man said, "I Can't Get No Satisfaction."

As we attempt to review the significant trends of popular music in the sixties and seventies, we note once again how it is that music holds up a mirror to society. The undeclared war in Vietnam, with all of the related domestic turmoil seen in campus draft-card burners and civil liberties protests, melded into the turmoils surrounding a new black consciousness (aided and abetted by black music, but also by Martin Luther King, Jr., and others). The calls for justice and equality included a long overdue women's movement—or movements. And there was Watergate.

Throughout this period can be seen the importance of Woodstock and its half-million faithful, and of Bob Dylan and Joan Baez and Simon and Garfunkel and their protest music reminiscent of Woody Guthrie's depression era folk ballads. There were significant messages in much of the counterculture rock and popular music of this period. Somehow, however, it is harder to assess the role and impact of highly merchandized "talent" such as The Monkees, or the bubble gum rock played by The Archies. The kindest thing to be said is that this music, for a while at least, sold records, received airplay, and took up space.

The Seventies and Beyond Woodstock ushered in the seventies, as did multitudes of other outdoor concerts celebrating natural and unnatural life. But those festivals of life had a flip side, death, which claimed several popular performers who had lived and played so very hard: Jimi Hendrix, Janis Joplin, and Jim Morrison of the Doors.

Perhaps we are still too close to this time period to put it in the kind of detached perspective the fifties and sixties permit. However, a few events in recent popular music history stand out. We have noted that some music reflected concerns over Vietnam and civil rights, while sharing the airwaves and record counters with bland pop sounds. At the same time, the seventies became what writer Tom Wolfe called "The Me Decade," a period not of throbbing social consciousness but rather of obsession with self, with success, with acquisitions. There was music to serve this self-consciousness. Some of it was loud, abrasive, and psychedelic heavy metal. Kiss, Alice Cooper, The Jefferson Starship, Grateful Dead, Grand Funk Railroad, Led Zeppelin, and, late in the decade, a gaggle of punk rockers probably fit this mold.

Romance is part of the obsession with self, and, as always, there was music to serve the mood. It showed up everywhere on the pop charts, with offerings by such "mainstream" country singers and gentle rockers as James Taylor, Crosby, Stills, Nash, and Young, Gordon Lightfoot, Carly Simon, Linda Ronstadt, Anne Murray, and Joni Mitchell. Country and pop seemed to merge during this period, so that it became impossible to stereotype the efforts of John Denver, Charlie Pride, Glen Campbell, Ray Price, Roger Miller, Dolly Parton, Kris Kristofferson, and the Australian songbird Olivia Newton-John.

Breakdancing—a new art form of the urban '80s.

Rock 'n' roll has not died, as many predicted back in the 1950s. Instead, it has branched out, cloned itself so to speak, with so many variations it has become difficult to keep them straight, let alone to gauge their ultimate contributions. That, some say, is a sign of health in the music business: there is so much coming and going these days that no single sound dominates the industry.

Some rock sounds have come and gone quickly. Others have been more enduring. There has been *heavy metal* or *acid rock* with its hard-driving sounds and frank lyrics; *chicken* or *bubble gum rock* with its softer, non-controversial messages; *disco,* typified by the Bee Gees' *Saturday Night Fever* sound in the late seventies; *soul rock,* as popularized by James Brown; *country rock,* popularized by Kris Kristofferson, Dolly Parton, and others; *folk rock,* by Gordon Lightfoot; *jazz rock,* by Chicago and Blood, Sweat and Tears; and *pop rock,* by Barry Manilow, Elton John, and Steely Dan.

The list of rock's progenies continues into the mid-eighties; *punk rock,* rude and nihilistic English working class music, typified by Sid Vicious and the Sex Pistols; *new wave,* essentially a cleaned-up, middle-class, Americanization of punk, typified by Duran-Duran and Boy George's Culture Club; *reggae,* a sophisticated Jamaican and Caribbean folk sound; *break dance music,* black urban music that essentially replaced street fighting with street dancing; *rapping* and *go-go,* a combination of black urban instrumental and "talking" music, a contemporary form of gospel music set to syncopated rhythm; *world beat,* truly international music, sort of "Afro-Latino-Caribbean-influenced-pop-rock"; and an emerging *cowpunk* or

Rock 'n' roll has cloned itself into a wide variety of sounds. It has not died, as many predicted.

rockabilly, a sound, being popularized by Jason and The Scorchers, the Blasters, and X, that preserves country music's down-home twang while capturing the raw energy of hard rock.

Can we safely generalize about contemporary rock by claiming that while no single style predominates, the music combines raw energy, electronically enhanced and amplified decibels, and flamboyant dress and demeanor that is conducive to video as well as audio performances? It may also be safe to claim there is a new sensuality, even sexuality in rock, ironically congruent with a new, liberated self-awareness on the part of female performers.[3]

Rock continues to separate generations, just as it binds them together. Teenagers who could not understand the prudishness of Ed Sullivan—who refused to show Elvis Presley below the waist—are now the generation who want warning labels on rock records and tapes. They are raising fears that rock music has subliminally embedded devil worship and that it blatantly and explicitly advocates sex and drugs rather than love and natural highs. The parents of that particular generation outraged their own parents by doing the shimmy, limbo, and Charleston, and, likely as not, their grandchildren will undoubtedly give rise to complaints that the popular music of the year 2000 is antisocial, depraved, and decadent. There is a certain time-binding capacity in all of this.

Lately rock has displayed a new social consciousness.

Meanwhile, rock is showing a surprising new face. Just as criticisms of rock's self-indulgence and antisocial tendencies were reaching a crescendo, and hearings were being held in Washington, D.C., rock suddenly began to emerge with an aura of respectability. Starting in late 1984, when Irish musician Bob Geldorf organized British pop and rock stars to help feed millions of starving Africans, rock seemed to enter an unprecedented period of social responsibility. "Band Aid" and "Live Aid" concerts reached all corners of the globe; the concerts and record sales raised some $92 million for the "Band Aid Trust," funds that are being spread across a variety of short- and long-term projects aimed at helping Ethiopians and other Africans became more self-sufficient.

Almost immediately after "Band Aid," American and Canadian musicians joined the move to focus the world's attention on the African famine. Forty-five top stars sang "We Are the World." The message of "USA (United Support of Artists) for Africa" was simple but effective, and resulted in millions more for charity. Later, with "Farm Aid," American rock stars put their talent to work in support of beleaguered U.S. farmers. Bruce Springsteen—"The Boss"—dedicated himself to saving the economy of his hometown when a major corporation decided to pull up stakes, and his guitarist Steven Van Zandt organized a troupe of rockers to produce an anti-apartheid album and video called "Sun City." U2 and Sting have written and performed songs offering personal reflections on such world issues as violence in Belfast and United States-Soviet relations, and other rockers continue to be involved in contemporary issues.[4]

These displays of social consciousness may be valid signs that music's opinion leaders recognize the need to use their abilities to help humanity. On the other hand, a skeptic might observe that almost without exception, when media institutions have come under serious public and governmental fire they have undertaken a public relations effort to improve their image, so rock music is just following the accepted pattern. Whatever the intrinsic or extrinsic motivation behind rock's charity, most observers would conclude that the world is probably somewhat better off as a result of the efforts.

Music Videos We have previously mentioned the flamboyant dress and demeanor of today's performers. Without question, these stylistic changes are tied to the rise of the music video. The typical video on cable's MTV (Music Television) channel will be seen by thirty million people, an audience larger than any but the top grossing motion picture, so competition for attention is intense. With exposure like that, it is little wonder that performers such as Cyndi Lauper, Madonna, Boy George, Prince, Michael Jackson, Billy Joel, Duran Duran, and others have had enormous influence over styles of dress, dance, song, and self-identity. (Who knows—maybe the Chicago Bears' 1986 rap video about the Super Bowl did more to raise the spirits of the Windy City than anything since the erection of the Sears Tower.)

Since August of 1981, when Music Television (MTV) was launched as a coast-to-coast, satellite-fed 24-hour-a-day cable program, there has been no turning back. MTV, whose FM signal permitted viewers to combine video with stereo, quickly became the nation's most popular cable offering. By 1986, its 30 million subscribers were an attractive target for upbeat advertisers, and the 300 or so music videos being played each day had changed the nature of popular music. Other networks and television outlets followed suit. Home Box Office and Cinemax use music videos to fill space between movies, special interest cable networks offer country and western, black, and religious/gospel videos, Ted Turner's superstation WTBS has "Night Tracks," and the commercial networks offer late night weekend video shows.

Videos are produced to coincide with the release of major (and minor) motion pictures and concert tours of popular groups, and to launch new artists into the public spotlight. Research indicates that the music videos have become as important as radio play in assuring the financial success of a recording. The record companies, which produce the videos and originally gave them to television as free promotional tools (and only recently have collected royalties from them), are obviously the major beneficiaries. Indeed, the entire video movement has salvaged the record industry from the severe sales slump that began in 1979.

By 1984, it became obvious to television, the recording industry, and Hollywood that something significant was happening in popular culture. One anthropological and critical study of music video has concluded that

MTV has become the nation's most popular cable offering. Video has salvaged the record industry.

it is a brand new form of mass communication, one that may alter the way future generations think of recorded music and how it is conceptualized. Eventually, the financially successful and creatively successful musician will be working in a combination of sound and visualization of sound, and the next generation of artists will visualize their music right from the outset.[5]

Without question, visualization is important to the consumer as well as to the producer of music videos. As *Time* magazine put it:

Increasingly, and perhaps irreversibly, audiences for American mainstream music will depend, even insist, on each song's being a full audiovisual confrontation. Why should sound alone be enough when sight is only as far away as the TV set or the video machine? Whole generations have had their brains fried with a cathode ray tube, a condition that creates a certain impatience and shortness of attention when limited to aural input. Posterity can rest easy—as Billy Joel points out, "Beethoven didn't have no videos, and he's been hanging in there"— but for rockers, popsters and soul brethren, video will be the way to keep time with the future.[6]

Music videos are affecting the other electronic media.

If video is the future, as many media commentators have observed, it will undoubtedly have an impact on other media. Radio, devoted as it is to music, is already sounding pale in comparison to the vivid images put out by music video. The aural medium has had to rethink its game plan, as it did a generation or so ago when television spoiled its programming monopoly. After all, if music without pictures seems boring, and radio loses its audience shares, the advertisers will be the first to notice.

The movies are also being affected by MTV and other video offerings. Top producers and directors are trying their hands in the music video business, where creativity, speed, and modest costs of production are enticements. In many cases music videos are being made to coincide with the release of Hollywood films: *Ghostbusters, Beverly Hills Cop, Top Gun, An Officer and a Gentleman, Against All Odds, White Nights,* and a whole host of others. The clever title track and video of *Ghostbusters* was said to have added 20 percent to the audience for the already blockbuster movie, a fact that did not go unnoticed in Hollywood.[7] Because three-fourths of today's movie goers are under age thirty, and have grown up with rock music, the marriage between video and film seems assured. On the negative side, however, critics bemoan the current glut of teen films that put more emphasis on the soundtrack than the story. Such films as *Flashdance, Footloose, Streets of Fire, Purple Rain,* and *Breakin'* are merely long-playing music videos, and exploit audiences who have been prepped on the three minute versions. The audiences, so far, seem quite willing to be exploited.

Video audiences are sophisticated, yet fickle.

The target audience for music videos may be the most sophisticated yet fickle group of media consumers in the world—American adolescents. For their entire lives they have been exposed to lavishly produced films, television programs, commercials, and musical extravaganzas. They expect their music videos to be produced with the same attention to detail. Yet, if we look at the first generation of music videos, those most popular prior to 1985, we see that the sophisticated productions had quite narrow appeal, aimed directly at the emotions of the teenager. The themes were repetitive: budding and broken relationships, the challenging of authority, sex, money, power, and even the threat of nuclear war. Adults were frequently portrayed in unfavorable lights, as overly materialistic, abusive, and unreasonable, while women were often seen as bimbos, prostitutes, or blatant sex objects.[8]

At mid-decade, for a variety of reasons, some significant shifts in the themes of music videos began to emerge. Videos became more sophisticated, perhaps because the faddish, 12-to-24-year-old target audiences had started to show the first inklings of boredom. (The same thing had happened a couple of years earlier with video games. After a meteoric rise in popularity that impacted on other entertainment, particularly buying and listening to music, video arcades began a slow fizzle.) But for videos, it became obvious that the medium had much greater potential than to merely create a sensory overload of noise, color, and light. The technology and insights were being put to work for music other than rock: jazz, ethnic, country, religious, and even classical and dance motifs.

In the still-dominant rock arena, videos were being used to introduce new themes about relationships, self-concepts, materialism, sexism, and the like. Consider some of the post-feminist videos, in which Cyndi Lauper convinced millions of viewers that it is OK to want to have fun and to be ideosyncratic, Madonna both flaunted and mocked sexism and materialism, Tina Turner proved that middle-aged performers can still cut it and be liberated, and Pat Benatar insisted that love is a battlefield in which women should prove invincible.[9] In three minutes one cannot help but rely on stereotyping to get a message across, but at least some of the recent stereotypes are challenging old assumptions. That, it would seem, is healthy.

Because of their pervasiveness, music videos have an almost instantaneous impact on popular culture. The moment performers such as Michael Jackson, Madonna, Boy George, Cyndi Lauper, and Prince release a video, their clothing and hair fashions become popular in millions of homes. Susan Flinker, a fashion writer, fashion video director, and author of the book *Hip Hair,* said that "Very simply, it's made very extreme fashion totally acceptable to anybody out there in the sticks, so that the outrageous clothing that would have taken maybe fifteen years to get visibility, now (is popular) within a week."[10]

The impact of music videos is seen on other fronts. For one thing, the distinctions between commercials and entertainment are growing fuzzy. Dozens if not hundreds of television commercials have appeared in music video formats. That should not surprise anyone, considering the fact that in many cases the most popular music videos have been made by directors whose previous experience has been in the advertising and commercial-making business. But the clips have been used to do everything from promoting the Olympics and Big Macs to selling fashion and crimebusting on television's "Miami Vice," creating some confusion about where art and programming let off and propaganda begins.

Music videos have introduced new themes, in new motifs.

Digital laser discs are
transforming the music
industry.

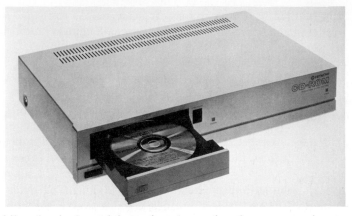

Meanwhile, classical musicians, dancers, and artists are starting to turn to music videos as vehicles to communicate to larger, post-*Amadeus* audiences. Once that happens, and the highly creative minds of master composers and artists devote themselves to combining traditional and contemporary media, a whole new wave of visualization and multisensory impact will be upon us. The fine and popular arts will have made a significant leap toward the twenty-first century, and we will look back at the eighties as a time when some of the most important traditions in mass communication took root.

The Future of Recordings

Compact discs—CDs—are causing a dramatic shift in the recording industry.

Digital laser discs or compact discs (CDs) have already begun to change the way we record and hear music, and are causing as dramatic a shift in the recording industry as the 33⅓ and 45 RPM records did when they first appeared. Based on computerized technology that transfers sound to a digitized system that can be played over and over without ever wearing out or losing fidelity, the CD has quickly moved from an expensive experiment to a basic ingredient in home entertainment systems.

In 1983, when first introduced on the market, CD systems cost $1,500. A year later, prices had dropped to $500, and are now readily available for under $200. Approximately 18 million of the shiny, 4½ inch discs were shipped in 1985, the industry's second full year of sales. That represents a phenomenal sales pace for a new communications tool.

Firms involved in production and sales of the new systems maintain that the CD will result in a fundamental upgrading of the music industry. More sophisticated recording and playback systems will demand higher quality performances, because audiences will actively listen to the music rather than use it as background for other activities. What the futurists are predicting is essentially a return to the early days of this century, when families crowded around their Victrolas and early radio sets, giving them their full attention.

In the meantime, music videos will continue to develop to their full potential. Home television systems equipped with stereophonic receivers are giving the video images ever greater impact, and, as noted earlier, the videos are improving in both quality and sophistication. The fine arts will join the popular arts in exploiting this new technology, to the mutual benefit of producer and consumer.

When the cassette tape recorder became popular—in miniature versions for solitary joggers or larger boom boxes for social gatherings—it brought on a dramatic change in the fortunes of the polyvinyl record manufacturers. Now, as CDs and music videos have started to dominate the recording industry, they will bring about a similar shake-up in production and distribution of both cassettes and records. Likewise, the makers of the playback systems are looking to new ways of serving the appetites of the buying public. After a period of turmoil, the marketplace is likely to settle in, but only for a short while, before the next set of innovations appears on the horizon. If we have learned anything about the recording business, it is that it is in a constant state of flux. We see no reason for that to change.

Summary

The recording industry has grown up as an auxiliary to radio. Like radio, it had its beginnings a century ago, and developed into a mass communications system in the early decades of this century. With occasional lapses, as for instance during the depression years, the recording industry has kept pace with the general move toward more media for more members of the audience.

Popular music, in particular, rock 'n' roll, has determined the financial success of the industry. Sophisticated production and marketing systems have been the industry's hallmark. The recent marriage of music to video, and high quality reproduction of sound permitted by compact disc recordings, deserve special note as cultural and economic phenomena.

Notes

1. Elizabeth Kaye, "Sam Phillips: The *Rolling Stone* interview," *Rolling Stone,* 13 February 1986, pp. 54-55.
2. Kurt Loder, "The music that changed the world," *Rolling Stone,* 13 February 1986, pp. 49-50, and David Zimmerman, "Rock's hall-of-famers will never fade away," *USA Today,* 23 January 1986, p. D-4.
3. Jim Miller, "Rock's new women," *Time,* 4 March 1985, pp. 48-57.
4. Sara Terry, "Ultravox singer sees a new age of responsibility in rock music," *Christian Science Monitor,* 13 February 1986, pp. 23-24; and Jay Cocks, "Come On, Let's Get Banglesized!" *Time,* 14 April 1986, p. 98.
5. John Gutman, "A marriage of sight and sound, video rock is on a roll," *Salt Lake Tribune,* 27 July 1984, p. M-3.
6. Jay Cocks, "Sing a song of seeing," *Time,* 26 December 1983, pp. 54-55.
7. Cathleen McGuigan, "Rock music goes to Hollywood," *Time,* 11 March 1985, p. 78.
8. Gutman, "A marriage of sight and sound."
9. Miller, "Rock's new women."
10. UPI, "Videos making big changes in pop culture," *Logan* (Utah) *Herald-Journal,* 14 March 1985, p. 9.

8

Television

The world sat in stunned disbelief, glued to a hundred million television sets, as the ill-fated space shuttle *Challenger* exploded shortly after take-off, killing all seven astronauts aboard. The much ballyhooed trip into space for an American high school social studies teacher had captured the fancy of a nation and world. All of us have been touched by a school teacher, so somehow her death brought about a different sense of shared loss than would the deaths of astronauts whom we may have respected and admired, but did not know. And it was television that most dramatically brought us the story, defined the situation for us, helped us express our grief.

At about the same time, in far off South Africa, television journalists were ordered to stop taking pictures of street rioting, to cease conveying to the world the struggles of an oppressed black majority demanding that the world take note of apartheid. Television pictures, the white government correctly inferred, were doing more to stir up the participants—and observers—than anything else in a century-long struggle for justice.

Two years earlier, in the same continent, the starvation of millions of Ethiopians and other Africans was so dramatically brought into the world's living rooms that our collective consciences were mobilized, and relief aid of an unparalleled scope poured forth. Drought and starvation were surely nothing new in Ethiopia, but our understanding of it had never before been so clear as it was once television focused its socially conscious lenses upon the tragedy.

When a TWA airliner was hijacked and several dozen American tourists taken hostage in a Middle Eastern power struggle, television not only brought the world the dramatic events, it helped shape the outcome of the crisis. The hijackers were offered press conferences and live televised interviews, and American network anchormen actually served as mediators in the unfolding drama. The world became the stage for the terrorists, the hostages, and the negotiators, and television carried the performance to all of us who vicariously experienced the events.

We could continue with example after example, all pointing to the increasingly influential role of television in our daily lives. In these cases, television significantly captured actual events for us, giving us shared meanings for situations, dilemmas, and world crises. But television operates with equal impact in other dimensions. In addition to its packaging of reality, it packages fiction, drama, culture, and the economy, to mention just a few. It has defined for us what it means to be a child or a senior citizen, a member of the majority or a minority, a responsible or irresponsible citizen or consumer. In every case, as we shall see throughout this chapter and elsewhere in this book, television has steadily increased its hold on us. Ironically, all this is occurring just as we are growing increasingly conscious of its reach and of our own need to be careful users of this pervasive medium.

What manner of medium is this?

The role television plays in American life in the late 1980s, some forty years after its commercial birth, can be better understood by reviewing its

TV packages reality, fiction, drama, culture, and the economy. Its hold on us is enormous.

statistical profile. There were 1,285 operating stations in 1987, of which 547 were commercial VHFs (very high frequency channels 2 through 13), 435 commercial UHFs (ultra high frequency channels 14 through 70), 111 noncommercial VHFs and 192 noncommercial UHFs.

More than 87 million American homes—98 percent of all residences—have television sets, and about 57 percent of them have more than one set. About 83 million homes—95 percent of television homes—have color television sets. That reflects an increase since 1970 of 27.3 million homes with television, a 23 percent increase in the number of homes with multiple sets, and a 56 percent increase in the number of homes with color receivers. (See Table 8.1.)

The average individual—child, woman, man—watches television for more than 30.5 hours each week, or four hours and twenty minutes per day. And, according to A. C. Nielsen statistics, the American household now averages seven hours and ten minutes a day in front of a set—a full hour a day more than the 1975 viewing levels. Households with three or more people in them watch more than sixty-one hours a week, ten more than the national average per household. Those statistics indicate that besides sleeping and working, television consumes the largest share of America's time, far more than any other leisure-time activity. One study showed that we spend 40 percent of our leisure time watching television, almost triple the time spent using *all* other mass media.[1]

The average child between the ages of two and five watches more than thirty hours a week, much of that time unsupervised. The rate drops by a few hours for older children, but one-fifth of our children watch forty or more hours per week, the equivalent of holding down a full-time job. By the age of twelve, the average child has already logged 12,000 hours of viewing time—the same number of hours he or she is expected to spend in the classroom between first grade and the completion of high school. Not

Table 8.1 The growth of television, 1946–1985

Year	Number of stations on the air[*]	Television households	Percent of homes equipped with TV	Percent of TV homes with more than one set	Percent of TV homes with color sets	Industry revenues (millions)
1946	6	8,000	.02	—	—	N.A.
1950	107	3,875,000	9.00	1	—	170.8
1955	482	30,700,000	64.50	3	.02	1,035.3
1960	583	45,750,000	87.10	13	.70	1,627.3
1965	596	52,700,000	92.60	22	5.30	2,515.0
1970	892	59,700,000	95.20	34	39.20	3,596.0
1975	952	68,500,000	96.30	43	70.80	5,263.0
1980	1008	76,000,000	98.00	49	80.00	8,800.0
1985	1194	85,000,000	98.00	55	90.59	19,900.0

Sources: Adapted from *Stay Tuned: A Concise History of American Broadcasting* by Christopher H. Sterling and John M. Kittross, © 1978 by Wadsworth Publishing Company, Inc., Belmont, California 94002. Reprinted by permission of the publisher and *Broadcasting/Cablecasting Yearbook.*

[*]Includes VHF and UHF, commercial and noncommercial stations

incidentally, by high school graduation that same child will have been exposed to 350,000 commercial messages—about 20,000 a year—and witnessed or vicariously participated in more than 200,000 acts of violence, 50,000 of which are murders.

Roper Organization surveys indicate that two-thirds of the American public turn to television as its major source of news, and more than half rank television as the most believable news source. Television surpassed newspapers in credibility in 1961 and as the prime news source in 1963; and according to the Roper studies, television has improved its position steadily since then at the expense of all other news sources. When asked which medium they would most want to keep if they could have only one, Americans since the first Roper survey in 1959 have been saying "television," rather than newspapers, radio, or magazines. In fact, since 1967, television has held more than a two-to-one advantage over its nearest rival, the newspaper, as the medium to keep when we are limited to one.

By the mid-1980s, videocassette recorders were beginning to have a major impact on the television marketplace. About 28 percent of American television homes were equipped with VCRs by the beginning of 1986, compared with only 13 percent a year earlier. The A. C. Nielsen Company said that the average weekly usage of those VCRs was two hours and 14 minutes of recording time and four hours and 18 minutes of playing time in 1985, with 60 percent of the recording being done while the set was turned off.[2]

The economics of television are staggering. Overall, the medium generates more than $20 billion worth of advertising revenues per year; that is slightly less than the newspaper industry, but far ahead of any other medium. Production and commercial costs are equally high. Producer-director Steven Spielberg's series for NBC, "Amazing Stories," was the most expensive half-hour program in history, with each episode costing between

Television's economics are staggering.

$800,000 and $1 million to produce. More typical is the $600,000 cost to produce each hour-long episode of a top-rated dramatic program such as "Dallas." Mini-series and sagas such as *Peter the Great* and *Amerika* have cost networks more than $25 million. The average 30-second prime-time network television commercial sells for $100,000, and a top show such as "The Cosby Show" commands more than $200,000. (That sounds like a lot of money to deliver a 30-second message to 20 to 25 million households, but sending 14-cent postcards to the same number of homes would cost ten times as much.) The larger the assured audience, the more willing advertisers are to spend money to reach them. Commercials for the final two-hour episode of "M*A*S*H" in February 1983 went for $450,000 per thirty-second spot, and advertisers have spent more than a million dollars a minute to reach the 100 million viewers for each of the last couple of Super Bowl games.

Most television programming, including news and commercials, comes from the three major networks—CBS, ABC, and NBC—who directly or indirectly, through affiliates, dominate 90 percent of the commercial stations. Lately, however, the programming monopoly long enjoyed by the three networks has been broken by independent programming syndicates, by cable and satellite-delivered channels, and by the proliferation of VCRs.

In this chapter we will focus primarily on network television, with appropriate mention of the changes it is undergoing due to the new forces in the marketplace—VCRs, cable, and the like. After a look at the history of the medium, we will discuss the way ratings work and then explore programming, the strategies employed by the networks, and the cost of preparing shows. In recent years research has revealed interesting facts about audiences, and this information is of particular interest to advertisers. In the last two sections, we will look at television news and some of the criticisms directed at the tube.

The Early History of Television

Zworykin and Farnsworth developed television systems in the 1920s.

Television has been around a lot longer than most people would suspect. The electronic discoveries of the late nineteenth and early twentieth centuries that gave us radio, motion pictures, and the telephone were coupled with various mechanical scanning devices for the transfer of visual imagery. A Russian-born American physicist, Vladimir Zworykin, who worked in the Pittsburgh Westinghouse laboratories, patented an all-electronic television system in 1923. At its core was a camera tube he called the *iconoscope*. In 1926 he devised a receiving unit he called a *kinescope,* a cathode-ray tube that carried an image consisting of thirty horizontal lines—a very sketchy picture by today's 525-line standards. Practical demonstrations of the system were made in 1928. Meanwhile, Philo Farnsworth was devising his own system, independent of corporation laboratories. In 1927 he successfully transmitted his first picture—perhaps prophetically, since it was a dollar sign! But years of squabbling over patent rights and financing meant there would be no commercial television for at least another decade, and by then the federal government had become inextricably involved.

Vladimir Zworykin holding the kinescope, or cathode-ray receiving tube.

In 1927 an experimental television program, in which Secretary of Commerce Herbert Hoover participated, was broadcast between New York and Washington, D.C. Soon several other broadcast stations were experimenting with television: WGY of Schenectady, New York, broadcast television's first dramatic programming in 1927, and shortly thereafter, the Schenectady and Purdue University stations were sending television images thousands of miles over shortwave radio frequencies. In 1930 RCA, which had taken over the radio research projects of Westinghouse and General Electric, demonstrated large-screen television at a New York City theater.

By 1937 seventeen experimental stations were operating, and starting in April 1938, television sets were available in department stores for interested Americans. The least expensive were American Television Corporation's three-inch screens for $125, with General Electric and RCA offering twelve-inch sets for up to $600. We should not forget that the dollar in 1938 went several times as far as it does today and that those depression-era dollars were being spent on experimental sets to receive experimental

broadcasts. The FCC in 1937 had concluded that television was not yet ready for public service on a national scale, and that there should be no commercial sponsorship.

RCA negotiated patent purchases with the independent Farnsworth, in addition to the millions of dollars it had already spent in its own laboratories, so it could market a complete television system by the late 1930s. RCA was not alone; by July 1939, fourteen different manufacturers were in production, including Crosley, DuMont, Farnsworth, General Electric, Philco, and Zenith. The FCC was called in to systematize broadcasting equipment. In 1941, after several years of heated debate among the manufacturers, the FCC decided upon a 525-line, thirty-frames-per-second system, which is still in effect. It also authorized commercial television operations to begin, and by May 1942, ten stations went on the air. Television was off and running. Significantly, during the same year that the FCC authorized commercial television, it also assumed jurisdiction over program content, instead of merely regulating the technical aspects of broadcasting. Its 1941 "Mayflower Decision" held that a broadcaster could not be an advocate and outlawed broadcast editorials.

In 1942 a wartime freeze was placed on television, and only six of the original ten licensed stations continued broadcasting throughout the war. In 1945, following the war, the FCC allocated twelve VHF channels to commercial broadcasting, at the most favorable spots on the broadcast band. This move was a boon to NBC, which had been urging its radio affiliates to seek television licenses, and a setback for CBS, which had gambled on FM development.

Rapid growth followed; the rush for new licenses, three competing schemes for color television, reservation of channels for noncommercial educational and UHF use, engineering standards, and the need for a national assignment plan for all channels made it obvious that television was developing faster than the FCC could regulate. So, on 30 September 1948 the commission stopped granting new TV applications in order to study the situation. This freeze, which the FCC promised would be brief, lasted through the years of the Korean conflict and was not lifted until 14 April 1952.

Chaos after World War II resulted in an FCC freeze on TV licenses from 1948 to 1952.

Despite the freeze on new stations, television grew at a fantastic pace between 1948 and 1952. When the freeze began, only 172,000 United States homes, 4 percent of the country's residences, were equipped with television. Only four years later the figures had jumped to 15.3 million and 34.2 percent, respectively. (See table 8.1 for TV's growth between 1946 and 1985.) Viewers in 1948 saw Milton Berle, a radio performer who had turned to television, inaugurate the "Texaco Star Theater," and Ed Sullivan host the first "Toast of the Town" show, which restored vaudeville to the American scene for a full, incredible twenty-five years.

Borrowing the time-restricted format of radio—the training ground for most of the pioneers of television—programming was made to fit segments of fifteen minutes, half an hour, or more. Most shows were performed live, without the security of "retakes" or the profit generated by reruns, which film and videotape offered. Live shows were preserved by making *kinescopes,* a film of the program taken directly from the screen. But these were of poor quality and were often unusable. Inadvertently, "I Love Lucy" helped to change that. When the show started in 1951, most television production emanated from New York. Lucille Ball and Desi Arnaz did not want to move from their home in Los Angeles, so they convinced CBS to allow them to film the program in Hollywood. Because they wanted a live audience, the show was filmed with three cameras, which cut down on interruptions and retakes. As the show became a hit, the value of the filmed episodes increased. Eventually "I Love Lucy" was syndicated, and reruns of the show can still be seen today. Now, of course, live television is rare. Most programs are produced on film or tape, weeks or months before air time, always with an eye to the potential for additional revenue from reruns.

"I Love Lucy" was the top-rated show in the mid-1950s and stayed there for half a decade. Westerns, such as "Gunsmoke," also corralled large audiences. Quiz shows, typified by the "$64,000 Question," appealed to the viewers' dreams of instant wealth. Television had become the preeminent entertainment medium, consuming material at an incredible pace and spawning imitators of every new show that became a hit.

The quality of programming in television has been an issue since the first critic evaluated a show. But for many, the 1950s are looked upon as the golden age of television. Live dramatic theater attracted some of the best-known playwrights, such as Paddy Chayefsky and Rod Serling. "Studio One," "Playhouse 90," and "Kraft Television Theater," among others, added substance to programming.

Before 1955 some shows were being broadcast in color, and by the mid-1960s most network programs were in color. The transition was relatively swift once the FCC had made a final decision about the technology to be used. In the 1940s both RCA and CBS developed color equipment. Because the CBS machinery was more advanced, the FCC found it attractive. The problem, however, was in the television receivers; the color sets designed by CBS would not carry black-and-white programs. After a complicated legal battle, the FCC approved the equipment developed by RCA, which could carry both black-and-white and color broadcasts.

The Networks

The three major networks—CBS, NBC, and ABC—dominate commercial television, as they have since 1948. Over the years the vast majority of America's commercial television stations have been affiliated with one or another of these webs. Prior to 1955, Allan B. DuMont, an independent producer of television equipment, competed against "the big three," but was forced out of business because he lacked the inherent advantages of the bigger networks. More recent competitors have specialized in sports, or news, or the like.

By 1937 New York and Philadelphia were linked by coaxial cable, which was required for long-distance hookups because television's line-of-sight signal rushes off into space as the earth curves (in the same way FM signals do, as we discussed in chapter 6) and because ordinary telephone lines are inadequate. Between 1946 and 1947 Boston and Washington were included in the linkage. Later, relatively inexpensive microwave relay systems were developed. By 1948 either cable or microwave linked the East Coast to the Midwest, and network programming became financially feasible. In 1951, before the freeze ended, coast-to-coast network linkage was ready.

With the technological apparatus lined up, all that was needed to make the system work was an audience. Prime-time programming—including Milton Berle, Ed Sullivan, drama, and sports—served as bait. Throughout the 1950s the costly quality drama of "Playhouse 90" served as loss leader. It cost more to produce than advertisers would pay, until larger audiences were in the fold. Once the audiences were hooked, however, television truly became a mass medium. According to some critics, the general quality of programming then began to slip.

The dominance of CBS, NBC, and ABC is based on their *owned-and-operated* as well as *affiliated* stations. The FCC permits a single owner to operate twelve television stations—as long as they do not reach more than 25 percent of the nation's homes. No more than five can be the generally profitable VHF variety. The networks during the 1940s and 1950s fought to establish their control of VHF stations—they still have not found it worthwhile to own UHFs—in the nation's major markets. All three networks now have about two hundred affiliates, ABC having closed the gap from which it had suffered since the 1940s, when it was the new kid on the block.

The affiliates, individually licensed by the FCC, carry the networks into less-populated corners of America. An affiliation agreement between a station and a network is a franchise to that station to carry that network's programming in its licensed area. Individual stations are not required to carry the net, but generally do so because the programming is generally superior to and cheaper than anything they can produce on their own. The net is their guarantee of an audience.

Affiliates

Some stations carry more network programming than others. All have the right to preempt network programming to carry other material. As previously discussed in chapter 6, affiliated stations usually are reimbursed by the networks for carrying the web. The industry standard for payment to affiliates is around 20 percent of the network's commercial rates for the time. To assure itself a greater number of affiliates, ABC began offering 30 percent on the rate card during the late 1970s, forcing NBC and CBS to raise the ante. In addition to this corporate windfall, affiliates have continued to sell local advertising for station breaks and *adjacencies,* or spot announcements before, during, and after network programming. That three-to-four minutes per half-hour can become highly profitable for local stations whose prime-time-network shows guarantee local advertisers a large captive audience. In fact, in some markets and for certain types of programming (the Super Bowl, major movies, and the like), enough local income is available that affiliates forego all network compensation.

Relationships between networks and affiliates are tenuous, with both sides keeping their eyes on the ledger sheet.

Originally, the networks required their affiliates to carry a certain high percentage of network programming, and specified certain programs that could not be preempted. By doing this the networks could guarantee a certain audience size to their advertisers. The FCC, however, outlawed these requirements, and stations now may choose not to carry the net whenever they wish, although that seldom happens. They may also pick up a rival network's programming if its affiliate in that area does not choose to use it. Affiliation agreements are a lot looser now than they used to be, and the networks' respective positions are less secure, in light of competitive and regulatory forces.

While the networks and the individual affiliates need each other, their relationship is not an entirely happy one. The affiliates are concerned about the development of each year's schedule since what the network offers will affect audience size and, hence, profitability in their own areas. Regional differences can cause problems, because some kinds of programming that are acceptable in the more liberal Northeast are less so in the South and Midwest. These interests must somehow be balanced by the networks, for whenever an affiliate does not carry the net, network ratings are adversely affected.

Animosities also arise because affiliates sell against the network, in competition with the net. From a major advertiser's standpoint, it is much simpler to buy the net. However, it is sometimes more scientific and economical to purchase time from individual stations in specific markets in order to carry out a certain marketing plan. For example, a snow tire manufacturer may decide there is no market in the sun belt. Aware of this, the individual stations, through their national representatives (reps), are constantly trying to lure advertisers away from the networks to buy time from them individually, which would mean considerably more profit to the station and no additional cost to the advertiser. One of the problems with national advertisers buying local rather than network commercial time has been the difficulty of proving whether or not the spot was actually run.

The network situation is further complicated by the development of a number of stations "groups"—that is, a number of stations under single ownership. Such powerful groups can exert considerable influence with the networks because they sometimes represent major outlets in large population centers (speaking in a single voice). For example, a large group, such as Group W (the Westinghouse organization of stations), can strike fear into network accountants' hearts by threatening to switch affiliations.

Public Television

All the complexity that plagues commercial broadcasting is multiplied in the noncommercial area. This area lacks the unifying theme of profit to keep it on course. Problems began during the FCC freeze between 1948 and 1952 when educators made strong presentations to the FCC concerning the potential of the new medium. As a result, the FCC initially set aside twenty-five channels for educational television (ETV), and the first of these stations went on the air in 1953 in Houston, Texas. ETV grew very gradually, and in due course, National Educational Television (NET) began to provide programming and distribution to the loose chain of educational stations.

Despite the enthusiasm of educators, ETV did not fulfill its promise. It was hampered principally by a lack of funds. Cut off from advertising revenues, it existed meagerly, receiving spasmodic contributions from municipalities, school districts, states, various agencies of the federal government, and from large foundations; but these contributions were never quite enough and never on an assured basis to permit anything more than a minimal subsistence. Furthermore, NET funds were lacking for either coaxial cable or microwave relay of programming, so ETV programs were delivered from one station to another by mail, which scarcely led to any degree of timeliness.

Into this picture came the 1967 Carnegie Commission Report to President Lyndon Johnson recommending the establishment of a public broadcasting service (PBS) distinct from ETV. One thing the Carnegie Commission Report would have provided was assured financing for PBS through a manufacturer's excise tax on television receivers sold. Although Congress implemented the report and established the Corporation for Public Broadcasting (CPB), it failed to provide the funding necessary to validate PBS as a television force capable of competing with the three major networks.

Public Broadcasting Service

From the outset PBS operated on a shoestring. By 1986, public broadcasting was receiving nearly a billion dollars per year from all sources, including the federal government (16 percent), states (29 percent), private businesses (16 percent), private individuals (23 percent), and a variety of other sources (16 percent). Although a billion dollars sounds like a lot of money, it represents only a tiny fraction of what the three commercial networks spend on programming alone. "On any scale of comparison," complained former PBS President Leonard Grossman, "public television in this country is subsisting on total revenues that make it the most desperately underfunded television system in the free world."[3]

Public broadcasting operates on a budget that is only a tiny fraction of commercial broadcasting's.

During the decade from the late 1960s to the late 1970s, the federal government's level of support of PBS's income more than doubled, from 12 percent to 27 percent. By the mid-1980s, its share was back to the mid-teens. As pointed out in chapter 6, the Reagan administration has advocated private and business support of broadcasting, and has sought the elimination of all federal support for public broadcasting by the year 1992.

To put the question of federal involvement in broadcasting in a broader perspective, we should point out that in many of the world's democratic nations, such as Great Britain, Canada, Australia, and New Zealand, public television is governmentally owned and operated. In Canada, for instance, federal support for television equates to $25 per year for each citizen. In the United Kingdom, the amount is $18, and in New Zealand, $38. By contrast, in the United States the federal appropriation for 1986 of $159.5 million amounts to about 67 cents per citizen.

With the 1967 Carnegie Commission Report, the foundation was laid for taxpayer support and control of American public broadcasting. A nonprofit corporation called the Corporation for Public Broadcasting (CPB) was established to develop educational broadcasting that would be objective and balanced when controversial, to facilitate distribution of the programs, and to assure maximum freedom of the broadcasters from interference with or control of program content or other activities. Unfortunately, CPB resulted in a bureaucratic maze which more often than not has been in conflict with PBS, its stepchild representing the individual noncommercial stations. Because the CPB members are appointed by the United

States President, politicizing became a factor, especially under Richard Nixon and Ronald Reagan. Their appointees influenced budgeting to minimize controversial programming such as "The American Dream Machine," which aired unflattering segments about Nixon, and a multi-part series on the history of the Vietnam conflict.[4]

Conservative commentators have lauded the Reagan administration's call for the end to federal support for broadcasting, saying the government had no business being in the broadcasting arena in the first place. Some have suggested that if the public and private businesses want to have public broadcasting, they should pay for it themselves rather than ask the federal government to do so. Pointing to documentaries and news specials they say have liberal slants, conservatives raise questions about why the federal government should support "this toy of the progressives." One has pointed out that public broadcasting has an influence on public policy disproportionate to the numbers of viewers and listeners, but that influence consistently tends to buttress the agenda of the left.[5]

Given the political controversies and the tenuous state of federal funding, public broadcasters have looked elsewhere for support. From the late 1960s to about 1980, support from increasingly strapped local governments and schools decreased by about half. Foundation support, especially from the Ford Foundation, dropped similarly. To offset these decreases, support from individuals tripled as a result of membership drives and those interminable on-air auctions. In the early 1980s, some PBS stations experimented with direct commercial sponsorship, but a two-year study concluded that advertising would never produce enough revenue to keep the public stations on the air.

PBS and local public TV stations have learned to scrounge funds from a variety of sources.

Since their inception, PBS and local public television stations have also gone hat in hand to corporations, which now finance nearly the same percentage of original broadcast hours as the federal government. This is an especially sensitive issue. AT&T, for instance, which is emerging as the nation's leading provider of electronic information and which stands to gain by earning a favorable public image, underwrites "The MacNeil/Lehrer News Hour." Likewise Prudential-Bache Securities underwrites Louis Rukeyser's "Wall Street Week." Several of the top corporate contributors are Exxon ("Great Performances"), Mobil ("Masterpiece Theatre"), Gulf, and

Atlantic Richfield, prompting critics to label PBS the "Petroleum Broadcasting System." PBS has adopted strict rules to bar underwriters from exercising any control over programming matters, including the selection of topics, but the potential for abuse exists nevertheless. For instance, an inflammatory docu-drama depicting the execution of an adulterous Saudi Arabian princess and her lover, sponsored in part by the Mobil Oil Corporation, drew a great deal of attention to PBS in 1980. The Saudi Arabian government expressed its displeasure to American officials; acting secretary of state Warren M. Christopher (along with Mobil Oil spokespersons) passed the protest along to PBS executives, who thought twice before airing "Death of a Princess."

In their search for programming independence, public television stations are encouraging the establishment of a billion dollar program endowment from private sources. Another source being explored involves having Congress set up an Independent Public Broadcasting Authority supported by a tax on the profits earned by commercial broadcasters. Former PBS and CPB executive John Wicklein notes that even a 2 percent tax imposed on broadcasters, who currently have free use of the broadcast airwaves, would generate some $400 million a year for public broadcasters. "In return for this tax, commercial broadcasters could be released from their present legal obligation to serve the public interest in their programming, allowing them to concentrate entirely on programs that maximize profit," Wicklein argues.[6] Variations of this proposal have gained some support in Congressional and commercial broadcasting circles, and it will be interesting to see what develops. As things now stand, federal support of noncommercial broadcasting is declining at the same time the government is lessening its controls over commercial broadcasters. What finally develops may be a far cry from the current economic and programming morass.

PBS Programs and Audiences

Essentially, PBS provides only distribution, programming, and services formerly performed by the controversial NET to its ETV affiliates, who are under no obligation whatever to carry the network. Stations may use the material or disregard it, as they wish, or they may tape it for later use. Most stations augment the PBS feed with inexpensive local and filler material, and home classroom subjects.

To date, PBS's major accomplishments include several English-imported series, including "Masterpiece Theatre" and "Civilisation," which it acquired only after all three commercial networks had turned it down (it was a gift from Xerox corporation). PBS's reliance upon the British programs has made more than one observer quip that public broadcasting would be in deep trouble if the British did not speak English. Other notable PBS programs, primarily homegrown, are "Sesame Street" and its offshoot, "Electric Company," produced by the Children's Television Workshop (CTW), originally foundation supported; "Over Easy," geared to the elderly, a demographic group long since abandoned by the commercial webs;

and science series such as "Nova," "Cosmos," "The Ascent of Man," "The Incredible Machine," and other National Geographic specials. Since the early 1980s, more than eight hundred colleges and universities have joined PBS's "Adult Learning Service," offering a wide variety of college courses to a quarter-million Americans. The programs are interesting and well produced, a far cry from many of ETV's talking heads. "The American Story," "Understanding Human Behavior," "Focus on Society," "The Business of Management," "Heritage," "Civilization and the Jews," "Congress: We the People," "The New Literacy," "The Constitution: That Delicate Balance," "The Brain," and "The Write Course" are among current offerings.[7]

Despite high-quality programming, PBS audiences remain minuscule.

Audiences for PBS remain minuscule, by network standards. Although PBS was supposed to be free of the tyranny of ratings in its operation, it has succumbed to an inevitable curiosity for evaluation. The results have not been encouraging. In the early 1970s, Nielsen gave it a rating of 0.4 percent of all viewing. Things have since improved, but not by much. Different studies of late indicate that somewhere between twenty and forty million Americans are exposed to PBS in an average week. But on any given night, even today, only about 5 percent of American homes will be tuned to PBS programming.

In the final analysis the persistent problem for PBS is that it is broadcast over the same air at the same time as CBS, NBC, and ABC. For a third of a century it has been the networks' business to find out what Americans want to watch and they have done this well; anything else simply cannot compete. Couple this with the growing demands on America's time made by videocassette recordings, movie channels, direct satellite broadcasting, and independent television networks, and it is little wonder PBS is struggling. A person can subscribe to *Reader's Digest* and *American Heritage* and read them both, but a person cannot watch CBS and PBS simultaneously.

The Ratings

Ratings are a form of feedback for programmers and advertisers.

The commercial cycle is comprised of a manufacturer with a product, a distributor, a consumer, and, generally, a financier. In television, the networks are the manufacturers, and programming is the product, which is distributed via affiliates to the audience-consumers, and advertising provides the financing. The manufacturer can judge the success of a product by the quantity purchased by consumers. But in television, the product is distributed without charge; there is no ring of the cash register. Instead, ratings are a form of feedback by which the networks and advertisers gauge the effectiveness of product consumption.

The ratings had their genesis in the days of early radio when the huge costs of production demanded an early warning system for measuring audience acceptance. Since the advent of television, technological improvements in computers and the refinement of public opinion polling techniques have developed the ratings into a near science.

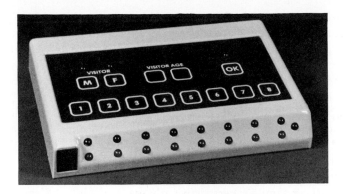

Using a sample of about 2,000 households, the A. C. Nielsen Company measures the program tastes of over 200 million Americans in 87 million television homes. Numbers of that magnitude occasionally provoke criticism. Even Nielsen, the largest of the half-dozen major polling firms for national television (in some 200 independent firms conducting ratings for radio and television), is less than perfect; for example, the Rocky Mountain states have long been underrepresented. But, in general, the measurements are accepted as valid by the networks. Perhaps more important, advertisers consider the Nielsen ratings to be official, so they become official.

For all their shortcomings, ratings constitute the only valid feedback in the TV business.

Each year the three networks spend millions of dollars for the Nielsen service, and receive some 90 percent of their audience feedback from that single source. Nielsen's ratings have long involved two feedback systems, the *audimeter* and the *diary;* recently, a new system, the so-called *people meter,* has been introduced. Storage Instantaneous Audimeters (SIAs) are tuning meters with recording circuits inside which are hooked to carefully selected sets in some seventeen hundred American households. The SIA silently measures all TV set usage within each "Nielsen family" household, recording when each set is turned on, how long it stays on the channel to which it is tuned, and all channel switchings. It even captures this information on backyard or patio sets, via a transmitter mounted on the TV set and linked to the audimeter, which is installed out of the family's view, in a closet or in the basement. Each SIA is connected by special telephone lines to computers in Dunedin, Florida, where they are electronically read at least twice a day.

The audimeters indicate whether the sets are turned on, but Nielsen also tries to find out whether anyone is watching. Thus diaries are placed in another twenty-three hundred households for thirty-four weeks each year. This sample of television homes, known as the National Audience Composition (NAC), is also equipped with a Recordimeter, logging the amount of time that each set is in use. The diary keepers are then expected to record daily on their Audilogs the same information about set and channel usage picked up by the Recordimeters.

Nielsen's

Nielsen uses the *audimeter, diary,* and *people meter* to rate programs. Complex and immediate ratings are possible.

Until recently, audimeter tapes were mailed to Nielsen every two weeks, but computer sophistication and programming decision-making dictates have grown apace, and now instantaneous feedback is possible. Currently, Nielsen is able to provide networks with complete reports of prime-time audiences on the second weekday morning following the programs; the networks and major clients have data terminals in their own offices connecting them to the Dunedin computers. In addition to these daily national ratings, Nielsen provides "weekly household ratings" on the second Monday following the end of each report week, and ratings in several other categories including ratings in top markets and ratings broken down by demographic characteristics.[8]

From the Nielsen family's perspective, the responsibility of each viewer representing the tastes of over 100,000 other Americans is sometimes overwhelming. One diary keeper told an ABC documentary team that he was so frustrated by the chore of recording all his selections that he started sending in marked-up copies of his *TV Guide,* and he felt such strain in representing 100,000 people he had never met that he soon quit altogether.[9] Others have admitted to upgrading their program taste while under Nielsen's surveillance out of concern that the rest of the viewers should be watching "better" programs than they naturally prefer. The Nielsen people claim they can compensate for these errors, however. For one thing, since Nielsen families are plugged into the system on five-year rotating schedules, there is little likelihood of systematically manipulating their inherent television preferences for the duration. For another, the diary and Recordimeter system is used to offset any systematic errors—such as sets left on while no one is at home—that crop up in audimeter homes.

Arbitron

Nielsen's nearest competitor in the ratings business, and the only other company accredited by the Broadcast Ratings Council, is the Arbitron Company (ARB), which measures both radio and television audiences. Arbitron relies primarily on weekly diaries kept by twenty-four hundred randomly selected households throughout the nation, augmented by recording meters placed in another twelve hundred homes. Arbitron families receive one diary for each television set and one diary for each person over the age of twelve for recording radio listenership. In addition, Arbitron has long employed a *telephone coincidental survey* in which viewers and listeners are queried on the spot, so to speak, about their current media consumption. Periodically Arbitron also uses personal interviews to ascertain more specialized data or to check on its own sampling procedures. Perhaps the biggest difference between the two major ratings services is that Nielsen's ratings concentrate on national or network programming and Arbitron's are concerned with more local audiences; numerous local stations subscribe to the Arbitron service, while Nielsen gets most of its income from the webs.

Of all the reports issued by the services, Nielsen's Television Index is probably the best known and most influential. Based on audimeters and, since the fall of 1987, people meters throughout the land, the NTI generates several types of data, the two most important being *ratings* and *shares*.

The *rating* for each program reflects the number of sets tuned to that show in comparison to the other 87 million television households that conceivably could be viewing it. Ratings are expressed as a percentage of all television households, regardless of whether the sets are turned on.

Nielsen gives "average audience ratings" or estimates of the number of households tuned to each network program during the average minute of the program, expressed as a percentage of all television households, and "total audience ratings" or estimates of households tuned to a particular program for six or more minutes. In addition, "persons ratings" offer such demographic data as sex, age, and family role of viewers of each program. In network television, a successful prime-time program has a rating of 17 or more, meaning that 20 million American homes are tuned in at any given moment to the show in question. The rating, then, reflects how many have been captured by a particular show, not necessarily in contrast with other shows but in contrast with everything else there is to do—eat, sleep, work, make love, or what have you.

The audience *share,* on the other hand, is a gauge of how a particular program stacks up against its other competition. Based only on the number of TV sets turned on, shares are the percentage of households watching a given program at any particular moment, compared with all the other shows they might be watching. This, according to network executives and advertisers, is a more crucial figure than the ratings, and it is easy to see why.

For decades broadcasters have hoped to capture at least one-third of all viewers, since—ideally, from the broadcasters' standpoint—the viewers will be watching one of the three networks at any given time. Such a situation, of course, is unreasonable nowadays, given the incursion of cable television, videotape playbacks, independent networks, PBS, and other options that have become increasingly available. Most American households (83 percent, as of 1986) have their choice of nine different channels, and some 30 percent of us can get more than twenty channels, so the goal of capturing a third of the viewers is elusive. On any given evening in the 1970s more than 90 percent of the television audience could be found tuned in to NBC, CBS, or ABC, but that is no longer the case. By the mid-1980s, after years of slow but steady erosion of their stranglehold on viewers, the three networks were pulling less than three-quarters of the viewing audience. The remaining quarter of the viewers were using their sets for other purposes. Meanwhile, the networks' combined ratings, of less than 50, indicated that fully half of the American audience had literally turned off their sets, and had found something better to do with their time. That figure,

Ratings and Shares

Ratings measure the households viewing a show, compared with all the potential TV viewing households.

Shares measure the households viewing a show, compared with households viewing other shows at a given time.

Today ABC, NBC, and CBS attract a decreasing percentage of TV viewers.

combined with decreases in HUT ratings—the percentage of Homes Using Television during any given period—reflects America's diminished infatuation with television.[10]

A share of 20 is now considered quite satisfactory, and anything over a 30 is considered an absolute hit. Thus, when "The Cosby Show" on NBC averaged a 47 percent share of the viewing public throughout its 1985 season, and hit a peak of 51 at one point, NBC had good cause to celebrate. The weekly audience of 62 million was more than the combined attendance (46.8 million) of every major league baseball game that year, more than eight times the combined attendance at all Broadway shows (7.4 million), more than the attendance for 1985's biggest movie, *Back to the Future* (50 million), and more than the number of voters it took to elect President Ronald Reagan in 1984 (53 million).[11]

For the last decade or so, it has not been enough to merely receive an audience totaling one-quarter to one-third of all viewers. Those viewers must be demographically suitable product purchasers. Nielsen now offers advertisers and programmers profiles of audiences, including their age, sex, education, socioeconomic status (ses), and so forth. Advertisers are enchanted with the young and relatively affluent, and with women, who comprise the majority of the viewing audience at all times during the day, drawing only approximately even to men during the late news. Networks charge the highest rates for shows appealing to young adults between the ages of eighteen and thirty-four, because that is the age group making the most use of kitchen, cosmetic, automotive, and drug products being advertised. The second most valuable audience, from an advertising perspective, includes viewers between the ages of eighteen and forty-nine. The assumption is that these groups are stocking up their households with what *New York Times* television critic Les Brown calls the paraphernalia of middle-class life.[12] As a reflection of their fixation on this audience, the advertisers spending the most money on television include Procter & Gamble, General Foods, American Home Products, General Mills, General Motors, Pepsico, Lever Brothers, McDonalds, Bristol-Myers, and Coca-Cola. The arena demonstrates America's conspicuous consumption of nonessential goods and services.

If it were just a matter of finding out how many people are watching television at any given time, the ratings services would not be embroiled in much controversy. However, driven by commercial considerations, the television industry has more than a passing fascination with its audiences. Because a single ratings point on a network prime-time show means between $61 million and $65 million in advertising revenues over the course of a year while a ratings point for a daytime show means $50 million in revenues, the networks and advertisers have intensified their search for accurate and detailed ratings data. They want to know not only what sets are turned on, but who is watching, how much time they spend watching, whether or not they watch the commercials, and how motivated they are to go out and purchase the products being advertised. As things now stand, it is nearly impossible to get answers to all these questions.

Things were simpler in television a few years ago, when television viewing was a routine family activity, and head counts told advertisers and programmers all they needed to know. Nowadays, however, viewing habits have changed, and so has their measurement. Multiple sets in most homes have splintered the audience, even within a given family unit. Alternatives to network programs—the aforementioned VCRs, cable offerings, etc.— have done likewise. Remote control devices permit, or should we say encourage, "zapping," or skipping across channels whenever commercials come on or when a program gets a little dull. Given all that zapping and skipping, neither the automatic meters nor diaries in Nielsen and Arbitron households are very likely to keep perfectly accurate account of America's overall viewing habits. This, of course, causes programmers and sponsors a mild headache. After all, those with a vested interest in television contents would like to believe that viewers are totally hooked on programs and held captive by commercials.

Surveys of VCR users tell advertisers and programmers that they are facing some additional problems that have heretofore gone unrecognized. For one, VCR viewers who are playing back prerecorded programs zap or fast-foward through commercials much more frequently than previously believed. For another, the Nielsen ratings count homes using VCRs to record programs as homes actually *watching* those programs, but surveys indicate that 75 percent of the recordings are made while no one is at home or viewing, and only 80 percent of those home-recorded shows are ever

Advertisers have intensified their search for accurate and detailed ratings. Meanwhile, the process has grown troublesome, with Orwellian overtones.

played back. Nielsen has no way of knowing who sees those shows, because playbacks on VCRs are not measured—we could be watching the shows we taped off the air, or a rental movie, or a video of Cousin Millie's wedding. It is a fair bet that only about one third of the commercials that have been taped on home VCRs will ever be watched by anyone, and Madison Avenue is not very happy.

To refine the art of measuring audiences, the ratings services have come up with some clever but controversial alternatives to their diaries and audimeters. One, called the "people meter," is being utilized by both Nielsen and Arbitron, with results that are both exciting and frightening— depending upon whom you ask. The people meter, designed by a British company called AGB Television Research, electronically records each individual family member's viewing behavior and correlates that minute-by-minute viewership with demographic data such as age, gender, income, ethnicity, and education. The system only works if all viewers punch in their code number when they start watching TV, and then punch out once they have finished. That way, it doesn't matter how much zapping or channel switching goes on—the computer will keep track of every bit of conscious and random viewing, by every member of the household. Naturally, the networks, cable companies, movie channels, and advertising agencies love the people meter. Whether people love it remains to be seen. No matter; it is now a fact of life. In late 1987 Nielsen began measuring its national audience with a single sample of two thousand households equipped with people meters. In September of 1988, the sample was expected to increase to four thousand homes. Meanwhile, Arbitron, in connection with Scan America, was planning to have five thousand people meter households on line by the 1988–89 television season.

"Refinements" of the basic people meter have Orwellian overtones. Sponsors are concerned that many viewers will not punch in their personal codes if they do not want anyone to know they are watching a particular show, and worry that many viewers may forget to punch out when leaving the room. So AGB and the ratings companies have dreamed up some alternative ways of keeping track of us, even when we do not want anyone to know what we are up to. One suggestion is to implant tiny electronic "bugs" in the navel of each viewer, or to strap individualized electronic bracelets on each viewer. The electronic impulses would automatically activate the recording meter when the "bugged" person enters a room where a set is turned on. Another proposal is to make all television two-way, with photoelectronic eyes built into them, so the sets themselves can keep track of who is viewing and who is not. Still another proposal, dubbed the "whoopee sofa," calls for furniture to be wired with infrared sensors that keep track of varying temperatures of family members' rear ends and therefore know who is in the audience and who is not.

These ratings systems, as sophisticated, bizarre, and obviously intrusive as they may sound, still do not answer the fundamental twenty-billion-dollar-a-year question: "Is anybody buying the stuff that's advertised?" Obviously, there is some relationship between the commercials on the air and the products that move off grocery and department store shelves, but even the best of today's research and ratings systems cannot demonstrate a direct cause-effect relationship. To get that elusive information, Nielsen, Arbitron, AGB, and a new company known as Information Resources, Inc., have designed a "magic wand." Families who are willing to be marketing guinea pigs will have a pencil-sized electronic wand built into their home TV meters. When they go shopping, they will remove the wand from the TV meter and wave it over the universal product code stamped on whatever packaged goods they buy. The computers will do the rest. "The rest," obviously, is to correlate the TV commercials and the purchases with all the demographic data built into the TV meters, so advertisers and marketers will know precisely what kinds of messages, under what kinds of conditions, have stimulated what kinds of consumers to make what kinds of purchasing decisions.[13]

From one point of view, these new ratings systems are exciting breakthroughs. After all, programmers and sponsors ask, why should they continue spending billions of dollars a year trying to reach random, invisible, and elusive consumers? On the positive side, programmers maintain that all these data will permit them to more carefully target potential audiences—giving high-brow, special interest programming to those few (few hundred thousand or few million) selective audiences who are highly likely to purchase the top-of-the-line, specialized products being advertised on the program. Entirely new and high quality shows could be created freed of the current tyrannical system under which anything not reaching twenty million viewers is considered a failure. At least, that is the argument of Madison Avenue.

Civil libertarians, on the other hand, express horror at the entire concept of people meters. They say that George Orwell was right when he prophesied a society that unwittingly gave up all semblance of its privacy, in the name of social efficiency. The year 1984 may have passed without the people meter being in widespread use, but the Orwellian *1984* may be at hand, thanks to a reasonable request by the television industry to learn more about its audiences.

A couple of final observations about the ratings. Currently, Nielsen and Arbitron conduct "sweeps" of local television audiences during the months of November, February, and May, on the assumption that these months reflect routine viewing habits. Even the casual viewer has come to recognize when *sweep weeks* are in effect, for it is then that local stations carry on multipart series of fascinating (if not always significant) news reports, and the networks lead off the weeks with blockbuster films and other specials guaranteed to draw an audience.

Sweep weeks and *black weeks* result in out-of-the-ordinary programming.

Within the rating systems there are also *black weeks,* periods during the year when no ratings are conducted by network agreement. Recognizing their obligation to provide a certain amount of public-affairs programming under FCC dictum, and also that such cultural programming loses large chunks of the audience, the networks have used black weeks as a device to satisfy the public interest, convenience, and necessity requirements for their affiliates (and wholly owned stations) at one fell swoop. Black weeks are filled with documentaries and cultural and public affairs programming, which the networks can offer with impunity because there are no ratings with which to be concerned.

Programming

Television programming is a highly competitive and incredibly expensive gamble always conducted with one eye on ratings. Prime-time programming is not so much concerned with the thought of developing an enormously popular program, although this helps, as it is with providing an evening's continuity across the prime-time hours and slotting shows to compete with offerings on the other two networks. Historically, one of the most closely guarded secrets of the networks was their forthcoming season's schedules. Prior knowledge of one network's schedule would permit the others to slot competing attractions against its most appealing offerings. Over the past few seasons, with networks willingly jumbling their prime-time offerings around, sometimes at a moment's notice, this secrecy has been less a factor. The end result has been a running corporate chess game of sorts, with programs and viewers acting as the pawns.

The continuity of programming is significant. A weak show will lose an audience not only for a particular time slot, but possibly for the entire evening. This is disastrous in ratings and, hence, in advertising revenue.

When a show fails in the ratings, it is often necessary to kill it outright. It is preferable to gamble with another program than to continue a proven loser, regardless of how devoted the supporters of the program may be. More often it is shifted to a different slot, in what is usually a sacrificial move. Certain viewing patterns are invariably destroyed by a move, which will further depress the ratings. In addition, the move is often to a slot against a highly rated competitor on another network. A few years ago, the critically acclaimed "Paper Chase" had a difficult time with the ratings. Instead of killing the show right away, CBS moved it several times. This allowed the network to recoup a portion of its investment in the program and to satisfy contracts it had with the performers. Eventually, however, failure was used as the excuse to drop the show. Complaints from loyal viewers may explain why star John Houseman and company later returned to the air in additional episodes filmed for cable TV.

A show's replacement, according to the theory of the *least objectionable program (lop),* does not have to be good, just less objectionable to most people than anything else on TV at that moment. Thus, it is possible for inferior shows to develop high ratings during prime time, provided the

The theory of "lop" says that the "least objectionable programming," not the highest quality, will get the best ratings.

alternative offerings by competing networks are worse. In view of the costs of television production, this is a key point. There is no need for a network to spend exorbitant sums of money to develop an expensive program for competition against a weak program when a cheaper one will suffice.

The preoccupation with ratings, of course, means that each network must attract as much audience as it possibly can and if possible outdraw the other two networks proportionately. Appeal to the largest possible mass is an appeal to the lowest common denominator and dictates a broad-based (usually noncontroversial) type of programming. And the program is more likely to succeed if a star is associated with the show. There is also a good deal of formula programming and imitation. Whenever one network develops a successful format, be it police beat, hospital, situation comedy, quasi-musical, mini-series, or real people exhibiting their eccentricities, others are sure to follow. A case in point was the rash of handicapped detectives who came in wheelchairs, or were blind, obese, obsequious, or senile, but always got their man. Even in the daytime hours, networks copy one another's game and quiz shows. Such repetitive programming is bound to offer a striking sameness to network programs. See table 8.2 for a list of the top-rated shows since 1950.

Typically, as the program directors for each network arrange their season schedules, they decide which programs will be retained from the previous year and which will be discontinued. They are then faced with placing their shows in the most appropriately competitive time slots, shooting largely in the dark. They are also faced with the problem of deciding what new programs they should put into the schedule, relying on proven formats and star availabilities. They then entertain a number of suggestions for new programming, new serials, new approaches to old problems, new and old movies, specials, and so on.

In the case of new programming, the program directors generally commission pilot programs from independent production companies for several of the most promising ideas. Pilots are expensive to make, because they are actually full-length films or videotapes of a new serial. They introduce the characters, the theme, and the treatment that will be followed through the entire schedule—unless, of course, audience reaction to the pilot dictates substantial changes. Because of their costs, many pilots will be shown several times as television movies and then sold to syndication companies for foreign showings on television and in theaters.

The majority of pilot films never become episodic series. Program directors and advertisers determine whether to contract with program suppliers for additional episodes on the basis of instincts, scientific pretests, and ratings of on-air pilots. Until recently a go-ahead meant production of a full year's—actually twenty-six—episodes. Nowadays, with competition and inflation taken into consideration, it is not unusual to order as few as four or six episodes, as the networks have come to think in terms of three "seasons" rather than one or two.

Next Season's Schedule

Planning a season's prime-time schedule is a corporate chess game.

Table 8.2 Top-rated television shows since 1950

1950–1951*	1. Texaco Star Theatre (NBC)	6. Gillette Calvalcade of Sports (Boxing) (NBC)
	2. Fireside Theater (NBC)	7. Arthur Godfrey's Talent Scouts (CBS)
	3. Your Show of Shows (NBC)	8. Mama (CBS)
	4. Philco Television Playhouse (NBC)	9. Robert Montgomery Presents (NBC)
	5. The Colgate Comedy Hour (NBC)	10. Martin Kane, Private Eye (NBC)

1954–1955*	1. I Love Lucy (CBS)	6. Disneyland (ABC)
	2. The Jackie Gleason Show (CBS)	7. The Bob Hope Show (NBC)
	3. Dragnet (NBC)	8. The Jack Benny Show (CBS)
	4. You Bet Your Life (NBC)	9. The Martha Raye Show (NBC)
	5. The Toast of the Town (CBS)	10. The George Gobel Show (NBC)

1959–1960*	1. Gunsmoke (CBS)	6. Father Knows Best (CBS)
	2. Wagon Train (NBC)	7. 77 Sunset Strip (ABC)
	3. Have Gun Will Travel (CBS)	8. The Price is Right (NBC)
	4. The Danny Thomas Show (CBS)	9. Wanted: Dead or Alive (CBS)
	5. The Red Skelton Show (CBS)	10. Perry Mason (CBS)

1964–1965*	1. Bonanza (NBC)	6. The Red Skelton Hour (CBS)
	2. Bewitched (ABC)	7. The Dick Van Dyke Show (CBS)
	3. Gomer Pyle, U.S.M.C. (CBS)	8. The Lucy Show (CBS)
	4. The Andy Griffith Show (CBS)	9. Peyton Place (ABC)
	5. The Fugitive (ABC)	10. Combat (ABC)

1969–1970*	1. Rowan & Martin's Laugh-in (NBC)	6. Here's Lucy (CBS)
	2. Gunsmoke (CBS)	7. The Red Skelton Hour (CBS)
	3. Bonanza (NBC)	8. Marcus Welby, M.D. (ABC)
	4. Mayberry R.F.D. (CBS)	9. Walt Disney's Wonderful World of Color (NBC)
	5. Family Affair (CBS)	10. The Doris Day Show (CBS)

1974–1975†	1. All in the Family (CBS)	6. Rhoda (CBS)
	2. Sanford and Son (NBC)	7. Good Times (CBS)
	3. Chico and the Man (NBC)	8. The Waltons (CBS)
	4. The Jeffersons (CBS)	9. Maude (CBS)
	5. M*A*S*H (CBS)	10. Hawaii Five-O (CBS)

1979–1980†	1. 60 Minutes (CBS)	6. Dallas (CBS)
	2. Three's Company (ABC)	7. Flo (CBS)
	3. That's Incredible (ABC)	8. The Jeffersons (CBS)
	4. M*A*S*H (CBS)	9. Dukes of Hazzard (CBS)
	5. Alice (CBS)	10. One Day at a Time (CBS)

1984–1985†	1. Dynasty (ABC)	6. The A Team (NBC)
	2. Dallas (CBS)	7. Simon and Simon (CBS)
	3. The Cosby Show (NBC)	8. Knots Landing (CBS)
	4. 60 Minutes (CBS)	9. Murder, She Wrote (CBS)
	5. Family Ties (NBC)	10. (tie) Falcon Crest (CBS)
		10. Crazy Like A Fox (CBS)

Sources: Craig T. and Peter G. Norback, eds., *TV Guide Almanac* (New York: Triangle Publications Inc., 1980), pp. 546–69; and *Broadcasting Magazine,* with permission.

*The television season from October to April

†The television season from September to April

Network "seasons" depend upon the ratings. As soon as the first ratings are in, never later than November, decisions are made to juggle or cancel programs. The second sweeps in February cause another shake-out, and March thus ushers in the third season. At the outset of the second and third seasons, new programs are matched up with established hits so viewers who do not bother switching channels will be lured into joining the audience. Mediocre shows are juggled to new time slots, sometimes in hopes they will find an audience, other times with the expectation that they will quietly fade away. Flops are replaced by new series. The networks introduce the second and third seasons with the same hype and ballyhoo traditionally reserved for the early fall season, despite the obvious fact that the new series they are introducing were not considered good enough to cut it in the first place.

Usually networks ask program suppliers to turn out twelve or thirteen episodes, enough for half a season. If ratings hold up, additional scripts and films will be ordered. If not, it is unlikely that audiences will ever see the episodes again, and the networks will write off the expenses as gambling losses. And the production companies, facing the same gambles, inevitably respond by cranking out clones of their already successful series.

Non Prime Time

Prime time, with its mass audience and lowest common denominator, is one situation in which television must find the way to appeal simultaneously to the sixteen-year-old Puerto Rican girl in New York City and the eighty-year-old North Dakota farmer and everybody between. The goals are somewhat different during non-prime-time hours.

Weekdays The daytime audience is overwhelmingly female: the home-makers of America—the purchasers of America. Daytime television is an incredible mixture of quiz shows, game shows, soap operas, old movies, and reruns. There is also a daytime audience of the elderly, the ailing, and the unemployed, but they are scarcely considered by programmers, for they have little purchasing power. Daytime television also tends to be inexpensive programming, as the networks conserve their dollars for the competitive prime-time hours. Game shows cost only one-twentieth as much to produce per half hour as prime-time shows, but they generate proportionately higher advertising returns. Much of daytime programming consists of packages made by independent producers, and often as not are contributed to heavily by advertisers. Prizes are contributed by manufacturers in return for product mention. Because costs to the network are relatively small, and the drawing power of these programs is high, daytime programming is taken quite seriously in corporate offices.

Children's Television Saturday morning offers another specialty, kiddie cartoon time, in what has become one of the most contentious arenas in the mass media. In days gone by, breakfast food manufacturers, toy makers,

Children's TV has blurred the distinction between entertainment and persuasion.

Program-length commercials have come to dominate children's television, to the chagrin of many watchdog groups.

and others hawked their wares on the commercials that accompanied these independently produced cartoons. Recently, however, the sponsors have taken over program production, and the result is known as "program-length commercials." The characters in forty or so after-school and Saturday morning programs are based on toys, dolls, games, breakfast cereals and other products—He-Man and the Masters of the Universe, Care Bears, GI Joe, Smurfs, Thundercats, Voltron, Transformers, Gobots, and the like. The multibillion-dollar-a-year toy industry has become so tied to the children's television that toymakers will not even consider introducing a new product unless it is tied directly to a children's television program. As a result, according to such critics as Action for Children's Television, children are confused over the distinction between program content and commercial speech.

In the face of challenges by Action for Children's Television, the national Parent Teachers Association, National Education Association, Dr. Benjamin Spock, and leading media researchers, a bill was introduced in Congress, "The Children's Television Act," proposing that the Federal Communications Commission require stations to air seven hours of educational programming per week. That 1985 proposal, plus an earlier one asking the FCC to reclassify the "program-length commercials" as paid advertising rather than entertainment programming, met a certain amount of indifference, if not hostility, from a federal government bent upon loosening its programming requirements for the broadcast industry.

Meanwhile, Mattel has sold some 125 million He-Man and the Masters of the Universe plastic "action figures," and weapon-laden Transformers robots have bumped the Cabbage Patch dolls out of the most-popular-toy slot. In 1985 alone, some $1.3 billion were spent in the United

States on toy guns, war toys, and war games that had been advertised on television. That figure represented a 600 percent increase in sales between 1982 and 1985. And it may also represent what one psychologist called "the most massive sale of war ideology to a generation of children in any modern democracy with the exception of Hitler's Germany."[14]

Tie-ins between toys and entertainment are nothing new. Superman cloaks, Davy Crockett coonskin caps, Peanuts lunch pails, and Sesame Street paraphernalia have been part of popular culture for generations. What concerns critics today, however, is the blatant manipulation of a child's love of cartoons into the general acceptance of—if not craving for—toys that celebrate deadly force and other anti-social values. They also cite the preponderance of these cartoon shows (1.5 hours a week in 1982, 27 hours a week in 1985, 42 hours a week in 1986) to the exclusion of "Captain Kangaroo" and other socially positive programs.[15] Substantive research on the true relationships between viewing such programming and children's actual behavior is still embryonic, so the debate among toy manufacturers, programmers, governmental agencies, parents, and educators most likely will continue for some time to come.

Sunday Mornings Sunday morning, a time of minimal viewing, has long been considered the "cultural ghetto" of television, but things are changing. Religious programs are quite naturally slotted into this period, as well as a considerable number of public affairs and quasi-cultural programs, to build up the public interest, convenience, and necessity portion of the weekly log. Televised religion ("televangelism") has grown to a controversial $500 million a year industry, as many video vicars have moved off the nets and onto their own satellite-fed channels, from which they reach millions of viewers and reap millions of dollars from direct donations. Pat Robertson, Jerry Falwell, Jimmy Swaggart, Oral Roberts, James Robison, Rex Humbard and Robert Schuller have become as recognized as the Mormon Tabernacle Choir—the latter a staple of Sunday TV and radio broadcasting for more than a half century. All in all, Sunday morning is no longer limited to amateurish broadcasts of local church services. A recent Nielsen study commissioned by the Christian Broadcasting Network revealed that more than 40 percent of the nation's television households watched one or more of the top ten syndicated religious broadcasts during the month of February, 1985. That audience, 61 million strong, is equivalent to the number of people who watch most prime-time network programs.[16]

CBS's "Face the Nation" and NBC's "Meet the Press" have been stable Sunday morning public affairs programs for decades. More recently, ABC's "This Week With David Brinkley" has also discovered an audience for news and feature stories during that time period, and the cultural ghetto hour has begun to show signs of economic health.

Sunday morning is no longer the "cultural ghetto" of television.

Costs of Programming

Television programming costs are enormous. Each of the three major networks spends some $1.5 billion annually on some fifteen hundred hours of programming, but collectively the networks still manage to make a profit. The least expensive programs, as noted earlier, are daytime game shows, which cost between $120,000 and $150,000 in license fees per week. These costs are low because advertisers pick up many of the expenses. Daytime serial dramas cost the networks an average of $550,000 per week, or $55,000 per half hour episode, fees that are also considered quite low when compared with the substantial advertising revenue derived.

Typical costs for half-hour network prime-time drama or action-adventure programs are in the $325,000 range, and hour-long shows run $750,000 to $800,000. (Pilots, as we have said, cost nearly twice as much.) The average price networks pay for a made-for-TV film is $2.5 million, a fee that gives them permission to show it twice over a four-year period. To purchase rights for a perviously released theatrical film, networks pay around $3 million, down substantially from the days before those theatrical releases were readily available, uncut, on cable TV or from local VCR rental stores.

The prices cited above are average, and the range is enormous. For instance, in the late 1970s "Battlestar Galactica" gained fame as the most expensive one-hour regular series ever produced in television, a million dollars an hour. More recently, Steven Spielberg's "Amazing Stories" became the most expensive half-hour show on the air, with each episode costing between $800,000 and $1 million. Mini-series such as *Peter the Great, Amerika, Shogun, The Thorn Birds, The Winds of War* and others typically cost $1.5 million per hour, with total costs in the $25 million range not unheard of.

Where does all that money go? It is quickly spread to writers, directors, cast, crew, sets, wardrobes, transportation, special effects, and publicity. For example, Aaron Spelling Productions' ABC hit, "The Love Boat," cost $100,000 per show for five producers (a line producer and two pairs of creative producers), $125,000 for the cast (each celebrity guest receiving $5,000 for an appearance), $40,000 for extras (faces in the crowd and walkons), $40,000 for writers (three teams, each preparing one-third of every weekly script), and $35,000 for music (a thirty-piece orchestra for six hours). Once miscellaneous expenses were included, the typical episode cost $350,000, not counting the $150,000 for the eighty to one hundred people needed to produce and film the series, the $40,000 in fringe benefits, and another $50,000 in postproduction editing. If we add to this the fee Spelling Productions paid Twentieth Century Fox for studio space and the fees to compensate writers and performers for the network's second run of the episode later in the season, "The Love Boat" was weighing anchor to the tune of $650,000 per episode. (Other programs face special economic problems. It would seem that shows such as ABC's "Foul-Ups, Bleeps & Blunders," and NBC's "Bloopers & Practical Jokes" would be low budget

TV production costs are enormous, and some of the financial arrangements downright mystifying.

Each episode of "Love Boat" cost $650,000 to produce, but earned only $550,000. The profits come from syndicated reruns.

operations, since most of their time is spent in showing previously filmed or taped bloopers and jokes. However, the half-hour shows typically cost $300,000 to $500,000 a week, with a large chunk of that going to the enormous number of fees paid to everyone who appears in any segment of the show—with extra fees going to those reluctant to have their foibles shown in public.)

ABC, by placing "The Love Boat" in the profitable 9 to 10 P.M. EST slot on Saturday nights, was able to justify selling six minutes of commercial time to national advertisers at $90,000 per thirty-second spot. Thus the network received as much as a million dollars for each episode for which it had paid Spelling Productions a mere $550,000 in "license fees." Did Spelling lose out on the deal? It would seem so, because its costs were $100,000 above its license fees. But Spelling and all other production companies know the real money is to be made when their successful series are rerun during the spring and summer, for second and third seasons, and in syndication or overseas. Millions of dollars are generated in this way, at no additional cost to the producers.[17] One television researcher maintains that it is very unlikely that a show will recover costs while in its first run before syndication and that deficit financing has been a staple of the industry for many years.[18]

The syndication market alone is worth more than $1 billion a year, with an economics all its own. The more chaotic the regular season programming, the less likelihood that any particular program will stay on the air long enough to amass a vault full of episodes for resale, creating a seller's market. Ideally, the producer of a half-hour drama or comedy will have a stockpile of one hundred episodes for sale to local stations who want to run the programs every weekday evening. If the show was successful, like "M*A*S*H," "Laverne and Shirley," "WKRP in Cincinnati," "The Jeffersons," "Taxi," or "Three's Company," there is no problem with balancing supply and demand considerations. However, with prime-time shows being canceled at the loss of a ratings point, fewer and fewer programs are now available for syndicated reruns. Those that are available are commanding astronomical fees, such as the $77,000 per episode one major market station paid for each episode of "Gimme a Break," a program that finished a mediocre 48th out of 101 shows in the previous year's ratings race.[19] "Family Ties" and "Cheers," which ranked 42nd and 35th in the 1983–84 ratings races, nevertheless earned Paramount more than $1 million *per episode* in syndicated reruns within a few short months of the end of that TV season.[20] With income expectations such as these, some independent producers are committed to making extra episodes of their programs even if they are canceled by the networks, so they will have enough on hand to syndicate.

Television Audiences

The network program directors are the principal gatekeepers of the television medium. Theirs is a different style of gatekeeping from that of newspaper or magazine editors, or even book publishers and film producers. Their style differs because they are concerned not only with audience appeal (specific or mass), as are the other media, but also with what their rivals are doing. They program as much against the other networks' offerings as they do for the audiences. Thus, gatekeeping has a competitive, three-dimensional quality and a new sophistication that is unique to television.

The preponderance of appeal to the lowest common denominator is reflected in much of the blandness and sameness that appear on television: the high incidence of screeching auto chases, gun fights, fisticuffs, drug busts, sickroom pathos, high society melodrama, family comedies, and young women in skimpy clothing. Theory holds that programming aimed at the so-called thirteen-year-old mentality will automatically reach everyone above that level. They will be capable of understanding it, and if the concept of the least objectionable program is correct, they will watch it. Those beneath this arbitrary level lack the competence and, more importantly, the purchasing power to be of concern.

Do viewership studies support these assumptions? Statistics cited at the outset of this chapter indicated that the average person in America watches television four-and-a-third hours per day, or thirty-and-one-half hours a week, while the average household has the set turned on for seven hours and ten minutes a day, or fifty-one hours a week. The time committed to watching television has been increasing gradually over the past two decades and shows little sign of reversing itself. Significantly, though, only one-third of the potential American audience partakes of two-thirds of that daily viewing. That third is referred to by the television industry as the steady, habitual audience, who turn on the set whenever they are near it and demonstrate little discrimination in choosing fare. Another group of viewers, considered occasional watchers, view frequently but without addiction. Finally, selective watchers are those who seek out specific programs but on the whole can take or leave television. Programmers, naturally, concentrate on the habitual and occasional viewers while paying only token attention to the desires of the selective group.[21]

During peak periods of the prime-time season, about 70 million viewers will be tuned in at 7:30 P.M., and are joined by 20 to 30 million others by 9 P.M. Viewing levels gradually decline to 50 to 60 million at 11 P.M. As Les Brown puts it, there are two ways of looking at these statistics: either to recognize, perhaps with dismay, that about half the population is drawn to television each evening, or to take heart in the fact that the other half of the population does not feel compelled to watch without good reason.[22]

What do we know about the viewers—that half of our population who are likely to be watching television on any evening? On the whole, viewers are older today than in years past. In 1970, two-thirds of the 187 million viewers were adults; by 1985, three-quarters of the 222 million viewers were over age eighteen.

We have already seen that women view more television than men. Also of interest is the fact that older men and women, those over fifty-five, watch more than any other age group, with older women being the heaviest viewers. By and large, young children view more than older children and teenagers; the age/sex subgroup that watches the least television is teenage girls, who view only twenty-one-and-a-half hours a week. The amount of viewing is greater for children who are black, who are from families of lower socioeconomic status, and who are lower in academic achievement and IQ.

Television is watched more in larger than in smaller families. Households with three or more members watch about sixty-one hours a week, ten more than the norm. The lowest viewing group is people who live alone. Even they, however, watch for an average of forty hours a week, which is more than America's average working week.[23]

Of those who describe themselves as super-fans of television, those with grade-school educations outnumber those with college educations nearly three to one. Overall viewership patterns, however, show that adult viewers with high-school educations watch more television than do those with grade-school educations, who in turn view more than those with one or more years of college. Similar patterns emerge when viewership is analyzed along income levels. The heaviest viewership is among middle-income groups, with upper-middle-income homes watching about an hour or two less per week. Those with the lowest income levels—who tend to live alone, to move frequently, and to make little use of any mass media—have the lowest level of television usage per household. At the other extreme, of course, lies another group of nonviewers. They are the individualistic intellectuals, frequently of high income, who consider television a waste of time.

Two of the most exhaustive studies ever conducted about television, by Gary A. Steiner and Robert T. Bower, offer substantial insights into America's fascination with the tube.[24] Conducted a decade apart, in the early 1960s and early 1970s, respectively, the national surveys systematically demonstrated that most viewers were satisfied with television, but that the better educated are somewhat more selective in their viewing and feel

Demographic studies tell us much about the changing nature of the TV viewing habit.

guiltier than other viewers about engaging in the television habit. Steiner's 1960s respondents indicated that television, more than any other invention during the previous generation—including new cars, refrigerators, washers and dryers, cooking appliances, stereo, radios, or telephones—had made their lives pleasant and interesting. Most felt their children were better off with television than without it. However, the Bower study a decade later found interesting skepticism among parents who were concerned about television's impact on their offspring.

The Steiner and Bower surveys indicate subtle shifts in America's attitudes toward its favorite invention. Other studies since the Bower survey reinforce the trends. Essentially, those trends indicate a growing disenchantment with commercial television. A decreasing percentage of "superfans" appear in all age and educational groupings; only females have remained faithful to the same extent they did during the early 1960s. Increased skepticism about advertising and television's disadvantages for children are evident. However, four-fifths of Bower's sample conceded that television plays an important educational role for their children, compared with fewer than two-thirds of the Steiner respondents. The increased number of viewers willing to pay for noncommercial television—30 percent in the Bower study, and 24 percent in the Steiner study—indicates growing support for pay television and alternatives to the long-standing attitude that "watching commercials is a fair price to pay for getting free TV."

The networks are aware of the diminishing stranglehold they once had on America and are seeking to compensate for it. As noted earlier in this chapter, they used to command the attention of 90 percent of the evening's viewers, but recently attract less than 75 percent. Looking ahead, media researchers predict that prerecorded home videocassettes will become the nation's leading entertainment medium by the year 1995, a situation that will surely change the face of both over-the-air and cable-delivered television.

Economic uncertainty has led the networks to diversify their interests, getting into other kinds of business. Some have acquired publishing outlets; others are buying into wholly unrelated endeavors. All have recognized the value of buying into or developing their own electronic media subsystems as a hedge against the economic chaos brought on by audiences moving into control of their own media fare—videodiscs, videotape recorders, and the like. (One interesting offshoot of the economic turmoil is that the networks themselves have become hot properties on the stock market, with all three either being sold or receiving offers to sell to other large corporate interests. In 1985 and 1986 alone, ABC was sold for $3.5 billion to Capital Cities Communications; RCA, parent company of NBC, was bought out by General Electric for $6.3 billion; and CBS fought off a couple of multibillion-dollar takeover bids.)

TV's massiveness may become the cause of its own demise.

Despite the fact that they are spending more time in front of the tube than ever before, Americans are spending less time than previously viewing traditional, network programming. The real question, of course, is why are Americans becoming ever-so-slightly disenchanted with television? Part of the answer undoubtedly lies in the sameness of commercial television. What was once enthralling has become a bore. Ironically, a part of the media specialization and fragmentation that television has forced upon books, newspapers, magazines, radio, and film is beginning to take its toll of television. People are turning to other things more suited to their individual interests, even though some of those things are done with the same television sets that have brought them everything from the "Texaco Star Theatre" to "The Cosby Show." In so doing, they have forsaken mainstream television. It may easily be that television's massiveness is bringing about its own demise.

In the United States advertising pays the media bills, more so in television than anywhere else. In recent years television has been taking slightly more than half of the advertising dollar spent on nationally promoted products. The biggest portion of this has gone to the three networks, which are uniquely equipped to offer national service, while individual stations in major markets have been receiving increasingly larger shares of those dollars.

National advertising uses television primarily because of its massive "reach." Relatively speaking, television saturation is probably no greater than that achieved earlier by newspapers or radio. But television offers new dynamic dimensions to advertising: drama, humor, dramatic effects, motion, color, and even stereo sound. The ability to create mood, excitement, and drama in connection with commercial products is a real asset; combined with vast audiences, it is overpowering. Advertising agencies have proven adept at conveying impressions, at probing the inner recesses of the mind, at developing ingenious appeals to the point where many of the commercials on network television are superior from production and entertainment standpoints to the supporting programming. This may not be entirely without design, as noted by some critics who insist that the real product of television is its audiences, who are delivered to the sponsors.

The costs of making a network commercial are astronomical. The commercial, after all, is the reason for the entire medium. It is a painstaking and time-consuming endeavor, utilizing all of an agency's expertise and genius, and all the production talent of studio professionals. Just to stage and film a simple thirty-second commercial in which an announcer merely faces the camera and holds up a product will cost over $25,000 once labor, editing, and duplication expenses are included. A $50,000 price is not unusual for network spots, while local advertisers are spending from $500 to $5000 to reach their more limited target groups. It is not entirely

Television Advertising

Commercials are often higher in production quality than the adjoining programs. This is not by accident.

Average children will see 350,000 commercials and 200,000 acts of violence on TV before they graduate from high school.

unusual for a half-minute network spot to cost more than $250,000 to get the message exactly right, to convey the perfect impression, to be "with it," and sell. At those costs—$8,333 per second—we are talking about the most expensively produced media fare in the world; a ninety-minute feature film would be in the $45 million range if similarly produced.[25] Advertisers expect to recoup a part of this investment by amortizing the cost over as long a period as possible. The more often a spot is run, the lower the unit cost on each run, even though actors are paid residuals for reruns.

So important is advertising to television that there was a time when the advertising agencies produced their own programs as well as commercials. The networks were little more than distribution channels for the agencies; they simply bought time for their clients and filled it as they wished. However, the quiz show scandals of 1959 brought an end to this practice, and the networks reassumed responsibility for programming. (See chapter 15, p. 581.)

Major advertisers can still sponsor an individual program or even a series but they have much less say about the content and scheduling of programs than they did in the 1950s. That does not mean that advertisers no longer exercise influence on the programming. Rather, because of the enormous expenses and risks involved in national television programming, there is by necessity a close relationship between the networks and advertisers.

Since 1970, most network advertising dollars have been spent on the purchase of *participating spots,* isolated commercial messages inserted into the programming. Technically, these are known as *participating network TV spots* when the national advertiser buys on a nationally broadcast network program, and as *national spots* when purchased individually on stations in major markets. *Local spot TV,* a third category of television advertising, is that purchased by local advertisers peddling their goods and services to hometown audiences.

Whether they are network, national, or local, such spots are not necessarily identifiable with the program. That is, they will run side by side with other advertisements for other products, and the program will not carry the advertiser's name as it would a sponsor's name. Under a provision called product protection, similar products promoted by competing spot advertisers will be separated.

Rather than sinking all of their ad dollars into the sponsorship of a single show, participating advertisers will purchase spots according to a *scatter plan*. The same message will appear during many programs, at different times of the week, and on different networks according to advertisers' wishes and the availability of the programs. Advertisers consider the scatter plan to be cost effective, giving them both reach and frequency—a combination impossible in print media, and far less satisfactory in radio.

Advertisers use *participating spots* and the *scatter plan* to reach their desired audiences.

The factors that enter into an advertiser's program choice are not determined by ratings alone, although ratings always have a bearing. The kind of audience and personal predispositions are factors, as are public relations concerns, such as when the cigarette companies would not buy any time before 9:00 P.M. on the presumption that children would not be in bed until then, or Sears' proclamation that it would no longer buy time on violent programs.

All commercials are not sold at the same price. There may be as much as a 25 percent variation in the cost of spots across a ninety-minute prime-time program, depending on how far in advance the advertisers buy, how frequently they buy, or the sorts of packages or deals they have made. Television advertising, while based on a rate card that in turn is based on the ratings, is still negotiable. Networks, however, do their utmost to protect their *cpm,* or *cost-per-thousand* rates, which is the amount they charge advertisers for exposure to a thousand homes (not viewers). This explains why Nielsen has been asked to conduct ratings on commercials watched (and find out why viewers are zapping or changing channels when commercials come on the air), and why at least one network has given free public-service-announcement spots (PSAs) to the Boy Scouts rather than cut the price of commercial time for Sunday afternoon professional football.

Since the mid-1970s, more than four-fifths of television spots have been thirty seconds in length, in contrast with the minute-long commercials popular in the 1960s. This explains why more commercials are on the air today than there were some years back, even though the total amount of commercial time per hour has held relatively steady at about six minutes during prime time, a standard agreed upon by the voluntary National Association of Broadcasters but not imposed by the FCC.

Research indicated that a thirty-second commercial has almost two-thirds the impact of a sixty-second one, so advertisers refined the art of making a pitch with forty-five words jammed into twenty-eight and one-half seconds of audio. In that time span scenarios are set, problems faced, and solutions found. Recently, behavioral scientists have found that message retention remains high when speech and action are compressed even further. This explains the proliferation of ten-second spots in which flashes of images and slogans are hurled through America's ozone night after night. The quickies were used first for station identification and program promotions ("promos") between shows or nested amid blocks of half-minute commercials. More recently they operate as a sort of booster shot or abbreviated reminders of the long form commercials in which the scenarios were originally fleshed out more fully. Individually, such an appeal may be effective, but the sum total, of course, is cacophony. The airwaves are cluttered, as are the brains of viewers faced with making sense of the messages. As advertising expert Otto Kleppner says, "the kaleidoscope of clutter and commercials produces confusion among viewers and a high rate of misidentification of brands."[26] The clutter will probably continue, however, because each spokesperson has the right to spend his or her dollars as he or she sees fit, and no commercial television station is going to refuse the income.

Since it is the network affiliates that actually reach the audience and since there is some selling against the networks on the part of their affiliates and the independents, the prime concern of a television station is for it to be considered part of a major market. The advertising agencies of Madison Avenue define a major market as 100,000 homes. That is the magical breaking point below which a station will not ordinarily be included in a national advertising campaign. For this reason, what once was the tallest man-made structure on earth was built in North Dakota to rise 2,000 feet above the sparsely populated plains. It is a television broadcasting antenna designed to fan out over the northern flatlands to reach 100,000 homes before its signal disappears over the edge of the earth.[27]

Television News

Each weekday evening, a third of America's 87 million television households, or 60 million viewers, are tuned to CBS, NBC, and ABC newscasts. Those 60 million viewers have made the news as popular as many other prime-time programs. To those audiences we have to add the tens of millions of other viewers who tune in to their local stations and to the 24-hour-a-day Cable News Network each day to keep up with what is going on in their world. As Robert MacNeil, co-anchor of the PBS news program "The MacNeil-Lehrer News Hour," puts it, "TV has created a nation of news junkies who tune in every night to get their fix on the world."[28]

Our addiction to TV news is reflected in findings from national Roper surveys, which indicate that two-thirds of America's adults receive most of their world news from television; that half the population says television is its *sole* source of news; that most people find television news more believable than any other source of news; and that a good percentage of them rely on television news because it is more convenient and requires less effort than any other media. Roper's conclusions have been challenged, as we shall

TV has created a generation of video news junkies. But news has not always been profitable for TV.

see, but they nevertheless paint a portrait of a nation enamored with television news, a nation that has made television news operations extremely profitable ventures.

It has not always been that way. For over a generation, starting in the mid-1950s when the evening news was a fifteen minute dose of headlines, news shows had been looked upon as respectable loss leaders. Because news used to cost more to make than it generated in advertising dollars, networks looked upon their unprofitable news operations as public service gestures, and affiliated stations considered them paramount in meeting their public interest, convenience and necessity requirements.

Since the late 1970s, however, things have changed at both the national and local levels. And there are signs that things are going to change even more drastically before the end of the 1980s. With the emergence of demographically attractive audiences for news programs a decade or so ago, networks began to get prime-time rates for the commercial minutes available on the newscasts. Advertisers were happy to cough up premium dollars to be on the network news; indeed, by 1985 they were paying triple the ad rates they had paid as recently as 1970. Over the same time period, the three networks went from spending an average of $3 million a week to around $15 million a week—or $750 million a year—to gather and report the news. The costs of producing the news may seem to be outstripping ad revenues. However, that is not necessarily the case. There is far more network news on the air today than ever before—prime-time news programs such as "60 Minutes" on CBS and "20-20" on ABC, plus morning and weekend and late night shows. And, because most of those shows take in more advertising revenue than they spend on production, network news operations on the whole have become the economic strengths of the networks. Only recently, in the face of threats from local news operations and other outlets, have network news programs begun to realize their goose has a finite supply of golden eggs.

At the local level, news operations for years have been carrying the stations. The evening and late night local newscasts have drawn ever larger audiences, and consequently, greater revenues from commercials. It is not unusual for a mid-sized television station to have a multi-million dollar a year operating budget, but to bring in twice as much in ad dollars. And the newest contender for news audiences, Ted Turner's satellite-delivered Cable News Network, after five years of struggling and some $165 million in red ink, turned the corner and became profitable in 1985. In the process CNN caused a significant ripple effect in boardrooms at both the network and local levels.

Since the late 1970s, television news seems to have experienced two major revolutions. One is primarily technological, the other primarily philosophical. However, both revolutions are linked by common economic considerations, and are difficult to discuss separately. Stated succinctly, the first revolution involves the deployment of satellites and mobile as well as

Cable News Network founder and sportsman Ted Turner.

Since the 1970s, TV news has undergone technological and philosophical revolutions.

computerized news operations that permit local stations and networks to report and transmit news immediately and with pizzazz. The second revolution, less visible but no less significant than the first, is a change in the contents and packaging of news that has been driven by a concern for ratings, by the desires to give audiences what they seem to say they want.[29] We will consider both revolutions, and their common motives.

The Technological Revolution

Prior to the mid-1970s, television news depended heavily on film, rather than videotape. Film was cumbersome, in more ways than one. It demanded bulky cameras, and elaborate processing and editing. It could be used only once. Reusable videotape, on the other hand, not only allowed for ease of editing, but when coupled with *ENG*—electronic news-gathering equipment—gave camera crews and reporters greater mobility. The time gap between when reporters got to a news event and when that event was presented to audiences became significantly lessened, and at times disappeared because of ENG. Sophisticated videotape cameras could be coupled to microwave relay units mounted on vans, permitting stories to be fed back to the stations for editing or even fed live, as on-the-spot coverage of breaking events. Crews bounced the microwave signals off buildings, off relay stations built on nearby hills or mountains, or even off helicopters. So long as the linkup was no longer than thirty-five miles or so, they could transmit a clear signal. In short, television developed the capacity to report today's news today—within limits.

Satellites, computers, and video ENG are changing news-gathering and delivery processes.

ENG may have revolutionized television news, but no more so than the communications satellite. Audiences have come to expect instantaneous coverage of events from every corner of the globe, thanks to networks' reliance upon a host of communications satellites hovering some 22,300 miles above the earth. Sports events, disasters, political stories and human dramas of every type have been brought to the world's living rooms since the mid-1970s. Networks and others have either rented transponder space or purchased their own communication satellites to pepper the globe with sound and visual images.

Entrepreneur Ted Turner became the first broadcaster to make use of satellites when he turned his Atlanta station WTBS-TV into a Superstation by sending the signal to a satellite ("uplinking") and then making it available to cable television stations across the nation ("downlinking"). He took the technology to a logical extreme when he created a 24-hour-a-day news service, Cable News Network, in 1980, and later, an around-the-clock headline news service. Operating on a budget of less than $2 million for a week of 24-hour-a-day news—compared with the $15 million weekly spent by each of the three networks for news that took up only a small portion of each broadcast day—and a young, eager-beaver, non-unionized staff, Turner very quickly became a force to be reckoned with.

As sexy as conventional satellite communications may be, the actual process of using the system remains cumbersome and expensive. Using *C-band* (the industry's term for conventional satellites) costs between $250 and $300 for a fifteen-minute "window" or open access, making it impractical for smaller stations. Because there is so much "traffic" coming and going between earth and the various communications satellites—business, telephone, military, and commercial broadcast traffic—C-band demanded that stations get a frequency clearance. That clearance assured them that their transmission from cumbersome, twenty-five-foot uplink dishes would not interfere with anyone else who was using the frequencies. It took anywhere from four hours to several weeks to get those clearances, so C-band was unlikely to be used for anything but major, anticipated news events such as elections or the Olympics. Thus satellite communication, in its early years, was seen as an impressive but expensive means of conducting only a limited amount of news reporting. It did little to help the local stations produce an improved news product.

A solution to that problem was presented by the FCC in 1983, when it approved a new satellite frequency range known as *Ku-band*. Ku-band is simpler and cheaper than C-band, permitting use of smaller, eight-foot satellite dishes that can be mounted on semi-trailer trucks or even behind vans. It has permitted stations to go live from anywhere in the United States, and to both send and receive satellite signals. A side benefit of Ku-band is its encouragement of stations to set up ad hoc networks, to pool resources and share expenses when covering news.[30] Groups introducing Ku-band service include CONUS and RCA Americom; the latter has offered free Ku-band receiving stations to any commercial television outlets that want them, in a move to expand the market and recoup RCA's investment. Many of the costs have been borne by the transponder purchasers, those special interest groups such as news syndicators who want to reach the broadcasters. As a result, more and more local stations are now plugged directly to national and international news services, and are not limited to the feeds provided by ABC, CBS, and NBC.

Before considering some of the ramifications of satellite communications, we should mention one more technological innovation that is having a major impact on television journalism. That innovation is the computerized newsroom, a system of word processors and editing systems that is changing the jobs of reporters and producers. In some ways the systems are similar to the VDTs employed in newspapers. They permit reporters and editors to write and correct their stories on desktop terminals, just as in newspaper newsrooms. However, the television computerized systems send the completed stories directly to TelePrompTers where they can be read on the air. As the stories pass through the central computer, the computer calculates the length of each one and precisely times the entire newscast to fit within its allocated time slot. If a reporter needs to make last second changes in a story, the computer keeps track of how many seconds of air time the adjustments are causing. Previously, there was a lot of guesswork and frenzy involved in changing a newscast on deadline, but the computerized newsrooms have made things run smoother.

C-band and *Ku-band* have been important to the satellite delivery of news.

The computerized newsrooms offer several other benefits. Computer generated graphics are resulting in more attractive displays for reporting weather, election stories, or other complex stories that need to be made more understandable. And, for the first time, television newsrooms have a system for filing, indexing, and retrieving information gathered by reporters. Older systems result in large, unwieldy piles of scripts, plus random notes and raw, unindexed video footage. Now, using their computers, TV journalists can systematically organize and store all their stories. In addition, they have access to newspaper files and other computerized data bases, and will be better able to keep up to date on national and world affairs.[31] Obviously, this should help stations overcome one of the major criticisms leveled at them—that they are a-historical, have little if any collective sense of what news has been covered and how it has been treated, and that they lack continuity in their treatment of major issues.

These innovations—satellites and computerized newsrooms—are beginning to have ripple effects in broadcast journalism. The new technology is much more than a set of expensive toys. In many ways it has redefined news for both local and network operations. And, not surprisingly, underlying these changes is the ever present economic motivation.

The new technology permits local stations to rely less and less upon the networks with which they have long been affiliated, and more and more upon CNN, regional and national Ku-band sources, and their own now-mobilized personnel for expanded coverage of news. Many stations are sending their own reporters and crews on foreign assignments—covering earthquakes in Mexico, political negotiations in Geneva, elections in the Phillipines, starvation in Ethiopia, guerrilla warfare in Latin America, and the Olympics in Canada.[32] Local stations affiliated with the networks—we might recall from our discussion about the radio industry—receive a share of the advertising dollar paid to the networks by national sponsors who wish to reach the largest possible audience. In days past, it was reasonable for affiliates to merely flip the switch and carry the web, socking away the small but steady revenues garnered with no effort on their own part. However, if they can sell the commercial minutes by themselves, they do not have to share the revenues with the networks. So it is in their own best interests to produce as complete a news package as they can. More and more stations are doing just that, pulling national and international news items from sources other than their networks. And, as one TV news veteran observes, "If, as now happens, the majority of the nearly 700 stations that carry the early evening network shows have actually provided most of the national and international stories before Rather, Brokaw and Jennings appear, what do the networks have to offer?"[33]

Technological changes are giving local stations more options than ever before, but are bringing panic to the networks.

The production of television news has begun to change because of a highly visible technological revolution, but such adjustments are probably not as significant as those being brought on by a much less visible revolution. This second revolution is one of philosophy, a change in the contents and packaging of news initiated by a concern for higher ratings and greater revenues. It is at base a change in how television journalists think they should be serving, or pandering to, their audiences. And this revolution has its beginning in the boardrooms, rather than the newsrooms, of America's television networks and stations.

Before television news was taken seriously as a source of information and influence—that is, before it started getting high ratings—newscasts consisted of talking heads and an occasional piece of film. In a way, it was radio news performed in front of a camera, with only tacit acknowledgment of the medium's visual dimensions. As we have seen elsewhere in this book, such a situation is practically inevitable; almost all news media steal their contents from existing media, until they learn how to do things their own way. Nowadays, as anyone not locked up in solitary confinement or marooned on a desert island can attest, the look and feel of television news have changed drastically. In *TV Guide's* words, television news is headed in a direction decried by traditionalists at all three networks. "The direction is toward commercialism—a dirty word to the traditionalists, but a fact of life for those who see TV news as what it really is: big business."[34] According to many observers, the news business has become show business, and this is not as it should be.

For evidence of the show business thrust, critics point to the astronomical salaries paid to the men and women who anchor newscasts at the network and local levels. They also fault the industry's emphasis on entertainment values, its concern with form rather than substance, in much that passes itself off as purely informational programming.

There is no doubt that television anchors and leading correspondents have become stars. They are among the most recognized figures on the face of the globe. In a recent national survey, 77 percent of those who were shown pictures of various public figures could identify the photo of Barbara Walters, compared with 70 percent who recognized Geraldine Ferraro and 68 percent who could name Vice President George Bush. Dan Rather, Ted Koppel, Mike Wallace and Tom Brokaw were all recognized far more readily than Secretary of Defense Caspar Weinberger.[35]

That recognition does not come cheaply. In fact, most of the anchors listed above earn a million dollars a year. Barbara Walters became the first member of the elite group when she was lured away from NBC in 1976 by a million-dollar-a-year offer from ABC to read the evening news and conduct interviews. (Half of her salary was paid by the ABC Entertainment

The Philosophical Revolution

In a search for higher ratings, TV news has revolutionized its contents and packaging.

division, which did little to quell critics' concern over a blurring of the distinction between news stars and other prime-time stars.) At the time of Walters's switch to ABC, network anchors were making about $200,000 a year. She created a quick adjustment in the business, giving Walter Cronkite and Harry Reasoner cause to demand a doubling of their salaries, to $500,000. Since then the stakes have grown higher.

When considering the following 1986 estimates, we do well to recall that a single ratings point on network evening news is worth somewhere between $60 and $70 million a year, so salaries and perks paid to network news personnel can be considered a basic function of the free enterprise system. Dan Rather, who in 1981 took the reins from retiring Walter Cronkite, "the most trusted man in America," reportedly makes $2.5 million a year on a ten-year contract. NBC's Tom Brokaw earns in the neighborhood of $1.5 million, and ABC's Peter Jennings and Ted Koppel make just under a million. Others in the million dollar range are Bryant Gumbel of NBC's "Today Show," Mike Wallace and Ed Bradley of CBS's "60 Minutes," and CBS's poet in residence, Charles Osgood.

By and large, women anchors are paid less than their male counterparts. Phyllis George was close to the million dollar range when she left CBS Morning News in 1985. Diane Sawyer makes $800,000 on "60 Minutes"; NBC's Jane Pauley earns $575,000, about half that paid to her "Today Show" co-host Bryant Gumbel; ABC's Kathleen Sullivan makes $350,000, and NBC's Connie Chung about $450,000.

Network salaries appear to be leveling off, especially at CBS, which reduced the size of its 1,250-member news department by 10 percent in 1986 after the network lost an estimated billion dollars in fighting off an unfriendly takeover bid by CNN's Ted Turner, and in 1987, when it was asked to swallow a $200 million budget cut. NBC and ABC news divisions faced similar pressures after their parent corporations changed ownership in 1985, and it seems unlikely that news anchors and star correspondents will be able to exert much bargaining pressure in the near future.

On the local level, salaries are sometimes as high as at the nets, based on the drawing ability of the stars and the size of the local market. Chicago pays best; Bill Kurtis earns $1.2 million a year at the CBS owned and operated WBBM, and his co-anchor, Walter Jacobson, reportedly earns a million. Their sports anchor, Johnny Morris, makes $700,000. Los Angeles, which used to pay the highest salaries, averages $400,000 for its anchors. It is another matter altogether at smaller markets, where anchors may earn about $25,000, and reporters only $15,000, depending primarily on their stations' reach and ratings.[36]

Lately something new has been added to the basic ratings-salary equation, and that is the agents or talent representatives who negotiate contracts for the ten thousand or so on-air TV newspeople. Agents became forces to be reckoned with in the 1970s when local anchors first turned into stars and personalities. The period of "happy talk" formats emphasized

"The Evening Stars" on TV news have become millionaires, with complex contracts negotiated by their agents.

showmanship as never before. It helped drive up ratings and, in turn, the marketplace mentality. Somehow it seemed only natural for the same agencies coordinating contracts for sports celebrities and movie stars to begin bargaining for TV newspeople's contracts. Nowadays, even though the happy talk format is on the wane, the emphasis on television news "talent" appears to be growing.

Among the things negotiated, in addition to salary, are perks, privileges, and ego-bolstering "status protection" clauses for anchors and star reporters. News stars fight over "guaranteed time," clauses assuring them a certain percentage of air time or a guaranteed number of appearances each week. They also negotiate the size of billing and photographs to be used in the station's promotional announcements and print advertising. Beyond that, the agents arrange stars' rights to limousines, wardrobe assistants, makeup artists, overseas travel, and editorial control over news broadcasts, including granting of exclusion from unwanted general-assignment reporting duties.

Broadcast executives call the talent representatives "flesh peddlers," and fear that the agents manipulate stations' and networks' profits by creating immoral bidding wars for talent.[37] In defense of the agent process, however, is the argument that anchors are like other television and film stars. They are valued, in the final analysis, for their ability to attract audiences, and contracts are necessary for those talents to be showcased to the greatest advantage of the network or station.[38] Unfortunately, however, as Barbara Matusow maintains in her insightful book *The Evening Stars,* the inevitable by-product of all these highly specific contracts is the agents' increasing intrusion into the editorial process. "If an anchor or reporter has a disagreement with the news director over an assignment, the agent, who in all probability has no grounding in journalism, may be called in to mediate. 'Talk to my agent' is an increasingly common refrain in local newsrooms."[39] This may portend a dangerous erosion of traditional institutional safeguards from within, as television news stars clamor for control over decisions that have always belonged to producers or editors.[40]

Media observers who grew up before television, or during its early years, tend to view television news as a kind of journalistic stepsister to the more legitimate news medium, the newspaper. They worry about the blend of showmanship and news. They are pretty sure they can tell the difference between form and substance, but they are worried about the majority of Americans who say that television is the most credible of all news media. That credibility, say the critics, is only because television management has spent a bundle of money hiring credible anchors who can "front" the evening news, giving the whole product a more believable, serious demeanor. Credibility is expensive, but television executives seem willing to do whatever is necessary "to counter the impression that the networks are greedy, immoral entities who contribute nothing to the public welfare."[41]

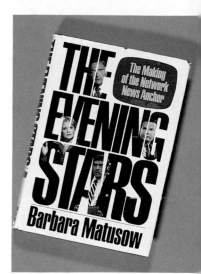

Is TV news paying too dear a price for credibility?

No matter how much they are pampered, or how much influence they may have on the form and contents of the evening news, anchors attract audiences. One *Journalism Monograph* study of network news viewing found that 41 percent of the viewers selected their news program on the basis of the anchor's personality and characteristics, 28 percent on the basis of channel, 9 percent on the basis of news quality or program format, and 22 percent on the basis of other or no particular reasons.[42] Other studies reveal similar, though not always as drastic, breakdowns.

TV News: Criticisms and Problems

Selecting an appealing anchor is only one of the commercial considerations made by network and local news operations. Some observers maintain that nearly all management and news gatekeeping decisions on television are made with an eye on the ratings. Consider the following:

To attract the largest possible viewership, news presentation formulas are drawn up with firm clarity and adhered to rigidly. Style, form, shape, time of presentation, length of treatment, and sequence of presentation are all determined by pecuniary considerations. The multiplicity of possible news items accumulating throughout a day is whittled down and winnowed out by selection, editing, and inference designed to entertain rather than enlighten. To fit within the limited time frame left after the commercials are accommodated, most occurrences are reduced to mere headlines. Such processes distort perceptions of the world and its events for those who have come to rely on television as their primary source of news. And there is evidence that this proportion of the populace is increasing.[43]

We select the above indictment not necessarily because it is unusually severe and caustic, but because in a few words it summarizes many of the criticisms being leveled against TV news, even though it may be overstated. Most television news reporters and editors publicly bristle at such comments, maintaining that news decisions are based on firm principles of journalism and a deep sense of obligation to the swelling ranks of television news viewers. Financial considerations are the concerns of network executives and station managers, and time and space and technological logistics are problems for the journalists, the gatekeepers maintain.[44] The truth, as in most such debates, probably lies somewhere between the extremes. Nonetheless, we can be certain that time, space, technology, and economics are the individual and collective tyrants in television news broadcasting.

Television news has justifiably been called controlled pandemonium. This is because, from the viewers' perspective, television news is conveyed in palatable and readily digestible chunks by seemingly unflappable reporters and anchors. What the viewers do not see is the madness that goes on behind the scenes, where producers, editors, floor directors, camera operators, technicians, and other off-camera personnel rush frantically to package the events of the day.

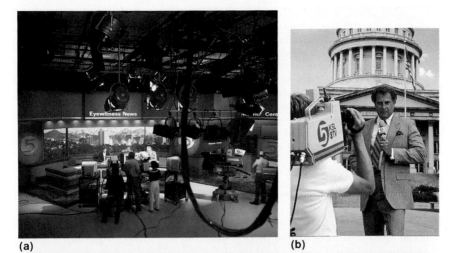

(a)

(b)

(c)

(a) KSL-TV News, live broadcast from the Salt Lake City Studios, with anchors Shelley Thomas and Dick Nourse, and weatherman Bob Welti. (b) KSL-TV reporter-anchor Dick Nourse does a stand-up report from the steps of the State Capitol. The cameraman is using a minicam. The story could be carried live, via microwave or satellite back to the studio, or trucked back to the studio for editing as part of a reporter package. (c) StarWest 5 and Chopper 5—state of the art equipment for today's major market TV stations. The truck can beam video and audio feeds from anywhere in the nation direct to the KSL-TV Salt Lake City Studios by way of a communications satellite. Chopper 5 is used constantly for news reports in nearby areas—weather, accidents, etc.—and can broadcast live by microwave relay or tape delay.

Newscasts are tyrannized by the clock. In twenty-two to twenty-four minutes (once commercial time is excluded) the network or local stations attempt to package the events that the reporters and camera personnel have recorded and the editors have judged newsworthy. This is a daily process that begins each morning when assignments are made to correspondents (at the network level) or reporters (at the local level). Because of the sheer volume of available stories and eager reporters, not all of the reports prepared during the day will be telecast. Most that do run will be edited to fit the exceedingly tight schedule.

Time, space, technology, and economics are tyrants in TV news broadcasting.

> The rigid limitation that time imposes on television journalists is not shared by their newspaper colleagues. Even though print journalists talk of the "news hole" in their publications, the space between the advertisements can be varied on any given day. Newspaper publishers can add a few pages if they feel the news warrants it. You cannot add a few television "pages," except in a crisis when all regular TV programming is wiped out to present unlimited coverage. On a day-to-day basis, TV schedules are rigid to the second. Everything must be compressed to fit, a process of elimination rather than inclusion, which forces constant compromise. No opportunity exists to add just one more feature or analysis for "only" thirty seconds more on a regular nightly newscast.
>
> Source: Av Westin, *Newswatch: How TV Decides The News* (New York: Simon and Schuster, 1982), p. 55.

Rare is the story that receives more than three minutes; more typical is the sixty-to-ninety second report, which equates to between two hundred and three hundred words, or a couple of newspaper paragraphs. As many television news critics—including Walter Cronkite—have observed, the script from a typical network newscast will not even cover the front page of a newspaper, so viewers should not assume they have meaningfully absorbed current events. (Such awareness, however, never stopped Cronkite from ending his newscasts with a sonorous "And that's the way it is. . . .")

Another tyrant is the visual aspect of television. In the 1950s the camera tended to focus exclusively on the newscaster who reported the news to viewers. It is not difficult to understand how this drab style got the name "talking head." And it is not surprising that today newscasts succeed or fail on the basis of their film or videotape coverage. This makes the camera an arbiter of what is newsworthy. If a camera crew can get to an event, shoot it, and relay the film or tape to the station in time for the newscast, then the event is newsworthy. If not, it isn't.[45] One metropolitan station's assignments editor says, quite simply, that the ironic rule of television is that "If you go to a story, it isn't there. If you don't go, it is."

As we pointed out earlier, electronic newsgathering equipment (ENG), coupled with microwave and satellite transmission from camera to station, has made the TV news process almost instantaneous in recent years. But the crews still must be at an event in order to record it. That truism means that events rather than underlying issues dominate television news, and that the predictable events—those to which reporters and cameras can be assigned in advance—stand the greatest chance of being covered. As one sociologist concluded in his study of national news coverage, anticipated stories provide a predictable supply of potentially suitable news, with access and availability made certain in advance.[46] The result is a variety of pseudo-events—political speeches, press conferences, and meetings—and other prescheduled activities of all kinds and shapes.

Veteran politicians and lobbyists have long known how to play to television's attraction for the predictable, the visual, and the dramatic. The eleventh commandment of television journalism, it is facetiously said, admonishes that: "Thou shalt not bore." Even as well-respected a journalist as Robert MacNeil, co-host of one of television's most serious news programs, has maintained that, "All television gravitates towards drama, and what passes for drama is often belligerence, people barking at each other, like soap opera actors, sounding vehement to make up for cardboard characters or too little rehearsal time."[47] Newcomers to the publicity business may initially object to such manipulation, but they either quickly adapt or are never heard from again. In their defense, it must be said that reporters and photographers have learned some lessons since the 1960s, when on more than one occasion they were bid to cover a demonstration and told it would be aired whenever they could get their cameras to the scene.

Part of the spatial tyranny over television is the location of camera crews. Because equipment and personnel are so expensive, networks and stations are limited in the number of crews they can support. Therefore the crews are placed where news is most likely to happen. For the nets, this means posting crews to Washington, D.C. for political news, New York City for economic news, Chicago for "heart of America" news (whatever that means), London for international news, and so on.

Location of camera crews is a temporal as well as a spatial tyranny. All else being equal—that is, barring earthshaking news events—to appear on the early evening newscasts televised from the East Coast, "reportable" events on the West Coast must occur not much later than noon and those on the East Coast not much later than early afternoon. As with newspapers, networks have to plan early in the day for the news outputs, and the closer it gets to deadline, the less chance there is for events to be covered. ENG and satellite communications have sped up the process noticeably over the past several years, but the sheer logistics of coordinating a mass of personnel and equipment naturally bias television news toward reporting those things that have occurred early enough to be packaged for later showing.

News teams are assigned for practical, economic purposes to major news centers, despite the fact that the networks like to leave us with the impression they are covering the entire country and world.[48] In a recent move caused partly by local stations' abilities to bring more and more of the world home to their viewers, the networks have reversed tradition and taken their coverage to the American heartland. CBS has even gone to the point of having Dan Rather anchor the news from various affiliate stations. The goal is to make its evening news more folksy, telling stories from the common person's perspective, even as it offers more background and analysis than the local stations generally do. This "populist television," which attempts to convey the mood of a time and place in legitimate journalistic fashion, has been fashioned by CBS News President Van Gordon Sauter, himself a product of middle America.[49] Meanwhile, all the networks, including CNN, have been doing some other creative brainstorming about how to expand their access to the world's news. Among the scenarios being played out is the use of cameras aboard orbiting satellites to cover events in space and from space, such as live pictures of a battlefront in the Iran-Iraq war, or raging forest fires in the isolated northwest United States.[50]

Although our discussion emphasizes network news, we cannot avoid mention of the significance of packaging in newscasting, which occurs at both the network and local levels. At most stations the newscast is the most profitable locally produced show, which means the station is willing to invest in the maintenance and improvement of the program. Nowhere in recent years have media consultants been as evident to the casual media observer as in local television news. National consulting firms such as McHugh and Hoffman, Inc., of MacLean, Virginia, and Frank Magid Associates of Marion, Iowa, have significantly affected television news in America. For somewhere between $10,000 and $50,000 (far more for the networks, of course), depending upon their market size, stations have been purchasing from the consultants insights into viewer preferences about anchorpersons, story types, tastes in weather and sports news, and even set design. The results of such studies, of course, are taken most seriously by lower-rated news operations, who are quick to fire an anchor, hire a new and more psychographically and demographically appealing one, and rearrange the furniture—the deck chairs on the Titanic, as it were. Formulas result: "Action News," "Eyewitness News," and "News From the Newsroom" are packaged as neatly and uniformly as McDonald's and Kentucky Fried Chicken food chains. Entertainment, not information, becomes the overriding emphasis, because entertainment is inherent in the medium and because the consultants, whose backgrounds are in marketing, not in news, are hired by management, not by news editors.

News consultants help "package" network and local broadcasts. Form often overshadows substance.

In summarizing how temporal, spatial, technological, and judgmental factors affect television news, critics Mankiewicz and Swerdlow have written that some informal unacknowledged laws of television news have developed:

Unattractive faces are almost never on camera in "good guy" roles; a fire at night will almost always be shown though an equally serious daytime fire won't; every news story must be complete within one minute and fifteen seconds, unless the program is doing an "in-depth" treatment, in which case one minute and forty-five seconds may be permitted. All this has led to an overriding law—The Trivial Will Always Drive Out the Serious. There are other limitations and strictures. A story with film, for example, will almost always take precedence over one that must be read from the anchor desk or reported in a "stand-up" on the spot. If there is film, the action film will almost always replace the film that includes only a conversation or a discussion—"talking heads" are to be avoided if there's any possible way.[51]

What does this mean to the viewer? Does television's preference for the visual mean that broadcast news is geared to the emotional rather than the intellectual level? Does the emphasis on action rather than thought, happenings rather than concepts, protagonists rather than underlying issues, mean that television news is devoid of explanation and background and is therefore only comprehensible to the already well-informed members of the public? Is it possible, as some researchers insist, that most television news reporting does little, if anything, to develop audiences' potential to analyze, think independently, or learn from grasping overall social patterns in the unfolding of events?[52]

Pundits and researchers often differ in their views about pros and cons of TV news.

Ease up, TV insiders admonish their critics. ABC News executive producer Av Westin admits that TV news is based on elimination rather than inclusion, and that time is the key factor that influences gatekeepers' decisions. Nevertheless, Westin says, "We have a great sense of responsibility to be fair, balanced, and accurate. We are not Communists trying to destroy America nor are we defenders of the status quo. TV news is show business, but it uses show business techniques to convey information rather than distort it."[53]

Critics and defenders of television news can take some comfort in recent studies that will help both camps prove their arguments about the medium. One set of studies indicates that viewers routinely misinterpret between one-quarter and one-third of any broadcast, whether it is news, entertainment, or commercials. Facts as well as inferences become muddled by serious and casual viewers alike, according to a Purdue University researcher. Another study, by the American Association of Advertising Agencies, reveals that the vast majority of television viewers, more than 90 percent, misunderstand at least part of whatever kind of programming they watch.[54] Such research helps support the arguments of television news critics.

On the other hand, a recent book-length study of how America uses and learns from television news demonstrates that public understanding of news information is considerably higher than previously believed—even though it is still lower than it ought to be if we expect an effective system of self-government. The study, entitled *The Main Source: Learning from Television News,* consisted of an innovative series of sophisticated research projects: surveys, naturalistic studies of people's news consumption, awareness, and comprehension, detailed content analyses of TV newscasts, and a broad synthesis of more than two decades of research from the United States and Europe.[55]

Questions raised by critics, and insights gained from empirical research, are provocative even if they are contradictory. In the meantime, many observers continue to accept the conclusions of the Roper surveys at face value: that television news is America's primary window to the world.

The world sat in stunned disbelief, glued to their TV sets, as the space shuttle *Challenger* exploded.

They may not fully appreciate what networks and local stations go through to keep that window clear, however. For instance, after the space shuttle *Challenger* exploded in January of 1986, the three broadcast networks broke into their regular programming within six minutes of the tragedy and stayed on the air continuously for five hours, without even taking a break for commercials. That commercial-free coverage cost the networks an estimated $9 million in lost advertising revenue that day. Not all of us were impressed with the sacrifice; the ABC switchboards fielded more than twelve hundred complaints about preempted soap operas.[56]

Television Criticism

TV has become the medium Americans love to hate. After all these years, we're still ambivalent.

Television seems to have become the medium Americans love to hate. Criticism of television is as common as the many nicknames we have devised to describe it—the boob tube, the idiot box, and the one-eyed monster, for example. TV has been said to mean "transcendental vegetation." And yet we continue to watch television, even as we complain about it.

In 1938, E. B. White looked into the future and predicted that television would be the test of the modern world in which "we shall discover either a new and unbearable disturbance of the general peace or a saving radiance in the sky."[57] Perhaps television has brought us a little of each.

Other writers during the 1930s and 1940s were also willing to take a stand regarding television. Many of them sided with Lee De Forest, who envisioned an enlightened populace dedicated to a strong home life: "Into such a picture ideally adapted to the benefits and physical limitations of television, this new magic will enter and become a vital element of the daily life."[58]

And it was magic to America recovering from the depression and World War II. People wanted to be entertained by the new medium and even the critics seemed optimistic. Television was often described in terms of its potential for good. Many saw it as an essential ingredient in giving children access to the world; in a short-lived series of newspaper ads, the American Television Dealers and Manufacturers tried to shame American parents into buying sets so their children would not feel ostracized by their friends or deprived of "sunshine for their morale." In 1952 the National Association of Broadcasters revised the television code to encourage innovative programming that would deal with significant moral and social issues. This did happen, of course, but more often the viewer turned to the likes of "Queen for a Day," a shlocky show in which the contestant with the most pathetic hard-luck story won cash and prizes.

Newsman Edward R. Murrow described the paradox of television:

This instrument can teach, it can illuminate; yes, and it can even inspire. But it can do so only to the extent that humans are determined to use it to those ends. Otherwise it is merely wires and lights in a box. There is a great and perhaps decisive battle to be fought against ignorance, intolerance and indifference. This weapon of television could be useful.[59]

> Sure, "Queen (For a Day)" was vulgar and sleazy and filled with bathos and bad taste. That was why it was so successful: It was exactly what the general public wanted. After all, the average American voted Warren G. Harding into office, reads the *Reader's Digest,* and made *Hercules Unchained* a smash movie. In the slightly amended words of H. L. Mencken, "Nobody ever lost money underestimating the taste of the American public."
>
> Source: Howard Blake, producer of "Queen For A Day," 1966.

Earlier in his 1958 speech Murrow had commented on the "decadence and escapism" of television and how the medium provided "insulation from the realities of the world."[60] The potential of television, said the critics, had yet to be achieved.

Those who wanted television to become a medium for enlightenment became increasingly discouraged during the 1960s. FCC Chairman Newton Minnow acknowledged that "when television is good, nothing is better," and then added, "but when television is bad, nothing is worse." He invited his audience to sit down in front of their television sets when their station went on the air in the morning and stay there without anything to distract them, keeping their eyes glued to the sets until the station signs off at night. "I can assure you that you will observe a vast wasteland," he commented, giving television a nickname that has stuck with it in some academic and regulatory circles.[61]

At one time television promised to be the saving radiance that even educators believed would unilaterally improve the learning of children. But in the past two decades, writers and critics have faulted television, asserting that it perpetuates dependency—or worse—in children. In the words of one of the nation's foremost television researchers, Dean George Gerbner, "It has profoundly affected what we call the process of socialization, the process by which members of our species become human."[62]

Recall the statistics cited in the introduction to this chapter, especially those that pertain to the number of hours children watch television. Consider the 350,000 commercial messages and the 200,000 acts of violence the average child will be exposed to prior to high school graduation. The evidence in support of Dean Gerbner's assertion is substantial.

Educator Neil Postman, a self-designated "media ecologist" who has written more than a dozen books on the subject, laments the fact that television seems to have changed forever the process of growing up, of being a child, of learning about life. It has eroded the dividing line between childhood and adulthood by making the world, warts and all, totally accessible to children. There are no more secrets, he complains; children learn about things vicariously they formerly learned from their parents or from direct experience, at appropriate points in their lives. Postman wrote in *The Disappearance of Childhood* that because television does not make complex

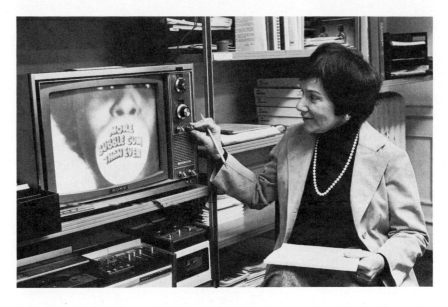

demands on either minds or behavior, it stunts intellectual growth at the same time it accelerates childhood.[63] In a more recent book Postman has argued that television has substituted immense quantities of useless information for knowledge. Instead of intellectual stimulation and vivid abstract ideas which need to be worked through logically, such as encouraged by print media, television has given us entertainment, Postman maintained in *Amusing Ourselves to Death: Public Discourse in the Age of Show Business.* He was actually more critical of television news and serious programs for being disguised entertainment, for not measuring up to their potential, than he was of "junk" television, which admits it is trying to do little more than entertain and divert our attention from the world's more serious issues.[64]

PBS anchorman Robert MacNeil, whom we have already cited as one who finds fault with the superficiality of television news, seems to agree with Postman in complaining that generically, television is guilty of shortening viewers' attention spans:

Everything about this nation becomes more complicated, not less. The structure of the society, its forms of family organization, its economy, its place in the world, have become more complex. Yet its dominating communications instrument, its principal form of national linkage, is an instrument that sells simplicity and tidiness; neat resolutions of human problems that usually have no neat resolutions.[65]

The criticism has become sharper, sometimes shrill; and the defenders have become less vocal. Newsman Sandy Vanocur has pointed out that "Television is in danger of becoming a whipping boy . . . a convenient scapegoat upon which we can . . . dump our frustrations."[66] Noted television critic Jeff Greenfield, after reciting the litany of evils television has supposedly caused—rises in crime, increased divorce rates, lower voter turnouts, falling scores on scholastic achievement tests, more sexual promiscuity, the collapse of family life, and the general decline of Western Civilization—accused television's doomsayers of playing fast and loose with historical evidence, or not looking carefully at the cause-effect relationships involved. "Television, powerful as it is, has shown precious little power over the most fundamental values of Americans," Greenfield wrote in *TV Guide*. "Given most of what's on TV, that's probably a good thing. But it also suggests that the cries of alarm may be misplaced."[67]

To accept without question the myriad accusations leveled at television would mean we are as guilty of shortsightedness as some say television is. We should not overlook the variety and vitality that network television now offers, nor ignore its continuing potential, especially now that it is being reformed by the impact of cable television and new technology.

Summary

Television has become part of the American way of life. More than any other medium it has infiltrated the lives of its audience. Television's almost instantaneous capacity and nearly total saturation of America's homes make it enormously time-consuming, averaging daily over seven hours per household.

Television is the result of technological developments of the nineteenth and early twentieth centuries. Network television began in the 1930s, and rapid growth of programming came about following World War II. Educational and public television has sputtered along for a generation, never reaching its full potential because of low levels of funding and the simple fact that it has to compete with the networks for audiences.

Ratings are important to television. They measure the total size of the audience and each network's (or station's) share of that audience during any given time slot, both of which are important to advertisers. Recently ratings have also delved into demography as advertisers devote more attention to the young and affluent. Women, too, who control most purchasing and who comprise at least half of all audiences, are very important to advertisers.

Programming is dictated by ratings as programmers at major networks jockey for position in the various time slots across the prime-time board. Selections are made for the presumed lowest common denominator of taste. The least objectionable programming (lop) concept contends that television is habit-forming and that people will watch regardless of what is presented, choosing only the least objectionable of what is on at any given time. Consequently, the gatekeepers of television have a complex task. They must second-guess their publics and at the same time program against what they think the other networks are scheduling for the same time slots.

Television production costs are enormous, but so are its profits. Each of the three major networks spends some $1.5 billion annually to produce some fifteen hundred hours of programming. The costs of producing shows range from about $12,000 for a half-hour game show to over $2.5 million for an average made-for-TV movie. Costs keep rising, and the networks more and more frequently attempt to extend the lives of their investments by running repeats.

Research into the nature of the television viewing audience offers many intriguing insights. On any given evening, half the American public is likely to be found in front of the set. Networks understand the demographic profiles of addicts, occasional watchers, and selective watchers, and they program accordingly. While most viewers are satisfied with television, there has been increasing skepticism about the experience in recent years.

Advertising is the dominant factor in television. Advertisers deal in minutes and homes. Advertising minutes have far greater potential for both frequency and reach into homes than do sponsorships. Advertising is sold on the basis of cost per thousand (cpm), based on exposure to each one thousand homes. At today's rates, commercials are probably the most expensively produced media fare. Recently, the half-minute commercial has dominated, but indications are that shorter commercials, including ten-second spots, will be used more frequently, although at some cost in effectiveness and viewer satisfaction.

We have become a nation of television news junkies. Network and local news programs have become profitable enterprises, often paying the bills for additional programming and operations. The three networks—CBS, NBC, and ABC—spend $750 million a year gathering and reporting the news. Ted Turner's CNN, with two 24-hour-a-day cable outlets, spends less than $2 million a week, but makes so much news available to so many American homes that the networks have begun to worry—and for good cause, as their share of the audience has declined in recent years.

News has undergone technological and philosophical revolutions during the 1980s. Innovations such as satellites, computerized newsrooms, and ENG (electronic newsgathering equipment) have brought about enormous changes in how the news is gathered and reported. Despite the hardware, news remains tyrannized by time, space, and technological factors. Meanwhile, the higher profitability of news has created changes in the cosmetics and showmanship of the industry, as stations and networks bend over backward to reach the largest potential audiences. Million dollar salaries are paid to anchors; most of the ten thousand on-air personalities have talent agents who negotiate their contracts. Traditionalists decry these profit-driven changes in the industry, insisting that an industry that keeps one eye on the ratings automatically takes on entertainment overtones.

The criticisms leveled at television are legion. It is accused of being bland, inoffensive, uninteresting, preoccupied with violence and sex, catering to the lowest common denominator while avoiding cultural uplift, and never realizing its full potential as a social instrument. It has been accused of almost everything currently wrong with the family, particularly children, and is said to be amoral and motivated solely by profit. Defenders suggest that we recognize the commercial nature of the beast and learn to use it wisely to serve the needs of society rather than the needs of the medium.

Notes

1. George Comstock, *Television in America* (Beverly Hills, Calif.: SAGE Publications, 1980), pp. 33-34.
2. "The Second 50 Years Of The Fifth Estate," *Broadcasting,* 30 December 1985, p. 72.
3. Tom Jory, "PBS And Corporate Underwriting," *Salt Lake Tribune,* 28 May 1978.
4. John Wicklein, "The Assault On Public Television," *Columbia Journalism Review,* January/February 1986, pp. 27-34.
5. Woody West, "Public Broadcasting: Pull the Plug," *Washington Times Insight,* 17 February 1986, p. 80.
6. Wicklein, "The Assault On Public Television," p. 34.
7. Neil Hickey, "Turn On Your TV, And Aim For A Diploma," *TV Guide,* 17 November 1984.
8. A. C. Nielsen Company, "Audience Research," eds. Craig T. Norback and Peter G. Norback, *TV Guide Almanac* (New York: Ballantine Books, 1980), pp. 70-74.
9. "ABC Evening News," 7 November 1977.
10. William MacDougall, "Why People Are Turned Off By Television," *US News & World Report,* 13 February 1984, pp. 49-50.
11. Sally Ann Stewart, "Cosby: The Undisputed King Of TV," *USA Today,* 5 December 1985, p. 1-D.
12. Les Brown, *Keeping Your Eye On Television* (New York: The Pilgrim Press, 1979), p. 12.
13. Harry F. Waters and Mark D. Uehling, "Tuning In On The Viewer," *Newsweek,* 4 March 1985, p. 68; and Neil Hickey, "Viewers Beware: Your People Meter Will Be 'Watching' Your Every Move," *TV Guide* 15 December 1984.
14. Rushworth M. Kidder, "Halting The Billion-Dollar March Of War Toys And War Cartoons," *Christian Science Monitor,* 16 December 1985, p. 23.
15. Ibid.; United Press International, "Group Says Children's TV Programming Suffers," *Logan* (Utah) *Herald-Journal,* 23 May 1985; Patricia McCormack, "Dr. Spock Joins Plea For Better Programming," *Logan* (Utah) *Herald-Journal,* 14 October 1985; and Peggy Charren and Robert Krock, " 'Program-Length Commercials' Turn Kiddies' TV Into Toy Store," *Variety,* January 1986, pp. 158, 192.
16. Robbie Gordon, "God's Medium," *Washington Journalism Review,* April 1986, p. 40.
17. Anthony Cook, "The Peculiar Economics of Television," *TV Guide,* 14 June 1980, pp. 4-10.

18. Muriel G. Cantor, *Prime-Time Television: Content and Control* (Beverly Hills, Calif.: SAGE Publications, 1980), p. 73.

19. Sharon Rosenthal, "$77,000 For 30 Minutes? *Gimme A Break!*" *TV Guide,* 8 December 1984, pp. 43-44.

20. "Special Report: The World of TV Programming," *Broadcasting,* 22 October 1984, pp. 54-82.

21. Brown, *Keeping Your Eye On Television,* p. 24.

22. Ibid.

23. Neil Hickey, "Do You Know Why Miniseries Usually Start On Sunday?" *TV Guide,* 7 December 1985, pp. 39-40.

24. Gary A. Steiner, *The People Look At Television* (New York: Knopf, 1963); and Robert T. Bower, *Television And The Public* (New York: Holt, Rinehart & Winston, 1973).

25. Jonathan Price, *The Best Thing On TV: Commercials* (New York: Penguin Books, 1978), p. 109.

26. Otto Kleppner, Thomas Russell and Glenn Verrill, *Advertising Procedure,* 8th ed. (Englewood Cliffs, N.J.: Prentice Hall, 1983), p. 127.

27. Martin Mayer, *About Television* (New York: Harper & Row, Publishers, 1972).

28. "The New Face Of TV News," *Time,* 25 February 1980, p. 65.

29. John Weisman, "Network News Today: Which Counts More-Journalism Or Profits?" *TV Guide,* 26 October 1985, pp. 7-13.

30. Barbara Matusow, "Station Identification: Network Affiliates Loosen The Apron Strings," *Washington Journalism Review,* April 1985, pp. 28-33; and Dennis Holder, "Local Coverage On Ku," *Washington Journalism Review,* October 1985, pp. 46-49.

31. Holder, "Local Coverage On Ku," p. 48.

32. Matusow, "Station Identification"; Holder, "Local Coverage On Ku"; and William J. Drummond, "Local TV News: Today The Bay Area, Tomorrow The World," *San Jose Mercury News West,* 16 February 1986, pp. 4-9.

33. Desmond Smith, "Is The Sun Setting On Network Nightly News?" *Washington Journalism Review,* January 1986, pp. 30-33.

34. Weisman, "Network News Today," p. 8.

35. *The People And The Press: A Times Mirror Investigation Of Public Attitudes Toward The News Media* (Los Angeles: Times Mirror Co., 1986), pp. 14-15.

36. Judy Flander, "TV's Top Dollar Talent," *Washington Journalism Review,* March 1986, pp. 19-21.

37. Neil Hickey, "The Newsroom Power Plays That Viewers Never See," *TV Guide,* 18 May 1985, pp. 6-9.

38. Barbara Matusow, *The Evening Stars: The Making Of The Network News Anchor* (New York: Ballantine Books, 1983), pp. 325-330.

39. Ibid, pp. 328-329.

40. Ibid, p. 330.

41. Ibid, p. 18.

42. Mark R. Levy, "The Audience Experience With Television News," *Journalism Monographs,* no. 55, April 1978, p. 7.

43. Robert H. Stanley and Ruth G. Ramsey, "Television News: Format As A Form Of Censorship," *ETC,* Winter 1978, p. 433.

44. Herbert J. Gans, *Deciding What's News: A Study Of CBS Evening News, NBC Nightly News, Newsweek And Time* (New York: Vintage Books, 1980).

45. Frank Mankiewicz and Joel Swerdlow, *Remote Control: Television And The Manipulation Of American Life* (New York: Ballantine Books, 1978), p. 99.

46. Gans, *Deciding What's News,* pp. 87–88.

47. Robert MacNeil, "The Mass Media And Public Trust," *Occasional Paper,* no. 1 (New York: Gannett Center For Media Studies, 1985).

48. Edward Jay Epstein, *News From Nowhere* (New York: Random House, 1973).

49. Michael Massing, "CBS: Sauterizing The News," *Columbia Journalism Review;* March/April 1986, pp. 27-37.

50. "Journalism in 1985: Bolder And Wiser," *Broadcasting,* 16 December 1985, pp. 54-58.

51. Mankiewicz and Swerdlow, *Remote Control,* pp. 97-98.

52. Stanley and Ramsey, "Television News," pp. 433-434.

53. Av Westin, *Newswatch: How TV Decides The News* (New York: Simon and Schuster, 1982), from the flyleaf.

54. Ron Powers, "Where Have The News Analysts Gone?" *TV Guide,* 8 October 1980, pp. 5-6.

55. John P. Robinson and Mark R. Levy, *The Main Source: Learning From Television News* (Beverly Hills, Calif.: SAGE Publications, 1986).

56. Richard Zoglin, "Covering The Awful Unexpected," *Time,* 10 February 1986, pp. 42, 45.

57. E. B. White, *The New Yorker,* 1938.

58. Lee De Forest, *Television: Today and Tomorrow* (New York: Dial Plan, 1942).

59. Edward R. Murrow, addressing the Radio Television News Directors Association, 15 October 1958.

60. Ibid.

61. Newton Minnow, addressing the National Association of Broadcasters, 9 May 1961.

62. George Gerbner, testifying before the National Commission on the Causes and Preventions of Violence, 1969.

63. Neil Postman, *The Disappearance of Childhood* (New York: Delacourte Press, 1982).

64. Neil Postman, *Amusing Ourselves To Death: Public Discourse In The Age Of Show Business* (New York: Elisabeth Sifton Books—Viking, 1985).

65. Robert MacNeil, "is Television Shortening Our Attention Spans?" *New York University Education Quarterly,* vol. 14, no. 2, Winter 1983, pp. 2-5.

66. Sander Vanocur, *Washington Post,* 1975.

67. Jeff Greenfield, "Don't Blame TV," *TV Guide,* 18 January 1986, pp. 4-6.

Film

With all the other mass media crying for our attention, why do we still go to the trouble of patronizing the movies?

Introduction

Film is a carrier of culture, but it is based on an optical illusion!

Film is based on an optical illusion, wherein the still frames of individual photographs are mechanically speeded up and projected until they blend into one another, creating the illusion of motion. Movies, motion pictures, film, and cinema all refer to essentially the same thing; but each has a slightly different connotation ranging from the commonplace to the aesthetic. This in itself gives a clue to the diversity of *film,* which is the term we will use.

Film is counted as a major mass medium because of the effect it has upon large masses of people over a relatively long period of time. In all its variations—the art film, the cartoon, the popular extravaganza, the industrial showcase, the educational teaching aid, the social documentary, and the new genre of the television "quickie"—it has unquestionably had a massive effect upon society. In this sense, as a carrier of the culture, a changing mirror of changing times, film is unexcelled.

However, in another sense it is scarcely a mass medium. Commercial films are not regularly issued and their specific audience sectors are not so clearly identifiable as they are with magazines and radio stations. While film does not foist itself upon its consumers daily, it has an advantage shared by no other medium. It commands attention, generally playing to captive audiences in a format over which the producer has complete control of emphasis, order of presentation, continuity, dramatic effect, and timing. These characteristics have made it over the years a superb medium for instruction and persuasion. A large portion of filmmaking is not commercial in the usual sense, but is devoted to sales, training, information, and other public relations purposes of corporations, governmental agencies, trade associations, and almost any other sector of society that has a message to communicate.

In chapter 5 we referred to film as an electric book, and rightfully so, because there are striking similarities between films, an electric medium, and books, a print medium. Principally, both are long-term undertakings; both are relatively expensive and hard to acquire; both embrace a certain unity and specificity of subject matter; and neither is supported by advertising. An additional parallel can be seen by comparing the commercial film and its desired box-office appeal with the trade book and its aspirations for making the best-seller list. The counterpart of this comparison is equating the noncommercial film, which is privately produced and contains a high degree of informational, educational, and persuasive content, with the textbook. The analogies are not exact but they do indicate a similarity of function between the two media. Also, the commercial film must excel at the box office within the first year or so, which is essentially the time frame for a best-seller, whereas the noncommercial film has a longer life dictated essentially by its purpose, just as a textbook does.

Another factor of comparison is that film, like the book, has a long and honorable history, stretching back into antiquity to drama. Film is, in effect, the mass production of drama and provides the same entertainment, educational, and persuasive purposes for which the dramas of ancient Greece were originally performed.

Most of this chapter will focus on commercial films. After a brief history of film, we will explore the business of filmmaking, especially production, distribution, and exhibition. We will also discuss the impact of film on audiences and how society has developed to modify those influences; nontheatrical films; the cross-cultural nature of commercial films; and the future of film as a mass medium.

Film is an electric book.

Historical Development

"Persistence of vision" was discovered and exploited.

The psychological principle on which film as a medium is based is known as *persistence of vision*—that the eye retains an image fleetingly after it is gone. For centuries inventors and magicians made use of this principle. Leonardo da Vinci's *camera obscura,* a darkened chamber in which an image was captured and focused, and Athanasius Kircher's 1671 *magic lanterns,* which attempted to project moving pictures, were early developments in the field. Throughout the 1800s, while the science of still photography was being refined, inventors were marketing hand-cranked "toys" that employed the persistence of vision principle—the idea that a sequential series of still pictures, when viewed or projected in rapid succession, would give the illusion of motion.

In 1877, two photographers synchronized twenty-four cameras to check a hunch of Leland Stanford, a railroad magnate. Stanford thought that at some point while his race horse was running, all four of its feet would leave the ground. The two dozen cameras proved him correct. Later, when the twenty-four pictures were projected in sequence on a machine with the unwieldy name of *Zoopraxinoscope,* photography and projection were united.

In 1889, Edison's kinetograph movie camera shot the first motion picture sequence of his assistant, Fred Ott, ''sneezing'' for the camera.

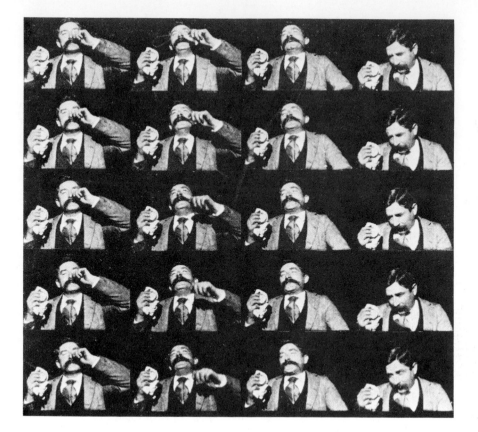

Thomas Edison gets much of the credit for creating the motion picture industry. Unfortunately for him, he forgot to get some patents.

With Thomas Edison's 1879 invention of the electric light providing a better source of illumination, mass viewing became possible. Within a decade, three other essential technical ingredients of the film were developed: a motion picture camera, flexible film, and the sprocket wheel.

Much of the credit for early film innovations goes to Edison and his gifted assistant, William Kennedy Laurie Dickson. They were intrigued with the possibilities of combining the newly devised phonograph with the flexible film being produced by George Eastman. In October 1889 Dickson managed to link the two in a demonstration on the *Kinetoscope* of a film in which Dickson's visage and voice were captured. As film historian Arthur Knight expresses it, what was in all probability the first actual presentation of a motion picture film also marked the debut of the talkies.[1]

Temporarily ignoring the sound dimension, Edison marketed the battery-powered Kinetoscopes and their minute-long peep shows throughout the United States. Within five years they were also in Europe, competing—unfortunately for Edison, who had not bothered to secure an international copyright—against numerous imitations and separately invented systems in England, France, and Germany. By 1896 the projecting Kinetoscope,

Vitascope, Biograph, and others were refined to project the films onto a screen so an entire paying audience, instead of only one customer, could view them at one time. Film was on its way toward becoming a mass medium.

Edison's Kinetoscope offered its penny-paying customers sensational but jerky pictures of locomotives rushing toward the cameras, fisticuffs, cockfights, Annie Oakley shooting down clay pigeons, hootchie-kootchie dancers, vaudeville acts, practical jokes, and sea waves. Not surprisingly, the novelty wore off quickly, even among the unsophisticated. When the "flickers" left the peep-show arcades and were projected before larger audiences in vaudeville palaces, the same cycle occurred. Initial wonderment at the technology was soon followed by boredom with the content. It was time for the fledgling industry to take a cue from the legitimate theater, and to do it one better.

Between 1896 and 1914, a French magician named Georges Melies "discovered" trick photography—stop motion, double exposure, superimposition, fadeout, fast- and slow-motion, and animation. He created over a thousand short films, including a fourteen-minute version of Jules Verne's *A Trip to the Moon* in 1902 that made use of professional actors, animation, many varied costumes and sets, and a detailed scenario.[2]

Melies's theatric artistry was recognized and copied by many contemporaries. One in particular, Edwin S. Porter, an employee at Edison Manufacturing Company, was especially impressed with the potential of film to tell stories. In 1902 he assembled and edited a collection of short films about firefighters into one longer dramatically narrative work, which he called *The Life of an American Fireman*. A year later, he improved upon his "cross-cutting" editing techniques in a production historians say revolutionized all moviemaking: *The Great Train Robbery*. In ten action-packed minutes, audiences were able to see a band of desperadoes hold up a mail train and get hunted down by a posse of cowboys. Tight editing and violations of normal time sequence characterize the film. At the film's conclusion, a gun was fired directly into the camera, horrifying many moviegoers who were adjusting to the new mode of storytelling.

Porter's 1903 *Great Train Robbery* is frequently cited as the first film to tell a story. A similar claim could be made for several other films because the narrative film was developed almost simultaneously in America, Europe, and Australia. But Porter can be singled out for assuring film a future that would transcend novelty and trickery. No longer could movies be mere reenactments of stage plays. They had become an art form in their own right.

The United States took the lead in filmmaking at the turn of the century, largely as a result of the huge waves of immigrants who arrived from Europe. For those people, mostly illiterate and unfamiliar with English, the

Turn of the Century

Novelty of the "flickers" soon wore off. New gimmicks for newly sophisticated masses were needed. Melies and Porter provided them.

The Great Train Robbery, 1903, was the first film to tell a story.

The Great Train Robbery, 1903, was the first motion picture to tell a complete story.

By 1908, the film industry's basic formula was set by the simple nickelodeon.

silent movies provided a rare escape from the drudgery of daily life. They flocked to see the one-reel thrillers, and their enthusiasm created a demand for more.

The earliest theaters were storefronts where half-hour showings ran continuously. In 1905 the first true theater was constructed in Pittsburgh and was named, appropriately, the *Nickelodeon,* because the price of admission was a nickel. Soon these specially designed pleasure houses spread across the nation. The period from 1903 to 1908 was the time during which the basic foundations of the film industry were laid, and the principles of wide—or mass—appeal, audience turnover, formula production, and character stereotyping became firmly established. The period from 1908 to 1914 was a time of struggle for the economic control of the new medium.

Early studios centered in New Jersey and New York formed a trust called the Motion Picture Patents Company with a subsidiary called the General Film Company. Members of the trust had exclusive contracts to purchase film and equipment, guaranteeing them a virtual monopoly. In addition, they sought to control production and to tie exhibitors' film releases to contracts on their equipment rental at two dollars a week, with no purchase possible.

Independents found the going difficult. They bootlegged materials from Europe or from friendly associates in the trust and they produced longer and more innovative films than their trust counterparts; but the independents were constantly disrupted by goon squads and harassing litigation initiated by the Motion Picture Patents Company. If that were not enough, the states of New Jersey and New York began taxing the studios.

The solution to the independents' problems was found in an isolated, small town in Southern California. Amid fruit orchards and attractive hills and valleys, which lent themselves to the filming of Westerns, the independents set up shop. For some of them, the move across country was one step ahead of the summonses being issued by New Jersey or New York sheriffs. Hollywood, the dream factory, offered inexpensive real estate, a seemingly endless supply of bright, sunny days ideal for the production of films, and a quick escape to Mexico if necessary.

Battles over production and distribution rights brought us to Hollywood.

The struggle with the Patents Company and the emergence of the feature length film gave shape to the industry and the three major segments that describe it: production, exhibition, and distribution.

Longer films cost more to produce, but the public seemed willing to pay more to see them. This provided an incentive to producers. At the same time the longer films began to attract a more educated and sophisticated audience and to compete effectively with the theater as an evening of entertainment. Film, as it developed, radically altered the nature of the legitimate theater, effectively removing it from the national scene, more or less restricting it to a purely provincial New York medium and, incidentally, using it as a proving ground for film musicals.

The Feature Length Film

Finally, the longer films reduced audience turnover and necessitated larger theaters; thus the ornate movie palaces of the 1920s and 1930s replaced the nickelodeons, and film-going became a middle-class diversion. The turning point in this chain of events can be marked by the 1915 production of *The Birth of a Nation,* a film as significant as *The Great Train Robbery* a dozen years earlier.

Griffith's *Birth of a Nation,* 1915, revolutionized the industry.

The Birth of a Nation was a three-and-a-half-hour historical epic produced at a cost of $100,000 to $125,000 by David Wark Griffith. Griffith, a Kentuckian and the son of a Confederate officer, selected the Civil War as the focal point around which he filmed his version of history—one in which the heroes were Ku Klux Klansmen. The eloquent and powerful film, a masterpiece of camera and editing techniques, was presented with full orchestration by theater musicians. The stereotypes it presented—particularly the "fallen" family of the South, carpetbaggers, and renegade blacks—caused race riots and mob behavior in many cities where the film was shown. Woodrow Wilson referred to the film as being "like history written in lightning." Indeed, Griffith proved in 1915 that film could have

The Hollywood Method

powerful emotional and propagandistic effects, at a time when the film industry was reaching wider and wider audiences who were coming to larger and more opulent theaters with greater expectations of entertainment and involvement.

The larger theaters had a voracious appetite. They had to be kept supplied with films that appealed to everyone, young and old alike. They could not as a rule be controversial, but they had to be good entertainment—that is, exciting, spectacular, sensational—and they could not depart too far from a proven mode. For example, Griffith's 1916 *Intolerance,* a massive sociohistorical study of human imperfection presented in four separate stories bound together by the theme of intolerance, was a commercial failure primarily because it was too novel, too complex, too pacifistic. Indeed, it was years ahead of its time and caused Griffith and his peers (including Charles Chaplin and Mack Sennett) to direct creative efforts along more tried-and-true lines in the interest of making a dollar. Recognizing the need for profit, filmmakers tended to foreshadow television's later offerings in blandness, tentative sensationalism, and formula emphasis.

After the Motion Pictures Patent Company failed in its attempt at control, power shifted to the studios, which created the vertically integrated film empires that ruled Hollywood for thirty years. Companies such as Metro-Goldwyn-Mayer (MGM) integrated production, distribution, and exhibition through their own movie houses. Distribution was the key to success, and *block booking* was the means used. Under block booking, exhibitors could not be selective; they were forced to contract for a number of pictures from a studio in order to get any major productions. If exhibitors refused, they got nothing and would go out of business. If they agreed, the exhibitors were assured a steady supply of films, many of which were second rate "B" pictures. At the time of booking, many of the films would be unproduced, and since they were being booked sight unseen, they were sold on the basis of proven factors, such as actors, actresses, plots or story lines, authors, directors, and sometimes even production costs.

As block booking became prevalent, the *star system* evolved into a type of insurance. The public became acquainted with the names and faces of performers by the 1920s and demanded to see more of their favorites. Among them were Douglas Fairbanks, Mary Pickford, and Rudolph Valentino. Cost controls became less important than the production budget, the size of which constituted another guarantee of success. There were numerous ways to spend money: on the stars who commanded fabulous salaries, on authors of best-selling novels for the film rights, on other proven writers to provide screenplays, on well-known directors, on an array of increasingly sophisticated technical innovations, and on promotion.

Film, by its nature, demanded promotion to call attention to its wares in the marketplace. In both a historical and legal sense it qualified more as a consumer product than as a medium. A 1915 Supreme Court decision held that the exhibition of films was a business, pure and simple, originated and conducted for profit, much like circuses and theaters; as such, it deserved no protection under the First Amendment. This attitude, coupled with the unquestioned fact that numerous film producers were growing increasingly rich, led to Hollywood's demand for press agentry and star promotion, and at the same time gave the fledgling public relations profession a substantial boost.

Once film became more and more organized as a major industry, it was inevitable that the artistic considerations of the medium would be subordinated to the practical. There was little room for experimentation, for art, and for departure from proven formulas. As production costs rose, Hollywood had to turn to Wall Street more frequently to finance its ventures. This move compounded the formula situation, for the bankers were far more apt to lend speculative money on a proven project than on an unknown quantity. In turn, there was an acceleration in the demand for best-selling authors, for expensive sets, for name stars, for proven formulas in the scripts, and for huge advertising and promotional budgets.

Popular mythology holds that sound came to films when Al Jolson turned to the audience in the middle of *The Jazz Singer* in 1927 and exclaimed, "Listen! You ain't heard nothin' yet." Like much folk history, this is erroneous. During the years after Edison's assistant William K. Dickson demonstrated talking pictures, experimental efforts were underway, including Warner Brothers' and Western Electric's Vitaphone process and Dr. Lee De Forest's phonafilm. The achievement of *The Jazz Singer* was the synchronization of sound with the feature-length picture. Al Jolson's 1927 utterance was prophetic, as audience desires for the *talkies* spelled the end of the silent era and a revolution in film technology and artistry.

Sound and the Depression

Hollywood's studio back
lots were worlds unto
themselves, even in the
1920s.

Sound changed the nature of
the industry, and made or
broke many stars.

Sound disturbed Hollywood because many of the stars of the silent era couldn't make it in the talkies. Whether it was because of accents, vocal pitch and timber, or the new methods of acting dictated by the single but omnipotent microphone, stars like John Gilbert, Emil Jannings, Lillian Gish, and Clara Bow were quickly out of work. Just as quickly, new stars appeared in the 1920s and 1930s, including Wallace Beery, Marie Dressler, William Powell, Myrna Loy, Gary Cooper, Clark Gable, James Cagney, Edward G. Robinson, Cary Grant, Spencer Tracey, Humphrey Bogart, Katherine Hepburn, John Wayne, and Bette Davis. The millions of dollars that had been spent promoting silent screen stars were lost assets, and the new, talking stars had to be created from scratch. Furthermore, the nature of sound changed the vehicles, and the story lines that had been appropriate during the silent era yielded to more dramatic scripts. New formulas were required, and the dimensions of sound required new and infinitely more expensive technology.

In 1927 some fifteen thousand movie houses were situated in America, but few were equipped for sound. Within two years nearly half the nation's theaters were wired, including almost all the metropolitan theaters, which generated the most income. At first Hollywood had counted on a dozen years or so for the changeover, but audiences stopped attending silent films—even though the average pre-1927 silent film was artistically superior to the first generation of talkies. By 1930, some 95 percent of Hollywood films were talkies, as the industry adjusted to the inevitable. Interestingly, not one major Hollywood studio folded during the changeover. They were more careful about the production and directing of the new films, which typically

Radio City Music Hall, although used primarily for live performances, shows how ornate theatres of the 1930s and 1940s could be.

cost one and a half times as much to make as their silent counterparts. And the involvement of big banks to finance the projects meant less artistic freedom, particularly during the years of the depression.

According to a former production chief of Paramount studios, the new era of talking films resulted in "more craftsmen, less teamwork, more complex organization, less pioneering spirit, more expense, less inspiration, more talent, less glamour, more predatory competition, and less hospitality."[3]

Sound had barely come to the screen when the Great Depression began, injecting new social factors into moviemaking. Radio was a relatively new and "free" medium. It provided home entertainment in the lean depression years, when even the price of twenty-five cents to see a movie was often prohibitive. While the excitement of sound delayed the full economic effect in the movie business for several years, the depression years for Hollywood, like the rest of the nation, were thin. Color technology, which had been held back by Hollywood lest the industry be disturbed again, made its appearance in 1935 (with *Becky Sharp,* starring Miriam Hopkins) as a counterdepression measure to offer something so vividly and dramatically new that America would spend money to see it.

Hollywood helped America escape from the Great Depression.

For those who could afford the price of admission, the magical movies were a way to forget one's troubles. They provided escape to *The Wizard of Oz* and to war-torn Atlanta in *Gone with the Wind*. Busby Berkeley became famous for making tap-dancing musicals with long lines of high-kicking chorus girls.

Table 9.1 Average weekly motion film attendance in the United States

Year	Average weekly attendance	Year	Average weekly attendance
1925	46,000,000	1960	40,000,000
1930	90,000,000	1965	44,000,000
1935	75,000,000*	1970	17,700,000
1940	80,000,000	1975	19,900,000
1945	90,000,000†	1980	19,644,000
1950	60,000,000	1985	20,385,000
1955	46,000,000		

Sources: Christopher H. Sterling and Timothy R. Haight, *The Mass Media: Aspen Institute Guide to Communication Industry Trends* (New York: Praeger Publishers, 1978), p. 352; Cobbett Steinberg, *Reel Facts: The Movie Book of Records* (New York: Vintage Books, 1978), p. 371; and *Variety,* 8 January 1986. Reprinted with permission of Variety, Inc.

*Sterling gives figure of 80,000,000 for 1935.

†Sterling gives figure of 85,000,000 for 1945.

While westerns, mysteries, and situation comedies were more or less stock fare, there was a certain faddishness to Hollywood productions in the 1930s and 1940s. A series of gangster movies in the early 1930s was followed by a rash of musical extravaganzas, of which *Flying Down to Rio,* with a chorus line on the wings of an airliner, may have been the all-time high. These extravaganzas were followed by a religious revival, and finally a blaze of spy and war films. Once a plot achieved any kind of success at the box office, every other studio rushed variations of that theme into production to capitalize on the trend.

A companion trend developed as the syndicated Hollywood columnist lent gloss to the scene, filling the nation's newspapers with gossip, sex, and the doings and misdoings of the movie colony. Several dozen magazines vied to outdo one another in dramatizing the scandals of "tinsel town," both on and off screen; Hollywood provided a kind of vicarious romance for a nation in the throes of depression.

In World War II, Hollywood helped carry the colors.

During World War II, the film industry also provided an additional form of escape and even a form of patriotism as it reached its full impact in the late 1940s, immediately following the war. The industry reached its peak in the period of 1945 to 1948, when 90 million Americans attended movies each week. (Attendance figures vary widely, from the industry's 90 million to the U.S. Department of Commerce's more conservative 66.8 million in 1948.) In 1949 the major studios—MGM, Twentieth Century Fox, Paramount, Columbia, RKO, and Warner Brothers—cranked out over four hundred celluloid fantasies. Thirty years later, fewer than half as many feature films were being turned out annually. Only recently, in 1986, has production equaled post-World War II highs.

The bubble of the film industry burst, primarily due to a 1948 court decision (*U.S.* vs. *Paramount Pictures*) and subsequent *consent decrees*. Sued by the Department of Justice for being in restraint of trade, the film industry reluctantly agreed to divest itself of its chains of movie houses. In essence, the empires were to be broken up, meaning the separation of exhibition from production and distribution. By the early 1950s the full impact of the consent decree was being felt; the studios lacked outlets for their potboilers, and the independent exhibitors demanded higher quality movies.

Other factors also entered into the decline of the studios, one of which was the formation of separate production companies by some artists. Actors, actresses, and directors banded together to produce films. Often they relied on the major studios for distribution of the final product, and sometimes for bankrolling. Increasing numbers of films were produced abroad, primarily as a result of tax advantages to the participating stars, and also because of cheaper foreign labor. Inevitably, some of the new independent producers came under the influence of the experimental and artistic techniques employed by foreign filmmakers.

Another blow to the film industry was dealt by a combination of national hysteria over communism and Hollywood's cowardice. During the years following World War II, right-wing groups scoured the countryside for Communist sympathizers or other "lefties." Scriptwriters and filmmakers who earlier had shown sympathy toward Russia, an American ally during the War, were hounded out of business—*blacklisted*. To protect their profits, studio chiefs turned against their former colleagues, and many of those named before the House Un-American Activities Committee were unable to find work for more than a decade. The blacklisting is regarded as one of the film industry's low points.

Then there was television. Television took audiences out of the movie palaces and placed them back in their homes. Hollywood could not compete with television on its own terms by providing an evening's entertainment, which forced the film industry to offer radically different techniques that were unavailable to the home viewer. Wide screens, stereophonic sound, and three-dimensional films were a part of the answer; so were the super-spectaculars and the so-called first runs playing at premium prices in selected movie houses. Another approach was the exploration of previously taboo subject matter: drugs, sex, homosexuality, and nudity.

But the strategies were not successful. Attendance dropped from a high of 90 million weekly admissions in the late 1940s, as indicated in table 9.1, to less than half of that a decade later, and was again reduced by half by the 1980s. As the market declined, so did the need for the big movie palaces; they were the first casualties.

The Postwar Decline

When the federal government made studios sell their theaters, and TV came along, the celluloid bubble burst.

Woody Allen, the front for a blacklisted television writer in the 1950s, is forced to testify before the House Un-American Activities Committee in *The Front*, a film based on real-life incidents.

The Impact of Television

TV was not all bad news for Hollywood. Cooperation and independent efforts came about.

Television was not entirely bad for the film industry; there were some hidden benefits. First, television freed the film industry from a slavish devotion to the lowest common denominator. It also freed film from a reliance on the pseudoguarantees of success: the star system, formula plots, huge promotion budgets, and the like. The breakdown of these accoutrements within the industry further encouraged independent production as a new breed of artistic directors began to take control. Film was and is a corporate art form demanding the talents of writers, directors, and actors. Where previously the economics of filmmaking emphasized the stars, the newer trend placed emphasis on the director in whose production the actors were more or less incidental. This, too, was a carryover from the influence of foreign films and the reputation of such impresarios (directors) as Federico Fellini.

The other big assist that television ironically offered the film industry was television itself. When television first began to make inroads on film's audiences, Hollywood fought television by prohibiting its productions from appearing on television. Beginning in 1955, however, the film industry discovered that television provided a brand-new market and source of revenue for old movies that were fully depreciated and had no other potential use for the future. Hollywood began selling old movies to television and in some recent cases has earned several million dollars a showing. The most noteworthy example is *Gone with the Wind,* for which CBS paid MGM 25 million dollars for multiyear exclusive rights.

The next step in this particular chain was the use of studio facilities, first for making television serials and later for making television films. The major studios had the technical hardware, the personnel, and the know-how to crank out television programs. More complex, however, was the television movie. Not all the contemporary films, especially those exploring the more permissive themes of sex and violence, were appropriate to family viewers. A new type of film was required—one made for television—which would have its initial showing over the air, and then later would be exported for foreign exhibition in theaters, where profits were almost guaranteed. Not only did the television movie provide an opportunity to tailor the subject matter to the limits of family viewing, but it could also produce films to meet television's exact specifications as to running time and commercials, something that Hollywood film, which was produced for other purposes, could never do. It was possible to interrupt the action on television movies at the exact time that a commercial would run without disturbing continuity. Further, television films were designed to run ninety minutes, less commercials, and would not require the extensive editing that often removed significant action from the Hollywood product adapted to television.

Conversely, the smaller audiences of theater movies were younger (three-fourths of the tickets purchased each year are by persons under age thirty), better educated, and more affluent than the audiences of television, and therefore they dictated different themes, like social justice, violence,

Table 9.2 The most popular films seen on television

Rank	Title	Date shown
1.	*Gone with the Wind*—Part I	11/7/76
2.	*Gone with the Wind*—Part 2	11/8/76
3.	*The Day After**	11/20/83
4.	*Airport*†	11/11/73
	Love Story†	10/1/72
6.	*The Godfather*—Part 2	11/18/74
7.	*Jaws*	11/4/79
8.	*Poseidon Adventure*	10/27/74
9.	*True Grit*†	11/12/72
	The Birds†	1/6/68
11.	*Patton*	11/19/72
12.	*Bridge on the River Kwai*	9/25/66
13.	*Helter Skelter*—Part 2*†	4/2/76
	Jeremiah Johnson†	1/18/76
15.	*Ben Hur*†	2/14/71
	Rocky†	2/4/79
17.	*The Godfather*—Part 1	11/16/74
18.	*Little Ladies of the Night**	1/16/77
19.	*Wizard of Oz*	1/26/64
20.	*The Burning Bed**	10/8/84

Source: *Variety*, 8 January 1986. Reprinted by permission Variety, Inc.
*Refers to movies made for television.
†Tied.

and sexual freedom. They also demanded, and received, different techniques of pacing and camera work. By and large, the theater audience was also more sophisticated, and more aware and appreciative of graphic technique, being more educated and having grown up with video from childhood. But with the critical successes of such TV films as *Roots, Holocaust, Shogun,* and *Masada,* the distinction between theater and TV films has blurred.

The Business of Filmmaking

Today, perhaps more than ever before, filmmaking is a business. The Hollywood studio system of old, which succeeded or failed largely on the basis of the star system and the instincts of studio moguls and directors for public tastes, has given way to conglomerate mentalities, where boards of directors and accountants as often as not make the final decisions. Given the fact that major studios such as MGM value their "libraries" at a billion dollars or more, it is easy to see why the giants of the industry seldom separate artistic from corporate decisions. Independent producers, working alone or using studio properties, must convince the studios, parent corporations (often insurance and investment concerns, banks, or oil companies), and, increasingly since 1980, private investors to take ever larger risks involved in producing a major motion picture.

Today, more than ever before, filmmaking is a big business. Corporate and artistic decisions are inseparable.

One of Hollywood's recent big-money losers, *Dune,* cost $42 million to produce.

Production

The average Hollywood film of 1987 cost $16 million to produce; it had to earn double that amount in rentals to break even.

And risks there are. As recently as 1972 the cost of an average film released by the Motion Picture Association of America (MPAA) was $1,936,000. By 1980 it had reached $10 million, and by 1987 stood at $16 million, far outstripping the rate of inflation. Hollywood's economic rule-of-thumb measure for success is that an average ($16 million) film's production cost has to be equaled by its domestic rentals—that is, how much money the United States and Canadian theater operators pay the producer for the privilege of showing the film. Today, films costing around $50 million to produce barely merit headlines, unless they bomb at the box office. Once print and marketing costs are added to production expenses, the "typical" feature film has to recover two to three times its total expenses in theater rentals before showing any profit. *The Cotton Club,* for instance, cost $51 million to make, and had to gross $120 million to break even; by early 1986 it had earned only $12.9 million in the domestic market. *Dune,* produced for $42 million, had a domestic income of only $16 million. (The most expensive film of all time, apparently, did not have to show a profit. It was the Soviet Union's *War and Peace,* whose estimated $100 million budget was underwritten by the government.)

According to *Variety* (the entertainment industry's trade magazine), as of 1986 only 216 of the thousands of films ever released had earned over $20 million in domestic rentals, and only 328 had earned $15 million. If we double those figures to include foreign rentals, reissuance at later dates,

New York City's evil spirits don't stand a ghost of a chance against ectoplasmic exterminators Ernie Hudson, Bill Murray, Dan Aykroyd, and Harold Ramis in Columbia Pictures' big money maker, *Ghostbusters*.

sales to television and home video (and doubling the totals is a generous guess—at mid-decade the typical film picked up only 40 percent of its total income from such sources), we are left with the startling realization that the overwhelming majority of films being released today will never make a profit. Over the past decade, the number of English-language theatrical films being produced worldwide has ranged between 175 and 515 annually; most of those films are made in America. It becomes obvious that a few (the *Star Wars* series, *E.T.: The Extra-Terrestrial*, *Indiana Jones and the Temple of Doom*, *Raiders of the Lost Ark*, *Jaws*, *Ghostbusters*, *Back to the Future* and the like), are going to have to keep the industry afloat, paying the bills for the big budget films (*Heaven's Gate*, *Cleopatra*, *Dune*, *The Cotton Club*) that will never show a profit.

Faced with rising inflation and an uncertain box-office attendance pattern, the MPAA has been running scared. Box-office take has shown a slight increase since 1980, but it has been somewhat artificial—the cost hikes in ticket prices barely offsetting the uneven attendance patterns. The barrage of Christmas vacation and early summer releases may draw record crowds and income, but the rest of the year the theaters continue to feel the pinch of television (particularly the cable and pay movie channels) and videotape rentals.

Nine companies—Columbia, Twentieth Century Fox, MGM/United Artists, Paramount, Universal, Warner Brothers, Disney, Orion, and Tri-Star—dominate the film industry, but they no longer control the projects from inception to distribution. Beginning in the 1960s, youthful directors

Nine companies dominate the industry, but they don't necessarily control their own projects from start to finish.

and producers who showed they could reach the younger audiences were financed by the studios to develop whatever kinds of films the studios suspected would make money. In some cases, with independent filmmakers such as George Lucas, Francis Ford Coppola, Martin Scorsese, Michael Cimino, Robert Altman, Woody Allen, and Stanley Kubrick at the helm, the tactic worked. Forget the star system, the filmmakers said; give us freedom to find a story, hire actors, develop an appealing movie. Since 1980, more and more of the independent filmmakers have decided to go it alone, even arranging their own financing through public and private investment agencies willing to take a high-stakes gamble in the glamour industry. Once they have completed production, they are likely to turn to the major studios for help in packaging and distributing their films.

Easy Rider was made by independent Dennis Hopper in 1969 for $370,000, and earned more than $40 million. *American Graffiti,* by George Lucas and Francis Coppola, cost a mere $780,000, but brought in some $145 million worldwide. Such phenomenal returns on investment convinced Hollywood to give independents their rein and cater to the youth movement. But the failures were as frequent as the successes. Cimino's reputation for artistic and financial genius, based on 1979's *The Deer Hunter,* was soiled considerably two years later when he bombed with *Heaven's Gate.* Between *Easy Rider* and *Heaven's Gate,* Hollywood—and America— had undergone several changes, and by the early 1980s the relationships between independent filmmakers and studios had also changed.

Studios successfully catered to filmgoers' tastes for disaster and violence and big stars/big budget films throughout the 1970s, with *The Poseiden Adventure, Jaws,* the *Airport* sagas, and *Earthquake* and *The Towering Inferno.* Then, toward the end of the decade, the brilliance of young filmmakers George Lucas and Stephen Spielberg resulted in what has become to date the most successful film genre of all time: the simple, sometimes galactic myth told with pizzazz. Computerized special effects, not superstars, helped make Lucas' *Star Wars* trilogy so phenomenally successful that he has organized a filmmaking company on his three-thousand-acre California ranch where he has been able to concentrate on cinema technology and assist others in getting started in the business. Computer-generated graphics—can you top this?—as introduced by Lucasfilms are changing the appearance and impact of films. Meanwhile, Spielberg's somewhat simpler stories (*E.T., Gremlins, Indiana Jones*) of good guys and bad guys continue to appeal, although the violence in them has given rise to a new barrage of criticisms and a reworking of the MPAA rating system.

Distribution

Since 1975, when moviegoers started standing in lines to watch Spielberg's thriller *Jaws,* producers have recognized the increasing value of media blitzes and marketing strategies to garner the needed return on investment.

Paid advertising, press releases and publicity, exploitative gimmicks, stunts, and the like, and special promotions involving tie-ins between the movie and a product and/or personality have been successfully orchestrated for mass appeal science fiction films (*Star Wars, The Empire Strikes Back, Return of the Jedi, Star Trek, Close Encounters of the Third Kind, E.T.: The Extra-Terrestrial*) and the adventure/disaster genre (*Jaws, The Deep, Raiders of the Lost Ark, Indiana Jones and the Temple of Doom, Gremlins*). Indeed, it is not unusual for the promotional budget to equal or even surpass the production budget. The rule of thumb throughout the industry is that it will take half as much to market a film as it did to produce it.

Basically, the reason such films are so heavily marketed is that in most cases, it is the distributor, and not the producer, who has taken the greatest financial risk in underwriting the film. In addition to providing the capital—either alone or in conjunction with other funding agencies—the distributor usually organizes the promotional efforts and deals directly with exhibitors in assuring favorable releases of the finished product.

Over the past decade or so, an increased number of tie-ins have aided the film business; not surprisingly in the world of mass communications symbiosis, other media benefit equally from increased consumer attention. The most typical tie-in involves a movie and a book, which increases the attention given to a title being promoted in both forms. Tradition holds that a good book might become a successful film, but contemporary media practice often finds a reversal of this process, starting with screenplays and ending up with best-selling books—frequently mass-marketed paperbacks filled with photos from the film, or even comic books, of which the best example might be the paperback versions of the various George Lucas and Steven Spielberg hits. Such tie-ins are forged with the hope that the popularity of one version will spur the other on to success. Not infrequently, television enters the picture, sometimes with documentaries on "The Making of . . ." or "Behind the Scenes of . . ." one or another mega-hit. Other times personalities from the films appear on talk shows, their visits carefully orchestrated by their film's publicity department. All in all, the complexities of media marketing demonstrate which tie-in combination—film, television, book, etc.—will return the largest number of dollars on the investment.[4]

Another aspect of a tie-in is the relationship between a film and magazines and newspapers. In the magazine world, there are dozens of publications devoted to film. And newspapers generally have entertainment sections that lean heavily toward the movies—quite naturally, since a good deal of their revenue comes from theaters advertising current showings.

Questionable distribution practices also improve attendance.

In addition to publicity and promotion schemes aimed at improving attendance and profits, the film industry has resorted to some questionable distribution practices. Selective contract adjustments, blind bidding, four-walling, and the older block booking methods are commonplace. Under *selective contract adjustments,* theater owners who cooperate with distributors in promoting and running their films are given preferential treatment. *Blind bidding,* which demands that exhibitors sign contracts for films before the films are released, has come under increasing fire on the state and federal levels. *Four-walling* became popular in the early 1970s when distributors rented the theaters directly from the owners, and kept all the box-office receipts for the limited engagements of their numerous low-budget films. Such films were frequently released or rereleased simultaneously across the country in carefully selected markets. They were accompanied by heavy television advertising campaigns, which often exceeded the cost of producing the film. Four-walling by the major studios was ruled illegal but remained a viable way for smaller independent firms, such as Sunn Classic Pictures (*The Life and Times of Grizzly Adams, Beyond and Back, Chariots of the Gods,* and biblical docu-dramas), to compete in the marketplace. *Block-booking,* the practice of forcing exhibitors to accept a producer's mediocre films along with the money-makers, is illegal but continues to be practiced in modified forms.

At the beginning of 1987 when *Variety* listed its annual "All-Time Film Rental Champs" in the United States and Canadian market, the value of promotional efforts made in recent years became increasingly obvious. Consider the mass-appeal films listed in table 9.3 and note how many of them are of very recent vintage. It is important to recall that the list does not include foreign theater income, sales to television networks, pay TV, syndicates, or home video—income which frequently equals or surpasses the domestic rental moneys.

Table 9.3 *Variety*'s all-time film rental* champs, of United States-Canadian market, 1987

Rank	Film	Release Year	Total rentals*
1.	*E.T.: The Extra-Terrestrial*	1982	$227,379,346
2.	*Star Wars*	1977	193,500,000
3.	*Return of the Jedi*	1983	168,002,414
4.	*The Empire Strikes Back*	1980	141,600,000
5.	*Jaws*	1975	129,961,081
6.	*Ghostbusters*	1984	128,264,005
7.	*Raiders of the Lost Ark*	1981	115,598,000
8.	*Indiana Jones and the Temple of Doom*	1984	109,000,000
9.	*Beverly Hills Cop*	1984	108,000,000
10.	*Back to the Future*	1985	101,955,795
11.	*Grease*	1978	96,300,000
12.	*Tootsie*	1982	95,268,806
13.	*The Exorcist*	1973	89,000,000
14.	*The Godfather*	1972	86,275,000
15.	*Superman*	1978	82,800,000
16.	*Close Encounters of the Third Kind*	1977 / 1980†	82,750,000
17.	*Top Gun*	1986	82,000,000
18.	*Rambo: First Blood Part II*	1985	80,000,000
19.	*The Sound of Music*	1965	79,748,000
20.	*Gremlins*	1984	79,500,000
21.	*The Sting*	1973	78,198,608
22.	*Gone with the Wind*	1939	76,700,000
23.	*Rocky IV*	1985	75,782,000
24.	*Saturday Night Fever*	1977	74,100,000
25.	*National Lampoon's Animal House*	1978	70,778,176

Source: *Variety,* 14 January 1987. Reprinted by permission Variety, Inc.

Rentals refers to dollars paid to the distributor, and is not the same as total box-office ticket sale grosses. Only the United States and Canadian ("domestic") markets are included here, which means the foreign market rentals and television or videocassette or disc sales, which sometimes equal or surpass the domestic rentals, are not considered. The dollar figures are not adjusted for inflation, biasing the table toward more recent releases. Adjusted rental figures would show *Gone with the Wind* holding down the top spot with more than one-third of a billion dollars in rentals.

†Two versions of this film were released.

The smaller audiences of the 1950s rendered the big palaces obsolete. Monsters like Radio City Music Hall in New York City, with its 6,000-plus seats, could not be filled, and the cost of upkeep was prohibitive. Consequently, many were torn down or were renovated and divided into two or more small theaters. The movie houses built during the 1950s were small, even tiny, and were located in shopping centers where parking and other services were available, or they were drive-ins where the customers brought their own seats.

Today, there are around 17,000 movie theaters in the United States. Fewer than 3,500 of them are drive-ins. On the average, an indoor theater has about 500 seats, compared to 750 in 1950. The newer movie houses reflect minimal construction costs and maintenance and often have multiple auditoriums. This allows a single operator to manage three or four minitheaters simultaneously from a central projection room, thus offering

Exhibition

Movies came to America's shopping centers, but now even the minitheaters are undergoing changes due to home video.

a range of movies to filmgoers. No longer is the audience limited to one selection: there may be three or four choices, and a moviegoer may be lured into attending three or four times during the same week.

In their never-ending struggle to lure the TV and home video audience back to the movies, film exhibitors in the mid-1980s started reversing their practice of constructing "mini-cinemas," which had averaged 150 seats. Today, the move is back toward the 500-seat theaters, and even toward some luxurious theaters that take advantage of the latest in neovideo. In addition, as the economy has forced a slowdown in shopping center growth, film exhibitors are returning to more traditional sites or expanding their present facilities.[5]

As one theater association president has explained it, it is time to get away from "cracker-box" theaters with "postage-stamp" screens, because those are already available in the home. To attract the ever-more sophisticated customers into the theater, exhibitors should pay closer attention to audience needs and behavior, provide clean and attractive locales, rely upon effective advertising and promotion, and offer a diversity of quality films.[6]

Many theater owners of late have resorted to their own creative marketing techniques. Some chains have exploited patrons' desire to "make a night of it" by adding a complete fast-food and bar service in the theater lobby and charging a set price for tickets and deluxe refreshments. Others have found they can stay in business by charging only a dollar or less on special nights each week (or every night if the fare consists of second run pictures). On the surface it would appear to be economic suicide, but the increased number of viewers helps pay the film rental costs, and any profits to be made occur at the concession stands. If we consider the fact that theaters make an estimated 600 percent profit on popcorn, 175 percent profit on pop, and 100 percent profit on candy, we can better appreciate the subtle economics of the film industry. The economic, psychological, and sociological phenomena of moviegoing become apparent as we find ourselves slurping and munching away in communal bliss, regardless of how thirsty or hungry we might be.

Most exhibitors make more at the concession stand than at the ticket booth.

Audiences and Their Guides

Film is a powerful medium because we are so willing to be manipulated by it.

There is a psychology of moviegoing that has long been recognized and exploited by governmental and educational agencies as well as by commercial interests. The experience of seeing a film emphasizes the vivid visual presentations in which images are already fully established, easily identified, and easily followed, even on the elementary levels. Viewers seem to attend films to enhance identification with film characters and to enjoy the aesthetics of the dramatic forms and vivid presentations. If these are the filmgoers' "needs," movies historically have met those needs through appeals to primary emotions and sentiments. The impact on audience members who attend films to be entertained is especially heightened. If they paid admission to laugh or cry, to be frightened, angered, sexually aroused, to reinforce their nationalistic or jingoistic pride, or merely to be removed from their everyday lives, their minds are particularly open to receive the producers' messages. Film is an unusually strong type of communication process, because the viewer is willing, even eager, to receive what the communicator has to offer.[7]

Whether for instructional, persuasive, or purely commercial purposes, film has unparalleled advantages as a mass medium. Complete control over emphasis, continuity, and effect—the use of special effects for dramatic purposes, the genius of editing, the ability to incorporate sound and music—are all designed for maximum purposeful impact on the audience.

Film tells a special story vividly, commanding attention. Further, it plays to a captive audience, and generally under ideal conditions. Theaters are designed to minimize outside interference (channel noise) and to concentrate all attention on the screen. Films have a pleasurable connotation, and whether instructional or not, they are regarded as play, not (at least by the audience) as social control.[8] Add to this the element of suspension of disbelief, and audiences are likely to be responsive to whatever message the screen offers them.

The connection between the message and the content is not automatic; for one thing people are complex creatures who do not always act in predictable ways. Also, the message or the content has to be in the film in order for people to respond to it, and for many years filmmaking in America was governed by a restrictive code. Finally, in order for a film to have an effect, people have to see it.

Film Controls

Historically, movies have not received the same freedoms, nor operated under the same restraints, as other media. Early in their history they were not even considered media of information, and were granted no First Amendment protection. That may have something to do with their strange history of self-regulation, which has ranged from an incredibly strict and controversial 1930 code to more realistic attempts at variable self-restraint beginning in the 1960s.

At the turn of the century and into the 1920s, we will recall from the discussion in chapter 2, the accepted theory was that mass media had great potential for manipulation and mobilization of society. Given this prevailing view, it is no wonder that strict controls were placed on the film industry after Hollywood established its reputation as a sin and scandal center during the swinging 20s. Threats for control—some real, some imagined—brought about efforts at internal house cleaning. The first was the 1922 creation of the Motion Picture Producers and Distributors of America (MPPDA), chaired by Will Hays, former Postmaster General and Chairman of the Republican National Committee.

In its formative years, the film business was subjected to very strict censorship. The MPPDA code prevailed from 1930 to 1966.

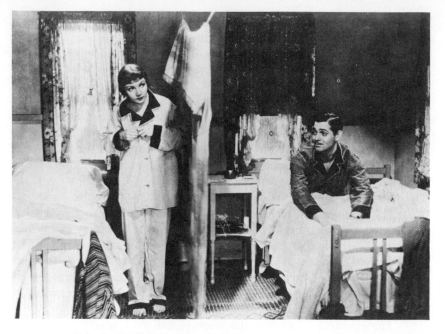

In 1930 the MPPDA adopted its Motion Picture Production Code. The Code's stipulations caused difficulties for filmmakers who were forced to submit their films to the MPPDA authorities before public showings were permitted. That policy, including the practice of insisting that any code violations be removed from the films before they could be released, came about several years after the code had been written. At first, it was intended to be a voluntary set of guidelines. However, when film attendance dropped off somewhat at the outset of the depression in the 1930s, filmmakers began taking liberties with the code in order to boost attendance. Public outcries that brought about the code in the first place were heard and felt again. This time it was the Catholic Legion of Decency and its pledge by eleven million Catholics to boycott offensive films. After 1934, the Hays Office (as the MPPDA was known) started issuing its seal of approval to acceptable films. Within a couple of years it was reviewing and approving almost all films exhibited in America (97 percent in 1937), sanctifying the "family film."[9]

Although undergoing minor changes, the code continued in effect well into the 1960s. For three and a half decades, critics blasted its attempts to create what they considered to be a highly untrue and misleading picture of life and the code's jellolike platitudes. It may have served its basic purpose, however, as it apparently played a role in reducing the scope and influence of state and local censorship boards and perhaps even warding off a federal motion picture censorship organization.[10]

In 1966, the Motion Picture Association of America (MPAA), as the MPPDA became known after 1945, developed its Motion Picture Code of Self-Regulation. The new code applied to production, to advertising, and to titles of motion pictures. It differed most significantly from the earlier code in its vigorous encouragement of responsible artistic freedom, and it attempted to establish a framework within which films deemed to be unsuitable for viewing by children could be so labeled.

MPAA set up the current rating system in 1966 so films could present more realistic portrayals of a changing time.

Movie Ratings Almost immediately, it became obvious to the MPAA that the revised code would not suffice. In the words of MPAA President Jack Valenti:

The national scene (in 1966) was marked by insurrection on the campus, riots in the streets, rise in women's liberation, protest of the young, questioning of church, doubts about the institution of marriage, abandonment of old guiding slogans, and the crumbling of social traditions. It would have been foolish to believe that movies, the most creative of art-forms, could have remained unaffected by the change and torment in our society.

The result of all this was the emergence of a "new kind" of American movie—frank and open, and made by filmmakers subject to very few self-imposed restraints.[11]

By 1968 the United States Supreme Court granted states the power to prevent children from being exposed to books and films suitable only for adults, and Hollywood immediately jumped on the "variable obscenity" and "variable availability" bandwagon. Instead of attempting to control the content of individual films, the MPAA decided to allow filmmakers to voluntarily submit their films to a rating board, where they would be rated on a four-point scale according to their suitability of viewing by children. According to Valenti, the only object of the voluntary rating system is to give some advance cautionary information to parents so they can make their own decisions about which films their children should or should not see.[12] And, as chairman of the rating board, Rutgers University professor Richard Heffner, says, "Our function is not to impose ideologies, morality, psychology or aesthetics, but to make an educated estimation of what most parents would think a movie should be rated."[13]

For sixteen years, the system included the categories: G for general audiences—all ages admitted; PG for parental guidance suggested—some material may not be suitable for preteens; R for restricted—children under age 17 must be accompanied by a parent or guardian; and X, no one under 17 admitted. (At first the MPAA had an "M" category for "suggested for mature audiences," but it caused some confusion and was replaced by the "PG" classification.)

Gremlins ground up in food processors and exploded in microwaves brought about the PG-13 rating.

G, PG, PG-13, R, and X ratings are supposedly based on a film's theme, language, nudity and sex, and violence. Theater managers do the enforcing.

In 1984, after a wave of public protest concerning the graphic violence in such PG-rated films as Steven Spielberg's *Indiana Jones and the Temple of Doom,* which he directed, and *Gremlins,* which he produced (in the former, the tearing of a heart from a living man, the flogging of children, and torture by immersion in boiling lava; in the latter, mini-monsters ground up in food processors and exploded in microwave ovens), the MPAA decided to add a fifth category of ratings: PG-13, children under age 13 admitted only when accompanied by a parent or guardian. Obviously, the new category has been inserted between the old PG and R ratings, and covers films a bit too rough—nudity, swearing, violence—for PG but not as rough as many R-rated films. With PG-13, Valenti said, the MPAA is not only cautioning parents to look closely at the film, "We're waving a flag and doing everything but tripping you as you go down the aisle to say 'don't let your young child go alone to this movie.' " Because some 60 percent of films made in the early 1980s have been given the R rating, thus putting them off-limits to the young teenage filmgoers, Valenti's group had some concern that a great many movies that would have received the R rating would stretch the definitions of permissible violence, swearing, and sex to earn the new PG-13 classification. "If PG-13 becomes the place to put soft R's rather than hard PG's, then we'll be in trouble," Valenti warned.[14]

Not everyone is satisfied with the MPAA's recent moves. Syndicated columnist Ellen Goodman, for one, was offended at *Indiana Jones and the Temple of Doom*'s graphic violence. Despite the fact that Paramount Pictures, which distributed the film, inserted a warning in its newspaper advertisements that read, "This film may be too intense for younger children," and despite the fact that even Spielberg himself admitted he would cover his ten-year-old child's eyes for twenty minutes during the middle of the movie, Goodman complained that:

Frankly, I am underwhelmed by Hollywood's attempt to modify its code instead of its behavior.

Rather than limiting sex and violence, they are increasing the ratings. Rather than improving the product, they are improving the warnings. It's rather like getting a broken car back from the auto mechanic with a new, improved description of its problems.[15]

Films are rated by a full-time, seven-member MPAA Classification and Rating Administration board. All the board members are parents, selected for two- or three-year terms. Film producers voluntarily submit their movies to the board for a rating. If they do not wish to receive any rating for their own films or ones they have imported, they are free to distribute the film without one. However, such films are automatically considered to be X-rated, and cannot be advertised on television or in many newspapers, and many theater chains refuse to carry them—all voluntary actions on the part of the media. (Bo and John Derek's *Bolero,* distributed without a rating in 1984, is a case in point.)

Many observers have felt that overly explicit sexually oriented films were the only ones to get the X rating, while graphic murder and mayhem were supposedly suitable for younger audiences. But Valenti insists that, despite popular opinion, all films are judged on four basic criteria: (1) theme, (2) language, (3) nudity and sex, and (4) violence.[16]

If filmmakers do not agree with the rating given by the MPAA, they are free to appeal the classification to a 22-member appeals board made up of theater owners, independent distributors, and studio representatives. The appeals, usually based on fear that an X-rated or R-rated movie will receive limited exposure, generally request and frequently receive a lower rating—PG considered to be the most acceptable. *Poltergeist* was originally rated R, but reduced to PG upon appeal. The review board also reduced the violent *Scarface* and *Angel Heart* from an X-rating to R-rating.

Ironically, Hollywood of late has come to consider G-rated movies as "death at the box office," because such movies carry with them connotations of goody-goody themes that will not appeal to mass audiences. Faced with sagging profits, Walt Disney Productions began making PG movies such as *Splash* under its "Touchstone" division. Other producers have been known to slip in an occasional obscenity or partial nudity to assure themselves of a "higher" rating.

Of the 342 movies rated in 1983, only eleven, or three percent, received a G rating; a decade earlier, fifteen percent received a G. Only one film voluntarily submitted for ratings received an X in 1983, compared with

G-rated films are considered "death at the box office."

twenty-eight in 1973.[17] For all practical purposes, then, the PG and PG-13 ratings are the ones most sought after, because they allow films to reach the lowest common denominator. As time passes, even the innocuous PG rating may fall into disuse, suffering the same stigmas attached to the G-rated films. After all, if becoming a teenager is a rite of passage, getting to see a PG-13 movie is part of that sociological phenomenon.

One final note on ratings. It is the judgment of the theater owner and ticket seller, and not necessarily the judgment of the parents, that prevails, for children are free to con an older teenager or adult into buying them a ticket and walking through a line to see whatever the youngsters desire. Like so many things in the world of mass media, the checks-and-balances system is not perfect.

Critics

Film critics play an important, but sometimes questionable, role in the industry.

Critics attempt to keep up with film; by devoting their lives to it, they develop a familiarity with it and an expertise that audiences could never hope to achieve. The public's time is too limited; their interests are too diversified. The critics serve as a kind of intermedia gatekeeper who use their opinion in an attempt to separate the wheat from the chaff for others, and to save them from exposure to the useless and meaningless. Like other media experts, critics have different goals. Critics writing for the *New Yorker* and *New York* magazines, and the *New York Times,* the *Los Angeles Times,* and many other metropolitan newspapers make honest efforts to evaluate the social and artistic merits of the various films to which they are exposed. Too often, however, as John Hohenberg points out, film criticism becomes merely an adjunct to the entertainment pages and a boost to advertising.[18]

Do critics make the difference between the success and failure of a film? Some producers of films that have been panned are certain that reviewers are to blame for poor attendance. Others say that word-of-mouth endorsements from individuals who have seen a film are more important than what a critic says. This is one reason why sneak previews are so popular. They also help a producer determine if there is negative audience reaction before committing funds to an expensive advertising campaign.

Sneak previews help establish a film's place in the market.

Michael Cimino, director of the $36 million *Heaven's Gate,* did not pretest the film via sneak previews. Negative reviews from newspapers in New York City, where the film was shown in November 1980, claimed that "Watching the film is like taking a forced four-hour walking tour of one's own living room" (*New York Times*) and that "Frankly, had the movie been filmed entirely in Russian without English subtitles it might have made more sense than it does in its present state" (*New York Daily News*). On the basis of that reception, Director Cimino and United Artists withdrew the film from circulation to attempt a $10 million reediting. Apparently Cimino believed in the power of the critics. But the reception the second time around was not much better, and the film disappeared quickly.

The little gold Oscar is worth millions of dollars to the winners.

The Oscars

One should not overlook the Academy Awards in a discussion of film evaluation and promotion. One night each spring two hours or more of television prime time are devoted to one of the most universally watched and extravagantly produced spectaculars on TV. The cost of the cast of this extravaganza alone would be prohibitive (if they did not appear gratis) when one considers the name stars, directors, composers, authors, and artists who are gathered by the hundreds to participate. While the Academy Awards may have begun as an effort to honor the artistic performances of the movie colony, some critics complain that with the advent of television they have become a promotional tool for old movies. They see it as an attempt to use television and the massive audiences it commands to publicize films that, by and large, have run their life course during the last year and to inject new life into them.

Hollywood's annual Oscar awards are actually commercials for the film industry.

There is no doubt that an Oscar in a major category or even significant nominations for the award can start a film back on the circuits again at first-run prices, often playing to far larger audiences on the "academy circuit" than it did the first time around. An Academy Award is not a guarantee of second-time success, but exposure of the film industry to 75 million people is almost bound to have some salutary effect on its wares. Industry estimates are that the gold statuette is worth $10 million in increased ticket sales for the "best motion picture" winner. And taking home an award almost always provides a boost to the career of the honored individuals, as well, making up for the embarrassment brought on by the barrage of self-serving hype in the trade magazines intended to influence the Academy's voting.

Other Film Types

Attention thus far has been concentrated on commercial or entertainment filmmaking, which is the entertaining and glamorous side of the medium. However, as in all media, film covers a wide range. There are cartoons—the ultimate in fantasy—which are often used as fillers but occasionally are developed into full-length movies, particularly by Walt Disney, who saw the childhood nostalgia of America's adults. There are also hard-core pornographic films, which are produced at bargain basement prices and marketed through an underground circuit. But these are the fringes of the film entertainment function.

Informational and persuasive films may have smaller audiences, but there are more of them.

Not to be forgotten are informational and persuasive films. Typically, these films play to smaller audiences than the entertainment films, but they outnumber commercial films by far. Some 16,000 informational and persuasive films are made annually in the United States alone. They are produced for governmental, industrial, agricultural, educational, cultural, and community organizations, such as chambers of commerce and visitors' bureaus. The number of such films has more than doubled since 1960. Together these films comprise well over a billion-dollar-a-year industry. In recent years many of these productions have used highly imaginative multimedia, wide-screen, and stereophonic effects to highlight their purpose and lend drama to the prosaic business of selling. Combining slides, film, and still photos in rapid flashes, using colored lights and multiple projectors, and most recently, adapting such neovideo as videocassettes and videodiscs, have almost created a new industrial art form. These productions offer widely expanded opportunities for employment, and absorb many of the university students who are training for film and video careers.

Newsreels and Documentaries

The information function of films started early with the filming of the inauguration of President William McKinley in 1897 and coverage of the Spanish-American war. As the film industry grew, newsreels became standard fare in the nation's movie houses. Produced weekly by such specialists as Pathe News and Fox Movietone News, they were a sort of mini-magazine on celluloid, concentrating as much on features as on hard news. At times, theatrics won out over objectivity and accuracy, the newsreel producers frequently contriving stories and faking scenes for dramatic impact.

Along with animated cartoons and live organ music, they provided a varied bill of fare in the movie houses of the 1920s and early 1930s. Like the picture magazines, they added a visual ingredient to the oral world of radio, the nation's principal mass medium at that time. In many larger cities, minitheaters exclusively showing newsreels found a ready market. Newsreels may have reached their heyday in World War II, bringing the sights and sounds of the war to domestic media audiences with an impact neither radio nor newspapers could match. Television, the catalyst for so many changes in media, did away with newsreels; coverage was easier and better done on the evening and late news and could be presented daily instead of almost a week late.

Related to the newsreel is the documentary, a film feature exploring in depth some aspect of society or the natural world. The earliest noted feature-length example is *Nanook of the North,* a natural history of an Eskimo family, which was produced in 1922 by explorer-filmmaker Robert Flaherty during the silent film days. Later, Flaherty produced an equally impressive documentary on life in Samoa, *Moana of the South Seas.* Since then documentaries have focused dramatic attention on some of the major areas of American life. During the administration of Franklin Delano Roosevelt, the documentary was refined as an instrument of domestic public relations—some might say propaganda. The depression period was captured by filmmaker Pare Lorentz with *The Plow That Broke the Plains,* a 1936 film exploring farming practices that led to the dust bowl, and *The River,* a 1937 testimonial to the Tennessee Valley Authority and flood control.

Recognizing the universal quality of film, the United States Information Agency (USIA) has made a number of films interpreting life and values of Americans for showing abroad. The efforts of the USIA have been hampered somewhat by the fact that these films run in competition with, and are far outnumbered by, Hollywood movies that show a different side of America. Thus, a credibility gap is established within the same medium. And the USIA films, no matter how well intentioned, do not operate in a vacuum. Other exposures—to American tourists and to United States advertising and products, such as Coca-Cola and McDonald's—create a confusion of American values difficult to understand.

The USIA is essentially a propaganda agency in the broadest sense of the term, and is restricted by Congress from domestic distribution of its materials, including film. One film, *Years of Lightning, Day of Drums,* a masterfully produced history of the thousand-day administration of John F. Kennedy to be shown to worldwide admirers, required an act of Congress to permit its exhibition in the United States. Another, Charles Guggenheim's *Seven at Little Rock,* which told the story of the integration of Little Rock, Arkansas schools, won an Oscar in 1964 without having been distributed to American audiences.

Newsreels and documentaries exploit the medium's informational function.

Propaganda and Education

We call it "information" or "communication," but other countries call it what it is: propaganda.

The propaganda potential of films has long been recognized. Documentary coverage of the Spanish-American War, which included dramatic shots of the aftermath of the explosion and sinking of the battleship *Maine* and Teddy Roosevelt's famous charge up San Juan Hill, may have been the first cinemagraphic propaganda. D. W. Griffith's *Birth of a Nation* and *Intolerance* were obviously more than dramas, capturing on celluloid the producer's values. The same can be said of the incredibly powerful products of early Russian and German filmmakers. Shortly after the Russian Revolution, the awakening of Soviet national pride and dislike of dictatorships were aided by such films as Sergei Eisenstein's *Battleship Potemkin,* a skillful and visually riveting account of mutiny by a ship's crew. In post-World War I Germany, filmmakers concentrated on stories of depression and helplessness, capturing the mood of the nation. Best known among the German films are *The Cabinet of Dr. Caligari, Nosferatu, Destiny, Metropolis,* and *Waxworks.* From both Russia and Germany emerged new techniques of storytelling, camera work, and direction, techniques that were to be employed by American filmmakers for generations.

American filmmakers were pressed into action during World War II to produce propaganda films in support of the war effort. Taking some cues—theatrical, not ideological—from German Leni Riefenstahl's classic Nazi propaganda films, *Triumph of the Will* and *Olympia,* Americans recognized the value of film to mobilize public opinion in wartime. Frank Capra's

Why We Fight and John Ford's *Battle of Midway* were unabashedly jingoistic efforts to motivate an initially naive domestic audience into accepting the need for global involvement and continuing deployment of troops. Capra's seven 50-minute documentary films tracing the rise of Nazism and fascism were commissioned by Army Chief of Staff George C. Marshall, who sensed that dramatically presented orientation films would motivate the hundreds of thousands of new and naive military recruits. An extensive series of psychological tests revealed that the films were highly successful in transferring factual information about the war effort, but somewhat less successful in changing soldiers' motivations, beliefs, and opinions. The research helped break the old belief in the media's hypodermic needle, stimulus-response effectiveness, and led to further studies on the subtle nature of media's role as agents of persuasion and motivation.[19] These films were originally underwritten by the Office of War Information, and led to expansion of American propaganda efforts into international film arenas— via the United States Information Agency.

Since World War II, other nations have recognized the propaganda potential of film. Film is sometimes used directly to agitate and mobilize public opinion to accepting radically new philosophies, as in the early days of Soviet Union, Red Chinese, and some Third World postrevolutionary nations. However, it is more frequently utilized to maintain public opinion and popular myths—"integration" rather than "agitation" propaganda— in most nations of the world. The United States is no exception, with its cinemagraphic releases reflecting cultural and institutional values.[20] For evidence, we need look no further than the recent spate of commercial films capitalizing on nationalistic pride—*Rambo, Missing in Action,* and even *Rocky IV*— and note the chagrin with which critics and officials from the Soviet Union have reacted to the "war mentality" reflected therein.

The Soviet Union itself has some 154,000 theaters—about 58 percent of the world's total and nearly ten times the number in operation within the United States. A prime reason for Soviet addiction to film has been that historically, especially before the advent of television, film was a key medium of propaganda and indoctrination. Today, it also serves as a prime social event, a break from routine.

Increasing numbers of educational films are being made for use in the world's classrooms, and new educational techniques are being developed that make heavier use of video materials, including both film and television. This is an interesting innovation that may modify the educational system considerably, but it is still too early to make any substantive judgments. The use of video materials may be one way to keep up with the acceleration of technological change, and with the visual orientation of the young. As this happens, of course, print will be de-emphasized and the entire structure of education will change.

Akira Kurosawa's epic *Ran,* an adaptation of *King Lear,* is regarded as Japan's most outstanding film report.

Film as a Cross-Cultural Medium

What's wrong with Hollywood's international diplomacy? Nothing, so long as we don't care how others see us.

One of the peculiarities of film, noticeable from the beginning, is that the language barrier is minimal. The immigrants who first populated the store-front theaters at the turn of the century were aware of this. Film has lent itself to cross-cultural transfer in a sense that no other medium has. Print demands language familiarity by its nature. The broadcast media, even television, are so heavily laden with verbal overtones that they hinder foreign comprehension. But film—like drama, its ancient ancestor—is universal. Dubbing and subtitling in native tongues can be done inexpensively, enhancing the universality. Film is also far more transportable than drama, and from its earliest history, became an export product with significant international effects.

From the beginning of filmmaking, the United States has dominated the industry. American producers had the early incentive to make an ever-increasing number of films commercially because of superior technology and large audiences, which found sufficient leisure to support film even around the turn of the century. Inevitably, these films found their way overseas to fill foreign movie houses with a United States product in the absence of a sufficient quantity of local films. Thus, film became an early communications export of the United States. Americans were visible on the world's movie screens to a degree that no number of emissaries or even tourists could ever achieve. But theirs was an unrealistic presence composed of fantasy and dreams. Exposed to the nations of the world were societal stereotypes and quaint moralities, a love of violence, and a sort of synthetic history, both past and present. An incredible wealth ran through all the pictures as celluloid strived mightily to portray the good life—a wealth bound to contrast unfavorably with the surrounding facts of life in much of the world. An unbelievable technology spread itself across the screen, generating a sense of awe, and the nations of the world thought they were seeing America.

Fantasy vs. Reality Abroad

Unaware of the American experience and lacking native perspective, other nations accepted American fantasies for fact. While American films were produced in America for the entertainment of Americans, as an escape from reality, there was no reason for other nations not to accept them as truth. This appears to be a pitfall of mass communications transfer, and it is quite possible that what originated in Hollywood as harmless entertainment, following a proven formula, emerged in Europe, in Asia, or in Latin America as a social document engendering awe, envy, and resentment.

It was not so much the treatment of American films—their improbable plots, their worship of sex symbols, and their cowboys and Indians—that caused this effect. It was the sheer overpowering volume of American export films, in the absence of other exposure, to a point where the films became America. Further, the more technically perfect the films became, the greater this delusion grew.

Moreover, it was the hundreds and thousands of "B" potboilers that offered this false view of America, rather than the big-budget extravaganzas. The very factors that characterized the old Hollywood—a large volume of production, the sacrifice of quality and technical perfection—were the same factors that betrayed the American image abroad. Rather than distorting the values of other lands, American films have distorted the foreign view of America, subtly and perhaps irreparably.

Foreign Influences Here

The transportability of film works two ways, and foreign films were not without influence in the United States. We have already indicated some of the early contributions of France, Germany, and Russia to filmmaking: Frances's Georges Melies with his trick photography and magical stories including the 1902 *A Trip to the Moon;* Germany's eerie tales of the macabre, most notably *The Cabinet of Dr. Caligari,* and the later Nazi propaganda films; and Russia's thematic montages and realism, manifest in Sergei Eisenstein's *Battleship Potemkin.* Also influential during the silent film days were the surrealistic works of Salvador Dali, especially his *Un Chien Andalou,* and the brilliant camera work of a French-German-Polish-Italian production, *The Passion of Joan of Arc.*

European film was less influential between the two world wars than it had been prior to World War I and again after World War II. Playing small "art" houses, foreign films gained a currency of modishness in the 1950s and 1960s. Principal among the foreign influences on contemporary film in America was the neorealism of the postwar Italian producers, such

Foreign films taught us many lessons.

as Roberto Rossellini and Vittorio DeSica. Their films dealt with the nitty-gritty and the hopelessness of life. French directors brought their "new wave" of films, which explored the actions and events of the lives of their characters—events that were often meaningless. Francois Truffaut became recognized as the leader of French "new wave" cinema with his influential classics *Farenheit 451, The Last Metro,* and *Small Change.* France's contribution was also one of technique—freeing producers from studios and taking them into the open—which even at the sacrifice of technical perfection brought a greater realism to the screen. English producers, meanwhile, discovered England's lower classes and their accents. In a sense they set the stage for the English revolution, which eventually erupted in gangs of Teddy Boys and the clothing styles of Carnaby Street and the Beatles. Indeed, a good deal of the American youth cult had its origin in English cinema.

Swedish films were and still are dominated by the genius of Ingmar Bergman who, more successfully than anyone else, introduced metaphor in film, raising cinema to an art form and exploring in the ancient Greek traditions the timeless themes of life, death, and truth.

From Japan came incredibly beautiful films, whose technical artistry and use of light, color, space, and misty camera techniques were entirely captivating. They brought the understated beauty of Japanese art to the screen. And, most recently, the history, geography, and cultural values of Australia have come into America's and the world's consciousness via film. All these forces found a refocus in the United States as its film industry underwent the throes of reorganization, and, to some extent, American films began to look for more complex themes, encouraging American independent producers to pay greater attention to the integrity and artistic merit of their work.

Foreign films, after a slow, almost "groupie" beginning in the 1950s and 1960s, became increasingly popular as the volume of production in the United States slacked off and was diverted to television. But popularity is relative. During a recent twelve-month period, only fifty-four imported films opened in New York City, the foreign film headquarters of America. There, in Los Angeles, in a few other major metropolitan centers, and in a smattering of university towns, foreign films are shown. Because the importing, printing, promotion, and distribution costs demand that a foreign film gross $400,000 in order to turn a profit, only a few entrepreneurs are in the business. Rarely do they earn as much as a million dollars in rentals for an imported film: *La Cage aux Folles* and *La Cage aux Folles II* from France, *Chariots of Fire* from England, and *Breaker Morant, Gallipoli,* and *"Crocodile" Dundee* from Australia are exceptions to the rule, having earned critical reviews and several million dollars apiece in American theaters. As Thomas Guback notes, the United States remains a curiously provincial nation when it comes to motion pictures. Public tastes formed and cultivated for decades by viewing predictable themes of films produced by the handful of American conglomerates are tastes not readily changed, and imported films have an uphill battle in the marketplace.[21]

The United States remains a curiously provincial nation when it comes to motion pictures.

The Future of Film

There have been few serious attempts to systematically predict the future of film, perhaps because most writers about film are caught up in the here and now, or perhaps because film has demonstrated a curious resiliency whenever it has been threatened in the past. Taking a cue from Jowett and Linton, who have devoted a complete chapter to the future of film in their thoughtful *Movies as Mass Communication,* we will look briefly at technological developments, demographic variables, and economic trends that may affect film in general and theatrical movies in particular.

Film's future is tied to technology, demographics, and economics.

As we have noticed in discussing the future of print media, and as we cannot help but notice when discussing electronic media, much of what happens to film is dependent upon the television industry. After their initial, mutual animosity, film and television appear to have struck a truce, with Hollywood producing made-for-TV movies and emptying its billion dollar vaults—for a price, mind you—to satisfy television's enormous appetite for time fillers. But Hollywood may be cutting its own throat by counting so heavily on television, especially since old movies are a nonrenewable resource and production.[22]

Burgeoning systems of in-home entertainment facilities such as videocassettes and videodiscs, which eliminate the necessity of "going out" to be entertained, have caused major shake-ups in film distribution and theater practices, if not film manufacturing. After years of lamenting that such neovideo spelled doomsday for the theaters, distributors and theater owners have apparently come to the conclusion that the media forms are symbiotic, not mutually destructive. Recalling their own experiences as moviegoers, and making use of research findings, the exhibitors have come so far as to begin selling videocassettes and videodiscs in theater lobbies. According to one estimate, rental and sales of videocassettes in various theaters were increasing theater owners' total revenues by 20 to 25 percent. As one Paramount executive put it, movies on videocassette are not regarded as a substitute for the popcorn, the shared emotion, and the enveloping sights and sounds of the theater-going experience; the regular moviegoers who own VCRs do not consider their VCRs to be in direct competition with the theaters, so why should the theater owners?[23] Besides, now that video rentals are exceeding box office takes, it's time the two media took their respective places in the entertainment marketplace.

Pay television, such as Home Box Office, Cinemax, Showtime, and other movie channels, is already absorbing large quantities of Hollywood's new movies (movies that are six to eighteen months old), and the trend is expected to continue. Improved technology of in-home units, including "projection television" sets producing six-foot-tall images, is bound to make staying home for movies almost as exciting an experience as heading to the theaters downtown or in suburban shopping centers. To compete with these innovations, theatrical films and movie houses are mounting technological campaigns to match those seen during the early television era, when three-dimensional and wide-screen extravaganzas were developed. In fact, 3-D was momentarily revived in the early 1980s.

Ornate theaters, complete with video and shopping arcades, bigger screens, 70-mm prints in lieu of the standard 35-mm size, incredible lifelike images made possible by filming at 60 frames per second rather than the current 24, and satellite-fed delivery systems from Hollywood to various theater chains are among exhibitors' ploys to lure the neovideo stay-at-homes back to the movies.[24] All of these cost money; even though filming at 60 frames per second rather than 24 frames would only add about 4 percent in film and lab costs to a movie's budget, exhibitors would have to spend up to $250,000 on projection and sound equipment to capture the intense imagery and heighten the audience involvement such film would permit. To make life even easier for the moviegoer, who might enjoy the communal experience once inside the theater, but who hates waiting in line to purchase a ticket, some theaters have already installed automatic teller machines that accept Visa or Mastercard.

Already, vastly improved sound systems have made their appearance. The Dolby noise-reduction stereo involves the rewiring of the theater at a cost of about $16,000. Another sound system, Sensurround, used low frequency sound to give *Earthquake* audiences heightened awareness of impending doom—but at a cost to theater owners. Drive-ins are experimenting with Cine-Fi, a system by which the sound track of the film is transmitted through the car's radio. Meanwhile, experiments with screen size and shape continue, in further efforts to give paying customers something they cannot experience at home.

Film futurists take some comfort in the knowledge that population trends indicate a bulge in the young, educated, and affluent. Some 45 percent of the world's population is now between the ages of twelve and thirty-nine, although demographers tell us that with the aging of postwar baby boomers, teenagers will soon represent a smaller percentage of the total mix than at present. In the past, persons under thirty have found films a satisfying way to spend their free time, and with increased leisure on the horizon, they may continue to do so. Should that be the case, Hollywood will try to satisfy the social inclinations of the younger group while tapping a broader market (one that includes those 25 to 40) by producing and distributing films for a combination of theater and in-home use. Already there are signs of this occurring, as filmmakers have broadly marketed "specialized" offerings, significantly reducing their investment risks. Cumulatively, commercial films may reach larger audiences even if theaters have

Filmmakers are realizing the wisdom of providing grist for newer media's mills.

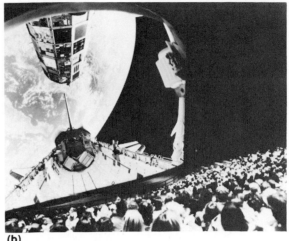

(b)

(a) The wrap-around, multisensory Omnimax theatre, the future of filmgoing. (b) Omninax patrons experience outer space in this film shot by space shuttle astronauts.

(a)

fewer patrons. To do so, however, more sophisticated marketing techniques, probably based on increased use of psychographic testing, may become more prevalent. The media blitzes accompanying recent Lucas and Spielberg films may be only the beginning.

One futurist has suggested that neighborhood theaters, even within apartment house complexes, will continue, but will use different distribution systems. He sees movies being sent electronically by cable and microwave, or even satellite, which would expedite delivery and reduce theater personnel.[25]

Others suggest that theaters will continue the path toward becoming total communications centers; not only where prerecorded films will be sold on tapes and discs, but where audiences will be more sensually involved in multimedia presentations. Crowd psychology will be exploited, and audience involvement in the film experience will be enhanced through the use of interactive feedback systems. One such system, experimented with as long ago as 1967 in the Czechoslovakian pavilion at Canada's Expo '67, allowed the audience to vote for different plot alternatives at various crisis points in the film. It necessitated seventy-eight different reels of film to accommodate the plot configurations the audience could conceivably choose. This technique has great potential for keeping audiences actively participating in their media.

Theatrical innovations, all efforts to more fully involve the audience, are expensive and tentative. In addition, they are frequently employed at some cost to the film itself, since audiences have grown accustomed to basic narrative and cinemagraphic formulas. Such formulas cannot be changed without affecting audience perceptions and moods, as Edwin Porter found in 1903 when his *Great Train Robbery* broke so many theatrical rules. North American audiences are too accustomed to feature-length fictional and narrative movies in which attention is consciously diverted from technique and focused on subject matter. Sophisticated technological innovations will be successful only if they can be incorporated adequately into existing storytelling techniques.[26]

When all is said and done, however, the key to the success of film will continue to depend upon the quality of the finished product. In the words of Robert Selig, president of the Theater Association of California:

If it is true that people—old or young—want to "go out," then the movie theater of tomorrow must be the "magnet" which provides all the elements of choice and resultant pleasure. All the statistics, comparisons, ratios and trends in filmgoing habits, VCR sales and cassette usage won't motivate a prospective patron towards a theater boxoffice. It always will be the "picture on the screen" and the total environment of the going-out-to-a-movie experience that will chart the final grossing figures, up or down.[27]

Summary

Through its antecedents, graphics and the drama, film stretches back to prehistory. Following the technological development of the electric light, roll film, and sprocket feed, and the application of the psychological principle known as persistence of vision, the curtain rose for the first halting one-reelers around the turn of the century. The first theaters were storefronts in major East Coast cities, where droves of immigrants, uneducated and illiterate, flocked for their only available entertainment in an alien land. Storefronts yielded to nickelodeons, which in turn yielded to the big movie palaces of the 1920s and 1930s, as a steady supply of films arrived from Hollywood.

Film was treated as a mass medium, catering to the lowest common denominator of appeal and sacrificing artistic considerations for economic discipline. The star system, lavish promotion and publicity, and huge budgets for formula pictures were all attempts to minimize the risk inherent in filmmaking during the years before television. The coming of television helped burst Hollywood's bubble, but the movie city was already in trouble as a result of the 1948 Paramount case wherein the industry agreed to divest itself of its theaters. Within a decade film's audiences dropped 50 percent, and within a generation, they dropped by half again.

In a pattern familiar to media historians, the new medium—television—was first viewed as a threat, and only later was it joined by the film industry, which sold old movies to TV at a considerable profit and eventually joined production forces. The two media found separate audiences.

Film's audiences were smaller, younger, better educated, more sophisticated, and more demanding than before. TV's audiences were represented by a truer cross section of America, the lowest common denominator.

Contemporary filmmaking, more than ever before, is a commercial business, controlled by economic decisions rather than artistic hunches. Three major elements constitute the business: production, distribution, and exhibition. The risks can be traumatic, even for the conglomerates who finance feature films, which today average over $16 million per feature. The industry has gone increasingly to publicity, advertising, marketing research, and gimmickry to gain a return on its investments.

Films generally play to captive audiences under ideal conditions in which the viewer is prompted to suspend disbelief. For these reasons, film is an ideal mass medium for instructional, persuasive, or commercial purposes. From the 1930s to the 1960s the content of American films was controlled by the code of the motion picture industry. Since then voluntary classifications have placed more responsibility on the theater owner. Critics help the viewer determine which films are worth seeing, while the Academy Awards are used by the industry to attract more patrons.

In addition to commercial films, the industry produces about 16,000 instructional and persuasive films each year. Other film types include newsreels, documentaries, propaganda films, and educational films. As a tangible consumer product, film became an early communications export. For better and for worse, it extended America's influence abroad; at the same time, foreign films with their unique production and artistic techniques influenced the American market and brought international understanding—and misunderstanding—back across the oceans.

To a large extent, the future of film is tied to the television industry, both in terms of product exchange and in consumer-controlled, at-home technology that is bound to affect theater attendance. Theatrical film is expected to become more sensory and more involving, particularly to retain its youthful, educated, and sophisticated adherents. Whether form or substance emerges victoriously is a key issue to be resolved, for new technology, like the 3-D and wide-screen innovations of the 1950s, may not suffice to demand loyalty from film fans who are already spending as much on video rentals as they are at the box office.

Notes

1. Arthur Knight, *The Liveliest Art: A Panoramic History of the Movies,* rev. ed. (New York: New American Library, 1979), p. 7.
2. Steven C. Early, *An Introduction to American Movies* (New York: New American Library, 1979), p. 6.
3. Peter Andrews, "The Birth of the Talkies," *Saturday Review,* 12 November 1977, p. 43.
4. "Media Tie-ins," *Publishers Weekly,* 11 April 1980, pp. 33–43; and Richard Reeves, "Lucas and Spielberg Strike Back for America," *Logan* (Utah) *Herald Journal,* 17 August 1982.

5. David Steritt, "Movies Try New Tactics in Battle With TV," *Christian Science Monitor,* 23 August 1982; and Thomas Guback, "Theatrical Film," in Benjamin M. Compaine, Christopher Sterling, Thomas Guback, and J. Kendrick Noble, Jr., *Anatomy of the Communications Industry: Who Owns the Media?* (White Plains, N.Y. Knowledge Industry Publications, Inc., 1982), pp. 244–45, 248.

6. Robert W. Selig, "Exhibs Think Big As Filmgoers Have Small Home Screens," *Variety,* 8 January 1986, pp. 34, 70.

7. Garth Jowett and James M. Linton, *Movies as Mass Communications* (Beverly Hills, Calif.: SAGE Publications, 1980), pp. 89–90.

8. William Stephenson, *The Play Theory of Mass Communications* (Chicago: University of Chicago Press, 1967).

9. Charlene Brown, Trevor Brown, and William Rivers, *The Media and the People* (New York: Holt, Rinehart & Winston, 1979), p. 139.

10. Harold L. Nelson and Dwight L. Teeter, Jr., *Law of Mass Communications: Freedom and Control of Print and Broadcast Media,* 4th ed. (Mineola, N.Y.: The Foundation Press, 1982), p. 397.

11. Jack Valenti, "The Movie Rating System," in *Mass Communication: Principles and Practices,* by Mary B. Cassata and Molefi K. Asante (New York: Macmillan Co., 1979). p. 289.

12. Jack Valenti, "The system works as guide for parents," *USA Today,* 14 November 1983.

13. Richard Zoglin, "Gremlins in the Rating System," *Time,* 25 June 1984, p. 78.

14. Gene Siskel, "The new PG-13 film rating: Is it effective?" *Salt Lake Tribune,* 29 August 1984.

15. Ellen Goodman, " 'Indiana Jones' spoils PG film rating," *Logan* (Utah) *Herald Journal,* 26 June 1984.

16. Valenti, "The Movie Rating System," p. 287.

17. Aljean Hermetz, "With return of Indy Jones comes debate over ratings," *Salt Lake Tribune,* 22 May 1984.

18. John Hohenberg, *The News Media: A Journalist Looks at His Profession* (New York: Holt, Rinehart & Winston, 1968).

19. Shearon Lowery and Melvin L. DeFleur, *Milestones in Mass Communication Research* (New York: Longman, 1983), pp. 114–147.

20. Jacques Ellul, *Propaganda: The Formation of Men's Attitudes* (New York: Alfred A. Knopf, 1965).

21. Guback, "Theatrical Film," p. 228.

22. William Fadiman, *Hollywood Now* (London: Thames and Hudson, 1973).

23. Michael London, "Movie exhibitors hear new video war strategy," *Salt Lake Tribune,* 15 March 1984.

24. Steritt, "Movies Try New Tactics in Battle with TV," and Jack Mathews, "Vivid images, visceral action," *USA Today,* 14 November 1983.

25. Wilton R. Holm, "Management Looks at the Future," in *The Movie Business: American Film Industry Practice,* eds. A. W. Bleum and J. E. Squire (New York: Hastings House, 1972), pp. 253–57.

26. Jowett and Linton, *Movies as Mass Communications,* p. 131.

27. Selig, "Exhibs Think Big As Filmgoers Have Small Home Screens," p. 70.

The New Electronics

10

Introduction

With network television, we became accustomed to looking at electronic screens for information and entertainment. By the mid-1980s, the average American spent four hours and twenty minutes a day looking at television. However, the audience for network television is getting smaller today. The average composite rating for the three major networks has fallen from over 90 percent in the 1970s to less than 75 percent today. One reason for this decline is that Americans can now get other programming and more varied information on their TV screens, or on TV-like computer screens, in the home.

Economic, technological, and regulatory factors suggest that we divide our examination of new electronics according to how the communication comes into our homes. Added programs and information may come to our home screens through clever uses of traditional TV signals, on communication signals from microwave distributors and satellite broadcasters, over coaxial cable, through television lines hooked to new computer technologies, and on prerecorded media such as videocassettes, videodiscs, and computer diskettes. In this chapter, we will examine these sources of information and entertainment.

Because the First Amendment bans the regulation of media content, most external control over the media is based on regulating the means of transmission. A program that might be judged to be obscene and illegal on network television, for example, may be perfectly legal over cable or on videocassette. We need to be aware that economic and social consequences flow directly from such technological and regulatory differences.

All of these new electronic media have uses and implications that we may only now be starting to realize. The use of new electronic technologies to place an unlimited range of information and entertainment on home screens is a trend that will continue into the next century. Technology exceeds adaptation, and new communication forms are constantly confronted before the old ones have been mastered. But by confronting them and trying to understand them, we are better able to try to shape their future as well as our own.

New Twists in Over-the-Air Communications

"Broadcasting" is no longer an adequate term to describe the incredible mix coming over the air.

While significant changes in electronic media have been brought about by new physical means of delivery (cables, phone lines, and prerecorded media), new twists in over-the-air communication have also added important new media services. Electronic modifications to traditional broadcast signals have led to subscription television, low-power television, and a user-controlled information service known as teletext. Multipoint distribution services (MDS) use microwave signals, which travel about twenty-five miles, to distribute programming and information services to high-population spots like hotels, apartments, and office buildings. A satellite broadcaster can skip both local broadcasters and MDS or cable operators to send a signal directly to receiving dishes on home rooftops.

In this section, we will examine these new twists in over-the-air communication. Satellite communication systems also figure prominently in fostering the revolutionary growth of cable systems, which we will examine in the section following this one.

Since the mid-1970s, traditional broadcasting increasingly has been challenged by new media competition born of the so-called new technologies. Its response has been to use technology itself to squeeze every service possible onto the regular broadcast signal.

New Twists in the Traditional Broadcast Signal

Subscription Television (STV) The first of these competitive responses resulted in *subscription television (STV)*. An STV channel simply relies upon subscriptions rather than advertising to support itself. It can restrict its reception only to subscribers by electronically scrambling its broadcast signal and attaching to subscribers' television sets decoders that unscramble the signal. As a pay service, its fare of first-run movies and special sports events is similar to that of a pay cable service.

Subscription television bills the viewer, rather than the advertiser.

In the mid-1970s, it was predicted that STV would win in competition with cable TV because the service could be started immediately, whereas the growth of cable depended upon the time and expense of wiring the country. Today, the future of STV looks bleak, thanks to a combination of cable TV, videocassette recorders, and video "pirates."

Between 1983 and 1986, the number of STV subscribers dropped from 1.5 million to only 250,000, as all but five of America's STV channels ceased operations. Up to 90 percent of STV subscribers have disconnected when cable TV has come to town. This is easy to understand when the cost of one STV channel is about the same as a cable service that offers dozens of channels. Also, for the price of a month's STV subscription, a viewer could rent at least a dozen movies on videotape, in the process getting exactly the films he or she desires rather than only the ones offered by the subscription service. Finally, signal theft by pirates who sell black-market decoder boxes that unscramble the STV pictures has continued to plague STV and other satellite-delivered television services.

Low-Power Television (LPTV) The most modest new twist on the traditional broadcast signal aimed at creating new media markets has been simply to reduce its power. The idea of *low-power television (LPTV)* is to limit reception of these new stations to a local market radius of ten to fifteen miles. A full-power TV station has a market radius of some eighty miles. In this way, for example, Providence, Rhode Island, could have a low-power signal on Channel 7 without interfering with Boston's Channel 7. Minorities, special interest groups (particularly religious groups), and existing media barons seeking to broaden their bases of operation are among the 40,000 or so who have applied to the FCC for LPTV licenses.

LPTV channels serve special interest or geographically limited audiences. The FCC is wading through piles of applications.

LPTV has had much appeal but rather limited success. It has taken root in small towns and out-of-the-way places. Alaska has nearly two hundred LPTV stations, about three times as many as the rest of the United States together. The Alaskan stations are operated by the state, primarily for weather and educational purposes in that vast state. Since 1981, LPTV licenses have been granted to almost three hundred applicants in the continental United States who have not followed through and built new stations.

Many potential operators believe that LPTV stations must also be carried on cable systems in order to be profitable. While LPTV markets are small, costs of television production are still high. Cable operators have shown little interest in carrying LPTV stations. They are required by law to carry full-power TV stations in their franchise areas, but they are not required to carry LPTV stations.

By 1986 the Federal Communications Commission had granted five hundred LPTV license applications, and another 27,000 were pending. The FCC still is trying to dig out from a flood of license applications from investors who see profitable future markets for LPTV services, but has revised downward its earlier estimates that as many as four thousand new LPTV stations eventually would go into operation.

Teletext The latest and most novel twist on traditional broadcast signals is that room has been found on them for more information. For both FM radio and regular TV signals, engineers have found ways to embed additional information that can be decoded and displayed as text and graphics on home screens. The result is a new user-controlled information service that challenges the creativity of editors and advertisers. This radically different service also challenges the ability of consumers to change their media habits and make decisions about what information they really want.

Teletext, this information system that piggybacks on a regular TV signal, may provide 100 to 200 pages of user-selectable information. Home viewers of teletext select the pages of information they wish to see by pressing keys on hand-held keypads. For example, a menu may appear on the screen indicating that sports scores are on page 100. By typing 100 on the keypad, the subscriber would have the latest scores displayed on the screen in a matter of seconds.

The kinds of information typically available are up-to-the-minute news, sports, weather, travel tips, financial reports, coming entertainment events, and advertising. More specialized services, such as stock and bond prices, are distributed to screens in homes and businesses on signal space similarly available on FM transmissions. The extra signal space for this information is called the FM subcarrier. On the TV signal, the extra space is called the vertical blanking interval.

Teletext offers information encoded on unused space in the broadcast signal; it is not a truly interactive medium.

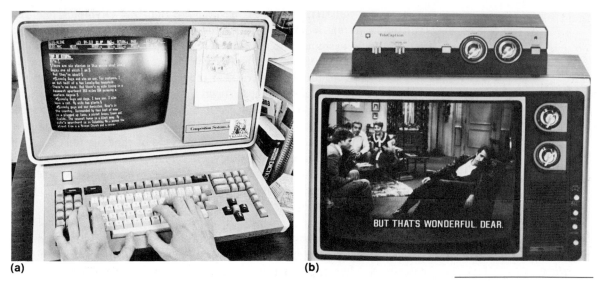

(a)　　　　　　　　　　　　　　　**(b)**

(a) A teletext operator preparing a news story for distribution.
(b) Closed caption TV for the hearing impaired is one use of the teletext system.

Teletext was first developed in the early 1970s in Great Britain, as was a related new information service, videotex, which we will discuss later.

Ceefax ("see facts") and Oracle (Optional Reception of Announcements by Coded Line Electronics) are teletext systems owned by the British Broadcasting Company (BBC) and Independent Television Authority (ITA), respectively. They have been broadcasting over the air to a British audience that surpassed one million by mid-decade; the only cost to the consumer is the purchase or rental price of the decoder.

Ceefax and Oracle have offered their teletext viewers choices of about one hundred different topics. They are indexed under broad categories, but most have numerous subtopics, including headlines, news, people in the news, and features. Instantaneous updates are included on breaking and sports stories, literally creating a deadline every minute for the BBC and ITA journalists. Weather and travel information include maps and schedules; consumer information includes food prices, recipes, science news, police news, and special information on education, farming, and gardening; finance news includes headlines, the *Financial Times* index, industrial news, reports from stock markets, exchange rates, and national and international finance. In addition, the teletext systems provide "reveal" buttons for "hidden answers" in puzzles, games, or educational topics seen on the screen.

Teletext has been far more successful abroad—especially in England, France, and Canada—than in the United States. Experimentation began in the United States in 1978, at Salt Lake City's KSL-TV. Several other stations, including KNXT-TV in Los Angeles and KMOX-TV in St. Louis, also tried out various teletext systems, with mixed results. KTTV in Los

Teletext has been more successful overseas than in the United States.

Angeles did well with its teletext coverage of the 1984 Olympics, and Ted Turner's superstation WTBS in Atlanta has offered teletext decoders to various cable systems that receive the superstation's signal. CBS, NBC, and Time, Inc., have all experimented with teletext systems, but economic and technological problems have held them back. One of the biggest obstacles has been the lack of an industry standard; rather than choose among the English, French, or an American (AT&T) teletext system, the FCC decided to let marketplace forces determine which technology would succeed. The decision has meant that no single type of sending and receiving equipment has become economically viable. By mid-decade fewer than five hundred American households were regularly receiving and using teletext, leading one writer to comment that teletext in the United States appears to be a solution looking for a problem.[1]

New Twists Using Microwave and Satellite Transmission

While the traditional broadcast signal is bringing us subscription television, low-power television, and teletext, a second cluster of new electronics is entering the home through microwave signals and satellite transmissions. In this section we will briefly consider the history and potential of two lesser-known and one well-known communications systems: Multipoint Distribution Service (MDS), Direct Broadcasting Satellite (DBS), and the burgeoning satellite-fed cable systems.

Multipoint Distribution Service (MDS) Multipoint Distribution Service is a variation of over-the-air television, using a microwave signal to transmit pay programs, generally to a master antenna serving a motel or apartment building, and occasionally to an entire city, but also to single family dwellings. This configuration of MDS hardware has given rise to the name "wireless cable." MDS operators can get into the business for about $250,000 to $500,000, the cost of a satellite earth station, allowing them to pick up pay-TV signals from elsewhere and relay them via microwave to rooftop receivers in a radius of twenty-five miles or so. Households or individual subscribers to MDS need a special antenna and downconverter, because the signal is transmitted over a special frequency that cannot be picked up by conventional television sets.

MDS uses microwave signals to transmit pay programs to a master antenna—usually in a motel, apartment complex, or office building.

Although MDS systems were operational many years ago, they got off to a slow start for the simple reason that they offered no reasonable way for an operator to earn money. MDS systems in the hinterlands could not boost over-the-air signals from distant television stations; they had to draw their signals from earth satellites. The advent of satellite-delivered programming by Home Box Office (in 1975) and other pay systems finally brought something marketable to MDS. The systems grew slowly; by 1985, between 500,000 and a million viewers in seventy different cities could watch one-channel MDS systems.

A study by the National Association of Broadcasters concluded that MDS probably will not be a major factor in the long-term war for paying customers, both because cable television is growing rapidly and because individual homes may sprout their own satellite receiving dishes and receive dozens of channels for not much more than the cost they are now paying for an antenna that picks up the single-channel MDS microwave signal. Even with a new version of MDS that offers up to four channels—*multichannel multipoint distribution service (MMDS)*—the package generally is not as attractive as cable or direct-to-home broadcasting. Where the latter are available, the MDS operators' primary function of receiving the satellite feeds and relaying them for a price to homeowners becomes obsolete. In major markets where complete cable systems have not yet been developed (generally because of the high cost of underground wiring) MMDS licenses have drawn considerable interest from existing broadcasters, cable companies, telephone companies, and even professional sports teams.[2]

One solution to the programming limitations imposed by MDS and MMDS systems has been posed by *Satellite Master Antenna TV (SMATV)*. It is a new wrinkle on an old system that has long been successful in metropolitan areas within reach of traditional over-the-air signals from area television stations. Whereas apartment and condominium complexes, hotels, motels, and the like used to have one large antenna that received area broadcast signals, then split those signals by directly wiring each room or housing unit to the master antenna, the FCC in 1979 opened the way for much more programming. It came in the form of large satellite dishes which carried the nationally distributed cable programs. Dish owners could add these programs—for a fee, of course—to the readily available over-the-air programs, and offer an attractive package to their tenants or guests. In all respects but one it is a cable television system; it differs only in that the hookups are confined to one privately owned building complex rather than an entire community, and therefore the system is free of the type of bureaucratic morass so many community cable systems have found themselves in.

Direct Broadcasting Satellite (DBS) The 1979 FCC decision referred to in the previous paragraph made it possible for anyone to set up an individual satellite receive-only antenna or dish without acquiring a special FCC license. That prompted newspapers, wire services, various businesses, and even private homeowners to investigate the potential of limited-usage receivers. Earth antennas for business use were made available in the $10,000 to $12,000 price range by the beginning of the 1980s, and dishes sprouted atop thousands of business establishments who were suddenly given access to the myriad of signals being transmitted from earth satellites.

The costs were prohibitive for most homeowners, however, so a new and improved system became necessary. With FCC permission in 1981, more powerful satellites were developed and put into orbit by the space shuttle, sending stronger signals to smaller and less expensive backyard receiving units. By mid-decade homeowners could tap into satellite signals for under $1,000. A joint operation by COMSAT and Sears & Roebuck went about the business of producing a $200 home dish only two feet in diameter, a real contrast from the ten- or twelve-foot dishes used commercially. There has been talk of satellites strong enough to send signals that eventually could be received by an antenna no more complex than a coat hanger. Theoretically, the system would bring the thousands of satellite signals to practically anywhere for practically no cost to the receiver.

Once all the technical questions were answered, DBS was expected to be alive and well by the mid-1980s. However, it was not to be. The anticipated $500 to $800 million price tag to send up a direct broadcast satellite that would reach only one-third of the nation put off even the most ambitious companies, and as of this writing not a single DBS system has become operational. The thousands of homes sprouting their own satellite receiving dishes continue to receive only those signals intended for cable or other systems, which has aroused the wrath of those systems underwriting the transmission costs. As a result, by 1986, HBO, the Cable News Network, and others had begun scrambling their signals, forcing home dish users to purchase decoders if they expected to continue receiving the programming. This, of course, has upset the home dish owners who had grown accustomed to a cornucopia of free programming. They insisted that the initial high cost of a receive-only dish justified their being able to tune in anything the satellites were capable of delivering, and a battle began to brew between them and the cable companies.

Satellite-Fed Cable Systems We have already hinted at the importance of the communications satellite. The experimental 1962 SYNCOM (Synchronous Orbit Communication) established the optimum height of a communications satellite to be 22,300 miles orbiting over the earth. Originally, communications satellites had the shortcoming of passing out of television's range in relaying their signals back to earth. This was corrected by placing a number of satellites in geostationary orbit—that is, orbiting at the speed at which the earth revolves so that they might always be directly above a given point. Thus, at least one satellite is always within range of any point on earth for reception, and the signal can be relayed from satellite to satellite to reach any other point on earth. Satellite-fixed orbit coverage is total.

The Communications Satellite Corporation (COMSAT), a quasi-public corporation, was established by Congress in 1962, with AT&T as the largest stockholder. Communications satellites, it should be noted, do a lot more than relay television signals. They transmit all kinds of other data, including computer information and telephone conversations.

Because satellites reach beyond national boundaries, it was inevitable that COMSAT would yield to the International Telecommunication Satellite Consortium (INTELSAT), a consortium of sixty-five nations engaged in international satellite communications. The INTELSAT efforts over the world's various oceans have resulted in transmission of a mind-boggling array of communications between and among nations.

In recent years, a large number of domestic communications satellites have been launched. Western Union's *WESTAR* and RCA's *SATCOM* rent space to many television cable systems and networks, and Satellite Business Systems (SBS)—a joint effort of IBM, COMSAT, and Aetna—has been set up for newspapers and other business ventures. Because CBS, NBC, and ABC were spending many millions of dollars a year to distribute their programs nationwide via land lines, the satellites offered an attractive savings. However, as the networks discovered, there were problems once they linked all their affiliate stations to satellite feeds: the local affiliates have gained access to all other satellite transmissions, including competing network and cable system feeds, and hence have less need for their own network offerings. The growing independence of local stations has helped explain the decline in network penetration, and means that national advertisers are not assured of a constant audience.

Satellite communication has been available since 1962. SYNCOM, COMSAT, and INTELSAT were the early birds.

Domestic satellites are being used for business and media communication at attractive rates.

The cable television systems have been the major beneficiaries of satellite communication. Earlier community antenna television (CATV) companies could only pass along signals received by super antennas placed atop tall buildings or mountains, which were still limited due to TV's line-of-sight signals. Home Box Office grew at an incredible pace after it went onto satellite in 1975, and served as an inspiration to other entrepreneurs who were seeking ever larger audiences. Satellite cable programming already includes more than fifty separate and diverse services. A partial listing of them indicates the broad range of programming and specialized audience appeal: ESPN (Entertainment and Sports Programming Network); the Arts and Entertainment Network; Cinemax; HBO; C-SPAN (Cable Satellite Public Affairs Network); Cable News Network (and CNN Headline News); Financial News Network; Home Theater Network; Lifetime; MTV (Music Television); National Jewish Television; Nickelodeon; PTL (Praise the Lord); Showtime; Video Hits One; The Weather Channel; American Movie Classics; Bravo; Cable Sports Network; Home Sports Entertainment; National Christian Network; The Playboy Channel; The Prime of Life Network; The Silent Network; Atlantic City Horse Racing; Catholic Telecommunications Network of America; Word of Faith Satellite Network; Country Music Television; The Nostalgia Channel; SelecTV; The Disney Channel; The Nashville Network; SIN (Spanish International Network); and numerous others.

In addition to servicing the cable television business, satellites have been a part of the newspaper, magazine, and wire service industries for years. Dow Jones & Co., Inc., has been using a satellite since 1975 to beam full-page facsimiles of the *Wall Street Journal* to remote printing plants across the United States. The *New York Times* in 1980 began whisking pages from New York City via satellite to Chicago for its new Midwestern edition, and in 1982 to the West Coast. *USA Today,* as discussed in chapter

3, has relied upon satellites to become the nation's first truly national newspaper—in slick four-color reproduction, no less. *Time* and other magazines have been making similar use of satellites for years. United Press International and the Associated Press effected significant savings on their AT&T land line bills when they began distributing their wire services to member and client newspapers and broadcast outlets in the early 1980s.

Not all reactions to the satellite boom have been positive. The networks, the National Association of Theater Owners, and others have fought and will continue to fight them because of the perceived and real threats to their status quo. The federal government has proceeded slowly in setting up regulations; and regulatory uncertainty, as we have seen, is a sure way of deterring a medium's development. At first, due to difficulties in measuring audience size, advertisers were slow to embrace the superstations or cable TV systems relying exclusively on the satellites. Also, satellites are costly (about $30 million), and have a short life expectancy (about eight or nine years), putting them within the financial grasp of only the giant firms. Because NASA was charging satellite owners $20 to $30 million to launch and track each satellite, the investments were substantial. Launching via the space shuttles resulted in some savings, and brought a frenzy of interested companies. However, following the explosion of the shuttle *Challenger* in early 1986, and shortly thereafter the failure of a French ARIANE rocket plus another American failure from Vandenberg Air Force base, the launching program was thrown into limbo for several years. Finally, satellites have been known to malfunction or to even get lost, as happened to RCA's *SATCOM II* late in 1979 when it disappeared somewhere in the ether.

The satellite boom has its drawbacks and critics.

When community antenna television (CATV, more commonly called cable) was first introduced in 1950, it was perceived as a way to bring clear reception to isolated regions. The television signal, which travels in a straight line, could not reach into the valleys of eastern Pennsylvania, where the first cable was installed. Now, however, cable has invaded the cities and suburbs in altered form. Cable still provides sharp pictures, but its appeal and its future are in the opportunities for programming, business data transfer, and interactive communication.

Over-the-air television is restricted to the available VHF and UHF frequencies, while many cable systems offer sixty-four clear channels (and some twice that), which increases the likelihood of more specialized programming, including all news stations, weather stations, sports stations, religious stations, and so on. In addition, some cable systems have the capacity to deliver specialized messages to individual viewers seeking business data, electronic mail delivery, banking, medical advice, and a wide variety of specialized information.

Cable Television: The Wired Nation?

Cable TV's new role lies in programming, business data transfer, and interactive communication. Its recent growth rate is impressive.

Table 10.1 Growth of cable television in the United States, 1952–1985

Year	Number of systems	Number of subscribers (add 000)	Percent of TV homes with cable
1952	10	14	0.1
1955	400	150	0.5
1960	640	650	1.4
1965	1,325	1,275	2.4
1970	2,490	4,500	7.6
1975	3,506	9,800	14.3
1980	4,200	15,500	20.0
1985	7,300	38,900	45.7

Sources: *Television Digest;* A. C. Nielsen: *TV Guide; Broadcasting Cablecasting Yearbook.*

Television of the future, the wired nation, and other optimistic terms used to describe cable convey an almost science-fiction quality to the potential of the medium. Depending upon who is asked, cable is either seen as the informational and cultural panacea for a nation of serious viewers, or just another expensive gadget for the nation's entertainment junkies. At one point, in the late 1970s, optimists predicted that eventually as many as one-third of the nation's households would be linked to cable. By 1987, those estimates had been vastly exceeded; half of the nation's households already had cable, and the percentage was steadily climbing. Nevertheless, pessimists continued predicting that cable's day had already come and gone, that the cumbersome and expensive coaxial cable was going to be supplanted by direct satellite-to-home communication systems and the burgeoning videocassette rental business. (See Table 10.1.)

The potentials of cable stem from the fact that it can both boost and import existing over-the-air broadcasts as well as originate its own programming. Because it is wired like a telephone, it can be as individually selective as a telephone and yet utilize telephone's two-way capacity. These obvious assets, combined with the computer, impart a Buck Rogers aspect to the future of cable. Furthermore, because it is video, it holds the potential for facsimile reproduction, which means it can duplicate written materials in the home—including newspapers, specialized reports, and the like. The computer capacity lends itself to information retrieval from memory banks that can either be flashed on the screen or reproduced in facsimile for the record.

Does it sound exciting? Well, there are problems. Costs, human nature, FCC regulation, competition, and economics have deterred cable's progress. Ironically, one of its great strengths has proven to be its stumbling block: cable is so versatile that it may easily be in competition with itself.

Consider, for example, that there are over 7,300 cable systems in the United States, serving some 19,000 different communities. Another thousand franchises have been approved, but have not yet been built. Each of

By 1987, half the American homes were wired for cable. But some were still forecasting cable's demise.

Cable imports, boosts, and originates programming.

Cable's growth has been stunted by costs, human nature, FCC rules, competition, and economics.

these is a little network that splinters the market to the point where none can develop an effective mass. When the audience numbers are not there, neither is the money for programming. Consider further that the majority of these new systems are not located in major population centers where the largest audiences are found. Rather, they are located in smaller cities and even rural communities where cable's ability to increase the fidelity of over-air broadcasting is an asset. In these areas cable's primary use is to provide interconnection with one or more of the commercial networks.

Only recently has cable begun to prove profitable in the largest cities. Back in 1966, the FCC prohibited cable from operating in the largest markets until UHF reached its potential. In view of UHF's erratic performance and the considerable amount of pressure under which it operates, the FCC in 1971 decided to permit cable operators to wire the largest markets. In these markets, however, broadcast competition is intense. For instance, in the early 1970s, when cable was allowed back into the marketplace, New York City had eight channels, and Los Angeles had eleven. Although cable provides residents of some cities a clearer picture, there was little incentive to add cable unless it could offer vastly superior or unique programming. But at a cost of some $300,000 to lay a mile of underground cable in a metropolis (compared with $10,000 per mile of stringing cable on existing utility poles in a rural area), cash-short cable systems had little capability of producing anything that would have major appeal. They were forced into situations of bidding with the networks for first-run movies, talent, and the like, but they lacked the funds and audiences. To become viable, the metropolitan cable systems needed a gimmick or two. Toward the end of the 1970s, they began to discover those gimmicks in pay-cable feature films and sports events. As noted earlier, satellite-fed programming was particularly beneficial to the individual cable companies. And, in the 1980s, large market systems started looking to their two-way capacity and appeal to special-interest groups as a way to carve out their place in the telecommunications environment.

Today, after several false starts, cable television has become a significant mass medium. Industry observers had long maintained that 30 percent of America's homes would have to be connected to the cable before it could become a medium of any real importance. Thirty percent, they noted, had been the magic number of television homes needed before national advertisers responded in 1949, and 30 percent was the number of color sets in use before advertisers began producing commercials in color. By that measurement, cable television became "significant" in 1982. Five years later, once cable penetrated half the homes in America, it was fully established.

Advertisers and investors have begun to show fascination with cable's potential. Despite the fact that most cable systems derive less than 5 percent of their gross revenues from advertising (the rest coming from monthly

**Advertisers
Take Note**

By 1982, cable reached 30 percent of American homes, and advertisers started taking it seriously. By 1987, it reached half the U.S. homes and was fully enfranchised as an ad medium.

subscriber fees), multimillion dollar ad packages have been put together for the largest cable networks. Overall, cable took in only $50 million in advertising revenue in 1980. Just four years later, the revenue was up eleven-fold, to $550 million. According to some estimates, the industry suffered a $200 million overall loss in 1982, but turned that around with a $400 million profit in 1983. In 1984 it doubled that profit level, to $800 million, and has continued a dramatic upward spiral. Much of that profit, of course, has not been based on advertising alone, but on increased levels of penetration into television households. And that increase has been coupled with a new enthusiasm among cable merchants, due to the fact that Congress has deregulated the industry. Starting in 1987, cable operators are permitted to raise their hookup and monthly charges to whatever the market will bear. This capacity to raise prices has come when overall construction costs and inflation levels are down, making cable an increasingly attractive commodity on Wall Street. By mid-decade, cable owners were trading properties left and right, with the average market value of a system jumping to $1,100 or $1,200 per subscriber, compared with $800 or $900 in 1980.

Slightly more than one thousand cable systems—15 percent of all systems—accept advertising on their "local origination channels," the channels set aside for community bulletin boards and locally produced programs. Altogether, the advertising revenues for cable remain small potatoes when compared with commercial network advertising. Until cable enters more of the major metropolitan markets, there remains some question about how viable it will be as a truly national advertising medium. However, as noted above, after years of losing money, the major cable companies were earning a profit by mid-decade.

At current rates, advertisers find it attractive to sponsor entire programs or major segments of programs, in ways reminiscent of network television in the 1950s. (The hawkers of records, tapes, albums, or magical kitchen products or health aids not available to retail stores, who so dominated the cable scene a few years ago, are giving way to more mainstream and seemingly legitimate commercial appeals.) Due to the nature of the medium, cable commercials need not be limited to the standard fifteen- or thirty- or sixty-second modes. Already there are experiments with commercial "mini-programs" of several minutes in length. They constitute fewer interruptions of the programs, on the premise that people paying for viewing cable expect something different from the din of commercial networks. In addition, there are "informercials" ranging in length from thirty minutes to four hours—talk shows (how to build a house, buy a boat, etc.), demonstrations of products, and so forth, all with clearly identified sponsors. Informercials capitalize on cable's unique relationships with specialized

Cable companies, by and large, are profitable these days.

audiences. However, the majority of cable systems have merely adopted the tried-and-true routines of the commercial nets, repeating the familiar pattern of new media being put to old uses.

Because cable audiences tend to be younger, better educated, and more affluent than noncable audiences, advertisers have something over which to drool. And the tastes of cable audiences, especially those on pay-per-view or interactive cable systems, are easier to pinpoint, since much of cable's programming is specialized. This helps create an advertiser's dream, in which audiences' demographic and psychographic characteristics are available and somewhat predictable, and in which there are fewer and fewer wasted impressions or ad messages reaching unintended audiences. Finally, given that cable audiences are smaller in size than network ones, the costs per commercial minute are lower, and at this time remain within the reach of smaller advertisers. Basically, cable has the same appeal to advertisers as do radio and magazines.

One way of knowing when a new mass medium is to be taken seriously is to notice the reactions of the established media. In the case of cable television, its time is obviously at hand. The networks and television programmers, including cinema programmers, have been reacting, either by complaining about cable's incursion into their territory or more significantly by quietly going about grabbing a piece of the cable action.

As the federal government took steps to increase cable's access to programming by granting cable operators the right to pick up as many signals as they want from distant cities and relay them to local subscribers, predictable reactions came from the National Association of Broadcasters (NAB), the Motion Picture Association of America (MPAA), and the Association of Independent Television Stations (AITS). The establishment argued that cable was getting something for nothing, that the infant industry had to pay its dues before starting to reap the heady profits possible in telecommunications.

Meanwhile, back at corporate headquarters, several communications giants responded to the unfairness of it all by joining the cable bandwagon, buying a piece of the action. Most of today's largest cable companies are subsidiaries of other media giants. Group W (Westinghouse), Cox Enterprises, Storer, Warner Amex, Times Mirror, S. I. Newhouse, Capital Cities Communications, Multimedia Inc., The (Chicago) Tribune Company, Scripps-Howard, Harte-Hanks Communications, McClatchy Newspapers, Time Inc., and the like are familiar names in print and electronic communications. All have invested in cable, not only as a natural outlet for programming, but as a hedge against the future when electronic publishing

Cable's audiences are younger, better educated, and more affluent than commercial TV's. Advertisers and programmers love them.

The Media React

Predictably, "old TV" senses a threat from the cable interloper.

If you can't lick 'em, join 'em!

or in-home production of news*papers* will complement other delivery systems. As an indication of the recent value being placed on cable operations, ABC in 1984 paid over $200 million for ESPN (Entertainment and Sports Programming Network), and in 1985, a consortium paid $2.1 billion for the massive Group W Cable systems.

Dow Jones, owner of the *Wall Street Journal* and other information services, has gone into the cable business via both land lines and satellite, bringing business news, stock quotes, and other information to customers upon demand. And American Telephone and Telegraph, the largest corporation in the world, has been fighting a running battle with the government and existing communications systems over its rights to enter the cable market as either an information provider or merely a hardware and electronics service. Others, too numerous to mention and moving too rapidly to track, have made cable television an industry worth following on the stock market. Keeping up with cable stocks is a dizzying proposition, but speculators are counting on the industry to continue its recent growth patterns for several more years before it settles into a pattern of stable, across-the-board profits.

Businesses eyeing cable are not limiting themselves to earning profits from entertainment alone. Market researchers expect cable revenues to be split equally among entertainment, business data, and other transaction services (mail, electronic newspapers, etc.) before the end of the 1980s. Local "loop market" revenues—including electronic mail, word processing, and teleconferencing—could become a $2 billion a year industry. Given the phenomenal growth rate of computers in the businessplace, the marriage of business and cable appears recession-proof.

Meanwhile, however, cable has tended to develop as a small duplicate of commercial television. Like most media that initially adapt older media contents to new media mechanics, cable has borrowed indiscriminately from commercial television's programming practices. A massive nationwide study in 1980 noted that 70 percent of United States cable systems were offering some form of local programming, and the percentage has increased since that time. However, before we conclude that this is a step in the right direction, we should look at the forms those local programs are taking. While 15 percent of the total available channels have been devoted to locally originated programming, much of that number consisted of message-service channels, classified advertising channels, stock-market reports, sports news roundups, and television listings. Few systems have offered their own newscasts, and much of the local programming has been done by volunteers and students. One of the major areas of local programming is coverage of local high-school, college, or community athletic events.

There has been a solid increase in the number of public-access channels (channels set aside for use by the public at a nominal charge), and additional political cablecasting on all levels.[3] Since the FCC forbids any program interference on the part of the cable operator with public-access

The marriage of business and cable may be recession-proof.

In a familiar pattern, the new medium borrows its content from the old media—TV services cable.

programming, the result has been some very strange programming—pressing, in many instances, against the boundaries of both credibility and taste. This has done little for the cable's public image and scarcely constitutes a sales tool for potential subscribers. However, it demonstrates that the open marketplace of ideas, as espoused by libertarians and our founding fathers, is alive and well in this new mass medium.

Cable is the basic delivery system for which people pay an initial connection charge and a monthly fee. Pay TV is a different matter; generally, it is a part of the cable delivery system and involves additional charges for viewing selected programs. It has been a controversial issue since the FCC started considering it in 1955. Supporters claim it provides movies, sports, culture, travel, nostalgia, and self-improvement (all without commercial interruptions) that even mass-appeal, sponsor-supported free TV cannot bring. Opponents counter that pay TV, seeking maximum profits, woos the same mass audiences as free TV and buys off the same audience-pleasing attractions, leaving the wealthier public to pay for the same entertainment it has always gotten free, and the poorer public with nothing. While the old arguments continue, the pay-TV business has been growing at exponential rates.

During the 1960s, pay television soaked up millions of dollars and shattered a lot of business hopes due to engineering, programming, and marketing difficulties. But having to endure rigorous regulation by the FCC and bitter opposition from broadcasters and motion picture operators, who sensed in pay TV a serious threat to their livelihoods, was only one of its problems. It was not until the late 1970s, about the same time satellite transmission became feasible, that the regulatory climate turned in favor of cable delivery and the pay-TV programmers. Once the FCC decided to allow cable companies to import distant programs merely by paying small royalties to the program owners and no payments to the stations or network airing the programs, cable TV and its stepchild, pay TV, were off and running.

In 1976 some 978,000 households received pay TV. The numbers increased to 13.5 million households by 1982, and 20 million by 1985. Today, two out of three homes receiving basic cable also receive some sort of pay service—either "pay per view" or, more typically, a monthly fee to grant access to all that a given pay channel has to offer. Developers continue to argue that the full potential of pay TV systems are only beginning to be exploited. Such growth seems inevitable, because pay TV is inexorably linked to the rapid development of cable, satellite, and in-home computerized communications, and to the federal government's spirit of broadcast deregulation.

Three methods for providing pay TV to audiences are *pay cable,* over-the-air pay TV, known as *subscription television (STV),* and *Multipoint Distribution Service (MDS).* The latter two systems were discussed earlier in this chapter.

Pay Cable

Pay TV raises questions about a nation of cultural "haves" and "have nots."

Only recently has governmental policy favored pay TV.

BLOOM COUNTY by Berke Breathed

Panel 1: THOSE PAY-CABLE STATIONS HAVE REAL NERVE SCRAMBLING THEIR SIGNALS, LET ME TELL YOU... / HOME-SAT-6 / FOR SALE CHEAP / 7-15

Panel 2: HERE WE SPEND $3000 ON A DISH TO STEAL THEIR PROGRAMS AND THEY STEAL THEM RIGHT BACK ...

Panel 3: YOU KNOW WHAT THIS IS? WELL I'LL **TELL** YOU WHAT IT IS ...

Panel 4: "JUST DESSERTS." / FASCISM. THEY'LL BE KILLING BABIES NEXT.

Decoder boxes unscramble the signals and keep track of the audiences, but video pirates constantly seek to short-circuit the process.

Pay cable enables families who receive the regular cable services to pay an additional monthly charge to receive special channels of programs such as those from Home Box Office and Showtime. Most companies charge a flat monthly rate of $8 to $10 to their cable customers who wish to receive the films, sporting events, or other noncommercial specials offered. One study showed that only 40 percent of cable subscribers elect not to take the available pay TV channel because of the added cost.[4]

Regardless of the receiving system involved—pay cable, STV, or MDS—pay TV customers have shown a preference for paying a flat monthly fee rather than a pay-as-you-view basis. Companies have found resistance on the part of customers billed via computer for each pay program they select. Some viewers consider it an invasion of privacy, knowing there is a decoding box hooked to their television set and telephone line, automatically recording and informing the pay TV company of their viewing habits. From the companies' point of view, an additional rationale against installing decoder boxes is the initial cost (an estimated $450 million to hook up two million homes in a large metropolitan area such as Los Angeles).

Of course, most customers probably would prefer not to pay at all for the privilege of watching pay TV. To the disappointment of pay TV companies, the number of viewers who get away with pirating the signals, circumventing the cashier, has been exorbitantly high. Arguing that the airwaves are free and that anything passed over those airwaves and through their homes and bodies can rightfully be intercepted, pay TV pirates have made millions manufacturing and selling special antennas and decoder boxes that beat the system. The pay TV industry has sought remedies from the courts, FCC, and Congress, and some states have responded by passing laws against the pirating of signals. Revision of the 1934 Communications Act and a federal privacy bill banning unauthorized reception of pay TV signals have been initiated, as the industry maintains it is illegal to intercept a private communication, just as it is illegal to eavesdrop on telephone calls relayed by microwave signals.

Sports and Movies Pay television has begun to be a force in some traditionally free-TV areas, particularly sporting events and movies. Networks sense in pay TV a serious threat to their lucrative arrangements with professional sports. It would take only 400,000 television homes willing to pay one dollar a week (a low figure) over a twenty-week professional football season to more than cover the income a single National Football League franchise is currently getting from the commercial networks each year. As the number of homes having pay TV increases, the potential income to the NFL teams grows even larger, causing the commercial networks and their advertisers grave concern. One of their worries, a socially responsible one, is that homeowners without access to pay TV will be denied their seemingly inherent birthright to spend their fall weekends ingesting pro football. Their other concern, less publicized, is that the loss of pro football and other major sporting events would be a devastating financial blow to the networks. As stated by one observer, "Every sport at every level will be touched, lawsuits are inevitable, competition for contracts will become fierce, and team owners will continue to face a string of major decisions."[5]

Meanwhile, similar battles rage in Hollywood. Film companies, which have had a longer-term affinity with telecommunications than professional sports teams have had, for quite some time have provided movies for pay TV. Home Box Office (HBO) (a Time, Inc., subsidiary whose stock zoomed after it began transmitting across the nation via satellite in 1975) has claimed to be the largest movie buyer in the world, both in volume and amount of money spent annually. With such competitors as Showtime, Warner-Amex Satellite Entertainment's The Movie Channel, Home Theater Network, Spotlight, and others in the movies-at-home business, Hollywood would seem to have every reason to make its hits available to pay TV, and even to crank out special movies for the insatiable video. Media mogul Ted Turner of Turner Broadcasting (Cable News Network, superstation WTBS, etc.) had one solution to the problem; in 1985 he paid $1.5 billion for MGM/United Artists, but said he would be happy to turn around and sell off everything but the studio's massive forty-six-hundred-film library, which he needed to fill the time on WTBS.[6]

Hollywood and independent filmmakers recognize the dollars to be made from cable and pay TV airings. As we mentioned in the earlier chapters on television and film, network and cable TV can deliver a larger audience than even a blockbuster film is likely to generate on its own volition. Therefore the film companies have little hesitation in asking for, and getting, millions of dollars for pay TV rights. Every year, it seems, the asking prices double, and the pay TV companies seem willing to pay whatever it takes to get the rights to air the films, usually about six months after they are released on videocassette and long before they go to commercial TV. Of course, now that those hits are showing up on pay TV and at videocassette rental and sales shops concurrent with their arrivals at suburban and small-town theaters, there have been many complaints about the encroachment.[7]

Sports and movies on pay TV please the customers but threaten the establishment.

An indication of the market's complexity is seen in HBO's struggle for independence from Hollywood by investing in its own films prior to production, locking up the pay rights for itself by means of a "prebuy" arrangement similar to that used by some commercial networks. The practice is risky, because the films may bomb at the box office and therefore be worth little for subsequent play on pay TV. However, as HBO itself has a paying audience in the tens of millions, it has reached the point of breaking even on many of its films without having to peddle them elsewhere.

A national survey of pay TV subscribers indicated that the single most important reason for paying extra is to receive movies not shown on regular television. Thus, because pay TV is growing at a dizzying pace, competition among companies for films and audiences is bound to continue.

Pay Cable's Future Most analysts see pay television as a growth industry despite current complications. Traditional Hollywood television producers, who have made their fortunes on syndicated reruns, are among those threatened by a system that should become an outlet for programming that is more adventurous than the familiar talk and game shows. Some of the more farsighted independent producers, however, have begun to latch onto basic cable and pay TV. They are feeding cable's voracious appetite by either developing original programs or making additional episodes of commercial TV shows that were canceled before enough episodes had been put "in the vault" to assure syndicated reruns (see chapter 8).

It might be said that the most enduring dream of pay TV remains unfulfilled. It has been unable to sell culture (such as opera, ballet, symphonic concerts, and Broadway plays) on a regular basis, despite the substantially increased presentations of such programs on public, noncommercial television. This, of course, prompts the question: why should we pay for such culture if the Public Broadcasting Service gives it to us free? Other key questions remained unanswered. First, the issue of two classes of society remains. Will those who are unable to pay for television programming, by reason of economic status or geographic quirk, be deprived of what increasingly is seen as an American necessity? (Not by accident, many welfare agencies consider the television set a necessity, not a luxury item.) Second, if we pay for the same programming we are now receiving free on commercial television, would we be subsidizing it by paying directly to the programmer, or would we be paying for it indirectly by buying advertised products? The ramifications most assuredly are not lost upon the networks and advertisers.[8]

New Mass Communication over Telephone Lines

Ironically, one of the fastest growing electronic means for sending information to home screens today is the telephone, whose invention in 1876 for sound transmission substantially predates both radio and television. The phone system today is being used to transmit text and graphics between computers. Operators of large information databases invite subscribers with

(margin notes)

Pay TV is growing at a dizzying pace, so competition for films and audiences is bound to continue.

The most enduring dream of pay TV remains unfulfilled; it still has a hard time selling "culture."

home computers to dial up their services and select information for display on their home screens. Besides up-to-the-minute news, weather, and investment information, the uniqueness of such systems lies in two-way services such as electronic shopping, home banking, and education.

The telephone has become one of the fastest growing media of mass communication.

All of the services described in this section could travel on two-way cable just as easily as on the telephone system. But nearly all American homes have telephone service, while less than half have two-way cable. Therefore for at least the next decade, the telephone communication link will be the predominant means by which these services come into homes.

Videotex is the term that originally was applied to two-way services with colorful graphics and text that came to television screens over the phone system. A special computer-like decoder was necessary to make this work. As such services are now becoming available on home computer screens as well as on television screens, the term is being applied to many other computer-based information utilities, which are gaining large numbers of subscribers.

Homes are being equipped at a rapid rate not only with personal computers but also with *modems,* electronic devices that allow computers to communicate over the telephone system with other computers. By 1990, almost half of United States households will have one or more computers, and half of these will be capable of telecommunication over phone lines. Computer owners with modems are being attracted to information services that come into the home through this means. Similar services are available through these information utilities as through videotex, although their screen displays with mostly text are currently less attractive.

Computer communication with modems also is having an impact on special group communication and on interpersonal communication. Professional groups are creating special electronic bulletin boards and databases that only their members can dial into. Ad hoc communication groups are forming around similar bulletin boards and informal networks created by computer communication hobbyists, today's computer equivalent of ham radio operators. Finally, individuals and groups are sending and receiving electronic mail, conducting computer conferences, and exchanging stories over computers equipped with modems.

We will now take a closer look at each of these categories of two-way systems that today rely mostly upon telephone links to reach your home screen. Whether you think of your home screen as a television screen or a computer screen may soon be inconsequential, as technologies converge toward a display screen that may be switchable between these functions.

Videotex

Videotex is the term applied to two-way computer services that supply text and graphic information to home computers and video display screens or TV sets. While most videotex services today are carried over telephone lines, they may be carried over cable, optical fibers, and satellite transmissions. They differ from teletext, which we discussed earlier, in the fact that they

Videotex is a two-way, interactive computer service.

The range of information services offered by the early Viewtron videotex system still is typical of today's computer-based information services. Electronic mail and on-line communication forums for special interest groups are among the more popular features unique to this new medium.

offer truly interactive communication between audiences and telecommunications sources, whereas teletext is still basically one-way communication delivered over the broadcast band.

To understand the scope of videotex, we should take a look at Viewtron, the first such commercial service in the United States. Viewtron was launched in late 1983 as an electronic subsidiary of Knight-Ridder Newspapers, Inc., in southern Florida. Three years and $50 million later, after it had gone nationwide, the Viewtron project was closed down when it became obvious that it had been ahead of its time. It had offered more services than its 20,000 customers were ready to use and to pay for. In fact, the greatest use of Viewtron had been by computer hobbyists who were using it for interpersonal communication instead of mass communication, as originally intended by the developers. The wealthy Knight-Ridder corporation had to admit it had been overly ambitious with the project.[9]

Viewtron's index of interactive topics included news, sports, money, education, games, shopping, messages, home and reference, and classified ads. Viewtron had more than 300,000 pages of locally produced information available in these categories. In addition to its basic topics, Viewtron tied into other national databases such as an on-line encyclopedia, a college selection service, stock price quotations, and airline schedules and prices. Among the locally produced information were restaurant menus with prices, lineups and results from race tracks, coming entertainment events with on-line ticket ordering, lessons for learning Spanish, a library reference service, and of course up-to-the-minute local news and weather. Obviously, the amount of information available in such a videotex system vastly exceeds the 200 or so page limit of a teletext system, because the videotex data base is as large as the system's master computer and other on-line computers to which it is linked. Being wired for videotex is truly a case of having the world at one's fingertips.

Users of videotex systems use a keyboard to tell the host computer what information they want. Users may press keys to make selections among options that appear on their screens. These options appear on index or menu pages, so called because choosing a listed option is similar to choosing an item from a menu. Users also may type keywords (for example, "election results") that take them directly to the information they are looking for, thus skipping menu pages otherwise needed to reach the information.

Once users have located a type of information they wish to see regularly, they may ask the computer to place those pages into an electronic "personal magazine." The personal magazine subsequently eliminates information searching and displays with a single keystroke information the users have indicated they wish to see regularly.

Videotex, like teletext, has not been an American invention. England, France, and Canada were the first to develop both systems. In England, the same basic technology used for Ceefax and Oracle, the teletext systems, was applied to a videotex system named Prestel (press and tell). Originated by the British post office in 1971, Prestel utilizes telephone lines, hooked to television sets, to make available some 25,000 different kinds of data, organized under such services as travel, weather, news, sports, entertainment, business, and other information, plus games, classified advertising, and so on. Prestel has been widely used in places of business, but, due to the cost of using the system, has received less enthusiasm among the general public. Likewise, the French system, Teletel, offers basic information on an interactive system. In addition, however, it has experimented with at-home banking and shopping, making use of credit cards with built-in microprocessors which connect to Teletel home terminals. In Canada, the videotex system called Telidon has been used primarily for educational purposes.[10]

In the United States, several videotex systems have been attempted, with mixed results. The most extensive and expensive—and hence the biggest failure—has been Knight-Ridder's Viewtron. In 1984 two other services went into operation: Gateway and Keyfax. Gateway was launched in Orange County, California, by Times Mirror Videotex Services. In the spring of 1986 the operation was curtailed, after having garnered only 3,000 subscribers and showing no real, short-term prospects of being profitable. Despite $1.2 million in advertising sales, the monthly income from customer use could not offset the Times Mirror's continued investment, which by some accounts had amounted to $30 million. Meanwhile, in the Chicago area, Keyfax continues to limp along. It is the product of Keycom Electronic Publishing, a joint venture of Centel Corp., Honeywell, and Field Enterprises, which publishes the Chicago *Sun-Times*.

In contrast to the massive investments undertaken by Knight-Ridder, Times Mirror, and Keycom, a more modest videotex system organized by the Fort Worth *Star-Telegram* has been much more successful. Startext, as the Texas videotex system is known, can be received by any personal computer without the special software that customers of other systems have to buy or rent. As a result, Startext had invested only $210,000 in hardware, and was breaking even with just 2,200 subscribers. It offers those subscribers access to news from the *Star-Telegram,* stock quotes, electronic classified ads, airline schedules, and other services for a flat rate of $9.95 a month. Home banking costs an additional $1 monthly for Startext subscribers, and $6 for non-subscribers.[11]

Newspaper companies have been involved in most of the American videotex operations.

From the above examples we can see that large newspaper companies back, or have some stake in, each of the American videotex ventures. While such ventures may provide new outlets for the kind of information newspapers gather, they also may provide investments that could protect such companies from shifts in how consumers may prefer to receive their information in the future.

The president of Knight-Ridder Newspapers, James K. Batten, explained: "Our logic was that if there was anything threatening for newspapers in these new technologies, we wanted to find it out sooner rather than later."[12]

This defensive strategy has been questioned by some media observers. One consultant told a large group of newspaper executives, "These new telecommunications businesses have absolutely no role in protecting the newspaper. Any link between newspapers and telecommunications is doomed to failure."[13] Some of the executives countered that the new technologies would not succeed unless newspapers brought to them traditional journalistic values of public service.

The prospects for the success of videotex depends upon whom you ask. Most observers now believe that videotex must merge with home computers in order to be successful. Even AT&T Information Systems, the supplier of the Sceptre terminals used by Viewtron and Gateway, expects to see videotex merge with personal computers. The director of venture development for AT&T still expects videotex in some form to be a $15 billion business by 1990, with penetration into ten million homes. He said in 1984, "We're at the dawn of a new era, but it's the Stone Age. We all need to dream a lot more about how to take this information medium and put it to use for consumers, finding the right mix of services for the right customers at the right price."[14]

Videotex services have tried to keep prices in the range of what people have been willing to pay for cable television services. While pricing strategies vary, the basic videotex service costs $10 to $20 a month without equipment. Leasing a terminal adds another $10 to $20 a month. Videotex vendors quickly discovered that $600 to $750 to purchase a terminal stopped many potential customers from subscribing, and monthly leasing of equipment has been incorporated into most marketing strategies. On some systems, phone line charges or data base use beyond the first five hours may add another $10 to $15 a month.

As an example of how fast things change in this market, software for receiving Keyfax on home computers became available in Chicago stores in November 1984. Only a year earlier, the director of advanced technology for the Times Mirror Co. had predicted that it would be another decade before videotex and home computers converged.[15] The videotex software for Keyfax, priced at $60, is available for Apple, IBM, Atari, and Commodore computers. As a result of this wider accessibility, Keyfax is expected to penetrate the market more successfully than some of the other videotex systems.[16]

The fact that other major companies are entering the videotex market may suggest its promise for the longer range future. Two joint ventures team up company threesomes possessing enormous financial clout. Trintex is the name of the videotex system being initiated jointly by IBM, CBS, and Sears. Also teaming up for a competitive videotex service are RCA, Citicorp, and J.C. Penney. These new services will be received on home computers. And they are national in scope, unlike the local orientation of Gateway, Keyfax, and Startext. With the participation of mass marketers like Sears and J. C. Penney, home shopping will be a major aspect of these services. The participation of Citicorp underscores the importance of home banking as a force driving the development of such systems. As we have seen, home banking has already proven successful in America and abroad. On a simple computer-modem communication system in California, Bank of America as long ago as 1984 had attracted ten thousand customers paying eight dollars monthly for home banking service.[17]

A Massachusetts Institute of Technology communications policy expert says that videotex will probably succeed. However it may not look like what we know as videotex now. Videotex in the early 1980s was envisioned with text and colorful graphics received in the home on a special terminal. Videotex in the future, more than likely, will be text with limited graphics received mostly in business offices on personal computers.[18]

Do we now need to change all of our statements about "home" screens to read "office" screens? What about the promises of life-style improvements that will be brought into the home when people have access to current information and, therefore, better control and use of their time? One videotex expert wryly observed, "When AT&T, IBM, Knight-Ridder, and Times Mirror all talk about how wonderful a given technology will be in the future, they leave themselves open to the suspicion that they haven't got a product that anybody wants today."[19]

The business community has shown interest in videotex as a medium for in-house communication. By mid-decade, more than two dozen private systems had been installed at companies like IBM, Equitable, and Buick.

Videotex may be too expensive for the average home consumer, but it has a real potential in the business world.

The description of future videotex as mostly text received in business offices on personal computers places it squarely in competition with several existing computer-based communication and data base services that have come to be known as information utilities. These and other information utilities are examined in the section that follows.

Information Utilities on Home Computers

By the time you read this section, the videotex media described above and information utilities on home computers may have blended into the same communications media. Potential videotex subscribers have been reluctant to buy special receiving equipment that can only display videotex. Americans, instead, are buying home computers that can be used in a variety of applications. This trend is forcing videotex operators into modifying their services so they can be received on home computer screens.

The separation of videotex from other information media on home computers is of some historical significance. In 1979, four years before commercial videotex was launched in the United States, two companies, The Source and CompuServe Information Service, began offering electronic news and information services to owners of home computers.[20]

In 1984 only 10 percent of United States homes had personal computers; by 1990, nearly half the households will be so equipped.

At the time, the potential market was small for such services. Nationwide, fewer than 50,000 personal computer owners also had modems, small electronic devices permitting computers to communicate over telephone lines, which are needed for receiving such services. In 1984, 10 percent of United States homes had personal computers, only a fifth of them equipped with modems. Only two years later, in 1986, 19 percent of the nation's homes had computers, and one out of ten had a modem—an indication that for the time being, the growth in computers was outstripping that of modems.[21] By 1990, nearly half of United States households are expected to have computers, with half of them having modems for communication uses. Obviously, this growing penetration of communicating home computers is creating a mass communications market for computer-based information services.

CompuServe Information Service, one of the more popular services, had 250,000 subscribers by mid-decade, while The Source had 80,000. In addition to CompuServe and The Source, information utilities for home computers include Delphi, NewsNet, and the Dow Jones News/Retrieval Service. While many other information utilities are now operating, these five have reasonably broad appeal and are representative of the range of such services. Dow Jones and CompuServe readily say their information services are profitable. Others decline to comment.

Most information utilities have up-to-the-minute news, stock quotes and other financial information, on-line shopping, electronic banking, and electronic mail.

Virtually all information utilities have up-to-the-minute news, stock quotes, and other financial information, on-line shopping, electronic banking, and electronic mail. They carry other on-line information such as electronic encyclopedia and abstracts from published journals and magazines. Education and games are also common features. Among the most popular features are those through which subscribers become active participants in communication with each other or with so-called "information providers." Electronic bulletin boards provide the means for individuals to discuss topics of mutual interest. The ability to self-publish on such systems may herald a new era of "electronic pamphleteering."

The wire services have created special news services designed to serve electronic information utilities. The Associated Press provides late-breaking news to CompuServe. United Press International is the source of news to The Source, Delphi, and NewsNet. A service aimed particularly at business, NewsNet also has up-to-the-minute information from the PR (Public Relations) Newswire. The Dow Jones News/Retrieval Service uses United Press International and also gets the latest news from subsidiaries of its parent, Dow Jones & Company, Inc. These subsidiaries are the Dow Jones News Service, the *Wall Street Journal,* and *Barron's.* The computer for the Dow Jones News/Retrieval Service is in Princeton, New Jersey.

CompuServe is owned by H&R Block, Inc., a company that helps people prepare their income tax returns. The CompuServe Information Service began as a way to use the excess computer power of a computer-time leasing service. Its computer is in Columbus, Ohio. The other information utilities were created by entrepreneurs who saw that computers could become a new medium of mass communication.

Telecomputing Corporation of America launched The Source in 1979, calling it "America's Information Utility." In 1980, it was purchased by the Reader's Digest Association. It operates twenty-three hours a day using a computer in the Washington, D.C. suburb of McLean, Virginia.

NewsNet was born in 1980 out of the death of the Philadelphia newspaper called the *Bulletin.* Its publisher, Independent Publications, decided electronic publishing was the communication business of the future. NewsNet has staked out a *narrowcasting* approach; unlike broadcasting, which casts a wide net over a broad geographic area, narrowcasting programs for a small, self-selective segment of the population. NewsNet carries more than 124 on-line newsletters in specialized subject areas. Whereas most information utilities are priced for the consumer market, business users of NewsNet may pay up to $120 an hour to read such newsletters as *Cellular Radio News, Sludge Week,* or *VideoGames Today.* The NewsNet computer is in Bryn Mawr, Pennsylvania.

The Dow Jones News/Retrieval Service began operations in 1981 specializing in business and financial information. It has had the fastest growth of any of the information utilities. Many observers say that an information utility must focus on the needs of a narrow market or audience in order to be successful. The Dow Jones service has done this. Its corporate news covers 7,000 companies in sixty industries.[22] It has financial facts and earnings estimates on 2,400 companies and financial disclosure information on almost 9,000 companies.

Some information utilities are *narrowcasting,* or appealing to small, self-selective audiences.

Besides its strong business coverage, the Dow Jones News/Retrieval Service carries movie reviews, sports scores, and weather reports for fifty major cities. Although home subscribers outnumber business subscribers, the use of the service is about equally divided between home use and business use.

Delphi was launched in February of 1983 by General Videotex Corporation of Cambridge, Massachusetts.[23] Being newer and smaller than some of the other services, Delphi tries to provide a friendlier and easy-to-use atmosphere for people at home using computers to communicate. Delphi subscribers can shop at home with Comp-U-Store, an electronic shopping service also available on CompuServe, The Source, and Dow Jones News/Retrieval Service. In addition, Delphi has an "electronic bazaar" where subscribers can sell, buy, or swap items.

Especially popular on Delphi is a subscriber directory in which individuals can post personal information and learn more about other subscribers who have posted information about themselves. This contributes to the friendliness and personal communication that occurs on Delphi. A key attraction of all of these information utilities is that they give subscribers a new power to communicate. The experience is quite unlike being a mere recipient of one-way communication as with traditional media such as newspapers and television.

One of the most intriguing new services is X-press, which offers news, stocks, and other information via satellite directly to office and home computers via local cable service. It is a partnership in which McGraw-Hill, Inc., TCI (a multi-system cable operator) and Telecrafter participate. X-press is being offered around the country and may be the forerunner of an entirely new area of effort for cable systems. Cable operators have been looking for a sexy service that will bring stability to their subscriber lists and reduce the high industry "churn," or tendency of people to routinely drop onto and off cable services. The service is important in our discussion because it marries the concept of videotex and cable, and illustrates the scrambling that is taking place in the industry to find the magic keys that will attract and maintain audiences.

Several of the information utilities have sophisticated computer conferencing facilities for small groups. The Source's conference system is called PARTICIPATE. Subscribers can participate with each other in discussions about a wide range of topics. Subscribers, indeed, can invite participation in a discussion topic of their own choosing. They have the privilege of closing the discussion to all except members of a select group, or they can open it up for all interested subscribers to participate in. The so-called discussion may last several days or several weeks. When you sign on to a conference, you are told what has been added since you last signed on. You may read what others have added and then make additional comments on your own by typing a new message for the conference. In a sense, it is a sophisticated form of intellectual graffiti.

Both The Source and CompuServe have had national experts lead "electronic conferences" on such topics as the nuclear freeze movement and educational computing. While traditional media can disseminate information about such newsworthy topics, only the new computer-based media provide for each easy participation by subscribers in active discussions of the topics.

CompuServe provides numerous special sections in which professional or special interest groups can have private communication and information exchanges for group members only. Public-relations practitioners are one of the public groups who conduct professional communication using CompuServe. A private group within this professional community, the National School Public Relations Association, also conducts organizational communication using CompuServe.

Besides these conferencing facilities, users of such information services may participate in less formal discussions through participation in on-line "bulletin boards." The Source's bulletin board facility is called POST and contains nearly one hundred individual bulletin board discussions on topics from "Aircraft" to "Zenith computers." In addition to bulletin boards for more than seventy special interest groups on CompuServe, there are three large boards called "Notice," "For Sale," and "Wanted" that all subscribers may use as electronic counterparts to traditional bulletin boards in public places.

On-line "bulletin boards" permit informal discussions, and "electronic mail" services permit direct communication across the miles.

Another popular form of informal communication on CompuServe is called CB Simulator, where subscribers may communicate with each other in "real time" by typing while they read what others are also typing. The words of several subscribers appear on a common screen image seen simultaneously by other CB Simulator users, who may be hundreds or thousands of miles apart.

Electronic mail, instantly delivering messages across the country, is a common feature of information utilities. You can send a letter by electronic mail even if your recipient does not own a computer on which to receive it. In this case, you compose your letter on your computer, indicate the address to which you want a paper copy delivered, then press a key to have your letter deposited in the United States Mail in the recipient's city. In 1984, estimates placed the number of electronic mail users at 200,000.[24] By 1986, the number was estimated to be 1.5 million, an indication of the service's popularity and potential for growth.

Dow Jones News/Retrieval Service provides an electronic mail facility through MCI Mail, an independent service. Electronic MCI Mail is available also to computer owners who may not subscribe to a general information utility. Delphi subscribers can send electronic messages, not only to other subscribers of Delphi, but also to subscribers of The Source, CompuServe, and two other independent electronic mail services, Dialcom and Ontyme.

The Source interconnects with the United States Postal Service's electronic mail facility, E-COM, and with Western Union Mailgrams. SourceMail is becoming so popular that businesses are putting their Source identification numbers on their letterhead stationary, just as many businesses include their Telex numbers, so that others can reach them through instant electronic communication.

Self-publishing is another special feature offered by The Source. While you must pay for the computer memory space to store your own story or feature, you may receive a royalty whenever your material is read by another user. The first electronic book, *The Blind Pharoah,* was published on The Source. Authors and entrepreneurs on this new medium have created weekly crossword puzzles, a computer-dating network, and at least two on-line magazines. *Real Times* is an electronic magazine devoted to owners of IBM personal computers. Another devoted to Apple computer owners is called *S.A.U.G. Magazine.* The Lutheran Church of America publishes an electronic newsletter on The Source that covers church and religious news. By mid-decade, The Source had some three dozen categories of self-published content.

While the lowest rate of $6 to $9 an hour for using the more popular information utilities may seem expensive, tens of thousands of Americans are using these new media to get news, to bank and shop, and to communicate with each other.

Besides subscribing to information utilities, more than a thousand Americans have set up their own computer-based communication systems, or electronic bulletin boards, offering their own information and communication services free of charge.[25] An era of "electronic pamphleteering" may be emerging as computer-based media are making mass communication affordable to ordinary citizens. Computer software for creating these two-way information services is available free through the public domain. Commercial packages may range from only $40 to around $300.

Most computer bulletin boards operated by individuals "narrowcast" to small and special audiences. There are bulletin boards for lawyers, doctors, and film producers. Some boards contain nothing but movie reviews, religious material, or adventure games. A number of retail stores have started their own bulletin boards for customer communication. Other boards have been set up by individuals for grass roots communication about political issues and pet causes.

Meanwhile, as home computer users are using their software and modems to become publishers, the traditional publishing houses are turning to the same information utilities to enhance their marketing strategies. For example, personal computer owners worldwide now have the luxury of ordering books from home, providing publishers and book merchants a potential market of millions of buyers.[26] The major outlet for book sales is CompuServe's Electronic Mall, which has a quarter of a million subscribers. The Source's subscribers use the Electronic Book Center for the

An era of "electronic pamphleteering" seems to be emerging.

Traditional publishing houses are turning to information utilities to enhance their marketing strategies.

Table 10.2 Publishers who have become most active in new electronic media

Company	Traditional products	Electronic products
Dow Jones	business newspapers, books	on-line news service, computer software
Dun & Bradstreet	business credit information, magazines, securities, ratings, airlines schedule guides	on-line credit information, computer software, on-line securities ratings, on-line airline schedules
Houghton Mifflin	books	computer software
Knight-Ridder Newspapers	newspapers, business information	home information services, on-line business information
McGraw-Hill	magazines, books, securities ratings, business information	electronic trading system, computer software, on-line securities information, on-line business information
Prentice-Hall	books	computer software
Scott Foresman	books	computer software
Scholastic	books, magazines	computer software
Times Mirror	newspapers, books	home information services, computer software

Source: "Publishers Go Electronic: An Industry Races to Relearn the Information Business," *Business Week,* 11 June 1984. By permission.

same purpose. Newsnet, Dialogue, and Westcom are other interactive computer systems that market books. All of the above permit customers to browse through electronic catalogs and order books, software, or other gifts from major publishing houses and book sellers. (See Table 10.2.)

Waldenbooks, Bantam, McGraw-Hill, The Christian Bookstore, Mercury House, and others have marketed their business, general fiction and non-fiction, technical, and religious books over the Electronic Mall and the Electronic Book Center. Although most of their customers are from the United States, the publishers have been receiving orders from Italy, Spain, France, Argentina, Brazil, Australia, and Japan.

The New Pamphleteers? Colonial America, where printing presses were within the means of numerous individual citizens, had many pamphleteers who spread their ideas by printing and distributing pamphlets throughout the populace. This early era ended as the technological means of mass communication became the province of corporations and large chains of communication companies.

Computer-based media nowadays may be putting the power to communicate with a large citizenry once more within the reach of individuals. This development in the new technological means of mass communication will contribute to the further decentralization of American society. It also represents one more means of communication that non-democratic, centralized societies will have to struggle to control in order to maintain their

Information utilities are contributing to the decentralization of society. Are we prepared for the consequences?

grip on internal information and communication. Only a vigorously democratic society can tolerate the unchecked growth of such a communications system. Even in the United States, voices have been raised asking whether this nation is prepared for the consequences of this new revolution in communication.

Before probing more fully into the social, economic, and political ramifications of the aforementioned media, let us consider the final category of new communications systems—ones with prerecorded information and entertainment.

Prerecorded Media

Consumer electronics, especially those offering prerecorded media, have changed drastically since the early 1970s.

Consumer electronics have changed dramatically since the early 1970s. Citizens-band radio gave a voice to the traveler isolated in an automobile. And about the same time, video games (especially one called Pong) launched the home-video-entertainment era and the brief but frenzied period of the video arcade. Advances (some may question the extent to which the CB and video games really are "advances" for civilization) have been swift. Today the consumer can purchase videocassette recorders, videodiscs, a personal computer, and an array of video and computer games or educational software to accompany many of the other telecommunications breakthroughs we have been discussing throughout this chapter. The home communications center, on the drawing boards and in the minds of science fiction buffs less than a decade ago, has become reality. It is not without its attendant problems.

Videocassette Recorders

Home video is experiencing a phenomenal growth spurt. It is nearly as large as the book publishing industry.

The birth of alternative television came about in 1975, when Sony introduced its Betamax, a home videocassette recorder (VCR). The Betamax quickly became one of the hottest selling items in electronics shops across America, despite the fact that it demanded a $2,300 investment: $1,300 for the VCR, and $1,000 for the Sony color TV set needed to play the Betamax tapes. Competition from other manufacturers, using a system called VHS (video home system), resulted in more sophistication and lower prices. Today VHS outsells Betamax about three to one.

By 1979 more than a million VCRs had been sold in the United States, and the sales pace quickened thereafter. In 1983, 4.1 million units were sold; in 1984, 8.3 million, and in 1985, 12 million. By some estimates, at the beginning of 1986 there were between 24 and 26 million American homes equipped with VCRs; there were 50 million copies of 6,000 different videocassette titles in circulation (with sales for 1986 expected to be in the 70 to 85 million range); 25,000 retail outlets were doing $3.3 billion worth of business selling and renting cassettes; and ten major publishing houses were producing a billion dollars worth of cassettes.[27]

Home video, a $1.25 billion business in 1983, had nearly tripled in less than two years, and was conservatively expected to double again by the year 1988. Putting this in perspective, we should recall from chapter 5 that at last check the 300-year-old book publishing business in the United States was a $10 billion industry, distributing its production through 12,000 retail outlets. Given its current growth rate, within fifteen years of its founding the home video business will equal or surpass the publishing business in volume, retail outlets, and dollars—and perhaps in consumer time commitments.

By mid-decade, the VCR had managed to penetrate some 30 percent of America's households, making it, like cable in 1982 and TV itself in 1949, a "significant" mass medium. The rapidity with which it achieved significance surprised most media observers. No other medium had so penetrated the marketplace in a mere decade. The amazing growth of the VCR reflects both the stepped-up pace of electronic communications and society's willingness to invest in an expensive diversion.

The typical VCR, available for as little as $300, plays six- to eight-hour videotapes, and offers computer programmable timers that can be preset to record selected programs on any channel up to two weeks in advance, plus fast forward, reverse, and pause controls. Since mid-decade, the hi-fi stereo VCR attached to a good amplifier and speakers has overcome one of television's major handicaps: a dinky three-inch speaker that could not even begin to capture the full range of recorded sound. (A half century ago, engineers decided that the FM band would be used for TV sound, giving television the capacity for excellent aural reproduction. However, most of the engineering developments in the meantime have been in picture quality, and sound has been left in the wings. The move toward stereo sound over the airwaves, by networks and local stations, has been slow coming. This means that, for most videophiles, a good stereo hi-fi VCR has been the first real opportunity to experience television's aural potential.)

Hand-held remote control devices permit viewers to manipulate most of the VCR's functions from across the room, without ever leaving the easy chair. Two conveniences especially admired by VCR owners are the pause button, which permits them to avoid taping commercials when recording off the air programs, and the fast forward button, which among other things permits them to zap past any remaining commercials when they are *time-shifting,* or watching prerecorded shows at a later date.

In addition to the standard features described above, VCRs can be purchased for making as well as viewing videos. A $400 to $1,000 color camera can be coupled with a battery-powered VCR to make up a complete portable video recording and play-back system. Aficionados can purchase ultra-low light cameras with full color viewers and built-in keyboard titlers allowing them to add written messages in a variety of sizes and colors to their tapes. Those interested in producing sports videos can buy a camera which superimposes a digital stop watch, along with the time, day, and date, on the recordings. The end result is not unlike the revolution in publishing described above; the amateur with an investment of less than $2,000 can be in the movie-making business.

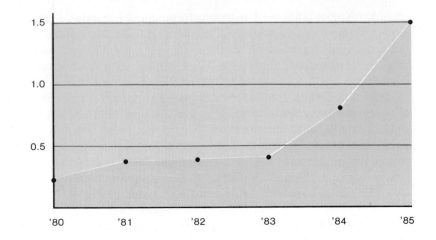

Video revenues earned by U.S. film distributors in billions of dollars. Source: Paul Kagan Assoc. Redrawn from *Time,* February 3, 1986, p. 65. Copyright © 1986 Time, Inc.

Home Video versus The Establishment The VCR's pause and fast forward buttons, so loved by consumers, have naturally become anathema to advertisers and network programmers, for the obvious reason that the commercials have not reached the programs' entire audiences. Recent efforts by Nielsen and other agencies to measure television audiences more effectively, to actually determine who is watching what show and what commercials at what time, have been controversial. Quite logically, advertisers do not want to pay for wasted impressions. Equally logically, audiences do not want their privacy invaded, and they see no way that sophisticated audimeters or "people meters" (see chapter 9) can gain the desired information about their viewing behavior without knowing more than the viewers care to reveal about themselves.

The controversy over "zapping" commercials is but one indication of the battles that have brewed between established media and the VCR. We have mentioned elsewhere that the VCR has had a major impact on television. Networks' share of the viewing audience has dipped significantly, and pay cable movie channels are engaged in never-ending conflicts with home video for the rights to early release of Hollywood films. Likewise, the VCR's appetite for rented or purchased movies has brought about a revolution of sorts in Hollywood. However, on both fronts—commercial television and the film industry—the legitimate initial fears of the VCR seem to have been mollified somewhat.

Few could have imagined, as recently as 1980, that video rental and sales outlets would crop up in every community, in every shopping center, in grocery stores and gas stations and convenience stores everywhere we turn. There are twice as many video retail outlets as there are bookstores in the United States, although the equation is somewhat spurious given that only a small percentage of the 25,000 commercial outlets for videos deal exclusively in cassettes. Indeed, the general attitude toward videotape rentals and sales is reflected in the fact that they are so readily available at checkout

VCR users who "zap" commercials are upsetting the ratings services.

stands—right up there with the razor blades, chewing gum, tabloid news-papers, magazines, and paperback books, where impulse buying is counted on to keep the economy rolling.

One of the film industry's greatest fears appears to be unwarranted. Despite the thriving video sales and rental business, and the fact that in some circles (especially high school and college) video parties have become one of the most popular social events, people are still going out to the movies. Surveys reveal that many VCR owners, who originally bought their machines so they could stay home and avoid the crowds and high costs of going to the movies, actually become more avid moviegoers after their VCR purchase. Apparently, the VCR helps expose them to an art form they had never fully appreciated. Overall box office take may have taken a slight dip since its 1984 high, but overall, the film industry has not suffered.[28]

Hollywood has learned to cooperate with and to some extent exploit the general craving for in-home viewing of movies. Most studios schedule the videotape release of their films within six months to a year of their first-run theater play, and other studios have done a lucrative business of renting and selling their older films. Walt Disney Productions has sold about 125 of its titles on cassettes, and MGM/United Artists marketed 300,000 copies of its 1939 classic, *Gone with the Wind,* at $90 a copy. In a marketing coup, Paramount discounted its popular *Raiders of the Lost Ark,* and quickly sold a half million copies at $24.95. Meanwhile, numerous other studios have followed suit. By 1987 some were peddling recent releases such as *Top Gun* at bargain prices because the videos included built-in, glitzy commercials. Between the first-run hits and the moldies, however, is the lucrative business for the "also-rans"—those Hollywood releases that did not do particularly well at the box office but which still have enough appeal to recover some of their production and marketing costs when released for VCR viewing.[29]

Home video's impact on Hollywood cannot be understated. In 1985, the watershed year when 12 million VCRs were purchased, 28 percent of the total revenue earned by feature films came from the domestic theatrical market, while VCR sales almost equaled that mark, with 27 percent.[30] (Once Hollywood sells a film to video, no matter how often it is rented out or shown, it earns no additional royalties. This has miffed producers and actors alike. They are calling for Congress to rewrite the copyright law to impose royalty fees on both videocassette recorders and blank videotapes. Retail outfits and consumers obviously do not mind the current arrangement.[31]) Comparing theatrical and home video figures is somewhat like comparing apples and oranges. The domestic theatrical market consists of the 300 or so new films made annually, while the VCR market contains many of the newer films plus thousands of older films resurrected from studio vaults. The latter supply, obviously, is limited. Once most of the "oldies but goodies" are trucked out, if home video's appetite for film is to be satiated,

Hollywood and the VCR may have reached detente.

Hollywood and independent operators will find themselves producing new releases at an unheard of pace. Before the coming shakeout, there will probably be an awful lot of shoddy films put on the market.

Speaking of shoddy films, popular myth has it that the largest and most profitable video market is for pornographic films. True, in the early days of the 1970s, as many as three-fourths of all video releases were X-rated. More recently, their share of the market has dropped to about 15 percent.[32] That drop in percent does not mean a corresponding decline in output. On the contrary, today there are more porno videos being made and marketed than ever before. Between fifty and seventy-five companies specialize in the production of X-rated videos. In mid-decade they were enjoying a 500 percent increase in sales volume over their 1980 figures, and were counting on a 100 percent annual growth rate for the next several years. The skin flicks are produced for an average of $150,000 (compared with $16 million for the average Hollywood film). Typically, about 20,000 cassette copies of a skin flick will be sold, giving the producers a $100,000 to $300,000 return over and above their investment. By means of comparison, major studios consider themselves lucky if their $16 million releases sell 50,000 video units. Thus "adult" videos represent an almost risk-free venture. So long as the courts rule that individuals can view whatever they want in the privacy of their own homes, the sales spiral of X-rated videos is likely to continue.

The only reason adult videos have a mere 15 percent of today's market is that so many other varieties of cassettes are competing for attention. Of the thousands of cassette titles available for sale or rent, a great many are how-to cassettes. They cover every imaginable topic, from how to improve your golf game, skiing, bowling, and hang gliding; to how to stay slim and trim (Jane Fonda's workout tapes have generated millions in revenue); to how to play musical instruments, invest your money, fix up your home, build a boat, buy a car, or take a vacation. Rock music, children's and adults' games, and thousands of business, educational, and other tapes are available through specialized distributors.[33] In sum, the VCR has become a mass medium, but an individualized mass medium. Its major function, it seems, is to provide home viewers with a cross section of traditional and specialized media fare, viewable and reviewable upon command.

Videotape may have a salutory effect on the increasingly expensive performing arts, such as Broadway and off-Broadway theater, ballet, opera, and symphony. Appeal of these arts has generally been restricted to around 4 percent of the population—an audience too small to justify network television coverage. Videocassettes (and cable TV) offer a potential whereby performances can be recorded and the tapes sold to aficionados for home consumption with the producer receiving a royalty on sales. With 30 percent of United States homes being VCR equipped, the possibilities for stimulating the economy of the performing arts are quite obvious.

Pornographic videos are still selling well. Their share of the market is smaller now, but only because there are so many other topics available.

The VCR has become a mass medium, but an individualized mass medium. Its traditional and specialized fare is viewable upon demand.

VCR Usage and Potential So far, in its infancy and adolescence, the videocassette recorder appears to have melded into the electronic media mix without causing a drastic shakeup. The day may come, however, when we stop using this new technology to perform old tasks and begin to recognize its new possibilities. As noted television critic Les Brown puts it, the VCR should be the hope for democracy, since democracy flourishes "when the consumer can get into the distribution system and actually make television."[34] When that day comes, when we start using the VCR to really create rather than merely play back video recordings, we will have gained some liberation from the electronic media's technological, artistic, and economic tyrannies.

The latest and probably most exciting form of television's progeny remains one of the most enigmatic. The videodisc has actually been around as long as television itself, since 1928. That year a London department store began selling a phonograph that played sound pictures in motion, with some in color. The "television gramophone" used standard wax disc records to project a coarse, low-resolution picture viewed through a perforated whirling disc. Although the novelty items were marketed for seven years, there is no record of how well they sold, and the fact that hardly anybody has heard of them today may be significant.

Videodiscs

Their successor, however, the contemporary videodisc, is being called by some observers the "omnibus medium." It can reproduce and effectively combine the best qualities of virtually all the other communication media: books and other printed materials; stills—slides, transparencies, photos, etc.; motion pictures; videotapes of all formats; digitized computer data; audio; and stereophonic sound.[35] In 1980, RCA President Edgar Griffiths said he expected the videodisc to start a new industry, "bigger than the broadcast industry, two-and-a-half times the record industry."[36] Four years later, RCA told the world that Griffiths had been overly optimistic. The giant electronics corporation pulled out of the videodisc market after having sunk $580 million in the futile production of its Selectavision, which had sold only 500,000 units.

Does RCA's disaster mean that the soothsayers were all wrong about the videodisc, that it has become the home electronics Edsel? Not necessarily. It is still too soon to determine whether the videodisc will become either the hottest new visual medium in history or electronics' most expensive flop.

Basically, a videodisc is merely a prerecorded disc the size of a 33⅓ RPM audio record. The gold or platinum discs run up to an hour per side when played on a videodisc player and viewed on a home television set. Like commercial records, the videodiscs are for play only. Home units cannot be used for recording.

The entire *Encyclopaedia Britannica* could be stored on a single videodisc.

As described, it hardly sounds like a revolutionary piece of technology. But its advantages over other electronic media are numerous. In one form—the only system on the market after RCA's withdrawal—laser-beam technology is used to read pictures and sound from grooveless discs. Since no stylus ever touches the discs, they never wear out. Fidelity is as good on the millionth play as on the first. The players include sophisticated features such as slow and fast motion, stop action, single-frame advance, and instant search and reviewability capacities. This laser system contrasts with RCA's Selectavision, a CED (capacitance electronic disc) system utilizing grooved discs, much like hi-fi records. The CED models, which sold for as little as $200, could be used for little more than showing home movies, whereas the more expensive laser-beam technology permits a wide variety of uses.

Unlike videotape, the discs provide rock-steady still frames in playback. When used to its full potential, a single disc is capable of storing a complete library of 360 books, each with 300 pages, assuming that each page would occupy one of the disc's 108,000 frames. Or a disc could be used to record 108,000 separate photographs or slides, or 13 billion bits of digitized information. Extremely high fidelity stereophonic sound is possible on the discs; a single disc can hold fifty hours of prerecorded music. Because the sound is recorded separately on two channels, movies on videodiscs have been recorded in two languages. Viewers can switch from English to Japanese in the middle of a word if they wish. Most intriguing of all, the digitized, laser-beam technology lends itself to instant access of any frame, image, or sound on the disc.

In the early going, many manufacturers explored unique applications of the videodisc. Telefunken/Decca designed a system using ten-minute discs for a Video Jukebox. Matsushita of Japan developed a combination video and color printing system called "Picture Paper," much like a television set with a color printing press spewing out sheets of hard copy upon viewer demand. MCA Incorporated (Universal Pictures and Decca Records) was producing thin flexible discs to be inserted in periodicals, indicating hopes that the market would be ready for videodisc catalogs, magazines, or talking encyclopedias.

The fact that the entire *Encyclopaedia Britannica* could be stored on a single disc has not been wasted on industrial visonaries. Libraries, concerned about the deterioration of their holdings, are contemplating transferring masses of documents to discs. Educators on all levels, including industrial and military trainers, are starting to recognize the enormous potential of the videodisc. (Army recruits train on interactive videodiscs that realistically simulate war conditions, without having to spend the $1,000 a round involved with firing live tank ammunition.[37]) Individual videophiles, who see the videodisc linked to a personal computer as an essential component of the home communications center, are also enthusiastic about the medium.

Sales of videodisc players lagged behind industry expectations in the early 1980s, but in educational functions the videodiscs have shown promise for such diverse purposes as language training, musical studies, and cognitive development of the mentally handicapped. Students can select whichever language they choose to learn, and move back and forth at will between their native language and the one they are attempting to master. Music students can not only hear chosen works in stereo but can also watch the performers, conductors, or musical scores. When interfaced with microcomputers, the videodiscs provide individualized, self-paced instruction for slow learners and valedictorians alike.

Researchers have demonstrated that the videodisc's effectiveness as a teaching tool is unsurpassed. Learning a specific task takes an average of 30 to 35 percent less time with the disc than with other methods.[38] A member of the House Science and Technology Committee, Rep. Judd Gregg (Republican, New Hampshire), closely watched developments in the videodisc field, then concluded, "I see interactive videodisc technology as revolutionizing the way we educate in this country. It's really moving education from the Industrial Age into the Information Age."[39]

Indeed, it is the videodisc's ties with the computer that have corporate and educational minds spinning. Phillips Research Laboratories in the Netherlands have developed a teaching system that combines the videodisc and the TV monitor to a home microcomputer with a built-in monitor screen. The microcomputer controls the videodisc and provides the graphics—such as the table of contents, questions, and so forth—on its own monitor screen. The student can follow instructions on the computer screen, respond to questions about the material, and request a repetition of any or all parts of each program. The student's progress can be made dependent on giving correct answers to the computer's questions. Another program, in experimental stages, allows teachers to design their own questions and instructions to go along with any prerecorded videodisc of their choosing.[40]

At Utah State University, instructional media faculty members have been working on a Videodisc Innovations Project since the late 1970s. They have developed a series of "videodisc vignettes" to supplement instruction in several academic areas. They are particularly excited about the industrial videodisc systems and microcomputers that allow students to physically interact with the program by merely touching the television screen in appropriate places. Doing so allows them to go from still-frame to lineal information in full color and motion almost instantaneously; to gain instant access to any frame on the disc; and to utilize the Matsushita "Picture Paper," giving them and their instructors hard copy printouts of any and all exercises attempted.

Industry jumped onto the videodisc bandwagon somewhat cautiously. The potential for videodisc had been recognized instantly, but high costs, lack of familiarity with the system among potential users, a low level

Videodisc's potential for business and educational use is probably greater than its potential for in-home entertainment.

of production capacity, and very long production turnaround were cause for some concern among the system's most likely customers.[41] By 1983, although mass marketers such as RCA and CBS were having trouble penetrating the home market, the obstacles standing in the way of industrial adaptation had largely been overcome, and the videodisc was on its way.

It may be significant that videodiscs have been on the American market only since 1978. The speed with which the industrial giants have proceeded to enter, back off, and reenter the competition is intriguing. Many problems still must be solved before the videodisc establishes its place in history, including technological refinement and compatability of systems; the cost to consumers; regulatory involvement over copyright and other issues; and a continuing supply of software (contents) to justify consumer interest. When these questions are fully resolved, the fate of the videodisc will have been settled in as short a time span as for any other mass medium in history. The growth rate of videodiscs represents a phenomenon. Much like the videocassette (whose financial success, particularly in playing rented tapes, has held back the home market for videodiscs), the newer technology is emerging on the scene far more rapidly than did any of the older technology, and with more rapid effects on the established media.

The growth rate of the videodisc represents a phenomenon.

Microcomputers and The HomeComCen

More than half the American work force is engaged in the postindustrial information processing era. The computer age is upon us.

The proliferation of home and office computers heralds the information revolution about which so much has been said. At its base is the refinement of miniaturized electronic components that reduced the cost of computer power by a factor of about fifty during the 1970s. Components that in 1950 took up an entire room have been reduced to the size of a cornflake, with no loss in productivity and with the promise of greater economic productivity and growth. In 1980 about 100,000 word-processing machines were installed throughout the country, giving offices and homes much greater efficiency in handling and storing written material than was ever before possible. By mid-decade one in every five American homes was equipped with a personal computer, and a majority of the tasks carried out in the nation's offices involved some crunching of words and/or data on an office computer. Indeed, more than half the work force is now engaged in what is called the postindustrial, information processing era.

With the new information revolution have come revised definitions of literacy and illiteracy. On one front, scientists and technicians are warning that within a few years it will be necessary to be able to program and operate microcomputers in order to get and hold a job, let alone enjoy the full potential of a new home communications system of entertainment and information processing. On another front, other scientists and technicians are making certain that the operation of computers is greatly simplified, so anyone—especially children—can adapt to them. With the prospect of tens of millions of user-friendly computers of all sizes being in use in the immediate future, the information and entertainment business is facing a certain shake-up.

Obviously, one task of the computer industry and business is to develop a system of interactive self-teaching programs for the microcomputers and word processors. Several colleges have instituted computer literacy requirements for graduation. Elementary and high schools have had to add computer science programs to their staple of reading, writing, and arithmetic. The unpleasant alternative will be a new version of functional illiteracy, a national debility we can ill afford. "Why Johnny Can't Program" might well become the most popular theme of supermarket magazines in 1990.

For the homeowner, the information revolution is taking the form of a complete home communications center, or HomeComCen. In its fullest form, as envisioned today, it will include a computer keyboard, a videocassette recorder and videodisc system, a television set (perhaps a wall-sized projection screen), an interactive cable, a home satellite receiving dish, a telephone interface, and a facsimile printer, perhaps with color capacity. All this hardware has been available since 1980. The basics (a regular-sized television but no home satellite receiving dish) could have been purchased for something over $5,000; the satellite dishes of the time would have brought the cost to $15,000. As technology progresses and mass marketing permits lower per unit costs, the same hardware should go for a more easily affordable $500 by 1990, according to some futurists.[42]

And what will we do with all this fancy hardware? Because of the economic nature of the beast, the HomeComCen will initially serve as an in-home adjunct to the financial world. As we have seen, parts of this system are being used for home banking and home shopping. Shopping over the HomeComCen is relatively easy. A consumer freezes the frame on an advertised (videocassette or videodisc catalogued) item desired, examines it, requests fuller data over cable's two-way capacity, and presses the purchase switch or touches the television screen at the appropriate place. This automatically debits the consumer's bank balance, credits the store's account with the purchase and the state with the proper sales tax, and starts the item on its way from the warehouse to the consumer's home. America's recent romance with plastic credit cards indicates a willingness to transact business without cash, and financial institutions have met little rebellion when the entire procedure is handled electronically.

The HomeComCen is changing the nature of broadcasting and the lives of Americans.

These new video options "will transform not only the face of broadcasting, but the lives of Americans as profoundly as the Industrial Revolution of the nineteenth century," according to Lionel Van Deerlin, former chairman of the House Subcommittee on Communications.[43] Even though there has been no national policy regarding the development and implementation of the hardware, it appears likely *neovideo,* as the systems are sometimes called, will have a dramatic impact on, and perhaps even abolish, over-the-air television broadcasting as we know it; will hasten deployment of interactive information retrieval and utilization; and will bring about a decentralization of the office. It has been suggested that the resultant back-to-home movement will become the single most significant social trend of the 1980s, and that automobile usage will decrease dramatically. Throughout the country, there has come a blurring of the lines between giant corporations and cottage industries. Corporate employees ranging from secretaries to managers are working at home on personal computers, "telecommuting" their work, only occasionally going into the central office. Meanwhile, numerous cottage industries have sprung up, with entrepreneurs of the information era setting up shop at home, even in the wilderness, from which they do everything from develop and market software; freelance or write articles and books on contract; manage far-flung holdings; buy and sell stocks, bonds, insurance, real estate, etc.; and generally carry on whatever businesses they can. One observer has pointed out that the merging of the new technologies (satellites, pay TV, fiber optics, interactive TV, computer communications) is going to change our lives drastically. As the United States enters the information age, it is the television set—not peanut shells or corn husks—that is the most promising candidate to lick the energy shortage.[44]

Impact and Dangers of Neovideo

Throughout this book, we have seen example after example of how each new medium of communication caused a disturbance among the existing media, and outcries from those convinced that the new media posed dangers to the economic, social, and political status quo. We have also noted that, after a shakeout period in which new media borrowed the contents of the old media before learning to take advantage of their own uniqueness, there has generally been a reconciliation. Almost invariably, the end result has been a symbiotic relationship among the survivors.

The new electronic media are assuredly not exceptions to these rules. It is still too early to speak in precise terms about the impact of all the media introduced in this chapter. We can note some early trends, however, and do some projecting based on preliminary evidence.

Earlier media have learned to coexist symbiotically. Will existing media adjust to neovideo?

Technical and Economic Factors

When traditional television was first altered, by cable and satellite and by being coupled to other media such as the telephone and computer, many members of the commercial broadcasting industries emitted death knells—their own. Interestingly enough, they were joined by members of the newspaper industry, even though for decades the two groups had been fighting one another for revenues and customers. For instance, early demonstrations of teletext and videotex made it appear that the stage finally had been set for the birth of the electronic newspaper and the death of advertising-supported television. TV set manufacturers, on the other hand, were delighted at the prospects of a nation—and world—dissatisfied with outmoded sets and eager to buy completely new models equipped to receive teletext and videotex. (This may explain the interest of RCA and General Telephone, both of whom are in the television set manufacturing business.) And while many newspaper people gnashed their teeth, the involvement of Knight-Ridder and Dow Jones and others can be interpreted as a farsighted sense of fiscal reality: when the electronic newspaper does come, who better than established newspaper companies to be in the forefront? It now appears likely that teletext and videotex will cause the most disruption to newspapers and commercial broadcasters who fail to become part of the action. Ignoring the new media will not make them go away, and wishing them ill will not keep them from capturing their share of the audience. Recognizing them as an inevitable part of the media mix, and planning accordingly, appears to be the safest bet for all concerned.

Looking further at the potential impact of teletext, videotex, low power TV, and other neovideo that have an interest in the news business, it may be that they will bring about a new definition of news. Perhaps of greatest interest to news audiences will be the recognition by newspapers and network television that news can and should transcend the time and space limitations now imposed upon them. Rather than define news in terms of the corporate gatekeeper's interests and limitations, news will be redefined by the consumer. Tomorrow's newspaper and television newscast may have to become more involving, more entertaining, offering greater depth and human interest to augment and offset the instantaneousness of teletext and videotex.

Financial support for the home information utilities will come from two primary sources: the consumer and the supplier. Additional charges to the homeowner, as with pay TV, will bear part of the burden. The remainder will be shared by the special-interest groups producing information, and by advertisers. Advertising, which has presented a vexing problem

The safest bet seems to be to accept neovideo as an inevitable part of the media mix.

Tomorrow's news consumers will redefine what is newsworthy.

in teletext and videotex's earliest stages, may have to become more informative and less manipulative than the network television commercials of today. (Meanwhile, commercials on cable, low power, and satellite delivered television might go the opposite route, and become informercials and mini-shows of their own.) Because teletext and videotex offer information only when requested, advertising on them will have to be appealing and informative to compensate for the intrusiveness factor. Indeed, the classified ad and *Consumer's Report* are perhaps the parents of tomorrow's new technology commercial. But, since it will cost far less to produce ads on some neovideo than for commercial television and print, we may find a less elaborate marketing system evolving.

There remains some question about which existing media are in the best position to move ahead with neovideo development. For instance, television and cable companies may be more likely than newspapers to get into teletext and videotex, only because the FCC has prohibited newspaper owners from acquiring broadcast properties in their home markets. Besides, it has always been somewhat difficult for outsiders to be granted broadcast licenses. This picture may be changing, with recent FCC moves to license hundreds of localized LPTV stations, to award franchises to minority and special-interest groups, and to deregulate and thereby enhance new delivery systems. But the very expensive gamble must be initiated by industry; the United States government is not likely to take the risks that Great Britain, France, and Canada have because it has less of a vested interest. Even when industry invests research and development monies into various neovideo projects, the FCC will have to adopt industry-wide specifications, lest the industry continue wallowing in long-term and somewhat self-destructive competition.

The enormous costs of wiring the nation with broadband or interactive cable must also be considered. Only a few of America's existing cable systems employ two-way cable, but lately it has become a requirement for cable companies seeking franchises in metropolitan areas. One of the most exciting developments in this respect is that of *fiber optics,* tiny flexible strands of glass capable of carrying beams of light or electromagnetic energy around curves and corners. A single 0.005 inch strand can carry 167 television channels, and the potential for each home to be wired for hundreds, even thousands of channels is not beyond speculation.[45] Mass production of the inexpensive technology may mean the wired nation is closer to becoming a reality, depending in part on the success of the direct-to-home broadcast satellites (DBS). However, it will still involve an expenditure of energy and a commitment equal to that of the two earlier wirings of America for electricity and telephones.

The question, of course, is whether America is as committed to a wider range of news, entertainment, advertising, and at-home shopping as it was to lights and conversation.

The half-inch diameter fiber optics lightguide cable is only one-fourth the size of the 1200-pair wire cable typically used throughout the Bell System, but it can handle four times as many phone calls as the larger cable.

Wilson Dizard, in his provocative look at neovideo, *The Coming Information Age,* reminds us that the telegraph, telephone, wireless radio, and other earlier communications machinery evolved slowly, strengthening economic productivity without disrupting social order in an expanding democratic society. However, he notes, in the present age of converging technologies and greater social complexity, the balance between economic productivity and social harmony becomes more difficult to maintain. The balance is threatened by a dilemma: technology as a productive force rolls on, while its contribution to social stability grows weaker.[46]

The balance between economic productivity and social harmony has become more difficult to maintain.

If the nation is committed to this rapid technological change, what are some of the ramifications?

Social and Philosophic Factors

Television has been called the most profoundly democractic medium in history. Today some are saying that the new electronics spell the end of that democratization process. Others disagree. Similar differences of opinion dominate discussions of society's new means of receiving information and entertainment, of the impact on established telecommunications, and of government's role in neovideo's development. Arbitrary conclusions are premature, but we do ourselves a disservice if we fail to consider the problems.

Marshall McLuhan's global village was supposed to be a by-product of commercial television. With its appeal to the lowest common denominator of society and with its instantaneousness, network television was to

Neovideo may be eliminating
our sense of community.

have given us a global sense of community. Whether it has is surely debatable. At the very least, it has given us some common social, economic, and political agenda items to consider. However, neovideo is very likely to eliminate even that superficial sense of community. Some suggest that the greatest danger of new electronic technology is a destruction of the common data base demanded by democracy. They insist that there must be a sense of community as well as a sense of individuality if democracy is to serve its mandate.[47] Their fear is that the new technology promises too much diversity in news and information for an increasingly disengaged citizenry.

What is the price of
information overload?

If we realize that even the most primitive videotex efforts could bring a one-hundred-page newspaper or magazine to the home in an amazing twenty-four seconds, we have to ask who among us has the capacity to consume that much information? Who *cares* to absorb that quantity and diversity? Given as much choice as teletext, videotex, and other neovideo promise, are we likely to shortcircuit due to information overload? Will the bewildering array of choices induce in consumers what Swedish social scientists are predicting, a clinical catatonia triggered by neurotic indecision over what to consume?[48] Or, as one scholar argues, are we likely to keep putting these new media to old uses, getting our daily doses of superficial community news in more rapid fixes, but never going beyond that, to fully exploit neovideo's potential? One result of this type of information production and consumption is a low level of job satisfaction among teletext and videotex journalists, who see themselves as rather uncreative and unchallenged automatons. If the technology does not lend itself to high quality journalism, it contributes little to the functioning of democratic forms of government.[49]

Could neovideo actually
enhance social integration?
Could it demand more active
involvement in
communications?

Forecasters suggest that individuals will grow socially alienated once they obtain information in the home rather than through interpersonal interaction. There are, however, credible arguments to the contrary. Among them is the suggestion that the availability of sophisticated communications services might enhance social integration due to the increased lines of communication through public-access channels on cables and in teleconferences. One might also argue that neovideo will be more demanding than today's television and will result in more active involvement, more participation, and more family or group discussion of media contents than ever before. This is predicted because the content of neovideo is actively sought out and is inherently more involving than passive viewing of conventional network television. At the end of a typical neovideo "program," there will be either a blank screen or a series of video questions, with the expectations of answers. Microcomputers in particular are demanding, and should increase intellectual activity. School-age children, among the first to be introduced to microcomputers, have taken to them with enthusiasm, playing the games, challenging the programs, creating new and unique images, displays, and sequences on the video screens and adding special effects to the preprogrammed packages. Family togetherness might possibly become an unexpected by-product of the new technology.

Theoretically, a strengthening of human bonds emerges as a natural response to a technologically complex society. At least, so argues prognosticator John Naisbitt in his best-selling *Megatrends*. His buzz words for this interaction are "high tech, high touch," and he has used numerous historical examples to prove his point.[50] Whether the "high touch" will naturally follow the infusion of neovideo is open to some question, however. A recent study of several hundred technologically advanced families suggests that there is no clear relationship between ownership and use of neovideo and the possession of any particular bonding or human values.[51]

Finally, there remains the question of society's priorities. We must consider whether this kind of investment in the future, requiring enormous risk capital, is where Americans wish to spend their money. Money, like time, is limited, and what is spent for this fantastic full-color, stereophonic, two-way, 3-D, multiple-channel, computer-accessed, ultimately flexible, and completely consumer controlled information system must come from something else. In the long run, it is not merely an economic question, but an economic-education-social-psychological-political one whose ramifications at present remain a mystery. Such systems are already here, and billions

Finally, it comes down to a question of society's priorities. Do we really want to invest in neovideo?

of dollars are being expended on their development. The kinds of judgments made this year and next, and the year after that, may prove irrevocable in terms of commitment to a path or a system.

Summary

The new electronics include an intriguing array of video technology that brings information and entertainment into the home. For economic, technological, and regulatory reasons our discussion of these electronics was divided according to how the communication systems deliver their goods to the home screens.

Traditional broadcast signals have been adapted to deliver subscription television (STV), low-power television (LPTV), and teletext. Microwave and satellite delivery systems bring multipoint distribution service (MDS), direct broadcasting satellite (DBS), and a variety of satellite fed cable services. Cable television, which has reached the point of being a significant mass medium in the United States, received a close look in this chapter, as did videotex and information utilities available to home computers. Finally, prerecorded media such as videocassette recorders (VCRs), videodiscs, and the HomeComCen were reviewed.

This chapter has uncautiously opened a Pandora's box. We have considered the current and potential natures of several forms of telecommunications. Serious questions have to be asked about their impact upon existing media and upon society as we know it. We speculate that the present and past hostility that established media have displayed toward television's progeny will ultimately succumb to a new state of integration, but not until there have been several more technological, regulatory, and economic battles fought. From today's perspective, we cannot predict with any certainty about society's ultimate response to these neovideo. We have suggested several alternatives, ranging from concern over America's loss of a sense of community to a newfound individualism that manifests itself in more caring, sharing family structures.

Notes

1. Lynne Schafer Gross, *Telecommunications: An Introduction to Radio, Television, and Other Electronic Media,* 2d ed. (Dubuque, Iowa: William C. Brown Co., Publishers, 1986), p. 185.
2. Ibid., p. 171.
3. "NCTA Report on Local Cable Programming," *Broadcasting,* 1 September 1980, p. 42.
4. *Broadcasting,* 25 August 1980, p. 111.
5. Kent Baker, "Cable TV Reshapes Sports," *The Press,* June 1982, p. 41.
6. "Fifth Estate's $30 billion-plus year," *Broadcasting,* 30 December 1985, p. 41.
7. Doug Hill, "Will the latest in home video empty the movie theaters?" *TV Guide,* 20 March 1982, pp. 21–22.

8. F. Leslie Smith, *Perspectives on Radio and Television.* (New York: Harper & Row, 1979), p. 418.

9. Barbara J. Friedman, "Videotex languor claims victims," *Presstime,* April 1986, p. 56; and Barbara J. Friedman, "Viewtron alters market goals," *Presstime,* February 1986, p. 36.

10. Lynne Schafer Gross, *The New Television Technologies,* 2d ed. (Dubuque, Iowa: William C. Brown Co., Publishers, 1986), pp. 175-77.

11. Barbara J. Friedman, "Videotex languor claims victims."

12. "Knight-Ridder's cutbacks at Viewtron show videotex revolution is faltering," *Wall Street Journal,* 7 October 1984.

13. Andrew Radolf, "Move into cable is not a way to protect newspapers," *Editor & Publisher,* 29 October 1983, p. 42.

14. "AT&T prophesies a $15 billion market for videotex," *Marketing News,* 26 September 1984, pp. 16, 20.

15. Aldrew Radolf, "Move into cable is not a way to protect newspapers."

16. Kathy Chin, "Keyfax service to hit Chicago," *Infoworld,* 8 October 1984, p. 16.

17. Efrem Sigel, "Videotex: into the cruel world," *Datamation,* 15 September 1984, p. 140.

18. "Don't judge tomorrow's videotex by today's data, expert says," *Presstime,* November 1984, p. 36.

19. Efrem Sigel, "Videotex: into the cruel world," p. 144.

20. Jan Owen, *Understanding Computer Information Networks* (Sherman Oaks, Calif.: Alfred Publishing Co., 1984), pp. 31, 36.

21. "High tech and the home," *Newsweek,* 17 February 1986, p. 5.

22. Brad Baldwin, "Dow Jones News/Retrieval," *Infoworld,* 30 April 1984, p. 54.

23. Russ Lockwood, "Electronic Oracle," *A+ Magazine,* May 1984, p. 104.

24. Russ Adams, "Cashing in on electronic mail systems," *Business Software,* June 1984, p. 24.

25. John Edwards, "Exploring private message systems," *Online Today* Vol. 3, No. 6 (June, 1984), p. 13.

26. United Press International, "Computers provide book market," *Logan* (Utah) *Herald-Journal,* 5 March 1986, p. 19.

27. Hy Hollinger, "Proliferation of videocassettes and barter TV syndication made a tumultuous 1985 for film biz," *Variety,* 8 January 1986, pp. 7, 238; and James Melanson, "Homevideo attains mass-medium status," *Variety,* 8 January 1986, pp. 141, 148.

28. Richard Corliss, "Backing into the future," *Time,* 3 February 1986, pp. 64-65.

29. Gail Bronson, "Videotapes give Hollywood a second shot at success," *US News & World Report,* 4 March 1985, p. 73.

30. Hy Hollinger, "Proliferation of videocassettes and barter TV syndication made a tumultous 1985 for film biz."

31. "Collect royalties on video-recorder sales?" *US News & World Report,* 13 February 1984, p. 48; and Lynne Schafer Gross, *The New Television Technologies,* p. 144.

32. Richard Zoglin, "VCRs: coming on strong," *Time,* 24 December 1984, pp. 44-53; and Howard Polskin, "Acting: awful; scripts: weak; sales: terrific," *TV Guide,* 30 June 1984, pp. 36-40.

33. Zoglin, "VCRs: coming on strong," p. 45.

34. Rushworth Kidder, "Video culture," reprinted from the *Christian Science Monitor,* 10-14 June 1985, p. 14.

35. R. Kent Wood, "The Utah State University videodisc innovations project," *Educational and Industrial Television,* May 1979, p. 31.

36. David Lachenbruch, "The coming videodisc battle," *Panorama,* April 1980, p. 59.

37. Associated Press, "Videodiscs catch on in military, business training," *Washington Times,* 30 October 1984.

38. Robin Lanier, "The interactive videodisc in action," *E-ITV, The Techniques Magazine for Professional Video,* April 1986, p. 37.

39. Associated Press, "Videodiscs catch on in military, business training."

40. "Videodisc," *Media Digest,* Spring, 1980, p. 4.

41. Robin Lanier, "The interactive videodisc in action," pp. 36-43.

42. *Broadcasting,* "Perils and prospects over the electronic horizon," in *Readings in Mass Communication: Concepts and Issues in the Mass Media,* 4th ed., Michael Emery and Ted Curtis Smythe, eds. (Dubuque, Iowa: William C. Brown Co., Publishers, 1980), pp. 223-24.

43. Desmond Smith, "What is America's secret weapon in the energy crisis? Your television set," *Panorama,* April 1980, p. 30.

44. Ibid., p. 32.

45. F. Leslie Smith, *Perspectives on Radio and Television,* pp. 433-34.

46. Wilson P. Dizard, Jr., *The Coming Information Age: An Overview of Technology, Economics, and Politics* (New York: Longman, 1985), p. 178.

47. Richard Gray, "Implications of the new information technology for democracy," in John W. Alhauser, ed., *Electronic Home News Delivery: Journalistic and Public Policy Implications* (Bloomington, Ind.: School of Journalism and Center for New Communications, Indiana University, 1981), p. 73; and Jay G. Blumer, "Information overload: is there a problem?" in Eberhard Witte, ed., *Human Aspects of Telecommunication* (New York: Springer-Verlag, 1980), pp. 233-35.

48. Neil Hickey, "Goodbye '70s, hello '80s," *TV Guide,* 5 January 1980, p. 12.

49. David Weaver, *Videotex Journalism: Teletext, Viewdata, and the News* (Hillsdale, N.J.: Lawrence Erlbaum Associates, Publishers, 1983).

50. John Naisbitt, *Megatrends: Ten New Directions Transforming Our Lives* (New York: Warner Books, Inc., 1982).

51. William R. Oates, Shailendra Ghorpade, and Jane D. Brown, "Media technology consumers: demographics and psychographics of 'Taffies,' " paper presented to the Mass Communications and Society Division, Association for Education in Journalism and Mass Communications Annual Convention, Norman, Oklahoma, August, 1986.

Part Four

Advertising, public relations, and news and feature syndicates are not media in their own rights, but essential support systems to the larger media whole. They are highly organized products of a complex and corporate communications system, but they have no audiences of their own in the sense that the mass media do. They work with other corporate entities through which they filter to people only indirectly.

It is necessary to differentiate between the advertising and public relations industries. Advertising deals in paid time and space by identified sponsors. It provides the economic incentive for most of the mass media, so that in addition to persuasion it provides the grease that keeps the mass media's wheels turning. It was, at last check, a $100 billion a year enterprise, subject to much debate and discussion for its omnipresence and impacts on individuals and the economy.

Public relations, unlike advertising, offers its persuasion free of cost to the mass media. Its messages are often newsworthy, filling many of the pages of print media and demanding the attention of

Media Support Systems

television's gatekeepers as well. The mass media rely on public relations far more than they like to admit to acquire legitimate news of events and organizations, for which they lack the necessary human and economic resources. Many PR efforts are hype, pizzazz, and pseudoevents, but it often takes these techniques to draw attention to legitimate issues of concern.

The news and feature syndicates serve as influential gatekeepers for the nation and the world. They are our eyes and ears to the outside, and they give voice to those whose messages would otherwise go unheard. Utilizing satellite and computer technology, wire services have shrunk the globe. Meanwhile feature syndicates package hundreds if not thousands of special interest features for omnivorous media.

The pervasiveness of these media support services, their influence upon the media they serve, and their indirect effect upon various audiences, publics, and each individual are the subjects of these three chapters.

Advertising

Commercial persuasion inundates audiences and passers-by.

Commercial persuasion inundates audiences and passers-by.

Introduction

Advertising pays most of the bills for the mass media; this has advantages as well as disadvantages.

A peculiarity of the American mass communications system is that advertising pays most of the bills. It accounts for around two-thirds of the revenue of the print media, and nearly all the revenue for broadcasting. Only public radio and television, and occasional special interest programming such as religious shows, manage to get by without advertising support.

Advertising's influence in the mass media marketplace has both advantages and disadvantages. On the plus side is the fact that the American people get an incredible variety of information, entertainment, and culture at minimal cost. A disadvantage is that nearly all of America's mass communications is heavily overladen with commercial or persuasive messages. Studies indicate that the average American is exposed to hundreds of commercial/advertising messages each day.

Commercial persuasion permeates American mass communications, as opposed to the political persuasion that is evident in most of the world's totalitarian countries. The private sector, seeking profits, has been able to come up with the massive investment necessary to develop America's diversified and sophisticated mass communications networks, while other nations that support mass communications by government financing have had to balance communications expenditures with many other kinds of national interests. Additionally, we should recall that government financing of the mass media inevitably brings with it governmental influence on the news and entertainment content. America has opted for "The Cosby Show" and "Miami Vice," just as the People's Republic of China has saturated its media with propagandistic opera and drama.

All advertising contains both information and persuasion. We shall be concerned here mostly with the persuasive function. The classified advertisements in the daily newspaper are almost pure information. So are most of the big supermarket specials. Their primary intention is to advise readers of the availability of a product—telling where, when, and for how much. The audiences either want it or they don't. Persuasion is another story, as we shall see.

To set one myth to rest at the onset, advertising does not sell, and cannot sell anything. Advertising predisposes, tips the scales if you like. In Latin, *ad vertere* means "to turn the mind toward."

The American Marketing Association (AMA) defines advertising as "any paid form of nonpersonal presentation and promotion of ideas, goods, or services by an identified sponsor." The AMA points out that advertising is a tool of marketing along with the product (and its packaging), price, distribution, and personal selling. Unlike public relations, the subject of our next chapter, advertising is openly and overtly subsidized information and persuasion, and its task is to present and promote far more than merchandise. Increasingly, as the Western world moves into a postindustrial era, advertising is used to promote ideologies and services. And, the AMA reminds us, that promotion is "nonpersonal"—it is "to whom it may concern." If advertising is effective, it is because the audience is receptive to it.

Other definitions stress the idea that advertising is controlled, identifiable information and persuasion by means of mass communications. Such definitions point out another distinction between advertising and other forms of promotion: whoever pays the bills to place the advertising in the media exercises control over how those messages are to appear. When you send a public relations release to the local paper, the editors and reporters become the gatekeepers. They determine whether, and how, your message gets told. But when you pay for the commercial space, you are the gatekeeper, and have far more influence in determining how the message looks, what it says, when and where it will appear, and who is likely to see it. In this sense, commercial advertising is more open and above board than public relations, and much more so than its totalitarian counterpart, propaganda, where persuasion more often than not is masked as news or its sources perverted or hidden.

Most of this chapter will be directed at persuasive advertising, which constitutes the bulk of the nation's $100 billion annual advertising budget. Following a discussion of the development of advertising, we will explore the role of advertising agencies and their function, including media buyers and the creative group. The next two sections will emphasize several advertising strategies that have been influential during the last thirty years and some reasons why advertising works. Because advertising is omnipresent, and is basically persuasive in nature, many people in the industry are concerned about deceptive practices. This concern has prompted efforts at governmental and self-regulations and attempts to make the business more professional. Nonetheless, advertising is still criticized frequently for its effects and costs, which will be the subject of the final section of this chapter.

Historical Development

Advertising is an ancient practice, but most of its early history is undistinguished. Following hundreds of years of usage, advertising began to grow very slowly only after the development of print technology. It has matured only within the last century, under the pressure of the industrial revolution and today's postindustrial economy.

From Wineskins to National Advertisers

During the illiterate times of bygone centuries, advertising was graphic and informational. The wineshopkeeper and the sandalmaker advertised their wares by hanging out a wineskin and a pair of sandals. To natives and travelers alike, these signs told where wine and sandals were sold. This was pure information, which indicated the availability of a product. However, human ingenuity never sleeps, and one day an enterprising wine seller hung out a larger wineskin. Not only was this bigger wineskin visible at a greater distance, thus automatically expanding its audience, but the size itself psychologically said that here was a bigger and better wineshop. In the course of events more people frequented the wineshop with the bigger wineskin. The sandalmakers followed suit. They made pairs of giant sandals and discovered that the sandals became conversation pieces and that folks came to their shops out of curiosity. Orders increased, and the correlation between exposure and sales became established.

The point is that the ordinary wineskin or pair of sandals provided pure information. However, once a bigger wineskin was built, an element of persuasion was injected: bigger is better. Then, as now, advertising presented varying combinations of information and persuasion, plus an important extra: something to catch the audience's attention—the bigger wineskin and the enormous sandals were both big pluses in the competition for attention.

The invention of movable type, as we discussed in chapter 5, led to increasing literacy, and it was natural that advertising, while retaining its graphic elements, should also take advantage of the new medium. Most of the early printers were also booksellers, who took quite naturally to advertising available or forthcoming titles in tracts and flyers. Because literacy was required to read a book, only the literate could read a flyer and only the potential customer could decipher the advertisement. This early example of advertising selectivity is a form that has achieved extraordinary sophistication in the past several decades.

Early advertising was local and selective.

By the early nineteenth century newspapers carried some classified advertising. One of the more intriguing ads appeared in several St. Louis papers during February and March of 1822.

TO
Enterprising Young Men

The subscriber wishes to engage ONE HUNDRED MEN, to ascend the river Missouri to its source, there to be employed for one, two or three years. For particulars enquire of Major Andrew Henry, near the Lead Mines, in the County of Washington (who will ascend with, and command the party) or to the subscriber at St. Louis.

Wm H. Ashley

Some of those who responded to the small ad became the most famous names in Rocky Mountain history, for they were the free trappers of the Rocky Mountain Fur Company who spawned legends and opened the West.

America's national experience has been tied into advertising from the very beginning. Many of the early settlers and investors in the colonies had been attracted by advertisements in the English newspapers. During the Civil War, advertising played a significant role in Northern recruiting, and certainly ads of all sorts were used during both World Wars, for war bonds, refugee relief, and enlistment: "Uncle Sam Wants You!"

Early advertising was local, serving local merchants and a local audience. But the industrial revolution changed this. It brought about the mass production of products, requiring mass purchases far in excess of what the local community could support.

By the 1840s railroads were an effective method for wider distribution of goods to an increasingly mercantilistic and literate society. The industrial revolution concentrated people in cities where the factories were, and which began to produce vast quantities of goods. The railroads hauled people and goods westward to build the nation and establish cities where the goods were both warehoused and sold in the uniquely American innovation, the department store. Printers were in the forefront of this westward movement, and often their simple but durable Washington presses accompanied the wagon trains, so that each new community was sure to have a newspaper, a vehicle for advertising.

In this picture, advertising began to come of age in calling attention to both the goods and the department stores. Given the nationwide competitive and free enterprise economy, the success of one new product prompted the development of almost identical products by competing manufacturers seeking their share of a proven market. The task of the advertiser became that of distinguishing between products that were more or less identical. More often than not, any distinctions lay in the advertisements rather than in the products. To the monied class or carriage trade, these advertisements were not only informational and persuasive, but a satisfying reminder of their financial status.

Ad agencies were in operation before 1850; they introduced the commission system.

The new emphasis on advertising began to take hold before the middle of the century with the establishment of a number of advertising agencies. In actuality, these early agencies were little more than publishers' representatives or space brokers. Often they would contract with a publisher for a certain amount of space in a newspaper, and then sell this space to advertisers for whatever amount the traffic would bear. There were no published rate cards in those days and no certified circulation figures. National advertising was on a happenstance basis, removed from the local familiarity that exercised a certain degree of control between a publisher and a merchant operating in the same town or city.

Early Agencies

In 1869 some order appeared with the publication of Rowell's *American Newspaper Directory,* establishing approximate circulations for the nation's newspapers. For the first time, advertisers had some idea of the coverage they were buying. At about the same time, advertising agencies came into being. Among the first to represent their clients with the newspapers were J. W. Thompson and N. W. Ayer & Son, often writing the copy and laying out the ads. For this they were paid a *commission* or discount by the newspapers, generally 15 percent. Publishers found the commission system useful to encourage agencies to sell space, while leaving themselves free to sell other space at the full price.

By the 1890s, the competition for attention was becoming intense. More and more manufacturers were selling vast quantities of more or less identical products. Purely informational ads no longer were sufficient, and more persuasion had to be injected to catch the customers' attention. Department stores engaged in open and sometimes ruinous price wars.

It had become clear by the 1890s that these nearly identical products had to be made distinct through advertising. Soaps are basically alike, as are beers, toothpastes, razor blades, and shirts.

Selling Techniques

Brand Names The first attempt to distinguish among identical products was through *brand name identification.* Using positive reinforcement on a more or less continual basis, an effort was made to establish the brand name as synonymous with the product or with quality. In England in the 1890s, Pears' soap was the subject of the first mammoth attempt to identify a name with a product. "Pears' Soap" appeared in English magazines and newspapers, on buses, light posts, board fences, and vacant walls until Pears' meant soap. The English did not ask for soap; they asked for Pears'. Coca-Cola, Kleenex, and Xerox are contemporary examples. It has gotten to the point that Kleenex means "facial tissue," and Xerox so describes "photo duplication" that the Xerox Corporation has spent millions of dollars clarifying the distinction, while at the same time firmly planting the name Xerox in consumers' minds.

By the 1890s, ads had to create any distinction among identical products. *Brand names* were established.

Slogans and Jingles While brand names were an effective advertising technique, something more was required to firmly establish product difference in the public mind. From the realm of political campaigning, advertisers borrowed the *slogan,* a catchy capsule summary of the product. Pears' Soap updated itself instantly with a catchy "Good morning, have you used Pears' today?" And it was with this greeting that at least two generations of English greeted one another every morning—marvelously effective, and superb word of mouth.

Slogans came and went, and it was not until the 1920s and the advent of broadcasting that new developments (and sounds) began to appear. The slogan gave way to the *jingle,* which was little more than a slogan set to music and rhyme. "Winston tastes good, like a cigarette should" is an evident and ungrammatical example. Coca-Cola's "It's the Real Thing" is almost a musical classic. It reminds us that often commercials are better than the surrounding programming, an event some suspect is not entirely by accident.

Slogans were catchy capsule summaries of the product's merits. *Jingles* were slogans set to music and rhyme.

The use of brand names and their accompanying slogans and jingles has carried on into the present. They continue to be successfully used even after more sophisticated persuasion techniques have been developed. Often brand names, slogans, and jingles are used in conjunction with other techniques, pyramiding on each other for maximum effect.

The Early Twentieth Century

Under attack, advertisers tried to enhance their "professional" image.

In the philosophical climate of the early twentieth century, advertisers were facing attacks leveled by muckrakers and concerned citizens. The industry responded with self-protective calls for professionalism and high ethical standards.[1] However, there was little change in advertising ethics from the turn of the century until after World War II.

Advertising agencies grew and prospered. The advent of radio introduced a new element into mass persuasion, and increasing complexities demanded expertise that few product organizations possessed. Creativity and advertisement testing became part of the agencies' stock-in-trade. Techniques for measuring the readership of magazine advertisements were superimposed on magazine circulation figures. These still exist in the form of the *Starch Reports,* which gives a clue to the exposure and drawing power of a print advertisement, taking into consideration such things as appeal of the artwork, layout, and copy readability.

The method of advertisement testing in the beginning was remarkably simple, and variations of it are still in use. Two different ads on the same subject were placed in a *split run* in a newspaper; that is, half of the edition carried one ad and the other half carried the other ad. The ads asked readers to return a coupon for a product at a reduced price, and the relative appeal of the two ads was gauged on the basis of how many people responded to each ad.

Following World War II a new spirit invaded advertising. Interest in psychological warfare and propaganda had injected heavy doses of the social sciences, particularly psychology and sociology, into the persuasion field. Computers, which first had a practical application during the war, greatly facilitated public opinion measurement, polling, and testing, and gave birth to more accurate ratings. Further, the lapse in civilian production during the war years had created a demand for consumer goods. As America's corporations turned their enormous capacity to peacetime manufacture, they generated gigantic pressure to move the goods. Advertising had a renaissance generated by new techniques, new products, and new audiences. Returning servicemen, reunited families, and a baby boom created an unparalleled demand for almost everything. Also, the genesis of television in 1948, only three years after the war, added new dimensions to advertising, as it did to most other phases of life.

Following World War II, advertising adopted heavy doses of social science.

Postwar Growth

Advertising was a $4 billion industry in 1947; in 1970 it reached $20 billion; by 1980 it passed the $50 billion mark; and only six years later it passed the magic $100 billion mark. These totals do not include the additional billions of dollars spent for packaging products, which is difficult if not impossible to separate from point-of-sale advertising. More than a half million people are employed in the various facets of advertising within the United States—in agencies, in the internal advertising departments of both industrial and retail establishments, and in the media.

Advertising, a $100 billion a year business, employs more than a half million people.

Newspapers still take the greatest share of the national advertising dollar, between 26 and 30 percent since 1960. Television takes between 21 and 22 percent, up from 13 percent in 1960. (See figure 11.1.) However, it should be noted that the newspapers' share is spread out among 1,687 daily newspapers and 7,600 weeklies, while television's percentage is concentrated among some 900 commercial stations and the three networks,

Figure 11.1
Advertising expenditures,
1986.
Source: McCann-Erikson, Inc.
From *Facts About
Newspapers '87, American
Newspaper Publishers
Association,* April 1987. By
permission.

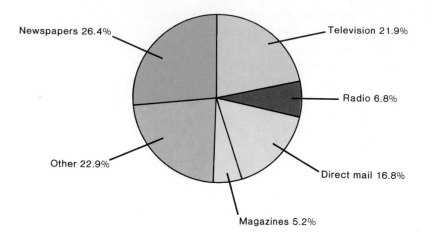

Newspapers 26.4%

Television 21.9%

Radio 6.8%

Direct mail 16.8%

Magazines 5.2%

Other 22.9%

Newspapers still get the
biggest share of the ad
dollar—but not by much.

plus cable. Further, newspapers tend to serve both national and local advertisers in about equal numbers, while network television concentrates on national advertising. Magazines handle about 5.2 percent of the national advertising volume, down from 10 percent in 1960. Radio receives 6.8 percent, about the same slice of the pie it earned in 1960. The remaining 39 percent of advertising goes for such things as direct mail, which gets a whopping 16.8 percent of the ad dollar, along with billboards, business and farm publications, and other miscellaneous outlets.

Direct-mail advertisements are designed to stimulate responses by return mail. However, the full impact of direct mail is often difficult to gauge because the persuaded consumer often has the alternative of purchasing the products at retail outlets. Attempts have been made to sell almost every type of product and service by direct mail—from insurance to antique autos. Some companies, such as the Spiegel mail-order house in Chicago, base their entire advertising plan on direct mail.

There are a half million people directly employed in the advertising business in America, and perhaps as many as a million more whose jobs depend directly upon advertising. In the first group are the 100,000 or so employed in 8,000 advertising agencies, plus those working directly with product manufacturers, in mass media advertising staffs, outdoor and direct-mail advertising firms, retail and wholesale outlets, and specialty companies. Among those million or so whose livelihoods depend upon advertising are all the other employees of commercial mass media, plus printers, clerical and sales staffs, paper manufacturers, and sign painters.

Advertising Agencies

America's advertising agencies vary from local one-person shops employing a principal and a secretary in smaller communities, to the creative boutiques of Madison Avenue employing unique individuals specializing in unusual effects, to the largest, highly organized, and even institutionalized variety.

Young & Rubican has been the nation's largest advertising agency, doing about $3.6 billion worth of business per year. In 1986 three other firms which had 10,000 employees and $5 billion in annual billings decided to join forces and become number one: Batten, Barton, Durstine & Osborne (BBDO) International; the Doyle Dane Bernbach Group; and Needham Harper Worldwide. Other billion-dollar agencies are the J. Walter Thompson and McCann-Erickson companies. Like other properties in mass media, successful advertising agencies are being bought and sold at a pace never before seen.

The major agencies tend to concentrate where the money is—on national campaigns for the nation's major clients—leaving local advertising to the smaller local agencies or to the individual advertiser, who is often served creatively by the local newspaper or broadcaster. Both newspapers and broadcasters provide ancillary services, either at a minimal fee or free, to local advertisers to help lay out ads or produce commercials.

Major agencies, doing a billion dollars or more each year in business, concentrate on national campaigns.

A persistent controversy in the advertising world concerns whether clients are better off with a small boutique where they can receive the creative attention of the firm's principals, or with the highly organized super-agency whose wealth commands superior talent and whose depth of services and personnel provide across-the-board coverage.

Agencies as Mediators

The advertising agency is a mediator between the advertisers (manufacturers or distributors) and the mass media (print or broadcast). To be successful, the agency must serve both interests.

Ad agencies are mediators between the advertisers and the news media. They tend to specialize.

It should be noted that agencies have certain client preferences. Some deal in automotive accounts almost exclusively; others in food and beverages, electronics, clothing and fashion, travel, or proprietary drugs or cosmetics. Through this kind of specialization, agencies are able to become intimately acquainted with the particular facets of their clients' distinctive pricing, marketing, and manufacturing problems, all of which have a bearing on the creation and placement of ads. The specialty situation is reinforced because an agency's success with a campaign for one client is quite likely to lead product competitors to that agency. Some agencies tend to favor print as their basic medium, while others concentrate on television or radio. As communications technology grows more complex, this specialization breeds a certain expertise in the medium that advertisers are likely to find attractive. Thus, the trend toward specialization that has been noted in the media has its parallel quite naturally in advertising, which serves the media.

Despite the increasing trend toward specialization, either in client type or media, all of the very big agencies are broadly based, serving a variety of clients in many lines of endeavor and maintaining the kind of in-house expertise in both product classification and media usage needed to meet any campaign. Quite obviously, only the very largest agencies can afford this kind of diversification.

Ad agency illustrators work from rough sketches to finished advertisements, taking special care to match colors and tones to give ads the desired effect.

However, size and institutionalization lead to a certain inertia and resistance to change. The smaller boutiques are neither so wealthy nor so ponderous and therefore have a lot more flexibility to keep up with rapidly changing approaches to organized persuasion.

Agency Functions

All agencies tend to have several functions: client liaison, creativity, production, placement, and housekeeping.

Because agencies serve both the client and the media, all embrace certain functions: client liaison, creativity, production, placement, and housekeeping. In a very large agency the *account executives* (AEs), often working under account supervisors, maintain a constant liaison with clients. They interpret the clients' problems to the agency's creative people, and then in turn explain the developed campaign to the clients. Since large agencies generally deal with large clients, the AEs usually deal with the clients' advertising managers or advertising departments—experts dealing with experts, each interpreting the other to their respective organization. Account executives are personality people; they are organizers who know how to work with others; they are the gray-flannel suiters.

The creative group is composed of art directors and artists, copy supervisors and copywriters, and layout people and graphics experts. Using a team approach, they develop the print ads, write the copy, determine how to use illustrations or photographs, decide on black-and-white or color, and make up the *storyboards* for broadcast commercials, which are static graphic depictions of what eventually becomes a live, moving commercial.

Production turns the layouts or storyboards into actual commercials in the form of matrices or camera-ready proofs for the press; into sets of expensive colorplates for magazines; and into tape, videotape, or film for radio and television.

Meanwhile, the media people select the media to be used by taking the budget and the composition of the audience into consideration and by choosing between and among communications vehicles: which newspapers, magazines, and radio stations; how much television on which stations or networks; or whether to use billboards or direct mail. They consult Standard Rate and Data Service (SRDS) and talk to the various national "reps" of groups of magazines, radio stations, newspapers, or broadcasters, always keeping in mind the ratings and *cpm*—the cost involved in reaching each one thousand members of the audience. One irony is that the agency media buyer who spends millions of dollars of client money annually is generally a woman just out of college, and the people she deals with most are highly paid national reps with years of media experience behind them, backed by reams of statistical data as to why their particular group is best in cost per thousand, in exposure, in purchasing power, or a variety of other complex demographic factors. Someone once said that anything can be proven with statistics, and the national reps do it daily.[2]

The housekeeping function of major agencies includes such things as accounting (a key function), research to develop the facts to be used in planning a campaign, personnel to take care of employees (sometimes thousands), and administration.

Some agencies—attracted by the huge commissions generated by major national automotive, food and beverage, drug, or cosmetic accounts—have tended to throw in a variety of other services in an attempt to attract and hold clients. These include public relations, public opinion sampling, market research, and promotional activities. The results of these extra services have not been spectacular. Since they are generally thrown in as an extra service, the agency tends to spend as little on them as possible. Thus, the job done is not generally of the highest caliber. "You get what you pay for" in advertising as elsewhere. Further, the lack of performance has tended to give many of these specialties a bad reputation among clients, particularly with regard to public relations.

Some agencies throw in a variety of other services to attract and hold clients.

A large advertising agency is a complex operation. Some of the major agencies, for example, have review boards composed of major executives who play devil's advocate with any planned campaign before it goes to the client. At the review board's request the client liaison and creative teams are forced to defend their handiwork before a critical panel of peers. The rationale for having a review board is that it is better to find the weak points of a campaign before the client or public does, millions of dollars later.

An integrated national campaign involves network television, which is supplemented by local television in certain markets. It involves major spreads in a dozen or so carefully chosen magazines. Also included are national newspaper coverage and supplementary radio, in both the major markets and the very minor ones that are reached effectively only by radio. Billboards on highways across the country and a sophisticated direct mail campaign to certain selected demographic audiences are utilized. The problems of a traffic department are mind-boggling, especially those of co-ordination and scheduling, which involve tying the media presentation together into a single theme, determining presentation priorities, coordinating with point-of-sale displays, tying in with public relations and promotional activities, and harnessing the energies of distributors and dealers and sub-sidiary agencies in fifty states and in a number of foreign countries with language differences. However, the big three automobile makers—Ford, General Motors, and Chrysler—go through orchestrated national cam-paigns every year when they introduce their new models. Most of the other manufacturers of consumer products contract similar campaigns whenever they launch a new product, which is often.

It is important to remember that whether the agency is a one-person shop, a creative boutique, or a major international agency, it must deal with both client and media in performing creative, production, placement, li-aison, and housekeeping functions. A one-person agency gathers all these functions together under one hat. The owner spends the day running from a meeting with a client to the drawing board to a conference with television sales reps, overseeing the shooting of a commercial on the side, hiring the models and photo crew, and keeping up with the bookkeeping at night. A big agency apportions these functions among a platoon of overlapping ex-perts. In either case, all such activities are performed under intensely com-petitive conditions. Madison Avenue is a jungle—and an ulcerous jungle at that—as agencies seek to steal each other's accounts and to hire each oth-er's personnel, and account supervisors leave to open their own boutiques, taking the agency's clients and creative people with them.

Ad agencies are intensely competitive.

Advertising Strategies

After World War II, advertising agencies moved into the public spotlight and became increasingly competitive. The 1950s were the years when Mad-ison Avenue became a household street and advertising people became ste-reotyped as manipulators of public opinion.[3] A rash of books appeared on advertising, among them Vance Packard's *The Hidden Persuaders* and Martin Mayer's *Madison Avenue, U.S.A.*[4] Both were popularizations born of public fascination with the manipulative techniques of applied social psychology.

The first manipulative technique to capture public attention was the *unique selling proposition* (usp) originated by the agency of Ted Bates & Company. The usp took as its starting point a fact that had been recognized since the dawn of the industrial revolution: that there was no real product differentiation among the various mass-produced products of competing manufacturers, so advertising must establish whatever difference there is to make the public buy one product instead of another. The unique selling proposition was based upon a characteristic of the product that could be dramatized, exploited, and made synonymous with the product on behalf of the manufacturer, *even though all other products shared the same characteristic.*

Nonexistent differences were exploited by advertising, and through the sheer power of public exposure were made to appear unique. For example, Bates originated the slogan "Cleans Your Breath While It Cleans Your Teeth" for Colgate toothpaste. This was usp, and Bates built an entire campaign out of it. The slogan was a natural selling device implying—but never coming right out and claiming—that Colgate was a superior toothpaste. Of course, even a cursory examination of the proposition suggests that all toothpastes clean the breath while they clean the teeth. There was nothing unique about Colgate, except that Bates had hit upon the slogan. By exploiting it he had effectively prevented any other toothpaste manufacturer from utilizing it. Even an attempt on the part of another manufacturer's agency to utilize a similar slogan would actually reinforce Bates's claim for Colgate. The usp is central to the advertisement.

Another Bates usp for Schlitz was that its bottles were "washed with live steam." This had a peculiar appeal in the 1950s when sanitary America was concerned about the hygiene of reusable glass beer bottles. Actually, the slogan was a redundancy (is there anything but live steam?), because all brewers washed their bottles with steam. However, Bates, by exploiting this point, had usurped it for Schlitz.

The next technique of pseudoscience applied to advertising was David Ogilvy's *brand image*. Ogilvy, who started his career as a chef at the Majestic Hotel in Paris, is testimony to the fact that an energetic imagination can bring success in many fields; he is also head of Ogilvy and Mather, one of the nation's largest agencies. As an aside, he claims he has had only nine ideas in his life, but they were apparently good ones. One of them was the brand image concept. Like usp, it recognized that advertising had to overcome the lack of essential differentiation among products. But it goes one step beyond usp and, in Ogilvy's words, "gives your product a first-class ticket through life."[5]

Brand image is based on snob appeal, to create an image for a product that puts it a little above the competition and hence makes it a little more desirable. The thought was that people would transfer that image to themselves as they bought the product. For example, Ogilvy hired Baron von

Unique Selling Proposition

Unique selling proposition (usp) dramatizes and exploits a product's basic characteristics.

Brand Image

Brand image used snob appeal to promote a product.

Wrangel for Hathaway shirts. The baron wore a black eye patch, the remnant of some romantic duel or exotic accident, and he exuded class. He was always fondling an exquisite antique ship model, or a beautifully wrought silver chased over/under shotgun costing thousands of dollars. And he always wore a Hathaway shirt. If you bought a Hathaway, you too would exude class, and all your neighbors would know it. You too would have acquired (presumably by osmosis) some of the baron's breeding, taste, and distinction.

Another of Ogilvy's creations was Commander Whitehead for Schweppes. Ogilvy tended to deal in personalities. Anyway, the commander's English accent simply dripped culture as he extolled for Schweppes' "schweppervescence," for "only pennies more"—that first class ticket, snob appeal. It worked.[6]

Motivational Research

Motivational research probes our inner selves; then advertisers play on our weaknesses.

The most extreme form of advertising's romance with the social sciences in the 1950s was Ernest Dichter's *motivational research (mr)*. Highly Freudian in its approach, mr maintains that people buy things for hidden reasons unknown even to themselves, and that these hidden reasons are more often than not sexual in origin. Through depth interviews mr sought to explore people's hidden motivations as to why they buy and then, utilizing these data, to build an advertising campaign. Psychologists conducted depth interviews and projective psychological tests, designed to induce interviewees to project their real feelings when they interpreted pictures shown to them or when they completed sentences. According to Dichter, the data gave marketers valuable insights for designing and selling products. Many advertisers and manufacturers remained highly skeptical.

Classic examples of Dichter's mr in action were campaigns for Chrysler Motor Company, Ronson lighters, and Ajax cleanser. When working with Chrysler, Dichter inferred through mr that the average American male, who at that time was the principal decisionmaker in automotive purchases, remained with his wife but secretly coveted a mistress. By equating a sedan with a wife, and a convertible with a mistress, Dichter concluded that a hardtop "convertible" would satisfy the inner needs of a man for glamour and romance, and at the same time, it would be safe and dependable like a wife. Once it was introduced, the hardtop convertible made history with its acceptance by an avid market.[7] It is unclear whether the market was actually motivated by the forces Dichter singled out, but the product was highly successful

Later, Dichter claimed similar success with Ronson lighters. He built his campaign on huge color reproductions in magazines and billboards of the lighter's flame, which in his view was an enormous phallic symbol appealing to men and women alike: men through wishful thinking, women from lust. Dichter went on to construct a television campaign for Ajax cleanser that utilized a white knight carrying a great lance astride a white charger and galloping down suburban streets, paying no attention to the traffic laws. The lance, representing a large phallic symbol, was designed to stimulate housewives of Hometown, America. Again, the product sales were impressive, but whether purchasers were motivated by Freudian or other factors was never really clear.

In the wake of Dichter's emphasis on psychological testing of markets, the late 1950s were also a time of considerable public consternation about *subliminal advertising*. According to its handful of supporters and legions of critics, subliminal advertising motivates consumers' unconscious minds. Even though the persuasive stimuli would not be picked up by the eye or ear, they would somehow be picked up and registered in the brain, where they would take effect. For instance, in one of the first efforts to demonstrate the power of subliminal suggestion, advertising researcher Jim Vicary inserted a few single frames of "Coca-Cola" or "Eat Popcorn" into movies at different intervals. A single frame would pass so quickly as a movie was viewed that the viewers' eyes would not be aware of having seen it, although their brains supposedly took note and motivated them to rush to the concession stands. Unfortunately—or fortunately, depending on one's point of view—the Vicary research project was badly flawed, in both design and execution, and the results he claims are highly suspect.

This is not to say that subliminal stimulation does not work. There is plenty of evidence, mustered through sophisticated equipment such as EEGs or psychogalvanometers (lie detectors), that the human brain picks up and reacts to sights, sounds, and smells that were never consciously experienced. However, this finding does not necessarily give substance to the claims

Subliminal Advertising

Subliminal advertising supposedly motivates consumers' unconscious minds.

Eye-catching and pulse-
quickening ads are
breaking many of
Madison Avenue's
taboos.

by many advertising researchers—and complaints by its critics—about the power of subliminal advertising. There is a substantial difference between having something merely register in the mind, and that "something" causing someone to go out and purchase a product or behave any differently than he or she would otherwise act. Despite decades of talk, and perhaps millions of dollars of investment by some advertisers, no one has yet proven that subliminal advertising is anywhere near as effective as old-fashioned, straightforward persuasion.

No one has yet proven that subliminal advertising is as effective as old-fashioned, straightforward persuasion.

If subliminal advertising has been successful for anyone, it has been Wilson Bryan Key. He has written three best-selling books lambasting the advertising industry for peppering its ads with embeds or hidden symbols, and faulting the film, popular music and magazine industries for scaring us to death and urging everything from devil worship to homosexuality. He has accused Ritz of baking the word "sex" into its crackers and Howard Johnson's restaurants for embedding orgies and beastiality in its placemats and menus. The titles and subtitles of his books are revealing: *Media Sex-ploitation; Subliminal Seduction: Ad Media's Manipulation of a Not So Innocent America;* and *The Clam-Plate Orgy And Other Subliminal Techniques for Manipulating Your Behavior.*[8]

Mainstream researchers in marketing and advertising laugh at Key's criticisms. Most of them at first blush claim that Key is imagining things, the way children see entire worlds passing by their eyes when they gaze at clouds. When pushed, however, many will admit that some advertisers and popular artists such as musicians and movie makers, seeking a competitive advantage, resort to using the types of stimuli Key discusses.

The Federal Communications Commission has concluded that the use of subliminal perception is inconsistent with the obligations of a broadcast license, and broadcasts employing such techniques are contrary to public interest. The FCC continued with the statement that, whether effective or not, such broadcasts are clearly intended to be deceptive. Psychologists and advertisers remain divided over whether such techniques are effective, so regulators have tended to ignore them. (In the mid-1980s a flurry of concern over devil worship in popular music led to public hearings in Washington, D.C., and some efforts to regulate at the local level. Meanwhile, even *Psychology Today* is cluttered with ads promising everything from improved sex lives to painless foreign language training brought about by subliminal stimuli, indicating that the jury is still out.)

Unemotional researchers insist there is no cause for alarm, that these hidden messages are totally ineffective at stimulating consumer behavior, let alone brainwashing a nation. The myth of subliminal stimulation persists in today's media environment probably because there remains a substantial number of folks who are intrigued by and perhaps even enjoy the idea of being led helplessly around by the nose, ear, eye, brain, or what have you.

Since the late 1960s, a new concept of specialization has come to national advertising, called *positioning*. A standard definition says that positioning consists of segmenting a market by (a) creating a new product to meet the needs of a selective group, or by (b) using a distinctive advertising appeal to make an existing product meet the needs of a specialized group, without making changes in the physical product.[9] Our interest lies more in the second half of the definition than in the first.

Earlier advertising was based on the demographic concept of the *lcd,* appealing to the lowest common denominator in the hope that differentiation in the advertising, regardless of product similarity, would attract a vague maximum response. This was really a sort of hit-or-miss approach. However, advertisers and their agencies watched as successful media specialization occurred—particularly in the radio and magazine fields where the various media pinpointed their appeal to specific audiences—and as a more

Wilson Bryan Key has sold a great many books in which he warns people about subliminal seduction. Mainstream researchers and advertisers don't buy Key's arguments.

Positioning

Through careful *positioning,* ads for nearly identical products carve out individual markets.

sophisticated advertising concept evolved. It was possible, they reasoned, to appeal to specific audiences even within the broad mass that newspapers or, especially, television represented.

Positioning recognized the inherent differences in people as individuals and the inherent impossibility of any product capturing the entire potential market. Advertisers were willing to settle for a share of the market, yielding a part of the pie to their competition and continually seeking to increase the size of their own slice.

Over the past several decades, there have been some classic examples of the repositioning of products through advertising. Back in the 1960s Volkswagen did not change its appearance, but modified its image from being an ugly German import to a cute "bug." Sterling Drug company changed Lysol from a disinfectant that was selling poorly in drugstores to a best-selling general purpose household cleanser/disinfectant marketed through grocery stores. Clorox, formerly marketed as a liquid bleach, has been successfully repositioned as a general disinfectant. Johnson and Johnson's Baby Oil, Baby Shampoo, and Baby Powder have been repositioned for adults who have "discovered" the respective products' versatility as make-up remover, a gentle shampoo for athletes and others who have to wash their hair daily, and relief for sweaty grown-ups who want to avoid chaffing and keep smelling sweet. California raisins have become a nutritional fruit snack that replaces the hollow calories of candy. Alka-Seltzer, the long-reigning remedy for overindulgence by slobs who couldn't believe they ate the whole thing, has repositioned itself as a cure for the stress experienced by health-conscious and success-driven yuppies. Many other products and companies have been saved from oblivion by recreating their images, if not their product lines.[10]

The positioning of specialized advertising is based on both *demographics* and *psychographics*. Demographically, it relies upon the computer or specialized mailing lists to identify and break out small, apparently disparate audiences of similar interest, or basic make-up. For instance, direct-mail campaigns in this day and age rely upon zip codes to give advertisers a "living picture" of potential customers. The zip code tells what type of a neighborhood people live in, what kind of car they are most likely to drive, where they will vacation, and whether they hold a white- or blue-collar job. Those data, and the inferences to be drawn from them, are more sophisticated than the old standard SES data—the socioeconomic statistics such as age, sex, income, education, and religion that people are likely to reveal to the Census Bureau or on surveys. In a sense, the "living pictures" contain psychological as well as sociological data.[11] Some advertisers use the term *geodemographics* to describe this expanded version of traditional demographic research.

Positioning is based on both *demographics* and *psychographics*.

Psychographics Such variables as audiences' values, moods, attitudes, opinions, life-style interests, and some other motivations are more subtle than those revealed in standard demographic surveys. When advertisers have a combination of these psychographic and demographic data, they are in a position to narrow the focus of their advertising campaigns and position the products more effectively in the marketplace. We cannot discuss psychographics without discussing positioning, for the two have become highly intertwined.

Psychographics has become the latest example of Madison Avenue's fascination with social science. As seen in the foregoing discussion of positioning, some advertisers have been making use of psychographic concepts for decades—even before the term "psychographics" entered the vernacular. However, the psychological principles on which much of that earlier advertising was based tended to be behavioristic or Freudian in nature. Many advertisers seemed to operate on the assumption that customers were somewhat sluggish, malleable, driven by needs and desires they barely understood (many of them sexual in origin), and in need of simplistic messages for many of life's complexities. Naturally, if consumers were depicted in this way, the solution was to psychologically condition them by incessantly lambasting them with insipid and banal messages. Unfortunately, even if the advertisers were employing fairly primitive psychological principles, many products advertised in this vein sold very well. (Think about the commercials or advertisements that you find the most offensive, and ask whether they might not have been created with this type of a consumer in mind. Then admit that the reason the commercials/advertisements were run so often was that they must have worked.)

Fortunately for most of us, advertisers' knowledge of psychological principles has been elevated lately. Since the 1970s, more sophisticated concepts of psychology have been widely applied to marketing and advertising campaigns. By and large, today's advertisers do not assume that the consumer is a passive sluggard, but an active, dynamic, and even obstinate participant in the marketing process. According to most advertising researchers, today's consumers increasingly exercise their free will, accepting or rejecting—usually for rational reasons—persuasive messages to suit their own needs, desires, moods, whims, opinions, attitudes, beliefs, and values.

Researchers took note of the fact that baby boomers and antiestablishment youth of the 1960s would be the primary consumers starting in the 1970s. The old acquisitive, conformist values pandered to by pre-1970s advertisers would have to give way to "individualism, experimentalism, direct experience, naturalism, appreciation of diversity, and taste," one researcher noted. "The central problem in advertising will be how to sell to values increasingly geared to processes, not things. Sales appeals directed toward the values of individualism, experimentalism, person-centeredness, direct experience, and some forms of pleasure and escape will need to tap

Psychographics has become the latest example of Madison Avenue's fascination with social science.

Today, most advertisers view the consumer as active, dynamic, and even obstinate.

Many of the detergents on supermarket shelves have been packaged and marketed in accordance with psychographic research. There's little difference among the products, but a great deal of difference among the packaging.

intangibles—human relationships, feelings, dreams, and hopes—rather than tangible things or explicit actions."[12] To understand and work with such consumer characteristics, advertisers needed sophisticated insights that came from research studies; creativity and gut instincts would no longer suffice.

Exhaustive research has gone into psychographics, and with interesting results.[13] Needham, Harper and Steers interviewed 3,288 consumers, asking each of them some 700 questions about their life-styles, basic attitudes toward life, personal preferences and habits, social and political views, and about the different kinds of products each one used. Tossing all this information into the computer, the company came up with ten different consumer prototypes. Five were male and five were female. Among them were Scott, the Successful Professional; Thelma, the Old-Fashioned Traditionalist; Dale, the Devoted Family Man; Cathy, the Contented Housewife; and so on. Another study, a 1970 one by Benton and Bowles, consisted of interviews with 2,000 housewives. Researchers emerged with six categories of consumers, including such prototypes as Outgoing Optimists, Conscientious Vigilants, and so on.

Of what value is this kind of research? One advertiser, working for Procter & Gamble, divided housewives into categories or "attitude groups" based on what they were looking for in a detergent. There were Practical Women, Convenience-Oriented Women, the Economy-Minded, the Traditionalists (known as "Mrs. Tide"), and the Experts. The Experts, for instance, were people who, in effect, set up a laboratory in the basement. They had soaps, sprays, hot-water and cold-water detergents, and bleaches.

Researchers imagined them rolling up their sleeves and playing the mad-scientist role each washday. Based on these insights, the advertiser concluded that the heaviest users of detergent were young middle-class women with large families, women who cared about their families but refused to be chained to the laundry room. Given that psychographic stereotype, the advertiser recommended that Procter & Gamble reposition Cheer, an established but poorly selling detergent, as an easy-to-use all-temperature product. A decade after Procter & Gamble took the advertiser's advice, Cheer had gone from seventh place to second among all detergents.

The most extensive use of this type of research has been thoroughly documented in an *Atlantic* magazine cover story: "Beyond Demographics: How Madison Ave. knows who you are and what you want."[14] For all its successes, life-style and values advertising has some problems. Take the positioning of soap and beer marketing, for instance. Some soap users are concerned with antiseptic cleanliness. Dial goes after this market, stressing its fresh smell, its cleansing qualities, and its ability to banish the curse of body odor. In so doing, it reaches out to the tens of millions of television viewers and magazine readers who are concerned, if not obsessed, with cleanliness and the fear of offending; it offers a haven to the insecure. Theoretically, it locks up this slice of the market. There are others in this vast audience, however, who are less concerned with hygiene than they are with their complexions. Dove emphasizes its one-quarter cleansing cream content, and gives hope to the consumers seeking to perpetuate the peaches-and-cream complexion of their youth.

> Life-style and values advertising has had some success, but it also has had its problems.

Meanwhile, one beer directs its appeal to the swingers. It is always being poured in happy surroundings, with people laughing, dancing, and singing. To the party people in the audience, it is a natural; it is their beer. To the wallflowers sitting at home, this beer represents companionship and gaiety, and holds the promise of popularity by association. Another beer appeals to the robust, active male. It is a he-man beer, consumed by active, energetic men who are working the rigs, running the trap line, hauling the nets, saving endangered river barges. Women are noticeably absent in the commercials; one sexist commercial with the boys off on a camping trip goes so far as to proclaim that "It doesn't get any better than this." However, 97-pound weaklings buy the beer in hope that some of the manliness will rub off on them. In effect, the positioning of such advertising directs itself to the mass within the mass.

But there are problems with such advertising. For an oversimplified example: what happens to the woman who cares both about personal hygiene and her complexion? Does she try both Dial and Dove, or use one while feeling guilty about neglecting her other "need," or in her confusion turn to another brand? What about the poor guy who wants to be both a he-man and a swinger? Does he become schizophrenic in his indecision about beer, or turn to hard liquor instead?

If the audience segments being appealed to are sufficiently large, there is no real problem. But stereotyping to a small, specific audience runs the risk of putting the product in a box from which it cannot escape, no matter how useful it might be for others. It is somewhat like good actors getting typecast in cowboy or villain roles, so they will never play Shakespeare.

Regardless of the problems posed by overkill and stereotyping, positioning appears to be the wave of the future. Its social scientific foundations are appealing in this day and age; its insights into consumers offer more specific guidance for advertisers than intuition and creativity alone. But positioning is a dangerous game fraught with the perils of overkill, which is expensive, and stereotyping, which places the product in a cage from which it cannot escape. Yet, the rationale of sharing the market is hard to destroy, and the presumed cohesiveness of the mass audience on which the lowest common denominator is based is hard to maintain.

Institutional Advertising

Institutional advertising takes on the image-building task of *public relations*.

Recent environmental concern points up another facet of advertising termed *institutional advertising*—sometimes called *public relations advertising*—which has been in existence for some time but now enjoys considerable emphasis. Institutional advertising does not seek to sell a product; rather it seeks to sell an idea or an institution. Many of America's major corporations, faced recently with a loss of credibility and public trust, as well as attacks from both consumer groups and the environmentalists, have taken to blowing their own horns in the mass media. Print and broadcast media are filled with their protestations of good citizenship—their environmental concern, their research for a rosier future, their emphasis on equal opportunity employment, and their stress on minority rights and in-house racial and sexual equality.

The nation's utilities, operating as controlled monopolies, have been in the forefront of institutional advertising for years. The "Bell Telephone Hour" of symphonic music, first on radio and later on television, was essentially public service, institutional advertising, designed to paint AT&T as a patron of culture. Gas and electric companies today focus ironically on energy conservation, seeking to slow rather than promote growth. Their generating capacity is outdistanced by demand, resulting in shortages and brownouts with an accompanying acceleration in public dissatisfaction.

The question arises, of course, as to why a utility operating from a monopoly position needs to advertise at all. The answer is simple. Private utilities do business on public sufferance; they are reliant upon public goodwill to maintain their privileged position. There is a considerable body of sentiment, both social and legislative, that holds that utilities should be publicly owned and operated, as many municipal utilities already are. Consequently, the volume of institutional advertising from the nation's utilities has traditionally been considerable.

Another curious form of institutional or "issue" advertising is being done by private firms with an ax to grind, or who are desperate to change their image. The Dow Chemical Company ran a series of commercials showing new college graduates saying such things as, "In two days I walk into a Dow laboratory and begin work on new ways to help grow more and better grain for those kids who so desperately need it. I can't wait." In case we might wonder why a company that has been cutting back its labor force would run such a campaign, we have to recognize that Dow's target audience was not the new graduate, but the 40-year-old consumers, business managers, and even legislators who recall Dow as the controversial company that produced Agent Orange during the Vietnam War and was the subject of nationwide college protests.

In other institutional "image" ads, Mobil Oil Company has taken strong stands on the responsibility of the news media; Seagrams became embroiled in controversy when networks refused to air its commercials showing that twelve ounces of beer, five ounces of wine, and one and one-fourth ounce of whisky contain the same amount of alcohol; the R. J. Reynolds Tobacco Company's magazine and newspaper ads maintained that being exposed to someone else's cigarette smoke is not as harmful as many people have maintained, and appealed to legislators and business establishments not to pass blanket "no smoking" policies.[15]

The Impact of Advertising

It is necessary, as far as advertising is concerned, to separate inconsequentials from matters of substance. Inconsequential items include beer and toothpaste, even motor cars, and a whole range of consumer goods that the general public tends to recognize as being basically alike. Even though consumers realize that it makes no appreciable difference which product they choose, they may have preferences. In such matters advertising may have an effect, to reinforce brand allegiance or to entice consumers to experiment with a nearly identical product. Matters of consequence, however, involve beliefs and attitudes ingrained in the person. These do not readily change, and advertising is relatively impotent against them.

Advertising, for example, has not proven particularly successful in political campaigning, which is perceived by the public as being a matter of substance affecting the welfare and future livelihood of each individual. It has been noted that advertising helps create name recognition, and may push an unknown candidate into the forefront during primary elections. However, in the general elections, after an extended period of campaign rhetoric, advertising, public appearances, and news reports, it is another story. While there have been some very well-financed political campaigns that have resulted in victory, we do not hear as much about the expensive campaigns that ended in failure. The results of political campaigns seem to hinge on factors other than advertising, and the best that can be said for political advertising is that it usually does no harm.

Advertising is great at focusing our attention on a new product or politician, but it has definite limitations when it comes to changing opinions on matters of substance.

'RELAX...IT'S ONLY THE REAGAN RE-ELECTION AD...'

Related to the consideration of inconsequentials and matters of substance in advertising is the fact that advertising is unsurpassed as a device for exposure—to call attention to, to introduce a product, concept, idea, or candidate. Once exposure is achieved, the purchase, acceptance, or election are in other undetermined hands hidden in the complexities of public opinion formation and change. Advertising is essential for product introduction. Once introduced, some will buy out of curiosity, but ultimately, success depends on the quality of the product. As they say, "Fool me once, shame on you; fool me twice, shame on me."

Advertising is unsurpassed as a device for exposure, for calling people's attention to something.

The Psychology of Advertising

In *Motivation and Personality,* psychologist Abraham Maslow defined a seven-stage hierarchy of needs. The needs progress from basic biological demands—such as hunger and thirst—to a complex psychological stage of self-actualization—a need to find self-fulfillment and realize one's potential. According to Maslow, needs low in the hierarchy must be at least partially satisfied before higher needs can become important sources of motivation.[16] For example, a missionary knows that there is little use in attempting to convert hungry peasants; religion settles better on a full tummy.

How does advertising relate to each stage on the hierarchy? While advertising seems capable of reinforcing basic awareness of—and stimulating us to satisfy—the lower order needs such as hunger, thirst, and safety, it is less capable of doing anything *significant* about our need for "belong-ingness," love, self-esteem, and order and beauty in life. It is highly unlikely that those who have had what Maslow calls "peak experiences" or transient moments of self-actualization (nonstriving, non-self-centered states of per-fection and goal attainment) have achieved this fulfillment through adver-tising. Advertising does an excellent job of alluding to each level of need, reminding us that we share common desires. But in terms of triggering these responses in us, advertising is less successful at marketing self-actualization than it is at emptying grocery shelves of Hostess Twinkies.

Former advertising executive Otto Kleppner, whose book *Advertising Procedure* has become a standard reference since its first edition was pub-lished in 1925, tells advertising practitioners they can trigger a multitude of significant motivations in consumers.[17] Some are physiological (like hunger, thirst, and mating—the satisfaction of which is essential to sur-vival), while others are secondary or social (the desire to be accepted, to succeed). The motivations to which advertising can supposedly appeal in-clude our tendencies to be acquisitive, to achieve, to be recognized, and to dominate. Kleppner does not maintain that advertising can actually fulfill all these motives and needs, but he argues that to be successful advertising must empathize with the goals, needs, desires, and problems of the people it is addressing. (This sounds like Kleppner had insights into psycho-graphics long before the term came into popular usage.) Even though we do not always understand our own motives for responding to advertising appeals, we tend to consume and display products that tell the world how we would like to have it think of us.

Advertisers spend billions of dollars each year tampering with Amer-ica's inconsequential beliefs, playing on brand loyalty and using unique selling propositions, slogans, and positioning to stress relatively meaning-less differences among products. Social psychologist Milton Rokeach says that when advertisers try to connect those brand changes (inconsequential beliefs) to other beliefs, they more often than not pander to our reliance upon authority figures to help us choose.[18] As a consequence, athletes, movie stars, and even old politicians expect us to transfer our positive images of them onto themselves or the products they tout. An appeal to change brands or try new products is based on playing with our negative self-concepts: exploiting our primitive phobias and neurotic anxieties about self-worth. Rokeach has taken strong issue with advertisers who picture humanity as being fundamentally irrational, guilt-ridden, and neurotic. The resulting advertising is exploitive, debasing, and insulting to human dignity, Rokeach maintains.[19]

Conclusions about advertising's ability to influence people, to stim-ulate motivations and trigger responses, appear to rest on the model of hu-manity adhered to by whoever is drawing the conclusions. If one believes humanity is readily manipulated, subjected to whims and motivations over

Advertising can affect lower level needs more readily than higher level ones. You can't peddle self-worth.

Advertisers spend billions of dollars tampering with our inconsequential beliefs.

Our views of advertising's impacts are probably based on our models of humanity—is the world manipulative or rational?

"Thirty-second nightmares"

Just when we thought we'd seen the end of social ostracism brought on by ring around the collar, or innocents being chastised for squeezing the Charmin, or having to share the agony of the hostess with spots on her goblets—along comes Joe Sedelmaier.

For a while, it seemed that the makers of commercials and advertising campaigns had learned that they can sell their products without appealing to our fears and phobias, our negative self-concepts. "Lighten up," we had told them. "We'll try a cleverly advertised product without being badgered to death or being made to feel like an idiot." Fortunately, many of them took our advice.

Sedelmaier, however, has gone one step further. He lightened up, all right, and appealed to our funny bone with a whole series of prize-winning commercials, especially for Wendy's Hamburgers ("Where's the beef?") and Federal Express ("YouknowwhatImeanPetePetecanIcallyouPete?" and "If this package doesn't make it to Peoria, it's your job!"). Sedelmaier, an independent commercial producer based in Chicago, has won more than seventy Clios—the advertising industry's Oscars—for his 30-second slices of life.

On one level, the commercials are hilarious and effective. Fiesty little Clara Peller and fast-talking John Moschitta, the "stars" of Wendy's and Federal Express commercials, brought America to its knees—and, not coincidentally, into Wendy's and Federal Express.

However, on another level, Sedelmaier can be criticized for appealing to the nation's collective fears and paranoia. Who hasn't been ostracized for requesting a little special service at a fast food joint? What business executive or secretary hasn't lived in dread of not getting the package delivered on time?

The Sedelmaier vignettes are stylish and forceful, but loaded with fear, dread, man's inhumanity to man, and the brutalization of corporate society. In Rokeach's terminology, they are taking advantage of our negative self-concepts and fears by tying them to inconsequential solutions. Sedelmaier, of course, insists that he loves humanity and that his little "slice of life" commercials are intended to get us to laugh at ourselves.

We could say that the jury is still out, but the fact remains that the advertising industry has practically knighted Sedelmaier, and the American public is patronizing the establishments he advertises.

Sources: Andy Meisler, "Playing to your fears—and your funny bone," *TV Guide*, 10 March 1984, and Deborah Papier, "He put the beef in TV's best ads," *The Washington Times*, 13 March 1985.

which it has little understanding and less control, all manner of mayhem may be laid at Madison Avenue's collective doorsteps. On the other hand, if one holds to a more democratic view, picturing humanity as rational and capable of making important decisions while being untroubled over inconsequential choices, a more forgiving picture of the advertising business emerges.

Since the muckraking years at the turn of the century, self-regulation of advertising has expanded almost as rapidly as the industry itself, resulting in a range of organizations covering all aspects of the business. In 1971 the three leading advertising groups—the American Association of Advertising Agencies (AAAA), the American Advertising Federation (AAF), and the Association of National Advertisers (ANA)—joined the most comprehensive self-regulating apparatus in advertising history, the National Advertising Review Council (NARC). Working with the Better Business Bureau (BBB), the advertising groups organized "to seek the voluntary elimination of national advertising which professionals would consider deceptive."

Self-regulation intensifies in an age of consumerism, but there are limits.

The BBB and NARC have no legal power; they rely upon "moral suasion" to terminate deceptive advertising. More than a third of the complaints brought by consumers, advertising competitors, or the BBB have been dismissed because the advertisers, when questioned, have substantiated their advertising claims. Another third have been resolved once advertisers modify or discontinue their claims. The remainder are dealt with administratively by the NARC or turned over to appropriate governmental agencies, such as the Federal Trade Commission (FTC).

As we consider the role of the various advertising associations in attempting self-regulation, we must remember that they are limited in their abilities in ways other than philosophy or differences of opinion over what constitutes professionalism. Should the associations attempt any actions that give the appearance of interfering with open competition in the economic marketplace, they will run afoul of federal antitrust laws. Indeed, that is precisely what became of the National Association of Broadcasters (NAB) code of ethics, as we shall see in later pages.

Advertising Organizations' Codes

In their own bailiwicks, large advertising organizations have established codes of ethics and made other attempts to become recognized as professionals.

Codes of ad ethics try to protect the innocent.

The Association of National Advertisers (ANA), whose members produce three-quarters of the commercials seen on national television, has established extensive guidelines for children's advertising. ANA's guidelines try to guard against any presentation that capitalizes on a child's difficulty in distinguishing between the real and the fanciful. The objective is to produce ads that are accurate in their representation of products and the benefits perceived by the child. The commercial must not mislead the viewer, in the specifics of copy, sound, and visual presentation, or in the total impact. If parts are sold separately, the advertisement should so indicate. If assembly is required, that too should be indicated. Finally, the average child for whom the product is intended should be able to duplicate product performance as shown in the ad.

Likewise the American Association of Advertising Agencies (AAAA) has given itself Standards of Good Practice. It conducts an investigation of each advertising agency wishing membership in AAAA, reading the agencies' contracts with clients to determine whether the agencies are behaving honorably. If accepted into the association, agencies are expected to produce no advertising that appears to contain:

False or misleading statements or exaggerations, visual or verbal. Testimonials which do not reflect the real choice of competent witness. Price claims which are misleading. Comparisons which unfairly disparage a competitive product or service. Claims insufficiently supported, or which distort the true meaning or practicable application of statements made by professionals or scientific authority. Statements, suggestions, or pictures offensive to public decency.

Better Business Bureaus

Since 1916 local Better Business Bureaus, outgrowths of local vigilance committees of the Associated Advertising Clubs of America, have also been a controlling force in advertising, though not strictly an internal self-regulatory one. Currently, there are some 151 separate bureaus scattered across the United States. They are supported by more than 126,000 different companies and businesses, which spend $19 million annually to guard against improper business activities, including deceptive advertising. In a recent year the BBB dealt with 6.5 million inquiries and complaints from businesses and the public, and checked out more than 40,000 advertisements for potential violations of truth and accuracy.[20]

Although the Better Business Bureau's code of advertising, much like the codes of many other trade organizations, lacks the power of enforcement, the BBB has become a bridge between governmental control and self-regulation. Its Printers' Ink model advertising statute has been adopted, in one form or another, in forty-five states and therefore has become the law of the land, guiding local and state governments in their efforts to prosecute deceptive advertising practices. And BBB's involvement in the National Advertising Review Council attests to its ability to influence practitioners from within the industry.

The Better Business Bureaus' codes and Printers' Ink statutes have a long and influential history.

Media Efforts

Media have their own ad standards to be socially responsible and to forestall governmental control.

Concerned that they may be in violation of federal, state, or local law, a great many newspapers, magazines, and radio and television stations have created their own standards that govern what advertising they will accept or reject. As legal scholars Nelson and Teeter point out:

The newspaper or broadcast station which permits dishonest or fraudulent advertising hurts its standing with both its readers and its advertisers. Publishers and broadcasters, who perceive psychological and economic advantages in refusing dishonest advertising, also appear to be becoming more cognizant that they have a moral duty to protect the public.[21]

Some of the resulting standards of advertising practice set up by individual media are highly complex and detailed. The *New York Times,* for instance, publishes a pamphlet, *Standards of Advertising Acceptability,* that includes generalized philosophical statements about credible advertising, and discusses the means it uses for self-checking and enforcing its standards. It also lists various kinds of unacceptable advertising and discusses "opinion advertising" of a political or philosophical nature, as well as the typography, format, and production of ads.

Louisville, Kentucky, newspapers have produced guidelines reflecting their strong concerns with sex discrimination. They list what they consider discriminatory terms frequently found in classified ads, along with their suggested substitutes. For example, *attractive, pretty,* and *handsome* are to be replaced by *well-groomed* and *presentable; barmaid* is to be replaced by *bar help* or the conjoined *bar waiter* or *waitress, foreman* to become *foreman—male or female, maid* to be *domestic help* or *housekeeping,* and so on.

Because nonlicensed news media are not legally considered to be common carriers, such as city- or state-owned utilities, they have enjoyed the right to carry on business activities with whomever they please. Despite recent thrusts by various interest groups and individuals seeking access to the advertising pages, the print media still retain the right to reject any or all ads they feel are inappropriate for their publications. Many, in the interest of carrying out public debates on controversial issues, have adapted their own concepts of the broadcasters' "fairness doctrine" (which will be discussed in chapter 14). They sometimes run ads they personally and institutionally abhor. In the process they may be operating in their own best interests, as they adopt one more technique of social responsibility that will help curtail public and regulatory incursions into their freedom.

Self-regulation in the broadcast industry is another story. The major networks, and most large stations, have adopted codes of ethics that are created by their standards and practices departments or divisions. At the national level, the standard of the industry for years had been the National Association of Broadcasters (NAB) code. In 1982, however, a federal court decided that some parts of that code violated federal antitrust laws, particularly ones concerning how many products could be advertised in a sixty-second time period ("piggybacking"), and how many total minutes of each hour could be turned over to advertising.

Broadcasting self-regulation is unique. When courts held the NAB code to be illegal, some rethinking was necessary.

The NAB's response to the court case was interesting. It chose to abandon its code altogether, suspending all of its television and radio advertising standards. Over the next few years, the NAB rethought its rights and obligations, and attempted to put together a new and legal code. Meanwhile, the competitive nature of broadcasting and the existence of individual network and station guidelines seem to have kept most operations in reasonably close alignment with the original NAB code. The entire experience has proven to be a fascinating example of how the federal government, in looking out for the best interests of its citizenry, caused a voluntary trade organization to abandon its operating principles, which had been put into operation to serve the same citizenry—plus, of course, to serve its own commercial self-interests.

Governmental Regulation

FTC, with its concern over truthful advertising, is but one of many governmental agencies checking the media.

Self-policing efforts notwithstanding, the government has become involved over the years in the regulation of advertising. Nearly forty federal agencies, and untold numbers of state and local ones, have gotten into the act. The most active and influential, insofar as most consumers are concerned, is the Federal Trade Commission (FTC). It is joined by the Federal Communications Commission, the Food and Drug Administration (FDA), the Post Office Department, the Securities and Exchange Commission (SEC), and the Alcohol and Tax Division of the Internal Revenue Service, among others. In all, more than thirty federal statutes contain advertising regulations. In addition, nearly all states have enacted statutes making fraudulent and misleading advertising a misdemeanor. In some areas, at some levels, enforcement is quite vigorous. In others, it is not.

The Federal Trade Commission

During the 1960s and 1970s, the FTC, increasingly sensitive to consumer pressure, moved with considerable success into the areas of *truth in lending* and *truth in packaging*. Fresh from its triumphs in these related fields, it undertook the more complex problems of *deception and truth in advertising*. Its mandate to grapple with these issues emerges from Section 5 of the Federal Trade Commission Act, which says: "Unfair methods of competition in commerce, and unfair or deceptive practices in commerce, are declared unlawful."[22]

Deceptive advertising presents a slightly different problem from earlier campaigns because another element is introduced. While campaigns in packaging and lending involved only the FTC and the individual offender, whether manufacturer or financial institution, deceptive advertising involves both the offending source of the advertising and the mass media that carry it. Thus, deceptive advertising as regulated by the FTC will affect all mass media, print and broadcast alike, and not merely the broadcast media falling under the jurisdiction of the federal government's FCC. Because advertising comprises the largest revenue portion of most of the mass media, a vigorous FTC cannot help but have significant repercussions on most of mass communications.

FTC oversees all media, using several means to control offenders.

As an administrative agency whose members are appointed by the President and whose budget is approved by Congress, the FTC tends to reflect the regulatory climate of Washington, D.C., at any given time. The 1970s may have been the FTC's heyday of activism, as it took on entire industries for their advertising practices, among them the legal, medical, and mortuary professions, and the makers of over-the-counter medicines, automobiles, and breakfast cereals. A beefed-up and over-eager FTC went so far as to propose a Trade Regulation Rule which would have required that all advertising to children under eight be banned, that advertising for sugared products be banned for children under twelve, and that dental health and nutritional advertising be paid for by the industry.

In part because such activism offended powerful lobbying groups, Congress tightened the reins on the FTC by passing what it euphemistically called its "Federal Trade Commission Improvements Act of 1980." It cut the FTC's budget, and changed its focus. (In 1985 only $2.85 million of the commission's $69.1 million operating budget was earmarked for the regulation of the $100 billion advertising industry.) Congress also told the commission it should no longer stew over whether advertising was "false and misleading," but should concentrate on regulating advertising that was out-and-out "deceptive," a condition that is obviously harder to prove. Overall, it seems that today's FTC is less sensitive to consumer pressures than it was prior to the Reagan administration. Due to ideological and budgetary policies the FTC has been reduced in its effectiveness as a federal regulatory agency. However, a look at its history and mandates for change should prove instructive.

From its founding in 1914 until the late 1960s, the FTC had been known as "the little gray lady of Pennsylvania Avenue" because of its reluctance to investigate and prosecute advertisers. After the 1960s and the rise of consumerism and challenges to advertising brought on by citizens advocate Ralph Nader and Action for Children's Television's Peggy Charren, among others, things began to change. It developed a series of guidelines on how industries should promote their products and services, and set up elaborate enforcement procedures. For instance, it was able to convince federal courts that advertisers were responsible for the overall impressions left by their ads; if the overall impression was false or misleading, the advertising campaign could be halted and the advertiser fined.

Other FTC regulations, and court decisions emerging from cases the FTC grappled with, dealt with questions of *clarity* (advertising statements must be so clear that even a person of low intelligence will not be confused by them); *fact versus puffery* (material misstatements of fact are not tolerated, but overblown and general opinions about a product's quality are); *questions of taste* (bad taste is not inherently deceptive or unfair advertising); *demonstrations* (depictions of product performance must not be misleading); *warranties* (under the 1975 Moss-Magnuson Act, they are not mandatory, but when they exist, they should clarify the limitations and tell consumers how to file a claim); *"free"* (if there are any conditions attached to receiving a "free" product, they must be clearly stated); and *lotteries* (the FTC makes it illegal to advertise lotteries in interstate commerce, and the United States Postal Service bans the use of the mails if a person has to pay to enter a lottery conducted by an advertiser.)[23]

The FTC tends to reflect the economic and regulatory climate of the times.

Advertisers should be able to prove their "factual" claims.

Documentation One of the most effective tactics taken by the FTC has been to require advertisers to fully document the claims made in their advertising. This has taken the form of submitting huge dossiers of statistics and scientific tests to the commission. If the advertisers fail to convince the commission that the ad campaign is fully justified on the basis of the documentation submitted, the FTC will usually issue a consent order asking them to stop making certain claims. Entire industries were brought under the FTC's microscopes during the 1970s, as noted above. In some cases the documentation was incredibly complex—such as a seventy-two-page packet of documents, some in French, to support Renault's gas mileage claims— which prompted the FTC to demand that advertisers supply a brief summary, in layperson's language.[24]

The guilty go to public confession.

Corrective Advertising Another tactic of the FTC has been an imaginative extension of the FCC's Fairness Doctrine, in the form of remedial or *corrective advertising.* In this scheme advertisers found guilty of employing false or deceptive advertising over a long and apparently successful campaign are required to use a certain percentage of their future advertising to acknowledge the earlier misrepresentation—in other words, to confess their sins.

In 1978, the United States Supreme Court upheld the FTC's authority to enforce corrective advertising, which was the first time the policy was fully tested in the courts. The case concerned Listerine mouthwash, which for one-hundred years had been advertising that it helped cure colds and sore throats. The FTC finally proved, based on current medical research, that colds are not caused by mouth bacteria, but by viruses entering the body through the nose and the eyes, which refuted Listerine's long-standing contentions. Warner-Lambert, which withdrew the advertising claims while the case was being adjudicated, was told to run language in $10.2 million worth of ads admitting that its product does not help prevent colds or sore throats, or lessen their severity.[25]

Other corrective advertising has been enforced against Ocean Spray Cranberry Juice and Profile Bread. Ocean Spray used to brag about the "food energy" in its juice; the FTC got it to admit that "food energy" was just another way of saying "calories." And the makers of Profile Bread, who formerly advertised that their bread would help people lose weight, had to disclose that Profile had the same calories per pound as other bread but was sliced thinner; at seven calories less per slice than regular bread, it was not really effective for weight loss.

Corrective advertising is probably not as effective as the FTC would have us believe; as often as not the damage from misleading ads has already been done by the time the corrective ads come out.

The FTC told Listerine not to claim it cured the common cold—but it didn't say anything about every other ill known to humanity.

Comparative Advertising Comparative advertising, or Brand X advertising, is the comparison of an advertiser's product with a competitor's. Under Brand X, the product is not named. But the FTC, after its documentation requirements, has been encouraging advertisers to name their competitors. The results have been especially evident in recent advertising wars being waged by the major hamburger chains and the soft drink industry. Those battles are being fought in print and on the air, but also in the courtroom.

The FTC says that if the documentation shows you are superior to your competitors, you should feel free to name them; the public has a right to know. However, in recent years a number of court cases have resulted from disgruntled manufacturers who complain that their product was subjected to unfair comparison. This fact alone has discouraged many advertisers from engaging in the comparison game.

Advertisers are encouraged to name their competitors, but there are good reasons not to.

There is another reason not to employ comparative advertising. The agency Ogilvy and Mather conducted a thorough study of the matter and said it would not recommend comparative advertising to any of its clients, because the only benefit it could perceive was to the competitor named in the ad. Running down the competition in the competitive world of marketing is probably a no-win situation.

Questionable Practices A couple of advertising practices that the FTC took a dim view of and largely eliminated were the so-called *mock-up cases,* and *bait and switch* advertising.

The mock-up cases concerned a shaving lather so efficient that it could shave sandpaper instantly on TV. It turned out that it was not sandpaper at all, but loose sand sprinkled on plexiglass that they were shaving with Rapid Shave. In the Campbell's Soup case, again on television, spots showed a particularly hearty soup, which got that way because the bowl was half full of marbles.

In another instance, the FTC held that advocates for toothpastes and other proprietary drugs wearing white lab coats on television conveyed the impression that they were doctors or dentists and hence the product was medically or dentally endorsed. In still another case, this one involving the Kroger chain of 1,200 grocery stores, the company was accused of making dubious use of survey research. Kroger claimed that shoppers in their stores could purchase 150 everyday items cheaper than in other chain stores. The FTC found fault with the methodology of their sample selection and survey form, and further pointed out that the survey did not include meat, produce, or house brands—three significant items to the food shopper.

Mock-up and bait and switch advertising practices are frowned upon.

Bait and switch involves advertising a product which is not available, or not available at the price stated. One Japanese car manufacturer advertised a low-priced model, but visitors to the showroom were told it was available on special order only, taking months to arrive from Japan. Comparable models on the floor were loaded with extras for $650 more.

In all of the above cases, the FTC took action against the advertisers, and got a message out to others who might be inclined to try the same techniques: don't do it. In the short run the advertising campaigns might move some merchandise, but in the long run the negative publicity that comes with being reprimanded by the FTC results in a net loss to the company.

In sum, as we consider the role and impact of the Federal Trade Commission, it seems that the agency is only as effective as the general mood of the society permits it to be. During periods of consumerism, the FTC has flexed its muscles, with impressive results. More recently, in a laissez faire deregulatory climate, the real regulator appears to be the marketplace itself. As conservative columnist James Kilpatrick once noted, "The federal authority has no writ to cure every imperfection in society. Some obligations ought to be left to individuals, for good or ill, if personal responsibility is not to be fatally undermined."[26]

Criticisms and Defenses of Advertising

Omnipresent advertising is the most talked-about aspect of U.S. business. Most observations are negative, not based on economic principles.

Advertising is an essential element in our contemporary economic system and is therefore an integral part of American life. Even though only 5 percent of the world's population lives within its borders, the United States generates 60 percent of the world's total advertising expenditures. Because advertising is so omnipresent, it is perhaps the most talked-about aspect of the national business scene. Merely being exposed to the hundreds of ads that bombard us each day has made each of us somewhat of an advertising expert. As experts, we have no shortage of opinions about advertising; not surprisingly, most of our opinions are negative. As consumers rather than economists, we tend to react personally to the offensiveness of commercial messages, frequently overlooking the fundamental economic "nature of the beast."

But even the economists are sharply divided in their assessments of advertising's true values. Some take the extreme position of historian Arnold Toynbee, who said, "I cannot think of any circumstances in which advertising would not be an evil." At the other end of the spectrum, President Franklin D. Roosevelt once mused that were he to start life again, he would most likely go into the advertising business. Such is the range of opinions about advertising.

Satirist-turned-advertising executive Stan Freberg once explained:

In talking with college students, I have asked them why they thought I went into the commercials business, and invariably *they* assume it was for the money. Or because "there's a lot of opportunities for advancement in that field," or "advertising is glamorous and exciting!" True, I have experienced tremendous stimulation, creating some rather unorthodox advertising campaigns, which in most cases were successful in making the sales curve jump, to say nothing of the client. I have also managed to successfully unnerve several hundred account executives along the way. In my day I have made gray flannel turn white overnight. That in itself is an exciting sport. But the thought that I might have been motivated to go into this business not as an advertising man but as a totally outraged *consumer* rarely, if ever, occurs to them. Indeed, the main thing that has held me in advertising is the thing that got me into it in the first place. That is the challenge of proving daily that advertising does not necessarily have to be dull, insipid, or irritating in order to communicate and thus sell the product.[27]

Over the years the criticisms and defenses of advertising have remained fairly consistent. Most of the criticisms have been aimed at the supposed psychological effects—namely, that advertising creates needs and wants and, in so doing, encourages people to emphasize materialism and live beyond their means. To the extent that these wants are left unfulfilled, it also breeds frustration and encourages aggressiveness and perhaps crime. While some evidence may be available to support these claims, there is little doubt that advertising is the product of a commercial and competitive society. It is the natural partner of the free enterprise system and it provides the distribution extension of mass production. Interestingly enough, it is the nonessentials such as soaps, snack foods, tobacco, toiletries, and soft drinks that get the most advertising as a percent of their sales. We are reminded by Maslow and Rokeach that advertising of such products does not actually create wants and needs, but it might encourage us to satiate inconsequential desires.

One of the dominant criticisms leveled at advertising is its cost to the consumer. Such costs are considerable. Because advertising costs are passed on, consumers are placed in the position of paying for the privilege of being persuaded. For example, more than $100 of the cost of the average American car represents advertising costs that the consumer pays when buying the vehicle. And the price of many cosmetics represents little more than the cost of the advertising that goes into them. Combined with the criticisms previously cited, this essentially means that members of the public are paying liberally to be urged to live beyond their means, to buy products they do not need, and—in the long run—to be spendthrifts.

At the beginning of the chapter we noted that advertising is an automated sales force, which is advertising's strongest defense. The existence of advertising has made possible a greater demand for goods and services. Translated into terms of mass production, this increased demand has led to economies and savings on the assembly lines and in purchasing, and may have rendered many products, including the automobile, far cheaper than they would otherwise have been—even with the extra advertising increment.

There are other assets. Advertising makes possible the most diversified mass communications system of any nation, providing unparalleled entertainment and considerable information and culture to the individual at a fraction of its true cost. Martin Mayer addresses himself to the psychological factors in advertising in his *value-added theory.*[28] He claims that

CATHY by Cathy Guisewite

advertising adds a vague dimension to a product. The woman who buys a particular soap with the belief that it will improve her complexion may be more satisfied in that belief—she may be a happier person—whether it does or does not; the assurance that comes from purchasing a product of believed quality—a Hathaway shirt, for example—provides a confidence to the individual that would have been unobtainable with a cheaper brand. According to Paul Samuelson, lipstick ingredients that cost pennies are transmuted by the alchemy of Madison Avenue into expensive cosmetics that essentially represent packaged hope, and what's the matter with hope?[29] In essence, Mayer and Samuelson are making a case for the psychological concept of confidence: we are what we believe and for some of us this is the added value of advertising.

There is little doubt that advertising increases the cost of some goods; that advertising encourages consumption, often at the expense of thrift; and that advertising often works on the public's psyches in various ways. But there is also little doubt that advertising supports the largest and most diverse mass media network in the world and, without doubt, the mass media system freest of overt governmental manipulation and propaganda; that advertising is a major employer and contributor to the overall economy; and that advertising, through increased demand, reduces the overall cost of many products.

Advertising is, therefore, both a blessing and a menace, and both these roles must be recognized if the operation of mass communications in society is to be understood. As in almost everything that we have discussed pertaining to mass communications, there are trade-offs. The bad is accepted with the good. It is not a question of either/or. It is a question of both, because under the existing system, which has had a long evolution and tradition, we cannot have one without the other, or we shall have a different system entirely.

It is advertising, by and large, that supports the diverse mass communications system within the United States. This relationship evolved largely as the result of a competitive, free enterprise economy where the private sector has provided the investment needed to expand the mass media in return for their use in selling goods and services.

Although ancient in origin, most of the growth of advertising has occurred during the years following World War II. Rapid improvement in media technology, interest in the social sciences, and a backlog of consumer demand all contributed to its postwar impetus. Today advertising is a $100 billion a year business divided into national and local advertising. Newspapers still get more of the nation's advertising dollar than any other medium.

Agencies can be one-person operations, creative boutiques, or large and institutional organizations. Regardless of size, all agencies perform five functions—client liaison, creativity, production, placement, and housekeeping—for which they are paid a commission or a negotiated fee.

Since World War II, strategies used by advertising have become more sophisticated. The first strategy developed was Ted Bates' unique selling proposition, followed by David Ogilvy's brand images, and various forms of motivational research. Subliminal advertising has been tried, with questionable success. Currently in favor is positioning, an advertising strategy that concedes part of the market to the competition. Through demographic and psychographic research, advertisers have learned to identify and reach specialized segments of the larger market. Institutional or public relations advertising has become a significant factor in recent years.

Advertising seems to be remarkably effective in selling consumer goods, inconsequentials that the public recognizes as trivia. It is much less effective in convincing the public in matters of substance, such as elections or higher level human needs or values. It is unequaled in calling attention to a new product or candidate. However, by its sheer pervasiveness, it has probably done as much as any other institution to propagate society's myths of progress and self-betterment: "A better world through advertising."

The industry itself, recognizing its questionable public image, has taken several steps toward effective self-regulation. Deceptive advertising has received most of the attention, primarily with an eye toward warding off formal governmental controls over the industry. Nonetheless, the government has exerted substantial influence over advertisers, particularly through the Federal Trade Commission. After a period of high visibility and success at controlling false and misleading advertising, the FTC has been reined in somewhat by a Congress that believes in deregulation or control by marketplace forces.

There is no shortage of criticisms of advertising, nor a shortage of defenses. It is attacked and defended mainly on psychological and economic grounds. Social critics and economists remain divided on advertising's shortcomings and contributions, but all recognize its integral role in contemporary America.

Notes

1. Quentin Schultze, "Comments on the History of Ethical Codes in the Advertising Business" (Paper presented to Association for Education in Journalism, Mass Communications and Society Division, Logan, Utah, 30 March 1979).
2. Martin Mayer, *Madison Avenue U.S.A.* (New York: Harper & Brothers, 1958).
3. Sloan Wilson, *The Man in the Gray Flannel Suit* (New York: Pocket Books, 1967).
4. Vance Packard, *The Hidden Persuaders* (New York: Pocket Books, 1968); and Mayer, *Madison Avenue U.S.A.*
5. J. A. C. Brown, *Techniques of Persuasion* (Baltimore, Md.: Penguin Books, 1963).
6. David Ogilvy, *Confessions of an Advertising Man* (New York: Ballantine, 1971).
7. W. Keith Hafer and Gordon E. White, *Advertising Writing* (St. Paul, Minn.: West Publishing Company, 1977), p. 244.
8. Wilson Bryan Key, *Subliminal Seduction: Ad Media's Manipulation of a Not So Innocent America* (New York: Signet, 1974); *Media Sexploitation* (New York: Signet, 1976); *The Clam-Plate Orgy and Other Subliminal Techniques For Manipulating Your Behavior* (New York: Signet, 1981).
9. Otto Kleppner, Thomas Russell and Glenn Verrill, *Advertising Procedure,* 8th ed. (Englewood Cliffs, N.J.: Prentice-Hall, Inc., 1983), p. 613.
10. Ibid., pp. 79–83, and Sharon Rosenthal, "Here's How They're Selling Alka-Seltzer to Yuppies," *TV Guide,* 22 December 1984, pp. 21–22.
11. United Press International, "Direct Mail campaigns pinpoint customers," *Washington Times,* 30 May 1985.

12. James Atlas, "Beyond Demographics: How Madison Avenue knows who you are and what you want," *The Atlantic Monthly,* October 1984, p. 51.
13. Ibid., pp. 49–58, and Peter W. Bernstein, "Psychographics is still an issue on Madison Avenue," *Fortune,* 16 January 1978, p. 78.
14. Atlas, "Beyond Demographics," p. 52.
15. John O'Toole, "Madison Ave.: Selling ideas, not products," *U.S. News & World Report,* 10 March 1986, p. 54.
16. Abraham Maslow, *Motivation and Personality* (New York: Harper & Row, 1970), pp. 80–92.
17. Kleppner, Russell, and Verrill, *Advertising Procedure,* pp. 324–327.
18. Milton Rokeach, *The Open and Closed Mind* (New York: Basic Books, 1960); *Beliefs, Attitudes, and Values* (San Francisco: Josey-Bass, 1968); and *The Nature of Human Values* (New York: The Free Press, 1973).
19. Milton Rokeach, "Images of the Consumer's Mind On and Off Madison Avenue," *ETC, A Review of General Semantics,* September 1964, pp. 261–73.
20. Kleppner, Russell, and Verrill, *Advertising Procedure,* pp. 569.
21. Harold L. Nelson and Dwight L. Teeter, Jr., *Law of Mass Communications: Freedom and Control of Print and Broadcast Media,* 5th ed. (Mineola, N.Y.: The Foundation Press, 1986), p. 650.
22. Ibid., p. 617.
23. Kleppner, Russell, and Verrill, *Advertising Procedure,* pp. 557–559.
24. Charlene J. Brown, Trevor R. Brown, and William L. Rivers, *The Media and the People* (New York: Holt, Rinehart & Winston, 1978), p. 400.
25. *FTC v. Warner-Lambert Co.,* 435 U.S. 950 (review denied).
26. James Kilpatrick, *Washington Star* (syndicated column), 16 May 1978.
27. Stan Freberg, "The Freberg Part-time Television Plan." In *Mass Media in a Free Society,* ed. Warren K. Agee (Lawrence, Kansas: The University Press of Kansas, 1969), p. 64.
28. Mayer, *Madison Avenue U.S.A.*
29. "American Issues Forum No. 8: Courses by Newspaper," February 1976.

12

Public Relations

Public relations is an enigma that defies simple definition. In its broadest sense it is an umbrella that includes publicity (with which it is too often equated), promotion, press agentry, and even advertising, plus behind-the-scenes promotional work in political campaigning, lobbying, sports, consumer affairs, education, health, social services, utilities, finance, and almost any other area one could imagine. Some years ago, one observer commented that public relations is a brotherhood of one hundred thousand whose common bond is a profession and whose common woe is that no two can agree on what the profession is. Today the profession has three times as many members, who still have not reached a consensus.

Public relations makes extensive use of the mass media, contributing considerable portions of media content in contemporary society. The public frequently confuses public relations with advertising. Part of the confusion can be lessened by trying to imagine our society without advertising and without public relations. We will undoubtedly find it more difficult to imagine society without advertising—the obvious, overt, paid-for, commercial persuasion—than without public relations—the frequently subtle, behind-the-scenes communications. Advertising uses paid time and space in the media, usually to deal directly with consumers, primarily to promote and sell goods and services, secondarily to promote and sell ideas. Public relations, on the other hand, uses free time and space in the media to promote institutions or ideas or candidates; it is more interested in reputation than in sales—more concerned about image than distribution of goods and services.

One PR practitioner once suggested that we celebrate National Public Relations Day by leaving blank in all newspapers and other periodicals, and by omitting from the broadcast media, any item with a public relations genesis. The ensuing volume of white space and silence in this impossible experiment would drive home just how much of our news, features, and entertainment is public relations, and how significant it is to our existence in contemporary society.

Public relations is both a condition and an activity—a noun and a verb. We say a corporation has good PR; that generally means we have a particular image of that corporation's image. On the other hand, when we say a corporation practices good PR, we are referring to a range of public relations activities such as writing news releases, giving speeches, and arranging special events. The activities typically fall into one of eight major areas: (1) opinion research, (2) press agentry, (3) product promotion, (4) publicity, (5) lobbying, (6) public affairs, (7) fund raising and membership drives, and (8) special events management.

Practitioners of PR have made numerous attempts over the years to define their livelihood. *Public Relations News* holds that public relations is the management function that evaluates public attitudes, identifies the policies and procedures of an individual or an organization with the public

Introduction

Public relations is hard to define, as it is such a broad umbrella covering many aspects of mass communication.

PR differs from advertising in its stress on reputation or image, rather than on sales or distribution. It is not paid for in the same way that advertising is.

interest, and plans and executes a program of action to earn public understanding and acceptance. The World Assembly of Public Relations Associations maintains that public relations practice is the art and social science of analyzing trends, predicting their consequences, counseling organization leaders, and implementing planned programs of action that will serve the interests of organizations and the public. These and other definitions all point to a similar goal: all functions labeled public relations are for the purpose of influencing public opinion.[1]

The irony of PR is that the industry devoted to improving images has an image problem of its own.

In this chapter we will attempt to distinguish public relations from other forms of mass communications. We will consider the irony of PR—the often negative image of an industry whose *raison d'etre* (reason for existence) is to promote a good image for others—and the industry's attempts to improve its own image. The history of public relations in America and abroad will be briefly considered, with attention paid to the movement toward social responsibility in business. Specialized roles of PR in contemporary society, and particularly within the political sphere, will be noted. We will also discuss the place of hype and organized persuasion in today's world.

Historical Development

The antecedents of public relations can be traced to earliest recorded history, including ancient biblical times. A 4,000 year old cuneiform tablet found in Iraq told farmers how to produce better crops; its function was similar to the PR bulletins distributed by today's agriculture departments.

Efforts to change public opinion are as old as recorded history. Greeks and Romans mastered the art.

In terms of publicity, lobbying, and even press agentry, we might look to ancient Greece, where the poets were the public relations people. Use of the poetic form in rhyme and meter facilitated memory, particularly among the illiterate, assuring that messages would be passed along in more or less the same form. Two poets, Simonides and Pindar, made a good living writing and selling odes of praise for those willing to pay. They were among the first press agents. The use of poetry to manipulate public opinion became so widespread in the Greek democracy that Plato in his *Republic* advocated the prohibition of all poetry—except that written for the government. Not only was this the first example of attempted governmental control of the mass media (such as they were), but it was the first advocacy of governmental public relations, which is, of course, a big business today.

In Rome, the same techniques of poetry and praise occurred with a number of Roman refinements. The Romans refined the poetic form, adding subtlety to public relations. Virgil's *Georgics* was, on the surface, a bucolic poem extolling the virtues of country living, the pastoral scene, clean air, fresh water, and a closeness to nature. Its purpose, however, was devious. Rome was overcrowded at the time, an early version of urban sprawl. There was not enough food to feed the population. The *Georgics,* commissioned by the government, was a public relations attempt to urge people to leave the city and take up rural residence, thus alleviating the population problem and providing more farmers to feed city residents.

Gaius Julius Caesar was a master of public relations who developed the first long-range public relations program. Early in his career his talent and ambition caused jealousies, and the Senate in effect banished him to Gaul to be in charge of an army, hoping that the people, with whom he was extraordinarily popular, would forget him.

Over the Roman roads that supplied the legions, Caesar sent back messengers regularly to Rome with his *Commentaries,* familiar to any beginning Latin student. The *Commentaries* were not reports to the Senate; they were reports to the people telling of his exploits. They were read and posted in the forum for all to see. They were written in the language of the people—punchy and alive. His famous "veni, vidi, vici" is a fine example— "I came, I saw, I conquered." The technique worked; over the long years of foreign battles—in Gaul, Spain, and Britain—he kept his legend alive until the day when he returned at the head of his victorious army. The people hailed him and made him emperor.

Matthew, Mark, Luke, and John were master public relations practitioners—or propagandists—with their gospels for the propagation of Christianity, in what Marshall McLuhan describes as a "conspiracy of communications." For the first time, he points out, a common language, Aramaic, combined with Roman order, the Roman roads, and safety of sea transport, permitted Peter and Paul and the apostles to spread the Word to a waiting world.

Throughout the Middle Ages and the Renaissance, public relations proceeded in a relatively informal way as troubadours spread messages from castle to castle and from town to town. As we pointed out earlier in this book, Gutenberg's converted wine press changed the face of the globe and ushered in the era of mass communications some five hundred years ago. Diverse points of view were published, and propagandists of all shades had outlets. By 1622, the Catholic Church, whose dogma had gone largely unchallenged prior to the invention of printing, found it necessary to close ranks. The "Sacra Congregation de Propagande Fide"—Sacred Congregation for the Propagation of Faith—was charged with spreading the faith through missionaries and other vehicles. That early innocuous use of the term "propaganda," which merely meant to spread or propagate ideas, gave way several centuries later to a more negative connotation.

Although public relations is sometimes called a uniquely American art, refined in the twentieth century, there is good reason to claim that PR has played an influential role from the very beginning of this nation.

By the time of the American Revolution, the power of public relations in swaying public opinion became obvious. The War for Independence was not the spontaneous uprising of downtrodden colonists against a tyrannical Crown. One way to look at it is as a long-range game plan engineered over nearly a quarter of a century largely by one man, Samuel Adams, who

Public Relations in America

18th and 19th Centuries

The Revolutionary War emerged from a lengthy PR campaign spearheaded by Samuel Adams.

THE BOSTON TEA PARTY DESTRUCTION OF THE TEA IN BOSTON HARBOR, DECEMBER 16, 1773.

hated the English. Adams formed Committees for Correspondence throughout the thirteen colonies, which traded and publicized cases of British brutality. He magnified a waterfront brawl into the Boston Massacre, and conceived the Boston Tea Party as a knee-slapper to dramatize excise taxation. With the help of Thomas Paine's fiery *Common Sense,* Adams fanned the flames of revolution. Catchy slogans and jingles, such as "Don't Tread on Me" and "Yankee Doodle," were part of the public opinion campaign.

Public relations did not cease when independence was won. The *Federalist Papers* by Hamilton, Madison, and Jay were a series of newspaper columns syndicated throughout the nation's press during the long battle for a strong federal government as opposed to a loose confederation. The Constitutional Convention, adoption of the Bill of Rights, and establishment of a republic could not have come about without a variety of lobbying and PR efforts.

Public relations activities leading up to the Revolutionary War, and those that followed it, differed significantly in one sense. The former serve as a type of *agitation propaganda,* stirring up public opinion, bringing about a willingness to fight for what seems right. The latter, on the other hand, are *integration propaganda,* activities that operate on a subtler and nonviolent level, intended to stabilize public opinion around certain "givens." Even a superficial analysis of the jockeying for public opinion in this period should indicate that public relations activities contain some elements of propaganda, and such propaganda is certainly not inherently evil or devious. As French social scientist Jacques Ellul argues, modern societies must have these elements of propaganda and public opinion formation if they are to survive and prosper.[2]

Public relations has much in common with both *agitation propaganda* and *integration propaganda.*

One of the two basic sources of American public relations as an organized practice can be found in political campaigns. When the founding fathers set up the electoral structure, they expected that natural leaders would emerge to be elected by the people. However, because political office meant preferment, it was not long before ambitious individuals began taking steps to assure that they would be counted among the natural leaders at election time.

In America, democracy and political campaigns made public relations inevitable.

Some of the early political campaigns were masterpieces of excitement. Catchy slogans, torchlight parades, brass bands, and beer busts were used in the effort to win votes. It was Andrew Jackson who changed the ground rules. An Indian fighter of repute and victor at New Orleans in the War of 1812, he campaigned on a populist ticket with a slate of electors pledged to him. His election in 1828 proved the merit of populist appeal, and the business of national campaigning was born.

The other basis of public relations was theatrical press agentry, of which P. T. Barnum (1810–1891) was the great exponent. Barnum recognized that people enjoyed being conned (directed). He was the author of the statement, "There's a sucker born every minute," and he exploited this principle throughout his life as a showman. He promoted an unknown singer, Jenny Lind, so successfully that people flocked to hear "The Swedish Nightingale." And he promoted a larger than average African elephant so effectively that the elephant's name "Jumbo" became synonymous with gigantic.

P. T. Barnum's "There's a sucker born every minute" reflected his concept of public opinion.

Barnum's buildup of General Tom Thumb, a midget masquerading as a Civil War general, was a masterpiece of promotion. Later, the production of General Tom Thumb's wedding to another midget, complete with a parade down Broadway in a tiny coach drawn by tiny horses, was another pinnacle of publicity. People knew that Tom Thumb was no Civil War officer, no hero of Antietam or Vicksburg, but the idea was so preposterous that they enjoyed it and were willing to pay money to hear his exploits and to see him strut in his little uniform and cockaded hat.

It is from this origin of public relations that much of its criticism stems. The theory was that anything was valid if it generated publicity. "I don't care what you say about me so long as you spell my name right" was a summary of the substance of press agentry. While political campaigning introduced specificity of appeal, press agentry introduced the outrageous as a device in the competition for attention. In the orchestra of public relations, press agentry represents the brass section; the aim of press agentry is more to attract attention than to gain understanding, since notoriety can be more useful than piety, thriving as it does in the box-office world.[3]

The years following the Civil War were the years of America's expansion, the winning of the West when much was overshadowed by the practical business of exploiting a continent. This was the time of the robber barons and the railroads, of "survival of the fittest" (widely misinterpreted from its biological origin in Darwin to a social doctrine), and of laissez-faire, its political and economic expression. The era had its public relations

Railroad magnate Vanderbilt's "The public be damned" reflected his big business's laissez-faire concepts.

In the face of the muckrakers and a credibility gap, big business turned to PR.

counterpart in William Henry Vanderbilt of the New York Central Railroad who, in 1882, expressed contempt for public opinion. When one of his advisers suggested that the public would not be particularly pleased with one of Vanderbilt's proposals, the magnate is said to have responded: "The public, why, sir, the public be damned."

The turn of the century brought reaction in the form of the muckrakers—authors and journalists—who saw business corruption and government collusion running rampant and set out to correct them, using the mass media of the time—books, magazines, and newspapers. Ida Tarbell's series "The History of Standard Oil;" Lincoln Steffens' "The Shame of the Cities;" and Upton Sinclair's novel *The Jungle,* an exposé of the meat-packing industry, are all examples of an unorganized campaign that ran a decade or so and resulted in some of the first social legislation. The muckrakers were the expression of a stirring social consciousness repulsed by the excesses of business and the timidity of government.

The Father of PR

Ivy Lee, the father of modern PR, had faith in people and urged business to be open and honest.

Business did not take this lying down. It turned to press agents to whitewash its reputation. However, the glaring discrepancies between business's whitewash and the all too visible inequities created an early example of a credibility gap. Business was in trouble. Into this picture in 1908 stepped Ivy Lee, the father of modern public relations. Lee, a press agent who dealt in truth, had an unprecedented idea. He published a declaration of principles and sent it to editors with whom he was associated. It said that he dealt in news and factual information, and that the editors were free to check any of his facts independently. Furthermore, if they felt his material would be better placed in their advertising columns, all they had to do was throw it away.

Ivy Lee had an enormous respect for the wisdom of the people. He felt that if they were given the facts they would make correct judgments. His basic tenet was that the public should be honestly informed of good news and bad. He proved this in the case of the Pennsylvania Railroad. The Pennsy had a wreck in which a number of lives were lost. Management's first reaction was to hush it up, but Lee pointed out that a wreck could not be hidden. With cars and engines strewn over the New Jersey landscape and bloodshed on the right-of-way, the public was certain to find out about it. Lee ran special trains for the press to the scene of the accident. He announced a system-wide survey of the company's roadbeds so that a similar wreck would not happen, compensation for the families of those who were killed, and hospitalization of the injured. In short, Lee converted a tragedy into a public relations triumph for the railroad, which was widely applauded by the press for the manner in which it handled the accident. Lee knew that it was far better to face bad news frontally than to let it linger and fester in secret until it destroyed credibility.

Lee's reputation as a public relations genius was not built upon this single incident, but upon a whole collection of similar campaigns. For instance, he greatly facilitated the tasks of reporters who attempted to cover

It took a massive public relations campaign to sell millions of dollars worth of Liberty Bonds during World War I.

a difficult strike in the coal industry, when he initiated the practice of offering handouts or press releases following each of the strike conferences. He was hired by the John D. Rockefeller family after the Rockefellers had developed a particularly unfavorable public image; in addition to being thought of as "robber barons" because of their enormous success with the Standard Oil Company, the family had viciously broken a strike at the Colorado Fuel and Iron Company. Lee worked on many fronts to show the public how socially responsible the Rockefellers were. The acceptance of the charitable Rockefeller Foundation, and political success of several Rockefeller children, speak to Lee's success.

The first decade of the twentieth century was a time of expanding mass media—proliferating newspapers, the development of the huge newspaper chains, stepped-up wire services, national magazines, and an embryonic movie industry—that breathed life into public relations. Its organized practice stemmed from this decade.

By World War I, public relations was sufficiently well established as a profession for President Wilson to call on it for assistance. He established the Creel Committee as an adjunct of government, so named because it was headed by George Creel, a newspaperman who was one of Wilson's friends. The committee was charged with controlling war information, helping to finance the war through the sale of Liberty Bonds, convincing the American people of the war's necessity and the need for families to conserve food and other commodities, popularizing the draft, supporting the Red Cross, and encouraging sympathy for the Allies and hatred of the enemy—a mammoth assignment that it carried out with considerable success.

World War I to World War II

Unlike the propaganda agencies of other countries, the Creel Committee advocated "expression" rather than "suppression." Rather than manipulating war news to keep the citizenry ignorant, it shared with America her military's successes and frustrations, and brought about unprecedented public and corporate involvement in the war effort.

The next milestone in the development of public relations as an organized practice was engendered by Edward Bernays. Bernays, who had been a member of the Creel Committee, set up practice in New York following World War I. Fresh from the Creel Committee's phenomenal success, Bernays became one of the best-known members of a fraternity that had grown convinced that public relations had enormous potential for mobilizing public opinion. In fact, Bernays' philosophy of public opinion manipulation is summed up in his phrase, "the engineering of consent."

Bernays wrote the first book on public relations, *Crystallizing Public Opinion,* which was published in 1923.[4] That same year, at New York University, he taught the first course in public relations. He was the first to apply Freudian and modern sociological perspectives to public relations (See chapter 2). Because he continued to contribute fresh ideas to the industry for some seven decades, in the eyes of many people he deserves equal billing with Ivy Lee as a father of public relations.

In 1930, with the beginning of the Great Depression, Paul Garrett became the first public relations director of the General Motors Corporation. When asked by GM how it was possible to make a billion-dollar corporation look small (wealth was suspect during the depression), Garrett replied that he had not the faintest idea. He proposed a program in which GM could use a part of its resources to provide high school and college educations for youngsters who otherwise would go without, and in which GM could make significant contributions to schools and municipal improvements in those communities where its plants were located. These were all windfalls to public agencies during the constricted years of the depression. Garrett reasoned that GM would receive public relations credit as a benefactor, in addition to which it would be making its plant communities more efficient and would also be developing a pool of educated executive talent for the future. Garrett referred to this as "enlightened self-interest," and his program was the first of the socially conscious programs of industry.

Public relations was hard pressed during the depression to serve business, which had been cast as a villain. Furthermore, because of its intangible nature, public relations was one of the first business services to be curtailed in a time of tight money. Labor unions and government, however, forged ahead in using public relations. Public opinion measurement, as developed by the Roper and Gallup polls, began in the mid-1930s, and proved to be invaluable during elections. Political campaign specialists also came into their own during this period. The husband and wife team of Clem Whitaker and Leone Baxter set up shop in 1933 in San Francisco, and in over a quarter century refined campaigning to an art form, winning all but a handful of the dozens of campaigns they managed.

The outbreak of World War II demonstrated once again the necessity of using PR in coalescing public opinion. This time around it was the Office of War Information, headed by radio newsman Elmer Davis, that became the government's PR arm. The OWI recruited millions of workers for the nation's new munitions factories, and helped smooth the transition from depression to wartime economy. It also promoted the sale of war bonds and the conservation and recycling of scarce materials needed for the war effort.

Psychological warfare, born of the social sciences, was refined in World War II. As described elsewhere in this book, propaganda films and nationalistic radio broadcasts became an essential part of the media mix. The messages were created to motivate the republic and frustrate the enemy; radio, for instance, was used both to remind Allied soldiers of home and the democratic values worth fighting over, and to convince the Germans and Japanese that they were engaged in a losing cause. (Naturally, the propaganda war was reciprocal, with both sides using similar tactics.) It is difficult to measure the actual success of such persuasive techniques, as we noted in chapter 2. Nevertheless, the public relations campaigns continued, on domestic and military fronts, and have become an essential ingredient in times of international tension.

In the years following World War II, public relations grew rapidly. These decades saw journalism and the social sciences form a loose union in organized practice. In spite of the growth, the field of public relations still lacks a concise definition. The problem is compounded by the image of public relations, whereby many people performing essentially public relations functions in government and industry and elsewhere are given different titles. The armed services lean heavily on the public information function as opposed to public relations. The terms *public affairs* and *vice-president of communications* are becoming increasingly popular, but these are actually euphemisms for *public relations* that, by seeking to avoid the stigma, merely emphasize it and create a crisis of credibility.

There are hundreds of thousands of public relations practitioners in America now. The Department of Labor says there are 172,000 public relations writers; other independent surveys of the field reveal that somewhere between a quarter and a third of a million people earn their living doing PR work. The discrepancies in numbers of personnel reflect different definitions of the field. As we shall see, the field encompasses a vast arena of responsibilities in an increasingly complex society, a society whose specializations and needs for public understanding and acceptance have made public relations a growth industry, with no letup in sight.

Public relations is carried out on international, national, regional, and local levels. Earlier in this century public relations people were generalists, jacks-of-all-trades. To a great degree this is still true in smaller agencies and communities. Under a single roof or hat they offer advice and counsel,

World War II and Beyond

Psychological warfare was greatly refined in World War II.

Public Relations Business Today

PR has international, national, regional, and local components. At each level, different tasks are carried out.

develop political campaigns, engineer product promotions, write annual reports and publicity releases, coordinate events, lay out brochures and produce audio-visual programs. They meet each new job as best they can, equipped with a thorough knowledge of the local community, an intimate working relationship with the local press (particularly the local newspaper, whose clippings give them a record of their efforts), and an ability to get things done on the local scene. In a sense, they are the in-house PR representatives of America's communities; they just don't happen to be on any single payroll.

Beyond the community level, there are still a number of "one-man shops" (increasingly, they are likely to be "one-woman shops"). They are small boutiques, but they tend to specialize, some in straight publicity, others in research and speech writing, still others in video scripts. Some do magazine layouts together with the necessary contacts in the magazine world. Recently a new breed of "image doctors" has put in an appearance. These wizards coach diction and take the hard edges off accents, and train clients in how to meet the press and public. (One group actually specializes in training political and business figures for appearances on "60 Minutes.") Some dress their clients and style their hair, choose their tables at smart restaurants, and tell them where to go, when, and with whom.

Much of public relations counseling is conducted on a local basis, for two reasons. First, a considerable portion of public relations deals with the community and community relations as a specialty—painting the organization as a concerned corporate citizen. Even the largest corporations have diversified their community relations into the communities where their branches and plants are located, and it is a local public relations counselor who serves these individual plants. Second, and more significantly, public relations is still print oriented in our electronic age. Newspapers have remained the principal vehicles of public relations practice because material that appears in print provides a tangible record of the public relations person's efforts. It is, thus, the local counselors who can best contact the local newspapers in their communities.

There are, of course, some major national and international public relations firms. Three dozen or so of them do at least a million dollars worth of business annually, with the ones at the very top doing more than $25 million. Best known are Hill and Knowlton, Burson-Marsteller, Carl Byoir & Associates, Ruder & Finn, and Harshe-Rotman & Druck, each of which has several hundred employees. Many of these top firms have offices in all the major United States cities and most major world capitals, operating internationally as general public relations counselors. In addition, many of the major advertising agencies have public relations divisions: J. Walter Thompson Company, Grey Advertising, and others.

At least fifteen hundred public relations counseling firms are in operation in the United States. All told, several billions of dollars are expended each year on public relations efforts. All PR practitioners work in one way or another for organizations: governmental agencies (whose public information officers probably outnumber public relations people in the private sector), business and industry, charitable and cultural organizations, educational institutions at all levels, hospitals, labor unions, churches, professional associations, and even the mass media.

All organizations have public relations whether they want it or not, just as all individuals have personalities; some are good, some are bad, and some are indifferent. Like individual personality, organizational public relations is capable of being changed by dint of hard effort over a long period. Those changes are hard to guage, somewhat like "measuring a gaseous body with a rubber band," in the words of one early writer. Public relations is necessarily long-term and is, in fact, a corporate personality dealing with publics rather than with other individuals.

All organizations have public relations, whether they want it or not.

Public relations may be either the "brushfire" or "fire prevention" type. Brushfire public relations moves from crisis to crisis seeking to put out the fires; it awaits imminent disaster before taking corrective action. It is widespread, but not particularly efficient. Fire prevention public relations seeks to foresee a crisis and avoid it. It is a continuing effort, constantly upgrading the organization, planning for the future, evaluating its own results, and utilizing this evaluation as updated research for future planning.

PR may be either the *brushfire* or *fire prevention* type.

A relatively new aspect of public relations called crisis management is a spin-off of fire prevention PR. Organizations seek to anticipate all the factors that could possibly go awry and then plan for them in advance. A key tool is the crisis plan detailing the actions which all affected personnel from top management down will perform. Often it includes prewritten press releases with blanks to be filled in, as well as a listing of where all principal people can be found. In the name of public relations, business is finally devoting its considerable organizational skills to the unexpected and unwanted which inevitably will occur. No one expected two Chevron tankers to collide in the night in San Francisco Bay, just as no one expected Tylenol capsules to be tampered with; only slightly more predictable was the suicide bombing of the American Embassy in Beirut, or the leakage of poisonous gas at the Union Carbide plant in Bhopal, India. Nevertheless, there were crisis plans in place for all of these incidents.

"Crisis management" is an important new facet of PR.

There is a kind of "iceberg" analogy to all this. Like the iceberg, whose visible tip constitutes only 10 percent of its mass, only about 10 percent of public relations is apparent to the publics in the form of news releases and news stories, films and promotions, speeches and mailings, and spot announcements and television programs. The rest is meticulous, grinding research and planning, painstakingly accomplished: deliberate statistical analysis, interviews, and long hours of reading and writing. It is far from the glamorous vocation that has been depicted—a misplaced stereotype of three-martini luncheons in exclusive bistros, glamorous travel, and lavish expense accounts.

Tools and Functions

PR is a two-way street, connecting management with its publics.

Good public relations is a two-way street. Not only does the PR staff interpret the organization to its public, but it gives feedback to the organization, interpreting the public's opinions, offering potential public reaction to various corporate courses of action. Unless it does the latter and unless management is both willing to listen and willing to provide the atmosphere in which this mutual exchange can take place, public relations is not doing its whole job and thus management is deprived of an invaluable tool in the decision-making process. (Gerald Ter Horst resigned as President Ford's press secretary not only because Ford pardoned Richard Nixon for his Watergate involvement, but because Ter Horst was not consulted in the decision and felt therefore that he could not do his job for the presidency.)

The tools used by public relations to reach its various publics are many. There are the mass media, of course, but there are also specialized media—public relations films, tapes, and publications, or house organs—that are used extensively to reach essentially captive audiences. Corporate annual reports are a form of public relations; so are addresses to groups, and seminars and conventions, which is why speakers bureaus, often supplemented by films and videotapes, are a stock-in-trade for major public relations operations. There is even a public relations wire service or "business wire" to feed PR stories directly into newspaper and broadcast newsrooms.

Public relations pays a good deal of attention to schools at all levels, providing educational films and tapes, and in some cases, course material. The idea is to condition the young with the thought that they can influence their parents and, then when they grow up, they too will become consumers and believers. The Bell System is particularly active in communications instruction of this type.

Public relations is not the same thing as publicity. Press clippings are not the true measure of effective PR.

Too often public relations is confused with publicity; in fact it has been identified with it. While publicity is one of the important tools of public relations, it is far from the sum and substance of the profession. Publicity consists of obtaining free space or time for promotional material in the press or on the air, and although this material masquerades as news, its purpose is often hidden. One of the advantages of publicity is that in a quantitative society it can be measured; the number of column inches obtained in the press and the number of minutes on the air can be totaled. Therefore a tangible record of publicity can be kept, which has led to reliance on the *clipping book* as a measure of public relations effectiveness.

PR, like advertising, consists of information as well as persuasion.

Like advertising, public relations includes two separate but related functions: information and persuasion. By far the easier to explain is the information aspect. This function also comes closest to public relations as a legitimate adjunct to the news, an auxiliary news source in its own right. Information, whether offered through the press or before groups or in house organs, keeps the various publics advised. It describes, delineates, and explains goods, services, events, and ideas; it seeks to break down the barriers of confusions and misunderstanding. It is this aspect of public relations that has proven most effective.

Public relations professionals often assert that their most important responsibility is to communicate their employer's side of the story to the public, since it frequently is not known. The PR function is important because few staff people in organizations have the skills and understanding necessary to communicate such messages.

The manipulative aspect of public relations, which has been called news management in government circles, is far more difficult. Persuasive news management is heavily bound up in the phenomena of public opinion formation and change. One method is the use of reinforcement, repetition of a theme until the public becomes supportive. It can backfire easily, accomplishing the opposite of its purpose. So little is known about public opinion that manipulative public relations has to be treated with extreme caution, as politicians are increasingly aware. Manipulative public relations tends to ignore the collective wisdom of the people. Abraham Lincoln saw this and expressed it succinctly: "You may fool all of the people some of the time; you can even fool some of the people all of the time; but you can't fool all of the people all of the time." He was talking about *public relations,* although the term had not yet been coined.

The most successful PR people are the ones who know that they are performing a useful information service in providing the media with news that could, in the ordinary course of events, be obtained in no other way. This is because none of the mass media have the physical or human resources, money, or time to fully cover all events. For some media, public relations people are needed to provide much of their news.

Although PR people do not go hat in hand a-begging, they are aware of the intense competition for attention. They know that they are in competition for newspaper space and broadcast time with a platoon of reporters, with the chattering wire services from around the world and nation, with the reams of feature material—syndicated or otherwise—and with every other press agent in town. They have established a working rapport with the media; the good ones have proven their integrity. For these pragmatic reasons, most public relations material in the mass media has a legitimate claim on news. It may be self-serving, but frequently it is news.

PR people know they are in tough competition for time and space in the media. Therefore they must deal in news.

Events and Pseudoevents Public relations deals in events. Outside of special areas of public relations, it makes extensive use of the mass media, particularly the press. The press deals in news, and events become the peg on which public relations can hang its client's hat. Events provide the vehicle to carry a story to the press and, at the same time, the rationale for an editor or news director to use it.

Events are what P. T. Barnum was thinking about when he marched General Tom Thumb's wedding procession up Broadway. Ivy Lee's Pennsy railroad wreck was an actual event which brought a favorable fallout and, not accidentally, improved the reputations of both Lee and the railroad. Events are what the White House staff members are thinking about when they list "photo opportunity": Nixon walking on the beach at San Clemente, or the Reagans embarking for Camp David by helicopter from the White House lawn.

Pseudoevents are PR fabrications whose exclusive purpose is to generate publicity.

There are events and *pseudoevents*. Events occur in nature and may be capitalized upon, as with Ivy Lee. Pseudoevents are manufactured out of whole cloth for the exclusive purpose of generating publicity, dramatizing an occurrence, or calling attention to something.[5] Sometimes the distinction between the two is frail, as when the presidential helicopter takes off within the range of camera and microphone. Sometimes the distinction is obvious.

An event, or pseudoevent, may range from a simple item such as the citizen-of-the-year award by the local Kiwanis Club to the dazzling production surrounding "Hands Across America," the $50 million fund-raising campaign in May of 1986 devoted to aiding the nation's hungry and homeless. It's one thing to buy a $20 plaque and inscribe it with the name of a local citizen who has volunteered time and energy to the local community. It's quite another to implement a monumentally ambitious plan to forge a human chain, six million strong, stretching from New York to California. Nevertheless, both campaigns are public relations, and both are newsworthy.

Both events can be categorized. We are a quantitative nation, so the one hundred-thousandth or one millionth item is intensely significant. The millionth visitor to the Picasso show at New York's Museum of Modern Art, the ten-thousandth blood donor this year in San Rafael, the six million people linking hands, etc. are newsworthy.

Awards and testimonials are good events, always worthy of a little ink. There is an organization in San Diego, California, whose sole purpose for being is to give an annual award to a prominent citizen. They do their thing to the attendant hoopla, then disappear for another twelve months.

The popularity of events cuts both ways. The media, being what they are, catering to the public, are ever watchful for something new, different, unusual, something people will talk about. Public relations people, on the other hand, have discovered that if the media will cover an event themselves, devoting their own time and resources to the event, the coverage will be more complete and colorful than if the PR person simply supplied the story and art.

Pizzazz and *hype*—terms to describe ultimate pseudoevents.

Pizzazz and Hype A couple of terms reserved for ultimate events, and the manner whereby they become ultimate, are *pizzazz* and *hype*. Pizzazz refers to events so outrageous by their scale and nature that they demand the media's attention. Hype refers to the merchandizing of these events or products in an artificially engendered atmosphere of hysteria.[6]

Recently, heavy doses of pizzazz and hype have been utilized for "Hands Across America" and the Statue of Liberty, in successful efforts to overcome the competition for attention that plagues an accelerating society.

"What is hype? An orientation."

Were the worthy schoolmaster on whom James Hilton modeled the hero of his novel *Good-bye, Mr. Chips* around today, he would probably be hosting a morning talk show entitled *Hello, Mr. Chips*.

Hype would have had its way even with this shy, essentially serious person, using all the contemporary techniques of marketing and packaging at its command—sales gimmicks, publicity stunts, promotional junkets, ad campaigns, and public relations—to inflate him into just another macrostructural phenomenon. You can bet your bottom buck there'd be Mr. Chips T-shirts, Mr. Chips rulers, Mr. Chips chalk and—*some* marketing whiz would be sure to come up with—chocolate Mr. Chipses.

"Hype." A word aggressively in tune with the times. Its very sound—sharp, shrieky, cheap, belligerent, predatory—alerts us: A hard sell is in the works.

Available to common usage as either noun or verb, hype can be most usefully defined as the merchandising of a product—be it an object, a person, or an idea—in an artificially engendered atmosphere of hysteria, in order to create a demand for it to inflate such demand as already exists. Its object is money, power, fame.

The provenance of "hype," like that of most coined monosyllabic words, is murky. Some say the word is derived from the hollow hypodermic needle; that in the narcotic addict's argot, "hyped up" meant "high." Others argue that "hype" is derived from "hyperbole," the Greek word for excess or exaggeration. Hyperbole is also a figure of speech not intended to be taken literally; the great Roman rhetorician Quintilian defined it as "an elegant straining of the truth."

There is nothing elegant about hype. And very little truth in it. But not all hype is unattractive, not all hype is bad. There are exaggerations that are unobjectionable, and some that are properly enhancing. Others, however, are outrageous and duplicitous.

Hype routinely debases language, the currency of communication, thus making the coin worth less; in a universe where *everything* is *"FABULOUS!,"* nothing is anything.

And most invidiously, hype manipulates taste as it vitiates our power to discriminate. An item may be inferior or superior—that, we often can't discern; all we can tell is whether the hype is first- or second-rate. The competition for our dollar is really a contest between one hype and another. Since we're buying the hype when we buy the product, hype is a product, too.

Hype differs from advertising—which it employs, along with public relations, as accomplices—in that it is not directly paid for. Its clear first principle is to attract as much *free* publicity as possible—news stories, magazine covers, talk-show appearances, gossip-column mentions, and editorials.

Perhaps the most nefarious feature of hype is that it is not bill-boarded as a species of advertising and therefore enjoys greater credibility. We distrust ads and press releases because we know where they come from; we don't always distrust hype because very often we can't see it for what it is. Its inner workings are as invisible as its results are visible.

Hype, like propaganda, is a conspiracy against the public. And the public not only tolerates it but unconsciously encourages it. . . .

Because we live in the age of hype, the process demands inspection, if not evisceration. How does a movie get to be a box-office hit? How does an ad campaign completely rejuvenate an industry? How does a book become a best seller? How does a rock star sell out Madison Square Garden? How does a hairdresser get to be as famous as the high-status people he serves? There's work involved. And hype is that work.

Source: Steven M. L. Aronson, *Hype.* (New York: William Morrow and Company, Inc., 1983), pp. 15–17. By permission.

"Hands Across America" was the brainchild of Ken Kragen, president of USA for Africa—the first of the massive mid-1980s relief drives that capitalized on media celebrities and pizzazz to awaken the world to pressing social issues. To organize the six-million-member human chain that stretched across 4,152 miles, Kragen solicited the help of 40,000 volunteers and lined up major sponsors such as Coca-Cola and McDonalds, plus about 1,000 other corporations. To focus media attention—as if that were still lacking—he got commitments from celebrities such as Kenny Rogers, who flew a group of famous friends to a remote southwestern desert, and Steven Spielberg, who "bought" a mile in the San Fernando Valley to fill with groups of handicapped and abused children. Other chairpersons included Lily Tomlin, Bill Cosby, Kareem Abdul-Jabbar, Pete Rose, and Jane Fonda, all of them well known for attracting the media. At the last minute Ronald Reagan changed his mind about participating, and received massive coverage for this unprecedented photo opportunity. "Ordinary" Americans who joined the fund-raiser were asked to pay $10 for the privilege of taking part. For that, they got a certificate. For an additional $25 they got a T-shirt, and for $35 a visor and a pin.

Prior to the linking up of the millions of hands, a three-hour program was beamed free via satellite to radio stations across the country, featuring songs like "We Are the World," "America the Beautiful," "I Wanna Hold Your Hand," and the new hit song written especially for the event, "Hands Across America." Tying the whole chain together was a network of 4,000 ham radio operators, stationed at roughly one mile intervals along the way.

At precisely 3 P.M. EDT, the ragged, motley line of volunteers raised their clasped hands from sea to shining sea and burst into song—and it was all captured on film, on live video, and by thousands of journalists.

A pseudoevent, engendered by hype and pizzazz? Perhaps. However, as promoter Kragen pointed out, it was organized with very serious objectives: to raise from $50 to $100 million for hunger and homelessness in the United States, and to send a very clear message to politicians, and anyone in and out of government, that hunger and homelessness are major issues that need to be addressed.

Only six weeks later, on 4 July 1986, the refurbished Statue of Liberty celebrated her 100th birthday with similar pizzazz. The reconditioning of the statue had been a national issue for years, with everyone from Lee Iacocca to the nation's schoolchildren collecting money. When the big day came, ABC TV paid $10 million for exclusive broadcast rights to the event, which ABC News President Roone Arledge called "part entertainment, part spectacle, and all genuine news."

ABC set aside seventeen and a half hours of air time over a five-day period for the party. On 2 July it offered a one-hour prime-time preview of the events; on the third it covered the arrival of the 35-ship international armada and later that evening devoted three hours to the opening ceremonies. They were not modest. They included President Reagan lighting Lady Liberty's torch; the presentation of the Medal of Liberty to a dozen naturalized citizens, including Bob Hope and Henry Kissinger; and a gala on Governor's Island with Frank Sinatra, Liz Taylor, and Mikhail Baryshnikov. Events on the Fourth included an international naval review and Operation Sail '86, with the largest-ever flotilla of tall ships. That evening saw a Boston Pops concert with John Denver, Whitney Houston, and Johnny Cash. Then came the biggest display of fireworks ever seen in America, followed by an hour long special on the meaning of liberty. On Saturday, the fifth, ABC continued its coverage, with Zubin Mehta and the New York Philharmonic in a Central Park concert, and on Sunday, the sixth, it wrapped things up with a sports salute by top American athletes, and showed closing ceremonies from Giants Stadium, with Patti LaBelle and Willie Nelson.[7]

The aforementioned events or pseudoevents attracted national and worldwide attention because they were unique, because they involved human drama, they were done on a large scale, and they drew celebrities. If those elements do not make something newsworthy, nothing will. Unfortunately, however, as we have already come to realize in our discussion of hype and pizzazz, there must be a limit to how much of this image manipulation the public can take, and how much will be effective. To return to the analogy of hype and drug addiction, how much is enough, and when do we realize that we have overdosed and slipped into a stupor? At what point do these full throttle promotional campaigns create a world in which slickness supplants substance, in which we have emptied our grab bag of superlatives

There's a limit to how much image manipulation society can tolerate.

(i.e., how many "stupendous!," "terrific!," "spectacular!," "unique!," and "greater than great!" events can we sit through?), and in which the celebrity becomes as common as the kid next door?

In a world overdosed on hype and pizzazz, is it time for a clever promotional expert to really stand apart from the crowd by trying some old-fashioned soft sell? Who knows? It just might work.

Audiences for Public Relations

In order to secure attention, public relations must pinpoint interest. This requires a clear definition of *audience,* or *public relations publics.* These publics range from the consumer to the government. Several publics are related to a business or industry: employees, dealers, suppliers, and competitors, for example. Other publics are part of the larger environment; among them are the community, educators, and financial interests. To a major corporation, many of these publics are discrete but interlocking. Each has a different interest in the corporation: employees in wages and fringe benefits, stockholders in dividends and appreciation of investment. Yet they all share an interest in the corporation; it is their rallying point. But the matter is not all that simple; an employee may easily be a stockholder and a consumer. The public relations people of the corporation must communicate with employees in different terms than they use in materials presented to the stockholders if they are going to hold the interest of each group. Defining these audiences, determining how to reach them most efficiently with the least overkill, and framing what to say to them is the job of public relations. At the same time, PR must try to measure the separate and collective pulses and give the readings back to management as guidelines to policy.

How does public relations reach these publics? It utilizes all forms of communication, not just mass media. It will prepare policy bulletins for employees and financial statements for Wall Street. Each year the public relations arm of the corporation will produce glossy annual reports for shareholders. It relies heavily on word of mouth. Public relations makes use of conventions, seminars, group addresses, and individual contacts with opinion leaders; it obviously employs mass media, from network television to community newspapers, with particular reliance upon specialized magazines.

Public relations has developed audience identification to a fine art, and ingeniously devises some imaginative means of reaching particular, discrete publics. Extensive use is made of demographics and public opinion polling. It is fair to say that both public relations and advertising, the organized forms of persuasion, are devoted to generating word-of-mouth coverage between individuals, for it is only through this means that a message perpetuates itself once it is let loose.

PR utilizes all forms of communication to reach its wide variety of publics.

By now we recognize that public relations is a combination of art and science. As such, it can be understood in terms of a series of processes. It is possible to model these processes, as we did in earlier chapters when considering the basic process of human and mass communication.

In their long-time standard work *Effective Public Relations,* Cutlip, Center, and Broom have offered a four-stage model of the PR process. That model, which has become a fundamental component of PR seminars and classes, suggests that effective public relations practitioners serve as listeners, counselors, communicators, and evaluators. The first stage is that of *research-listening,* which involves probing the opinions, attitudes, and reactions of various publics toward the organization, along with learning as much as possible about the oganization's problems and potentials. Next comes the *planning-decision making* stage, in which public attitudes, opinions, ideas, and reactions are brought to bear on the programs of the organization; the process enables the organization to chart a course in the interests of all concerned. Third is the *communication-action* phase, when the actual "PR-ing" takes place. The chosen course of action is put into effect, and the organization makes certain all those who may be affected, and whose support is essential, are clearly informed and involved. Finally, to complete the circle, is the *evaluation* stage, in which the results of the PR program and the effectiveness of the techniques used are assessed. Obviously, in an ongoing program, the evaluation stage leads right back to updated research for future planning.[8] Cutlip, Center, and Broom suggest that all four of the phases are of equal importance. Even a casual observer of the public relations scene will have to admit that, in most cases, the third stage—the actual program of promotion—receives far more attention than any of the other three, which might explain why so much PR appears random and ineffective; it tends to be one-way communication.

It is the reversal of this model, and the emphasis on public relations as a two-way street, that has led Grunig and Hunt to develop a new theory of public relations which they call the "two-way symmetric model" in their book *Managing Public Relations.*[9]

To go back a moment, they note that press agentry developed by Barnum is pure one-way persuasion, refined and updated by Bernays and very much with us yet. Ivy Lee contributed information to the public relations scene, albeit still one-way communication; it is still highly evident in governmental, military, and not-for-profit public relations.

With Paul Garrett's "enlightened self-interest" for General Motors in the depression, we find the beginnings of two-way public relations, and the recognition that public reaction can and should influence corporate policy. This resulted in PR directors sitting in America's boardrooms, if not actually serving on the boards, where they could serve as the eyes, ears, and voice of the public. As time moved on in the post-World War II years, and the refinement of polling techniques coupled with computer magic began

A New Model of Public Relations

The art and science of PR can be depicted as a communications model.

Traditional models show PR to be a one-way street, with information and persuasion flowing from the company or source to the public.

New models of PR show it to be a two-way street.

to offer rapid, sophisticated feedback, more and more organizations began to lean heavily on evaluation to chart their courses. This is what Grunig and Hunt refer to as "asymmetric or unbalanced two-way communication."

The unstated implication in the asymmetric model is that an organization is using its feedback to further its own ends, and that the only purpose of the research is to discover the most effective way of marketing the product, selling the idea, raising the money, or settling the strike. Essentially, while the communication may be two-way, the basic corporate goal is and always will be one-way.

In the symmetric model, on the other hand, which is still too new to have a viable history, the research, feedback, and evaluation are used to try to effect genuine accommodation to differing viewpoints. It recognizes that there is daily change in the air, and that the corporation must change to survive and learn to live with its publics. The symmetric model sees the corporation and the public operating in a mutually agreeable and mutually acceptable world in which both can advance toward new goals, and even change their identities, while retaining confidence in each other and without looking around to see who is gaining.

The Grunig-Hunt symmetric model calls for new theory and even a new definition of the public relations process. It suggests that the late twentieth century demands new vision for corporate survival and profit.

We have heard much recently about "social responsibility" as a desirable corporate goal. Milton Friedman has remarked that there is a social responsibility to make a profit. This is no facetious one-liner, but a profound economic truth. There is the responsibility to the employees, the customers, suppliers, shareholders, and community. Hundreds, thousands, and perhaps millions of people may be reliant upon the corporation, and all will be affected if the corporation does not make a profit and continue to exist. Lee Iacocca demonstrated this dramatically as he resurrected the Chrysler Corporation.

There are dangers in an asymmetric program of "social responsibility." The DuPont Corporation discovered through a public opinion poll that by and large the corporation was disliked locally. To correct this, using the asymmetric model in conjunction with the currently popular concept of social responsibility, it embarked on a public relations campaign to prove itself to be socially responsible, to be a good community citizen. At the conclusion of the campaign, it conducted an evaluation and discovered that yes, the community felt the DuPont Corporation was socially responsible, and disliked it even more. The community apparently did not care about social responsiblity, whatever that is. A more careful analysis of its image problem, and a sincere effort to contribute to the community's welfare, may have had a different effect.

All this is by way of saying that, from a public relations standpoint, there may be a deeper bottom line, and that America's corporations had better find it. Grunig and Hunt's "symmetric model" of two-way interreaction points a way toward the future.

Earlier we pointed out that at the outset public relations people tended to be generalists, and that, to a great degree, such is still the case in smaller agencies and communities. Given the increasing complexity of the twentieth century and the need of so many different institutions to maintain favorable relationships with their respective publics, it is only natural that public relations has become more and more specialized. At this juncture we will consider several of those areas of specialty: political public relations, lobbying, and not-for-profit public relations. There are, of course, other areas we could review, but a look at these specializations offers some insights into the diverse nature of the field.

It is not surprising that a nation founded in PR should find PR such an integral part of its political process. Nowhere else are the results of opinion molding so quickly and conclusively known, and nowhere else are they so accurately measured, as they are at the ballot box.

Political campaigning and public relations activities have been inseparable throughout American history. The campaign of William Henry Harrison, an Indian War hero, was based on a slogan from his most famous battle: "Tippecanoe and Tyler too" (for his vice-president). Harrison's campaign featured torchlight parades and giant barbecues. Andrew Jackson, as noted ealier, was the first populist president. Much later Mark Hannah, as campaign manager for Grover Cleveland, sat his candidate on his own front porch in Ohio and brought America to him to shake his hand. By the trainload they came, to return home and tell of how they were personally greeted. Hannah developed special appeals for special groups and was the first to emphasize audience identification—not everyone reacts to the same stimulus, and different appeals should be used to reach farmers, coalminers, railroad workers, lumberjacks, traveling salesmen, and homemakers.

Franklin Delano Roosevelt became one of the masters of public relations, both before and during World War II. He first saw the capacity of radio for the molding of public opinion. His "fireside chats" offered often during the depression brought his soothing, sonorous voice into America's living rooms and calmed America's fears, while holding out hope for the future. FDR was facing enormous obstacles, with big business and most newspapers speaking against his massive changes in federal policies, his "New Deal." But his personal charm and adroit use of the fireside chats swayed public opinion, and demonstrated to business, the military, education, and other establishments the value of building public trust.

By the 1950s television was becoming a force to be reckoned with in shaping public opinion. President Dwight Eisenhower was the first to allow televised press conferences, and used them effectively to demonstrate his affable personality. He hired Robert Montgomery to coach him in playing to the cameras; it was the start of a whole new type of politics.

Specializations in Public Relations

Today PR is becoming more and more specialized.

Politics and Public Relations

Politics and PR have been inseparable throughout American history.

Today's politicians must
master the art of television.

Ronald Reagan, "The
Great Communicator,"
delivering one of his
Saturday morning radio
addresses. He has
successfully used the
medium to go straight to
the people, controlling
his own image.

John F. Kennedy displayed the same inherent understanding of television that FDR did of radio. He saw the enormous potential of video for projecting images rather than issues, and through his 1960 TV debates with then vice-president Richard Nixon emerged as presidential timber. Once in office JFK used television to talk to the nation disarmingly, exploiting his personal charisma while prompting a sense of trust and optimism about the future.

All presidential candidates and incumbents since have followed JFK's clues, some with more success than others. Lyndon Johnson, who succeeded JFK, and Richard Nixon never seemed at ease before the cameras. That dilemma surely did not help Johnson gain public acceptance during the trying days of the Vietnam War, nor did it help Nixon during Watergate.

The fortieth president, Ronald Reagan, himself a former film actor and television series host, is certainly aware of the power of both radio and television. Through his weekly radio talks he, like FDR, can go directly to the people. Through his carefully orchestrated television appearances, including press conferences and "photo opportunities" (when he is walking past the camera crews and reporters, visible but out of conversation range, and therefore not really accountable), he can control the images he wants to project. Some call him "The Great Communicator"; others, "The Teflon President." Both labels recognize his public relations skill in manipulating the electronic media and, in turn, public opinion.

From William Henry Harrison to Ronald Reagan, presidential politics have indicated the extent to which public relations has infiltrated government and show how adroit media usage constitutes the means of projecting an image. The public relations image is not necessarily a mirage; it can easily be a reality (as it was for FDR and JFK) that needs only the proper lens to project it where it can be seen by everyone. Choosing and using that lens is a part of public relations.

Political Campaigning Political campaigning differs from organizational public relations in several important respects. From the practitioner's standpoint, it is seasonal. Political campaigns occur only at certain specified intervals. The campaign specialist, therefore, is faced with a feast or famine situation, which is one of the reasons for the high costs of campaigning. Second, political campaigning is a crisis situation wherein all activity is always concentrated into too short a time frame. Everyone knows about the harrowing schedule of a politician on the campaign trail: the twenty-four hour days, the jet travel, and speeches in Pittsburgh, St. Louis, and Los Angeles all in one day. The public relations people who arrange, schedule, and promote all this activity have equally appalling schedules. They worry about the press, the advertising, the crowds, the advance arrangements, the timetable, the speeches, the sources of money, the budget, the press kits, the infighting, the polls, and the candidate's blunders. Years of public relations activity are crowded into a couple of months.

Political campaigning, too, is the one form of public relations that
operates within a known time frame with a known cutoff date—election
day. Campaigning is also the one form of public relations that is accurately
measurable—at the ballot box on election day. The results are tangible, to
be sure, but some method is needed to measure effectiveness during the
process. Too much is at stake to leave campaigning to intuition alone. Public
opinion polls became an extremely popular tool to judge the effectiveness
of a campaign, uncover weaknesses to be minimized, and indicate strengths
to be capitalized.

Finally, political campaigning differs from more traditional public relations, in that there are usually only two candidates in a general election,
and only two points of view in a ballot proposition or a bond issue: candidate
A or candidate *B,* yes or no, for or against. Thus, the entire spectrum of
public opinion must be compressed into only one of two choices. In society
as a whole there is a wealth of personal opinions, interests, and tastes. But
this is not so in politics, where general elections force a choice between two
alternatives, neither of which may exactly meet anyone's criteria. As often
as not this choice is made on the basis of the lesser of two evils rather than
on genuine conviction. This being the case, the political campaigners must
seek to enhance the stature of their own candidate while seeking to discredit
the opponent. A vote taken away from an opponent counts as much as a
vote attracted to the cause.

Political campaigning demands
special and intensive PR tools.
Its results are clear-cut. The
candidate or platform wins or
loses on election day.

Political campaigning has, with its charges and countercharges, occasioned enormous criticism aimed at public relations, advertising, and the mass media. This might be due to the fact that, of all the forms of public relations, political campaigning is the most visible, the most clearly identified, and the most concentrated. A national presidential campaign takes on the aspect of the Roman circus. There are those who claim it provides a catharsis, an emotional safety valve every four years that permits the public to keep on a more or less even keel the remainder of the time. In defense of political campaigning, we should note that it is a no-holds-barred activity in which truth and misrepresentation are intermingled on both sides, and from whose massive exposure the public can come to a reasonably well-informed conclusion.

Lobbying

Lobbying is another persuasive technique allied with public relations. In fact, a number of firms that specialize in political campaigning have developed lobbying specialties because of their contacts in the legislative halls. The term "lobbying" came from the coatrooms, or lobbies, of the legislative chambers, where interested parties "button-holed" their representatives. Lobbying is concerned with influencing government, bills to be passed or defeated, appropriations to be made, "pork barrel" projects to be obtained for localities, or bridges, dams, roads, post offices, courthouses, or other public projects to be built.

Many of the labor unions and the major business and professional associations—for example, the National Association of Manufacturers (NAM), the American Medical Association (AMA), the National Association of Broadcasters (NAB), and the National Education Association (NEA)—employ lobbyists. These trade and professional associations have found it expedient to watchdog legislation, to encourage passage of certain measures and to discourage others as they affect the goals of their respective members.

Lobbying results in an imperfect balance of public opinion and legislation, but it gives people a voice.

The rationale is simple. All citizens have the right, perhaps the duty, to contact their congressional representatives to indicate their feelings on proposed legislation. This is the principle of representative democracy. In a highly organized and complex society, lobbying merely extends this individual right to an organizational right, as the organization is composed of many individuals with a common interest. As the First Amendment, conceived initially as an individual right of access, has evolved into an organizational right to print or broadcast, so within the Constitution the individual right of representation has been extended to serve society.

Lobbying is enforced by the so-called unwritten law; lobbyists are expected to be advocates, present their points of view, but are not expected to misrepresent or lie. Should they be caught, even once, the word passes quickly and no one will ever talk to them again about anything. They become pariahs. Since it takes many years to develop the know-how, knowledge and contacts to be a successful lobbyist, with its attendant reward in

money and perceived power, there are few willing to jeopardize their status for a short-term advantage; of course, the ones who try it do not reach the top.

Lobbying is largely individual people dealing with other people. It makes little use of the mass media, and while many general public relations people have come from working in the media, most lobbyists have come from the law, or from the halls of government. Former legislators and political appointees—such as President Reagan's controversial aide, Michael Deaver—frequently become lobbyists.

Most of the public relations tasks we have discussed so far have been applied to corporate or profit-making institutions, or to political campaigning and lobbying. Obviously, many of the same tools can be used in nonprofit enterprises. However, because of their unique place in contemporary society, with their overriding need to earn public acceptance and good will, the nonprofit institutions deserve special mention in our overview of public relations.

Not-for-profit Public Relations

Thousands of not-for-profit institutions have a strong need for public relations: the nation's charities, museums, galleries, zoos, community theatre, research foundations, universities, libraries, churches, hospitals, and even government at all levels, including the military. It has been estimated that one out of every six professional workers in America is employed by a nonprofit organization, as is one out of every ten service workers.

There is probably more activity in public than in private PR sectors, because in a great many cases the only "product" is an intangible, an idea. Generally, it is called public information, and the public information officers (PIOs) are charged with the responsibility of informing the public about the organization, its works, and its needs. Their tasks also include raising funds, broadening and maintaining volunteer participation, winning public acceptance of new ideas and new concepts (many of which are highly controversial), effectively marketing programs and services, and developing channels of communication with the disadvantaged who are cut off from society's mainstream.[10]

A quick overview of the special public relations problems in several nonprofit arenas should prove instructive.

Military PR Consider the PR tasks of the military. In a democracy, the armed forces need a good bit of public relations to justify the defense budgets. Caspar Weinberger, President Reagan's Secretary of Defense, once commented that there were 2,400 public information officers in the Department of Defense to serve two newspaper reporters assigned there. While this misses the point that the 2,400 PIOs have far wider constituencies than those two journalists, it does indicate the emphasis that government places on public information. It has been this way throughout most of American history, particularly in the twentieth century. The military's PR problems have grown acute in these post-Vietnam years. Public debate continues over calls for increased defense spending while budgets are being cut on the domestic front. It has not helped the Pentagon and military suppliers' PR cause to be accused of price gouging—the $500 hammer and so forth.

Nowadays every branch of the military, including the reserves and National Guard, make extensive use of PIOs to reach both the internal publics (the servicemen and women, actives, reserves, and retirees, plus civilians working on the bases) and external publics (the general American public, which has not always been favorably inclined toward the military, plus international publics, the legislature, news media, industry, etc.). The Air Force's Singing Sergeants and its flying Thunderbirds, the Coast Guard Band, and all the other highly visible components of the military's PR apparatus demonstrate the services' awareness of how important it is to leave a good impression on the public.

One of the military's key publics is the news media, and relationships between the two groups have always been testy. In the wake of the news media's uproar over not being permitted to accompany the first wave of military to reach Grenada in 1983, delicate negotiations were held. They resulted in a set of agreements about what information should be made available, and how the press should pool its resources, in times of military crisis. Like most such things, it was not completely satisfactory to either side. But that is the nature of public relations.

Hospital PR The nation's hospitals are another interesting public relations case study. When we stop to consider that almost all of their customers would rather not be there, and that being hospitalized is costly in terms of time, finances, and discomfort, it is little wonder that hospitals have unique PR problems. Few of us understand why the costs of medical care have skyrocketed so much more rapidly than any other services. However, at the same time, we expect more comprehensive care than ever before, from highly trained specialists. And we do not appreciate it when those specialists appear to be aloof, complacent, and indifferent to our individual needs.

Unless hospitals offer good service that displays a genuine concern for each individual patient's welfare, the PR battle is lost. This explains why health care public relations is such a growth industry, and why most hospital employees, from the orderlies to the chiefs of surgery, are part of the PR battle plan. Increasingly, hospitals are hiring full-time public relations directors. Their responsibilities are likely to include overseeing employee relationships, coordinating volunteer groups, arranging tours and speakers bureaus, conducting public opinion surveys of patients and handling patient complaints, and working with insurance companies to ease the financial burdens. They may be responsible for producing news releases and house organs, and are likely to coordinate the hospital's softball team.

Hospitals need to display genuine concern for their patients, most of whom are not happy about having to be there in the first place.

Public Education PR In public educaton, from kindergarten through college, there is also a great need for public relations work. Schools, like churches, used to be stable, tranquil institutions that received little public attention. That is no longer the case. Schools have become one of the most controversial and highly visible institutions in the nation. National studies by foundations, the Department of Education, and individual researchers, plus headlines in the daily press, have declared a variety of crises in education: drugs, vandalism, and violence in the schools; lower academic achievement scores and calls for "back to basics"; busing and desegregation problems; teachers' strikes; failures of school bond issues; and the like.

More than half of America's local tax dollars—property tax, sales tax, etc.—are spent on education, and in this era of consumerism the public has begun to demand its money's worth. Accountability is the new catchword in education, and the need for careful public relations has never been greater.

Individual schools, parent-teacher organizations, school districts, and state education offices have by and large come to recognize the need for improved images and more open communication with the taxpayers and their children. Recognizing the problem and doing something systematic

Schools have been tossed into the spotlight of publicity, and their need for careful PR has never been greater.

about it are not necessarily one and the same. Solutions have ranged from individual teachers or principals who make sporadic phone calls to local newspapers announcing "something interesting" happening at the school, to full-fledged, ongoing public relations campaigns that involve two-way, symmetric communication.

Good public relations practices in the public education sector include open door policies wherein the community is invited to the schools to observe special as well as everyday activities; open school board and PTA meetings at which significant issues are fully aired; full use of public advisory committees; ongoing surveys of community opinion; house organs and other internal PR activities, including personnel work; news releases, press kits, and ongoing ties with local news media; community service activities by school groups, including teachers; etc.

Schools have an enormous range of external publics to satisfy. The interests of parents, taxpayer groups, service clubs, patriotic groups, civic groups, industry, churches, alumni, athletic boosters, labor unions, legislators, government administrators, and teachers' unions must be met. Obviously, such heavy demands for communication cannot be handled randomly if the schools wish to be seen as important and fully functioning institutions in society.

Colleges and universities have unique PR problems.

At the college and university level there are additional and unique PR problems, and thousands of PR practitioners trying to address them. Traditionally, a college news bureau was expected to handle the entire PR task by sending out news releases that described research and service activities, scholarship and honor roll students, and special events such as commencements. Recently, these offices have added audio-visual components to their external PR work, producing electronic news releases that are intended to meet the needs of regional broadcast outlets. And, in a period of specialization, the PR offices are likely to employ experts who can tell the stories of the various academic centers—science, agriculture, engineering, music, medicine, law, and all the rest.

A separate office usually handles similar tasks for the school's varsity athletics, concentrating on football and basketball teams, the revenue makers. The Sports Information Directors (SIDs) have always taken on additional responsibilities, such as arranging the press guidebooks or brochures, setting up the press box for games (and arranging for food and drinks—somebody started the tradition of feeding sports writers, and the tradition has never been broken), and working with coaches and athletic directors in handling details on recruitment, fund-raising, and the like. A fascinating case study in public relations, one with which most sports fans are familiar, is the "making" of an all-American athlete or the Heisman Trophy winner.

In addition to the news bureaus or public information offices, colleges and universities have a battery of PR activities being carried out by the college development and alumni offices, where fund-raising and continued good will are the hallmarks. Increasingly, higher education has lobbyists who join the presidents and provosts in their regular treks to the state capitals during budget hearings, where the ivy tower types have to compete with gray flannel suiters for scarce public dollars.

Public relations by itself cannot, and should not, solve the kinds of problems being faced by the military, hospitals, schools, or other nonprofit institutions, any more than it is the panacea for all the ills of the private sector. The problems are more severe than just image problems. They need, and in the majority of cases are receiving, scrutiny at all levels of management. However, unless management recognizes the value of incorporating mutually beneficial, two-way symmetric communication to the task, the problems are unlikely to be resolved. And that communication is as likely to be informational as it is persuasive in nature.

One of the basic problems surrounding public relations as a practice is the difficulty of measuring its effect. Advertising can be measured at the cash register. But public relations is a necessarily long-term intangible, so completely interrelated with everything that an organization does that it is difficult to separate the effects of public relations alone. This, in turn, has contributed considerably to the questionable reputation that public relations has. It is ironic that a field that professes to mold public opinion, to change organizational images, and to accomplish corporate and political miracles should itself have such a poor public image. There are a number of incompetents in the field calling themselves public relations practitioners. As their substandard work multiplies, the reputation of the profession as a whole becomes damaged. There are neither standards (such as passing the bar for lawyers), nor a stipulated course of study (as for doctors), nor licensing (as for architects, engineers, and accountants) for the practice of public relations. Anyone can hang out a shingle.

Recognizing this, the Public Relations Society of America (PRSA) has embarked upon a program of accreditation for its members. In order to receive full accreditation, members must have a minimum of five years of public relations experience. In addition, they must pass an eight-hour examination in public relations principles and techniques, history and ethics, and they must also undergo an oral examination by a panel of three accredited peers. The exam is discriminating. Typically about 40 percent of those who take the test fail; many keep trying until they succeed. About one fouth of the 12,000 members of PRSA are currently accredited, and are permitted to put the initials APR after their names.

The hope of PRSA is that gradually business and government and other organizations will come to recognize that PRSA membership and accreditation mean that the PR person has satisfied at least the minimal qualifications of experience and knowledge and can be expected to bring a certain amount of professionalism and expertise to any job. Such a person should be the best choice for an assignment. PRSA accreditation *should* mean that the individual is a competent practitioner who has satisfied professional requirements. The PRSA program has been slow to develop; there is much to overcome.

Self-Regulation of Public Relations

The Public Relations Society of America has worked to raise PR's public image, and with mixed results.

> We are reminded that several years after resigning from the highest office in the nation, Richard Nixon admitted that he had misled the world about his role in Watergate. His memoirs explained that his response to the Watergate crisis was tactically incorrect, since, in his own words, his first response to Watergate was "that it was just a public relations problem that only needed a public relations solution."

To instill professionalism in public relations at the ground level, PRSA sponsors student chapters on campuses across the nation. There are some five thousand members in the 145 PRSSA (Public Relations Student Society of America) chapters in America. Many of these students serve working internships with PR agencies during their senior years, and a significant portion of them step into PR jobs upon graduation. Interestingly, more than three-fourths of the PRSSA members are female, and their entry rate into the marketplace has been very rapid.

Meanwhile, the International Association of Business Communicators has been making a considerable impression in PR circles. As the name indicates, it tends to restrict itself to the private, profit arena. It, too, has a rigorous accreditation program. Overall membership in IABC is similar to that of PRSA—about twelve thousand. However, only about four hundred, or 3 percent of them, have received full IABC accreditation. In mid-1987, IABC and PRSA began exploring the possibility of a merger. The move would prove cost-effective and might strengthen the overall profession of public relations.

Recently, particularly as a result of the political campaigning excesses for which public relations must share a part of the blame, there have been movements in several states toward public relations licensing. Some communities have instituted stringent regulations of the public relations practice as it affects lobbying before public agencies and political campaigning. It is quite possible that PRSA's movement toward professionalism, which is a form of self-regulation designed to forestall governmental control, may fail as being too little and too late. If widespread regulation and/or licensing of public relations does come about, it is inevitable that in the name of social responsibility it will be another form of communications control. Government regulation may not be entirely undesirable, but public relations is so interwoven with the entire fabric of mass communications that control cannot help but have side effects on the content, utilization, and future development of the mass communications complex.

Public relations, though relatively new as an organized practice, is as old as humanity. Nevertheless, it is enigmatic, and difficult to define. Its historical genesis can be seen in ancient Iraq, in the poets of ancient Greece and Rome, through the Dark Ages, and in early American history. As a modern profession—though we use the word *profession* with some reservation—it is essentially an American product, having its origins in political campaigning, theatrical press agentry, and the Industrial Revolution.

The use of public relations in the United States was highly visible beginning with the American Revolution and the plots of Samuel Adams, and continued through the establishment of the republic with the *Federalist Papers*. PR was a natural for political campaigning, as Andrew Jackson and others quickly realized. P. T. Barnum was the father of press agentry, while the age of "survival of the fittest" was represented by its PR counterpart "the public be damned." Public Information, widely used by government, military, and non-profit institutions, began with Ivy Lee, the "father of public relations" in the early 1900s.

Edward Bernays added an emphasis on persuasion and "the engineering of consent" after World War I, while Paul Garrett, first public relations director for General Motors, conceived "enlightened self-interest" during the depression of the 1930s. Garrett's model, still widely used and refined in the age of computerized public opinion polling, can be referred to as "asymmetric two-way public relations." This simply means that while the public interest is a consideration, all public relations is essentially devoted to enhancing the profitability of the corporation.

Recently James Grunig and Todd Hunt, in *Managing Public Relations,* have detailed a new theory of "symmetric two-way public relations" in which public or audience input is actually used to change and mature the corporation or institution, rather than merely improve its image.

Most public relations is conducted on a local basis, although there are national public relations firms. As many as a third of a million practitioners are working in the United States; a great many of them are part-timers, amateurs thrown into the mill for the nation's thousands of charities, community organizations, schools, and what have you. Some PR practitioners conduct "brush fire" public relations—meeting crises as they arise. Other forms—like "fire prevention"—utilize research, planning, and evaluation in an attempt to foresee and forestall potential trouble spots. The industry serves organizations either internally as a part of a company or externally—as independent counselors.

One of the major tools of public relations is publicity, but it is not the only tool. Its functions are informative and persuasive, dealing with events and pseudoevents; the latter are created for the sole purpose of generating publicity. Hype and pizzazz accompany pseudoevents, and sometimes even legitimate events, in efforts to concentrate the attention of the media and general public.

PR is like an iceberg. Only about one-tenth of public relations activity shows in speeches, releases, broadcast time, pamphlets and the like. The remainder is painstaking research, planning, and evaluation, the cyclical process through which quality PR operates.

Public relations serves the press as a legitimate source of otherwise difficult-if-not-impossible-to-get information. The press serves PR as an outlet for its messages. The rationale is that no medium has the resources needed to fully cover every event, let alone the significant nonevents, occurring within its circulation area.

An irony of public relations is that the craft itself suffers from a poor public image. The PRSA, through a program of accreditation, is working toward increasing professionalization in an area where there are no legal standards.

As an organizational or corporate personality, public relations exists, whether good, bad, or indifferent, and through the application of appropriate techniques is capable of changing for better or worse.

Notes

1. Scott M. Cutlip and Allen H. Center, *Effective Public Relations,* 5th ed. (Englewood Cliffs, N.J.: Prentice-Hall, 1978), pp. 8–12.
2. Jacques Ellul, *Propaganda: The Formation of Men's Attitudes* (New York: Alfred A. Knopf, 1965).
3. Cutlip and Center, *Effective Public Relations,* p. 9.
4. Edward Bernays, *Crystallizing Public Opinion* (New York: Liveright, 1961).
5. Daniel J. Boorstin, *The Image: A Guide to Pseudo-Events in America* (New York: Harper & Row, Harper Colophon Books, 1964).
6. Steven M. L. Aronson, *Hype* (New York: William Morrow and Company, Inc., 1983).
7. Matt Rouch, "Liberty's Fourth: 5-day ABC-TV fest," *USA Today,* 22 May 1986.
8. Scott C. Cutlip, Allen H. Center, and Glenn M. Broom, *Effective Public Relations,* 6th ed. (Englewood Cliffs, N.J.: Prentice-Hall, Inc., 1985), pp. 199–200.
9. James Grunig and Todd Hunt, *Managing Public Relations* (New York: Holt, Rinehart & Winston, 1984).
10. Cutlip, Center, and Broom, *Effective Public Relations,* 6th ed., p. 518.

News Services and Syndicates

13

Introduction

There is no way the average newspaper, radio, or television station could bring us the quantity of national and international news and specialized features we get every day if the local news staffs had to do it all themselves. On their limited budgets and with considerable time pressures, the local reporters and editors scarcely can generate enough material to keep us informed about our own communities, let alone the doings of Washington, New York, and overseas. In addition, we can hardly expect homegrown journalists to possess the myriad talents we look for every day in our media: the incisive understandings of federal budgets, major league sports, and international diplomacy; the ability to explain how to fix our autos, purchase a home, invest our dollars, tend to our health, and mend relationships; and the artistry to create comic strips, crossword puzzles, and gourmet dinners. Fortunately for the local media and for our eclectic tastes, the mass communications network has come to incorporate a wide variety of auxiliary news and feature services.

In this chapter we will consider some of the basic services found in our local media, with primary emphasis on the press associations, or news services, as well as the feature syndicates. Such services have been with us in one form or another since the middle of the nineteenth century. They have a respectable history as packagers of information. Their existence came out of the costly business of news gathering that, as it became more competitive, also became prohibitively expensive for any single news operation to bear. Essentially, the news services are a pooling of costs and efforts—joint ventures.

Not too long ago a discussion of news services would have been limited to the doings of the Associated Press and United Press International (the two primary news services) and the various feature operations that mailed or wired their wares to subscribing news media. Things are no longer that simple. The complex new electronic technology, as discussed in chapter 10, has joined forces with traditional news services and syndicates, offering many information services direct to individual subscribers. As a result today's typical mass media consumer has a far broader spectrum of nearly instantaneously delivered news and features, just as print and electronic media have expanded their networks to gather and distribute global information. Today, more than ever, the news media have to rely on these auxiliaries for a major portion of their content.

This chapter begins with a historical treatment of the major news services, then considers their current operations and problems and contributions to the news media at home and around the world. In addition to the news services' contributions in shrinking our globe, their role as so-called "communications or cultural imperialists" must be considered. To that end, we briefly cover the recent calls by developing nations for a "New World Information Order."

Media Support Systems

Next comes a discussion of the pooling efforts made by various print news conglomerates—the smaller news and feature services organized to serve the interests of member papers and others which care to subscribe.

The final section of this chapter considers the history and impact of specialized feature syndicates that contribute a generous dose of nonnews materials to the media. A cornucopia of columns, comics, horoscopes, how-to-do-it pieces, and dozens of other general subject areas are provided by hundreds of competitive feature syndicates today.

The effectiveness of supplementary news and features cannot be underestimated. Take newspapers, for instance. Some have calculated that news service news and photos constitute 15 percent of the average daily paper. When you consider that 60 percent of a newspaper is advertising, the news service contribution is greater than any other single source. At least 10 percent of the overall space—up to a third of the news hole—goes to various syndicated features and another 10 percent to specialized departments—such as society, business and finance, and sports. That leaves only about 5 percent for local news. That 5 percent, of course, includes locally produced photographs and graphics. These percentages indicate how much of the average paper is "canned," or prewritten and delivered—at a cost—to the newspaper.

Likewise, the broadcast media make very heavy use of these services. Smaller radio and television stations wishing to carry regular newscasts, but unable to employ specialized reporters, rely almost exclusively on the services. Broadcast networks and groups have effectively moved out to establish their own pooling systems, collaborating for visual and aural materials with other broadcast services around the world. Some observers have suggested that this type of arrangement may become the dominant news method of tomorrow, given the direction broadcast news is going (see chapters 8 and 10). That being the case, the services deserve a closer look.

News Services: Historical Development

From the beginning, the press in America served a dual purpose. It was isolated and local; it was also the principal public communications link to the old world. Even in colonial times, newspapers in the scattered population centers were eager to receive news from England and France as rapidly as possible, and through transatlantic shipping they had far better and more regular contact with Europe than they did with their sister colonies. It became common for publishers to send sloops out to meet incoming vessels and to return with the news from abroad, sometimes days before the slower oceangoing ships could beat their way into harbor and dock. Curiosity and economic advantage put a premium on such news. In due course, as the populations increased and newspapers became more competitive, relays of horse couriers sped foreign notices, the news of the world, first from the tip of Long Island and later from as far away as Nova Scotia, to the waiting presses in New York. The costs of riders, horses, oarsmen, and relay stations were great, and rivalry among the competing couriers was intense.

Curiosity and economic advantage have always motivated news media to be the first with news from distant places.

Telegraph operators
helped shrink the globe
in the mid-1800s.

In 1844 Samuel Morse invented the telegraph, which heralded the dim beginning of the electric age and the communications revolution. Newspapers were quick to see the advantages of telegraph, which meant that they could import distant national news from Washington and elsewhere. However, telegraph costs were high (involving stringing lines and hiring operators), and it soon became apparent that having six operators telegraph six stories about the same presidential message to six editors at more or less the same time was senseless.

The Associated Press

The AP was founded in 1848
as a cooperative news service
with members.

The Associated Press was formed in 1848 as a cooperative endeavor among six New York newspapers. Costs were drastically reduced, and each of the papers was assured that it would receive the same news at the same time. The Associated Press was, and still is, a membership organization, governed by a board of directors elected from among its client-members. It expanded gradually, acquiring new members in different geographical areas on an exclusive basis. Thus, AP membership was an extremely valuable asset to publishers, for it meant that they alone in their territory had access to outside news. Journalism history is replete with examples of faltering newspapers selling for high prices solely because of their AP membership.

However, as other newspapers were added to the AP membership roster, they reflected a wide variety of differing political and social viewpoints. Prior to the Civil War, Northern newspapers tended to be abolitionist; Southern newspapers were proslavery. Partisan differences at that time were even stronger than they are today, and the politically oriented press often conveyed this partisanship in the news as well as in editorial columns. The AP quickly discovered the necessity of reporting the news objectively—that is, reporting the facts alone, without color or opinion. If AP member newspapers wished to inject subjectivity into the news accounts, they were free to add whatever insights (or biases) they wished to the fundamentally neutral wire service accounts of political and social events. As a considerable portion of the news content of each newspaper gradually became wire service copy, newspapers began to adhere to objectivity as a fundamental operating procedure.

By the Civil War, the wire service was fairly well established, and the eastern portion of the nation at least, where most of the action took place, was "wired." Apart from being the first modern war in the sense of its use of long-range weapons and massive firepower that de-emphasized individual combat, the Civil War acutely demonstrated the attraction of war as a news source. The major newspapers across the nation sent correspondents to the battlefields. Most of this news flowed to the nation's editors, in the North and South, over the wires. However, telegraphic performance in the din of battle was uncertain. Lines could be cut by artillery barrages or sappers could snip them purposely. Therefore, correspondents developed the technique of sending the most significant news first in bulletin form and following up later with the details. Thus *journalese* and the *inverted pyramid* style of writing were born in the heat of battle, whereby the lead to a story—who, what, when, where, why, and how—was sent first and the balance of the story followed in order of decreasing significance until the story was told or the wires went out, whichever came first. The journalistic principles of objectivity and the inverted pyramid were the historical contributions of the wire services, forms to which the wire services adhered almost slavishly for more than a century, long after many newspapers adapted newer (or reverted to older) narrative and interpretive forms of writing.

During the years following the Civil War, the years of laissez-faire and "survival of the fittest," America's efforts went to rebuilding and to expansion—the winning of the West. Technological innovations mounted one upon the other, and the field of communications was no exception. The older cities grew, and more and more people flocked to new cities that sprang up along the railheads across the continent. All demanded newspapers to

Service to diverse members via telegraph brought about objective reporting and *journalese*—the *inverted pyramid.*

serve them. The railroad and the Industrial Revolution, serving one another
cooperatively, made national distribution possible, which led to both advertising and national magazines. The transatlantic cable, which sped world
news to the waiting presses, was completed in 1866, and international coverage became a part of the budget allowance of the Associated Press.

AP Shares the Wires The growth of telegraph led to the formation of the
great newspaper empires of Edward W. Scripps, William Randolph Hearst,
and Joseph Pulitzer. It permitted them a centralized control over basic editorial policy; it assured that the treatment of international and national
news would be similar in all their papers; and it made possible national
advertising on a local basis.

Scripps was the first to break loose from the AP's domination of the news industry, because several of his papers were denied world news by the AP's exclusive agreements. He founded the United Press (UP) in 1907 primarily to serve his own newspapers, but he sold the service to other papers that wished to use it, thus defraying a part of its cost. Such a system was logical because a part of the copy generated by the UP was local copy for Scripp's own newspapers and fed onto the wires when appropriate. Hearst followed suit in 1909 with the International News Service (INS), characterized from the beginning by the sensational Hearst style. Neither the UP nor the INS ever really rivaled the AP in coverage. They were stopgaps for major papers offering different viewpoints on world and national events. They made outside coverage accessible to many newspapers and communities locked out by the AP's exclusive franchises.

UP and INS differed from the AP in both policy and organization. While the AP was established as a nonprofit cooperative venture to serve its members, the services provided by Scripps and Hearst were private companies aiming to make a profit by selling information to clients. During its early years, the Associated Press had split into several factions, due to disagreements of its members over policies and organizational hassles. Until 1915, members were not allowed to use rival wire services, and established AP papers could veto membership requests filed by competing papers. In 1945 the United States Supreme Court forbade the latter practice as an unfair restraint of trade, but by then the growing strength of UP and INS rendered the question moot.

Throughout most of the first half of the twentieth century, the AP was the largest wire service, UP was second, and INS was a distant third in number of bureaus and reporters. AP had established a reputation for objective, accurate, and factual writing. UP's forte was the human-interest, personalized journalism we associate with its founder, Mr. Scripps. INS, in the Hearst mold, became noted for its aggressive (and ofttimes sensational) coverage of national and world news, and for high-quality writing. In 1958, due to economic pressures, United Press and the International News Service merged, calling itself the United Press International (UPI). The new wire service was able to compete on a nearly equal basis with the AP, and United States and world news media benefited from having a choice between two large, aggressive services.

AP and UPI Today At last count AP claimed more than 15,000 customers worldwide—newspapers, magazines, and radio and television stations. Three-fourths of America's daily newspapers, and half of its radio and television stations, are members. Its 2,600 employees at 198 bureaus across the globe (123 domestic, 75 overseas) turn out 14 million words which reach more than one billion people daily.

Scripps founded the UP in 1907, and Hearst launched his INS in 1909.

In 1958, UP and INS merged to form UPI. The new service was better able to stand up to the AP.

UPI serves more than 13,000 clients worldwide, including 8,000 overseas newspapers. In the United States it serves some 600 newspapers, 2,300 broadcast stations, and 875 nonmedia subscribers. Its 1,500 full-time employees working at bureaus across the globe (over 800 of them in United States bureaus) write some 13 million words each day. UPI has 231 bureaus worldwide, giving it a larger number of bureaus than the AP. However, on the average its bureaus are smaller than the AP's, and it typically serves smaller newspapers and broadcast stations than does the older service. Although the number of UPI newspaper clients has dropped slightly in the past decade—a period of economic chaos for the younger service—UPI's worldwide reach has been expanding, particularly via cable TV.[1]

The two services are in intense competition. Probably nowhere else in the mass communications business, including the television networks, is competition so intense to be first with the news. Remembering that among the nation's dailies and radio and television stations (and in the thousands of media outlets outside the United States) there is a deadline every minute, a minute's delay in flashing a wire service story may determine whether the AP or UPI version is printed or broadcast. In what must be some sort of record, for example, an AP reporter filed a story on the attempted assassination of President Reagan a mere 53 seconds after the shooting. The economics of the matter are simple. If an editor, looking back over a year's national and international coverage in the newspaper or on the station, discovers that one of the wire services has been used predominantly, the editor may easily decide to discard the other one. On a big newspaper, such as the *New York Times,* that's a million-dollar-a-year decision; on a small daily, it may be only a $100 or $150 per week decision. Like so many other mass media operations, news services are paid on the basis of how many people are in the audience.

Because the cost of news gathering increases with inflation and because the intense competition from domestic, foreign, and other supplemental services is driving the sales prices down, all is not well with the major wire services. UPI, supposedly a profit-making organization, lost money for thirty-five years in a row—as much as $7.7 million annually—before declaring bankruptcy in 1985. The story of UPI's recent economic woes are instructive.

In 1982 the Media News Corporation of Nashville, Tennessee, bought the troubled agency from the E. W. Scripps Company for $1, after Scripps grew weary of writing off the burdensome annual tax loss. Media News Corporation staked the future of UPI on cable television, direct broadcast satellite service, low-power TV, videodisc, videocassette, and computer data bases, and promised to make UPI profitable once it moved fully onto the electronic bandwagon. Old-time newspaper journalists winced upon hearing that the "wire service" they knew and loved was turning electronic. UPI's new young owners, Douglas Ruhe and William Geissler, were pleased to acquire a service endowed by its previous owners with state-of-the-art equipment, including a $10 million computer, communications and technical center in Dallas, digital newspicture darkrooms in New York and Brussels, and some five-hundred video display terminals in its various bureaus around the world.

Between 1982 and 1985, amid complaints of mismanagement and of moving too quickly to "go electronic," UPI became enmeshed in a series of funding crises. Twenty million dollars in debt, it laid off hundreds of employees. Those who remained took a 25 percent pay cut. UPI was left with a simple choice: either stop business altogether or declare bankruptcy and await a sugar daddy who could reorganize it and bail it out. The former alternative did not settle well with media observers, who feared that if the AP were left as the only major news service, the result might be a complacent monopoly. Fortunately, UPI was able to choose the lesser evil. After thirteen months in Chapter 11 bankruptcy it was sold for some $41 million to Mexican newspaper mogul Mario Vazquez-Rana and a minority partner, Texas businessman Joe Russo.

The newest owners announced in mid-1986 their commitment to strengthening the editorial personnel and acquiring new communications equipment. Vazquez-Rana, who has an impressive record of turning losing companies into financial successes, told his subscribers that UPI "shortly will embark on what I am determined to make the brightest chapter in its proud history." The Mexico City businessman, seeking to quiet concerns about foreign ownership of a major United States-based agency, pledged to subscribers, "I have no interest or intention of using the news agency to further any personal ambition, to favor any person, any cause, any region or any country." He planned to pour millions of dollars of operating capital into the operation, money drawn from his sixty-two profitable Mexican newspapers. Reporters laid off during the bankruptcy, plus new staffers, were being put to work expanding UPI's delivery of in-depth features, special investigative reports, international financial news, sports, show business, and coverage of culture, arts, and celebrities.[2]

UPI's new owners in 1986 promised to breathe new life into the service.

Between 1982 and 1986 the struggling news service saw five different corporate presidents attempting to turn the tide, improve staff morale, and stem the flow of dropped clients. In late 1986 alone, despite Vazquez-Rana's promises of upgraded performance, some forty newspapers opted not to renew their UPI contracts. Among them were the *New York Times* and the *Daily News;* the *Wall Street Journal* discontinued taking all but the statistical stock quotation services. The *Times* said it would rather rely upon the AP and its other services, and would put its million dollar annual UPI bill into improving its own reporting and editing staffs.[3]

High distribution costs have plagued the news services, and they have turned to the new electronic technology for survival. In the early 1980s UPI's annual phone bill was $13.5 million, and AP's was $16 million. That was the price of leasing telephone or land lines from AT&T ("Ma Bell") for delivery of news and pictures to the news media. Little wonder then that the wire services were anxious to transmit high speed verbal and pictorial data by means of communications satellites. By mid-decade more than half of the wire services' American newspapers were taking satellite-fed delivery directly into their newsroom computers. AP and UPI owners still maintain two sets of delivery capabilities—high and low speed—to disseminate the news over satellite and land lines, respectively, but are looking forward to a 100 percent high speed delivery system by the 1990s. Making the systems fully operational should put a substantial dent in the budget problems the news services have been facing. And the new technology will render the term "wire service" obsolete, although traditionalists will probably continue to use the term.

News services have turned to new electronic technology to cut costs and gain efficiency.

The News Service Operation

The news services traditionally have operated a remarkably efficient two-way, dovetailed system of both news gathering and distribution. In the distribution link most news service copy originated in New York for domestic consumption, then traveled over trunk lines to the news services' major metropolitan subscribers. At various relay points across the country, its contents were scanned by *wire filers* who took the appropriate items, major stories, and material of a particular regional or local interest and forwarded them onto the state and regional circuits for use by smaller clients. The system worked much like a transcontinental freight being broken up in Kansas City, with some cars continuing on to the coast and others being sidetracked to Omaha or Dallas.

In days gone by, news gathering worked in reverse of its distribution. Many news service bureaus, often one-person affairs located in local newspaper offices (AP bureaus) or independent shops (UPI bureaus), consisted of a bureau chief or a small group of gatekeepers who gathered stories from the local community and filed them on the wire. As the material moved in reverse, the wire filers had to decide whether to kill the item or pass it along to New York for national distribution. In the event of a major newsbreak in an out-of-the-way place, the wire services would send reporters to cover it from either New York or the regional bureaus. In smaller communities where there were no bureaus, the wire services hired *stringers,* generally reporters for the local papers who moonlighted for the news services and were paid by the column inch or for each story used. In the smallest towns, services used housewives and high schoolers, paying them a flat rate for each story they submitted.

At the end of the pipeline, in the newspaper office, the news service copy emerged on a battery of teletype machines, the number of which varied depending on how many of the services the newspaper subscribed to—whether a single national wire or a combination of national, regional, sports, business, and other wires. In the earliest days, the news was transmitted by a Morse code operator who, on a good day, could transcribe thirty-five words per minute. The teletype increased transcription to sixty-five words per minute, resulting in some savings to the newspaper, since it no longer had to pay a Morse code operator. Eventually, in the 1960s and 1970s, the teletype machines printed justified copy (copy with even margins, as in this text), ready to run, together with perforated tape on the teletypesetter (TTS) ready to be fed directly into the typesetting machines. The tape reduced composing room time considerably and permitted the services to operate a little closer to deadlines than previously, but still delivered news at only one word per second.

Thanks to computers and other new technology, all of this has changed in the past several years. Since the late 1970s, wire service news gathering and distributing methods, and the speed with which they are carried out, have been altered almost beyond recognition.

Since the late 1970s, wire service news gathering and distribution methods have been altered drastically.

Today, a system of nationwide distribution circuits is still in effect, with relevant stories being channeled to appropriate bureaus and media. However, thanks to the data storage capacity of large computers, an editor in South Succotash, for example, who desires to run a story about the eradication of hydatids in Tasmanian sheepdogs, no longer has to wait for the news service to announce when the desired story will be released. Instead, the editor merely calls up the news service's central computer in New York or Dallas and requests the entire story. In other words, the local editor has more control of information. And when the information arrives, it does not dribble along at the sixty-five-word-a-minute limit necessitated by leased telephone lines and slow teletypesetters. The new high-speed teleprinters and satellite beams have brought about the computer-to-computer exchange of stories at the rate of 56,000 words per minute. Even though few readers in South Succotash will desire much information detailing the problems of Tasmanian sheepdogs, if they did and if the editor could process it and all the other data appearing on the VDT screen, the stories could be printed instantaneously whenever desired.

Editors connected to the UPI's "demand" service have the opportunity to order up on a given day those stories deemed of interest to the local community. A more typical pattern, however, is the news services' traditional system of suggesting to their subscribers what they believe are the major stories of the day and the editors' traditional approach of allocating space for only those key stories designated.

A day in the life of a wire service is divided into cycles. Twice daily the wire services draw up a *budget* of news, choosing ten or twelve stories that, in their estimate, are the most important news events of the day. The services then wire or beam a summary of this budget to editors across the country, so they will know what to expect and can plan accordingly in making up their papers. The service then runs through the major stories, one by one, interrupting the cycle with bulletins of other major events and continual "updates" on the running budget stories. As updates and news leads are released, their written presentation is such that they can be fitted directly into what has been received previously with only minimal disturbance to material already set in type. After the cycle has been completed,

News services provide news *budgets* during each daily cycle. Local editors are active participants in the two-way gatekeeping process.

it begins again. On a fast-moving story, there may be as many as a half-dozen updates and new leads within a cycle carried by the "A" or major news wire. Other news items are relegated to the "B" wire for secondary news and major features, or on the sports wire, business wire, or racing wire. Because of the two-way nature of the gathering-distribution system, a client is able to query the wire service for data of special local interest, such as how the local representative in Congress voted on a certain bill. In this example, the query would be relayed to the Washington bureau, where one of a hundred or so staffers would ask the Representative's office or their Capitol Hill reporter, and the private answer would be flashed back to the client.

The world headquarters of both AP and UPI are in New York City. UPI's central computer is in Dallas, as previously mentioned; AP has half a dozen or so regional computer centers, where much of the gatekeeping and wire filing take place. No longer does a story move from the small bureau to the "B" wire where it is judged worthy of "A" wire treatment or handled as second-class information. Now it goes from the small bureau to one of the AP's regional computerized centers or to UPI's central computer, from which it is made available to clients.

Both AP and UPI operate separate radio and television wires. After 1935 UPI began supplying radio stations with the same wires as newspapers. This required rewriting for the less formal, more immediate audio style of the air. When AP began serving broadcasters in 1940, there was a competitive incentive for both AP and UPI to develop radio wires written in the audio style, so that newscasters could take the wire service copy directly from the teletype and read it on the air—*rip 'n' read*. The radio/television writing style is punchy, immediate, and pays less attention to spelling and punctuation than does newspaper style. In essence, it plays for the ear, not for the eye. In addition to the broadcast news stories, UPI maintains a daily radio pickup service for its clients whereby they can tune into UPI and carry a national news budget directly onto the air or tape it for future use. Both UPI Audio and AP Radio provide complete newscasts over leased telephone lines to their clients and members. In addition, both services offer television stations a steady diet of still and motion pictures for use in newscasts.

Most nonlocal photographs found in America's daily newspapers are courtesy of the news services. Pictures are transmitted anywhere in the country via special circuits and satellite, and are reproduced instantaneously by facsimile machines in each newspaper. New techniques of electrostatic transmission electronically deliver glossy photos on dry paper, ready for printing.

News Service Beats The news services are no different in their news gathering operations than the typical newspaper or broadcast news operation, as we described in earlier chapters. However, the news services function on

AP and UPI are among the services subscribed to by broadcasters, many of whom still *rip 'n' read*.

News services provide photos to newspapers. Satellites and other new technology are used.

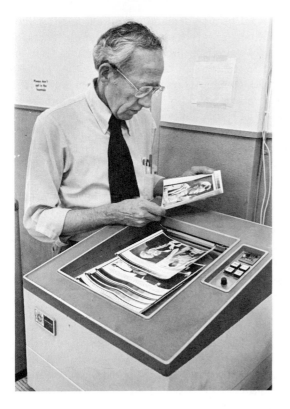

a larger scale. They, too, maintain beats with the presidency, on Capitol Hill, in the major executive departments, at the Pentagon, in the major state capitals, and on Wall Street. Internationally, they have major bureaus staffed in the leading capitals of the world—London, Paris, Rome, Berlin, Moscow, Tel Aviv, Cairo, Tokyo, and Buenos Aires—several of which have regional responsibility. For example, Tokyo carries most of the coverage for the Far East, and Buenos Aires—where coverage of the British and Argentinean conflict in the Falkland Islands originated—is the focal point for Latin American news.

The press associations have their own versions of the *tickler file,* to remind themselves of regularly scheduled events that demand coverage: the World Series, the Super Bowl, the Indy 500, the national political conventions and elections, holidays, the winter and summer Olympics, and so forth. News of such significant events is in great demand by their clients and is awaited breathlessly by tens of millions of Americans. Gathering such news produces huge logistical problems of moving dozens of personnel around, shifting assignments, and temporarily leaving the more routine beats minimally covered to concentrate on the big event that, experience has shown,

News services cover the world as a good newspaper or broadcast station would if it had the resources.

will often develop an unexpected turn or demand a continuing effort. For the unexpected, press associations also rely on their far-flung network of hundreds of bureaus, the news gathering facilities of their thousands of clients, and their countless stringers.

The major areas in which the news services concentrate correspond closely to the major departments of the daily newspapers. Concentration is on government and politics, on sports, and on feature or society news.

The Washington bureaus of press associations are staffed by more than one hundred people covering every aspect of government on a fairly regular basis. Even so, they are sadly understaffed considering the enormity of the federal bureaucracy and its hundreds of agencies. Consequently, a considerable portion of their time is devoted to watching the other media: the television networks, the bureaus of the big dailies, the press and broadcast groups, the major syndicated material, the national columnists, the foreign press associations, as well as the big lobbies and news magazines. These are all potential subsources of governmental material in addition to tips from the interested, both inside and outside of government. The approach is a good deal like that of the city editor of an afternoon paper who scans the A.M. coverage to be sure nothing is overlooked.

Politics is great copy because it represents the raw material of controversy. On the national level it becomes the epic struggle of two titans playing for real prizes—power, prestige, opportunity, and position. Political campaigns are gladiatorial in nature. Their ability to generate long-range emotion makes them indispensable to the American scene, and money in the bank for all the news packagers.

Sports shares with politics some of the elements of competition. Here again giants are seen struggling for supremacy. Everyone picks favorites and experiences a vicarious thrill in watching opponents vie. In a sedentary and white-collar nation increasingly removed from physical contact and from direct thrill and danger, athletes are surrogates, performing skills for contemporary society and doing its exercising.

The status of the participants in sports, government, and politics as news sources is yet another factor to be considered. In a nation of 235 million persons, only a tiny fraction of them can be in the news. Some of them rise to the surface at society's officially prescribed gathering points—the various news beats. Others (athletes, politicians, and officials) occupy positions that are almost by definition newsworthy. Society itself, in one way or another, has chosen such persons for preferred treatment. In response to society's demand, the news packagers heavily treat their affairs, both public and private, because society's curiosity is ravenous. This formal and informal selection of individuals by society also constitutes a sort of self-screen device of news. In a circular process, news is what society says it is. And that is what the wire services will relay to the newspapers and broadcasters for replay to society.

The widespread reach of only two major press associations occasions a similarity to the news that is striking. It is quite possible for a person to breakfast in New York while reading the *New York Times* and then fly to Los Angeles for lunch and find the same story verbatim in the *Los Angeles Times*. Anyone who has driven across the country has also been exposed to a progression of individual local stations airing the same radio wire material, differing only in the accents of the announcers. Such sameness is heightened by the similarity of both the AP and UPI approach to the news and their objective style, which allows little leeway for individuality of treatment.

This situation places awesome responsibility and power in the hands of the relatively few editors and wire filers of the press associations who are, in fact, the gatekeepers of the nation's news. Because only a dozen or so major stories are included on the daily budget cycle, it is these stories, by and large, that will be carried on the country's front pages and over the daily radio and television news programs. Of course, there are hundreds of other stories available, if the local papers' or stations' wire editors are willing to move through the feeds to seek out pieces they think might serve their audiences. However, the bias is always in favor of those stories the wire services highlight in their budgets. Even editors who are not particularly enchanted with a story are under great pressure to carry it, not only because the news services have highlighted it, but because the editors know that other papers and stations are likely to play up the budgeted items and thereby place them on the public's agenda.

AP and UPI could control the
news agendas of the nation,
but several forces limit the
threat.

Radio news commentators are perhaps in an even more restricted position, having to use the radio wire material because there is no other news available. There is speculation that as the financial pressures become greater on the news services they may be forced to curtail some of their coverage, despite their best intentions. This would reduce their budgets and ultimately limit the availability of information. Other auxiliary news and feature services may pick up part of the slack, but more likely in specialized rather than general news coverage.

Several factors operate to balance the potential threat of news services' power:

1. their historical dedication to news, their determination to uncover it, their long-term success in doing so, and their sharply honed news judgment;
2. the intense competition between associations, which involves constant watchdogging of each other, and consequently minimizes the opportunity for purposeful bias;
3. the increase in the number of correspondents and stringers employed by or representing the various newspaper or radio groups, the television networks, the several syndicates, and the larger metropolitan newspapers, which enhances competition and provides a greater variety of both news stories and treatment;

4. in the case of AP, at least, the cooperative nature of the venture, including the fact that the service is dependent on its members to supply a large part of the news it provides to other members; and

5. the fact that more and more interpretive reporting is finding its way into the news services, as a response to the demand of subscribers, who discovered that the old mode of total objectivity was no longer adequate for their better-educated and more sophisticated audiences.

The mainstay of the news services is still objective reporting, but this is augmented now with an increasing number of interpretive, background, and even opinion pieces.

Gatekeepers for the World?

AP, UPI, Agence France-Press, Reuters, and TASS dominate the world scene.

Lest we have the erroneous impression that the Associated Press and United Press International are the eyes and ears of the world, we shall consider several other international wire services and the controversy surrounding the way the world's news is mustered.

At present five news agencies dominate the world scene. In addition to the American-based AP and UPI, the major watchdogs are the Agence France-Press (AFP) in Paris, Reuters in London, and the Telegrafnoie Agentsvo Sovetskovo Soyuza (TASS) or Telegraph Agency of the Soviet Union. Virtually every nation has its own news service, often controlled by the government. For example, there are the Yugoslavian agency Tanjug, the West German Deutsche Press Agentur, the Japanese Kyodo, the Spanish Fabra, the Italian Stefani, and mainland Chinese Xinhua (formerly New China News Agency). Some of these are significant in the world flow of news, but when conversation turns to the major agents of world communication, the "big five" come instantly to mind.

AFP was organized at the end of World War II as an autonomous public body operating under the watchful eye of a board of directors comprised of French publishing representatives. With one hundred foreign bureaus, it services thousands of newspapers and broadcast stations either directly or through joint agreements with national news agencies in individual countries. Like the AP, it is a nonprofit enterprise.

Press associations in the British Commonwealth owned Reuters prior to 1984, when it sold stock on the open market for the first time. Reuters originated in the 1850s when Paul Julius Reuter began delivering financial news between the Prussian town of Aachen and Brussels, Belgium, by using carrier pigeons. In the ensuing decades, the service expanded its scope, and supplies its clients in 112 countries with electronically transmitted timely news plus data about currency exchange rates, commodity prices, stocks, bonds, and even the availability of tanker space. The recent moves to financial news and information-retrieval services made it profitable for the first time in four decades, and when it began selling some of its stock to the public in 1984 its once disgruntled majority owners reaped a paper windfall of nearly $1.5 billion. Because of the news service's new wealth, it has expanded its news coverage. The staff of 612 reporters and editors in London and at ninety-two bureaus provides complete and thoughtful coverage of the Middle East, Africa, and British and Commonwealth countries.[4]

TASS is known as the official Soviet news agency. It was founded in 1918 as Rosta, and became TASS in 1925. It serves print and broadcast news outlets within the Soviet Union and abroad. TASS is known not only as the voice of Moscow for the rest of the world to hear, but as the government's ears as well. Foreign correspondents in one hundred nations gather local and international news, interpreting the world's affairs according to official policy. The entire operation is under the control of the Central Committee of the Communist Party, and shares the Committee's goals of furthering the policies and ideologies of the Party. As gatekeepers, TASS correspondents and editors are highly influential in gathering and disseminating information and ideology.

A New World Information Order? If TASS is considered by the Western world to speak in the singular voice of Soviet policy, what can be said about the other four international agencies? Over the past decade or so, they have come under increasing attack from non-Western nations for their tendencies to see and report the world through Western, capitalistic lenses. All things considered, AP, UPI, AFP, and Reuters on a daily basis take the subtleties of world events and filter them through a gatekeeping system not compatible with much of the world's interests. Because of their enormous scope, these Western agencies gather news from one Third World nation, process it in Paris, London, or New York, and then file it off to other Third World nations. In the meantime, Western knowledge of the outside world is coming almost exclusively from these wire services, and the rest of the world learns of the West's doings by the same means.

When we combine the influence of the wire-service gatekeepers with the fact that Western powers own and control most of the world's information delivery systems, it is little wonder Third World nations view the situation with alarm. Crying out against what they call cultural imperialism or communications colonialism, they have raised in the United Nations and elsewhere numerous calls for a *New World Communications Order,* or a *New World Information Order.* They have requested independence and equity in access to global communication resources so that their own views, values, and developmental efforts will be reported more fully. They have sought the right to restrict the flow of information across national borders, and have requested power to license journalists and impose an international code of journalistic ethics.

When the United Nations Educational, Scientific and Cultural Organization (UNESCO) issued its report on these problems in 1980, Western journalists were predictably upset. The so-called MacBride Commission

Non-Western nations resent Western news domination; they have requested a *New World Information Order.*

Western journalists tell
the Third World's stories,
but some call it "media
imperialism."

Western journalists tell the Third World's stories, but some call it "media imperialism."

suggested that all news media, including the international news agencies, would be expected to promote governmentally established social, cultural, economic, and political goals. The commission's report was not adopted, despite intense debate, but UNESCO supported a three-year study to deal with questions of media imperialism. By the mid-1980s, UNESCO's efforts had been matched by the Prague-based International Organization of Journalists (IOJ)—a predominantly socialist group—and a Talloires, France, gathering of representatives from more than fifty Western or democratic media organizations. The three groups emerged with widely divergent solutions to the questions of media colonialism. Despite the fact that all three were in favor of journalistic responsibility and accountability, they differed in terms of who or what agency should guarantee or enforce these goals. Due largely to frustration over UNESCO's tendencies to adopt numerous restrictions on the press such as licensing of journalists and what many saw as local censorship (along with its incessant political bickering and inefficient use of funds, 25 percent of which came from America, plus UNESCO's strong support of a New World Economic Order), the United States withdrew from UNESCO in 1985.[5]

Debates over media colonialism or imperialism, or what have you, will continue so long as the international news agencies monopolize the flow of information between and among the world's developed and developing nations. Given the tendency of any of us to see the world as we have grown

There seems little hope for a "quick fix" to perceptions of media colonialism.

accustomed to seeing it, there seems little hope for a quick or easy settlement to the issue. When we are in a position to open and close the world's information gates, we assume awesome responsibilities.[6]

In the newspaper field, following the precedent established long ago by Scripps and Hearst, many groups are finding it advantageous to organize smaller wire and feature services principally to serve their own newspapers. Material written for one newspaper, in some instances, can be used by others in the group, thereby achieving some savings and lowering the unit cost. Alternately, some newspapers have syndicated their materials for mutual use, as, for example, the *Los Angeles Times-Washington Post* syndicate. Both papers profit from unparalleled coverage of the nation's capital and of southern California at no additional cost, and enjoy the output of some of the world's best foreign correspondents. Once this pooling occurs, whether on a group or independent syndicate basis, the next logical step is to offer the material for sale to other newspapers, defraying a portion of the original investment in the material.

Individual newspapers also get into the act. The *New York Times,* the nation's newspaper of record, offers outstanding comprehensive material and unsurpassed coverage of the nation's financial capital to its hundreds of clients worldwide. The *Chicago Tribune* and the *Chicago Sun-Times* offer material from, but not limited to, the nation's heartland; the *Toronto Telegram Services* covers Canada. In each instance, the service represents extra income from an investment already made. Using the same economic principle employed by UPI, the suppliers put a teletype in the hands of their subscribers, billing them on a sliding scale according to each subscriber's circulation. Columnists and key sections (sports, business, politics, and entertainment) from the host paper are made available on a same-day basis to whichever papers have subscribed. In this respect newspaper groups and private services have an advantage over the news services in that they utilize pooled information directly and derive income through subscriptions besides, while the revenue of the news services must come from media sales alone. Newsmagazines also have gotten into the packaging business, getting more revenue mileage out of their basic coverage. *Newsweek* is notable among these.

Many of these services also tend to specialize their coverage. The Women's News Service speaks for itself. Others in the list of a hundred or so specialized services include the National Black News Service; the Chinese Information Service; entertainment, sports, and religious services; and a plethora of others delivering their sometimes propagandistic messages wherever they will be paid for and used.

Related to the wire services are the various city and regional news services. These are local, private services that follow the press association principle. The city news services maintain reporters at the major beats within a city (city hall, courts, police stations, and so on), gathering news for the small dailies, the weeklies, and the radio and television stations within their

On a smaller scale, various news media pool their resources.

Private businesses often
subscribe to the Dow
Jones Financial News
Service.

area. They sell these data to local media, relieving them of the expense of maintaining reportorial staff, and yet enabling them to keep current on local happenings. The more enterprising city news services also operate a radio pickup wire for local stations, sometimes on a telephone "beeper," and a few shoot film for television.

These services in the larger cities are generally supplemented by the so-called *business wires*. Business wires are really public relations wires, and they operate in reverse of the other wires. Instead of charging the media, they maintain teletypes in the newspaper, radio, and television outlets, and they charge their sources for carrying the material. The service assures rapid and complete distribution of public relations material. The more enterprising public relations people reinforce their story with a phone call advising the editor that the copy is coming over the business wire.

More legitimately deserving of the "business wire" title are the highly successful services provided by Dow Jones and the Commodity News Service. Dow Jones for years has provided newspapers and private offices with a financial wire, carrying the substance of the *Wall Street Journal*. Recently, in its effort to stay in the thick of the information explosion, Dow Jones has begun offering its pool of financial data on an interactive basis to homeowners or anyone else who has access to a microcomputer and telephone. Its success stories include working with "electronic newspaper" delivery systems in Florida, Texas, and elsewhere; in academic institutions; and in the nation's boardrooms, large and small. (See chapter 10.)

Meanwhile, the Commodity News Service caters to a wide gamut of specialized market-analysis needs. From its Chicago and Kansas City headquarters, it distributes Farm Radio News and Grain Information News reports on crop reports and farm-market conditions; it distributes the Lumber Instant News for the lumber industry, providing information about wood prices, forest conditions, and national economic trends that would impinge upon the industry; and it distributes other specialized market reports to interested groups.

All in all, the diversity of opinions and information made available by the wide variety of auxiliary news services bodes well for the media. Not only do the participants receive a broader spectrum of information and opinion than they could generate on their own, but their readers, listeners, and viewers are spared the sometimes monotonous sameness of AP and UPI copy.

Feature Syndicates

In the 1880s Samuel McClure, who published *McClure's Magazine,* decided to capitalize on his investment in "nonnews." He recognized that the growing newspapers had a need for "untimely" material of a human-interest or feature nature. Material that made interesting reading, but did not have to make the next edition, could be held for a slow news day. McClure put out fifty thousand words a week to newspaper subscribers on such topics as fashion, homemaking, manners, and literature. Thus, the *feature services,* or *syndicates,* were born, as packagers of the untimely. Newspaper publishers did not have the money to produce this kind of material themselves. However, it could be purchased at a fraction of its cost on an exclusive basis in their territory, it was a filler bargain; publishers could save some reportorial salaries while filling their newspapers with readable content.

As the news services are basically packagers of the information content, the feature syndicates have become the packagers of entertainment content, both graphic and verbal. Anywhere from 10 to 35 percent of the average editorial content of newspapers is composed of this pre-packaged or "canned" entertainment. It includes the advice to the lovelorn columns, comics and cartoons, daily horoscopes, crossword puzzles, word games, coin and stamp collecting columns, chess features, a wide range of how-to-do-it pieces, fashion, interior decorating, cooking, dressmaking, and nostalgia. Book, art, drama, and film criticism, and a variety of humor, gossip, financial, and political columns are also included. (See table 13.1.)

A century old, feature syndicates fulfill media's vast appetites for nonnews.

Newspaper editors try to achieve some sort of balance of features depending on their evaluation of the mix of their particular readership. The larger newspapers try to achieve political balance in their national columnists, offering both liberal and conservative viewpoints. While ostensibly these are presented to permit readers to make up their own minds between two divergent viewpoints, they are actually a surefire audience formula in that readers will read the one because they agree with it and will read the other to get mad, which is a comparable "entertaining" emotional reaction.

Table 13.1   Classifications of feature syndicates

1.	Advice	26.	Gossip
2.	Antiques	27.	Health (medical)
3.	Astrology (graphology, biorhythm)	28.	Hobbies
4.	Automotive	29.	House & home
5.	Beauty		A. Decorating
6.	Books		B. House plans
7.	Business (economics & finance)		C. Do-it-yourself
8.	Children's features	30.	Humor columns
9.	Coins & stamps	31.	Legal
10.	Comics	32.	Maps & charts
	A. Strips	33.	Miscellaneous columns and features
	B. Sunday pages	34.	News services
	C. Cartoon panels	35.	Outdoors
11.	Commentary		A. Camping
	A. Political		B. Fishing
	B. General		C. Hunting
12.	Computers	36.	Patterns (sewing)
13.	Consumerism	37.	Pets
14.	Contests (circulation & promotion)	38.	Photography
15.	Ecology	39.	Photo services (pictures)
16.	Editorial act	40.	Polls & surveys
	A. Caricatures	41.	Puzzles
	B. Cartoon panels		A. Crossword
17.	Education		B. Quizzes
18.	Entertainment		C. Word games
	A. Films	42.	Real estate
	B. Music	43.	Religion
	C. Radio/television	44.	Retirement (senior citizens)
	D. Theater	45.	Science (technology)
	E. TV program listings	46.	Serializations
19.	Etiquette		A. Nonfiction
20.	Family (child raising)		B. Fiction
21.	Fashion	47.	Short stories
22.	Fillers	48.	Sports (spectator and participant)
23.	Food & wine	49.	Teenage features
24.	Games (backgammon, bridge, chess, etc.)	50.	Travel
25.	Gardening (farming)	51.	Women's pages

Source: From the table of contents, *Editor & Publisher 60th Annual Syndicate Directory,* 27 July 1985. By permission.

There are hundreds of feature syndicates listed in the annual *Editor & Publisher* directory of syndicated services, ranging from small, one-person outlets to giants like the Newspaper Enterprise Association and King Feature Syndicate with their $100 million operations. Because all syndicates bill their clients on the basis of circulation published, widely syndicated columnists like Ann Landers (1,146 newspaper clients worldwide and a readership of 85 million) and Art Buchwald, or cartoonists like Garry Trudeau ("Doonesbury"), bring in hefty six-figure incomes annually.

All told, more than 2,500 different features are available for purchase from some 350 different syndicates, although only eight services dominate the industry. The eight which provide 90 percent of the nation's comics, editorial cartoons, columns, features, puzzles, games, book excerpts, weekly "one shot" items, and graphics and features packages are King Features Syndicate, United Media (which includes United Feature Syndicate and Newspaper Enterprise Association), News America Syndicate, Tribune Media Services, Universal Press Syndicate, The New York Times Syndication Sales Corporation, Los Angeles Times Syndicate, and The Washington Post Writers Group.

There are 350 different syndicates, but eight of them dominate the industry. They help newspapers be all things to all people.

Larger newspapers, with over 100,000 circulation, purchase about 100 to 150 syndicated features a year, while smaller papers, with circulations of 20,000 to 50,000, typically use thirty to fifty features. Depending upon their circulation size, competitive status, and the number of features they purchase, papers typically spend between $15,000 and one million dollars a year for the services. The cost per newspaper is not great—as little as $5 per week for a feature purchased by the smallest newspapers, and up to $500 for the top markets; the average is in the neighborhood of $15. This explains why over the course of a year the average editor will have to cope with many dozens of syndicate salespersons, each claiming to offer exclusive rights to the next Peanuts-type surefire success feature. As in most commercial aspects of the media, competition is intense, and the gatekeeping process never ends.[7]

Feature syndicates invest as much as $100,000 in producing and launching a new feature; the average is said to be around $25,000. Syndicates themselves are besieged with thousands of would-be creators annually. Larger syndicates typically receive between 3,000 and 4,000 unsolicited comics and the same number of text submissions a year. From that cornucopia, they typically take on only three or four of each.[8] The normal split between the author-creator and the syndicate is fifty-fifty, which helps explain why those thousands of unsolicited comics, opinion pieces, special features, and what have you keep crossing the desks of syndicate editors.

Success breeds imitation, and once a newspaper has gambled and won circulation or interest by publishing a given feature, it is unwilling to see competing or local or regional newspapers reap rewards without having to gamble. Thus exclusivity contracts have long been the norm in the features business. A newspaper signs a contract guaranteeing it exclusive rights to publish a given feature within a given radius—anywhere from fifty to 200 miles—or assuring itself that it will face no competition for the feature in any county where the paper circulates to 20 percent of the households. There may be other variations to these contracts, but the upshot of it all is frequently considered a form of censorship and restraint of trade by papers denied access to popular features. Suburban newspapers in particular have been angered at major metros for buying exclusive rights to the most attractive features and then not using them regularly. Recent legal challenges by the smaller papers have resulted in a more open marketing system among the syndicate services. This is as it should be, the smaller papers insist, if American journalism expects the public to take seriously the industry's claims about free access to information.[9]

A glance into the future of syndicates shows an industry with enormous growth potential, even though the inevitability of consolidation may result in fewer services dominating the market. Audience diversification, competition for attention, and the tendencies for newspapers to experiment with total information and entertainment packaging means there will be a need for more, not fewer, features available through the syndication services.

The future of syndication is exciting; there's plenty of room for growth.

Summary

The press associations, AP and UPI, are packagers of information in probably its purest form. They are the result of newspapers pooling their resources to offset the increasing costs of gathering news. The AP was founded in 1848, after the invention of the telegraph, by six New York newspapers to share wire costs. It was a membership organization, which soon admitted other newspapers on an exclusive basis in their respective areas. An AP membership became a very valuable asset. Early, AP developed the technique of reportorial objectivity in an attempt to answer the differing social and political viewpoints of its many members. Later, during the Civil War, the concepts of journalese and the "lead" were born to get the most important information out first in case of telegraph failure.

The telegraph also made possible the introduction of national advertising into newspapers and assisted in the centralized control necessary for the establishment of the great newspaper empires of Hearst, Scripps, and Pulitzer. Scripps ended AP's nearly sixty-year monopoly on national and international news by founding the United Press in 1907. Two years later Hearst started the International News Service for the same purpose. Both organizations found it profitable to sell their services to other newspapers in order to defray costs. Neither was an effective competitor for AP until they merged in 1958. Today, AP has 15,000 members worldwide, and UPI has 13,000. They are in constant, intense competition to be the first to bring the news to their clients. Somewhere out there, AP and UPI subscribers are fighting a deadline every minute of the day, and to run second means to be left out of the paper or off the air; to risk subsequent cancellation.

AP and UPI have turned to satellite transmission of data for more rapid and less expensive communication with subscribers. Modernization of the services includes reliance upon centralized computers and "on demand" access to news by editors scattered across the nation. The services continue to draw up daily news budgets, suggesting to editors the most important stories to be carried during the day. Both services cater to the special needs of their radio and television subscribers, offering satellite and wire transmission of photographs.

The news services handle news beats in much the same way that metropolitan newspapers do, with general assignment and specialist reporters covering the news on beats when news is most likely to occur. Politics, sports, society, and business are the wire services' stock-in-trade. With their budgets, the news services wield enormous influence as national agenda setters; the same versions of the same stories are repeated coast to coast, hour by hour, in the nation's subscribing media. Fortunately, the news services' potential for abuse is limited by their own integrity and hunger for news, sense of competition and cooperation, growth, and response to subscribers' demands for more interpretation and analysis.

AP and UPI share the world's news-gathering chores with a great many other agencies, the largest of which are AFP of France, Reuters of England, and TASS of Russia. The four Western agencies have been faulted for "communications colonialism," as they gather and report the world's news through capitalistic eyes. Calls for a New World Information Order, with developing nations taking a larger role in shaping their images around the world, are creating some difficulties for the Western agencies.

Today, an increasing number of smaller newspapers and broadcast groups have organized news services to serve their own papers in addition to making these services available to others. Also, some of the nation's major papers and broadcast operations, alone or in concert, have made material available on a syndicated basis, which increases the diversity of news available in the marketplace. City news services condense municipal news for the smaller media. The so-called business wires are used for public relations purposes where the source pays for inclusion and the newspaper receives it free. More significant are the financial and business services provided by Dow Jones and the Commodity News Service. Married to new technology, the business services are available to homes and small companies instead of just to large media subscribers.

Finally, there are the feature syndicates, the 350 or so organizations devoted to providing America's newspapers with nonnews, with the untimely content that fills so much space and attracts so many readers to the press. Comics, columns, crossword puzzles, games, and other self-indulgent fare have become big business. Successful features bring their individual creators six-figure incomes and keep syndicate sales personnel knocking on editors' doors with bargains they can't, but often should, refuse.

If the wire services gained favor among the nation's newspapers during a period when objectivity was stressed, then feature syndicates are receiving patronage under differing circumstances. Last century an avoidance of opinion satisfied the relatively small range of opinion differences represented by wire service clients. Today, increasing affluence, a much higher educational level, and a diversity of interests and the proliferating media to serve them are requiring a higher degree of diversification than historically provided by the wire services. Objective news is not enough anymore; reflective news is demanded.

Into this breach come the feature services, which, in one form or another, have proliferated to serve a diversity of public tastes and the multiplying departments that the daily newspaper evidences. The features bring magazine content to the press. Society has again demanded fragmentation and specialization from even such monolithic enterprises as the two major news services. In a familiar pattern, media consumers are receiving the mix of hard and soft news, substance and fluff, news and entertainment, and even the persuasion they have requested.

Notes

1. "Pulling Wires," *Time,* 6 May 1985, p. 65.
2. UPI, "New Owners Take Over Reins at UPI," *Salt Lake Tribune,* 12 June 1985, p. A-5; Don Kowet, "UPI's Prince With a Mysterious Past," *Washington Times Insight,* 16 December 1985, pp. 62–64; and UPI, "UPI owner says news service to expand," *Logan* (UT) *Herald-Journal,* 15 June 1986, p. 6.
3. Margaret Genovese, "UPI," *Presstime,* December 1986, pp. 6–9.
4. Janice Castro, "Reuters' $1.5 Billion Bonanza," *Time,* 6 February 1984, p. 58; and William A. Henry, "Reuters' Hot Financial Flash," *Time,* 11 June 1984, p. 78.
5. Stuart James Bullion, "Truth, Freedom, and Responsibility: Seeking Common Ethical Ground in International News Work," *Journal of Mass Media Ethics,* Spring/Summer 1986, pp. 68–73.
6. Oliver Boyd-Barrett, *The International News Agencies* (Beverly Hills: SAGE Publications, 1980).
7. Marcia Ruth, "Syndicates Serve Spreading Smorgasbord," *Presstime,* November 1986, pp. 16–24.
8. *Ibid.,* p. 20.
9. Marcia Ruth, "Newspapers Are Asking for Less Exclusivity," *Presstime,* November 1986, p. 22.

Part Five

This final section of the text considers how the mass media are regulated, how they are permitted and encouraged to fulfill their functions of informing, educating, persuading, and entertaining society. In chapter 14 we treat the media's relationships to government; in chapter 15 we treat media's self-regulatory activities.

The divergent theories of press and government are considered in these two chapters. The logic of authoritarianism and its inequities, the hopes of libertarianism and its flaws, and the call for social responsibility and its ramifications are analyzed, along with numerous other perspectives on the proper relationships between media systems and their constituent societies and governments. As we will see, most of the history of controls over mass media is a history of compromises, sometimes more subtle than at other times. The path to these compromises is not always clear, but we are able to follow it to some logical conclusions.

Today's mass, diverse, and complex society cannot be kept in check by informal and unwritten codes, as in days gone by. Instruments of mass communication are as massive and complex as the society in which they function. Media's efforts to earn a profit and serve society are not terribly dissimilar to other corporate

The Media Environment

enterprises. However, journalism is the only business to have been singled out in the United States Constitution for special privileges. Government was explicitly told to pass no laws abridging freedom of the press. Nonetheless, American history has demonstrated that such guarantees are not absolute.

As government and media have grown more complex, in tune with the complexity of society, certain precautions have been necessary. Laws of libel, privacy, obscenity, and the whole of broadcasting and advertising controls were instituted to protect society. Informal controls also went into effect. Codes of ethics, in particular, served to protect the media from bearing the burden of additional governmental regulation.

The greatest control of all may be that exerted by the body politic in the formation of that vague reality called public opinion. Media simply are not free to print, broadcast, or produce anything that comes into their heads. They must conform to the expectations of their respective and often highly selective audiences. Ultimately, then, the imperfect balance among public taste, governmental sanction, and individual and corporate media behavior is reached, but not without sustained and severe differences along the way.

Regulation and the Law

14

Introduction

Who and what permit media to carry out their various functions, and even to be irresponsible?

We have already sketched numerous explanations of how and why the media do what they do in society. Let us take a look at how and why the media operate as they do within the political and regulatory environment. Panoramic views of media's appropriate roles in society must include details about who and what permit the media to exist, to inform, to make a profit, to entertain, to persuade, to challenge conventional wisdom, and even to be irresponsible.

In the first half of the chapter we will discuss several theories of the press and the regulations and controls that apply to the print media. Next we will look at the unique regulations applicable to the broadcasting industries. Unlike the print media, especially the newspaper, they have had to operate under the watchful eye of governmental agencies created solely to keep them in check, to assure their responsible behavior.

The chapter also considers the adversarial relationship between journalists and government, and explores various means used by vested interests to exercise control over the press. To conclude we will investigate the difficulties arising from media coverage of court trials and other areas in which the First Amendment comes in direct conflict with the rights and responsibilities of other individuals and institutions.

Theories of the Press

All media reflect the political and social philosophies of the countries in which they operate.

Media operate under two basic philosophies—*authoritarianism* and *libertarianism*—and their several offshoots.

The mass media, as leading scholars have pointed out, always take on the form and coloration of the social and political structures within which they operate. It is only natural for the media to reflect the system of social control in which relations between individuals and institutions are adjusted.[1] Tentative explanations of the different systems have been offered, but the 1956 treatise by Frederick Siebert, Theodore Peterson, and Wilbur Schramm, entitled *Four Theories of the Press,* has remained one of the most durable. In general, they demonstrate that media operate under two general philosophic modes, *authoritarianism* and *libertarianism,* or their respective offshoots, which we will discuss momentarily, *Soviet totalitarianism* and *social responsibility.* Other scholars have expanded the categories, describing the uniqueness of media operating in *revolutionary* cultures, in third world *developmental* systems, and in modern European *democratic socialist* conditions.[2] (See table 14.1.) As we consider all of the various categories of media systems, we come to the conclusion that they are merely subtle refinements of the two basic categories, one stressing control and the other stressing freedom.

Men, as well as nations, tend to be authoritarian or libertarian. Of course, they are all somewhat schizophrenic, but basically they are disposed toward either a well-structured, disciplined world view with definite rules and an ordered society, or they are disposed toward an open, experimental, nonrestrictive society with a minimum of rules and controls. Governments are designed on the philosophical base of one of these two basic orientations.[3]

Scholars continue to quibble over the lack of precision in the designations of media systems, and over whether all presses are inherently socially responsible—"responsible" to whatever social, economic, and political system in which they may be operating. This means that what is acceptable journalism, entertainment, and persuasion in the American system could be highly irresponsible media performance under a different system. The very fact that the world's media systems are so different from one another, and that even within one nation the system of checks and balances changes from one day to the next, makes the study of such controls fascinating.

From the earliest times rulers attempted to control their domains. Beginning with ancient cave dwellers, through the growth of mass cultures in Mesopotamia, the Holy Roman Empire, and on to today's multifaceted cultures from Asia, Africa, Latin America, and most of the civilized world, the general history of civilization demonstrates a pattern of oppression, of subjugation of the people to the control of leaders.[4] In ancient Greece, Plato advocated rule by *philosopher-kings,* assuming that even an enlightened society should be kept in check by the wisest and strongest leaders rather than through open and shared government. In the late years of the Middle Ages, the Italian political advisor Machiavelli suggested, in *The Prince,* that leaders should exercise any and all options available to maintain their authority. (From Machiavelli we get the expression "The ends justify the means.") Throughout history, more people in more societies have lived under authoritarian regimes than under any other form of government.

After the printing press came into use during the fifteenth and sixteenth centuries, English monarchs (especially Henry VIII and Elizabeth I) devised new strategies to exert their authority. We noted in chapter 3 that the English government imposed restrictive censorship on printers through the use of licensing, taxation, and seditious libel laws. *Licensing,* which was started by Henry VIII, was a form of direct censorship (prior restraint), since there could be no publication whatsoever without the permission of religious or secular authorities. Having paid an exorbitant fee or bond for the privilege of printing, few printers were willing to lose their investment by irritating the authorities. Besides, they quickly learned that those printers who cooperated with the authorities received numerous side benefits, such as exclusive and relatively lucrative rights to print authorized books, pamphlets, and posters.

Taxation was the second means of controlling the press. In 1712 Parliament passed the first Stamp Act, a tax on newspapers and pamphlets, on advertising, and on the print paper itself. This basically amounted to a tax on knowledge. It was selectively enforced to punish scandalous and licentious publications, to force publications to register with the government, and to bolster the treasury. You will recall from your studies of American history that the refusal by the American colonists to pay the 1765 Stamp

Authoritarianism

Authoritarians tend to exercise prior restraint by means of licensing, taxation, and seditious libel laws.

Licensing: the means by which only those media operating in accord with governmental policies will receive permission to publish or broadcast.

Taxation: a means of controlling the press by forcing publications to register with the government. It was a tax on knowledge.

Table 14.1 Five theories of the press

	Authoritarian[1]	Communist[1]	Revolutionary[2]
Developed	In 16th and 17th century England; practiced in modern dictatorships in Latin America and elsewhere	In the Soviet Union after 1917; currently found in Eastern Europe and the Communist nations	In the Soviet Union before 1917, in independence struggles of Third World, and wartime occupied nations
Out of	Philosophy of absolute power of monarch, his/her government, or both	Marxist-Leninist-Stalinist thought, with mixture of Hegel and 19th Century Russian thought	Writings of Lenin and experiences of early nationalist movements and resistance movements
Chief purpose	To support and advance the policies of the government in power; to service the state	To contribute to the success and continuance of the Communist social system, especially party dictatorship	To overthrow existing national government or free the state from foreign dominance
Who has right to use media?	Whoever gets royal patent or similar permission	Loyal and orthodox party members	The people who oppose existing authority
How are media controlled?	Government patents, guilds, licensing, sometimes censorship	Surveillance and economic or political action by government	Uncontrolled by government or existing society
What is forbidden?	Criticism of political machinery and officials in power	Criticism of party objectives as distinguished from tactics	Not governed by official proscriptions on content
Ownership	Private or state	State or party	By revolutionary organizations
Essential differences	Instrument for effecting governmental policy, though not necessarily governmentally owned	State-or-party-owned; closely controlled media existing solely as arm of the state or party	Essentially a transitional theory

Source: By permission, University of Illinois Press and Robert G. Picard.

1. Authoritarian-tending
2. Indeterminant tendencies
3. Libertarian-tending

Table 14.1 *Continued*

Developmental[2]	Western[3]		
	Libertarian Adopted in England after 1688 and in U.S.; influential in Western and pro-Western states	*Social Responsibility* In the United States in the 20th century	*Democratic Socialist* In 20th century Western Europe
In 20th century non-industrialized, non-communist Third World nations			
Marxist thought combined with communication-for-development views; Schramm/Lerner/Pye	Writings of Milton, Locke, Mill, and philosophy of rationalism and natural rights	Writing of W. E. Hocking, Commission on Freedom of the Press, media codes, and practitioners	Modern Marxist thought combined with writings of classical liberal philosophers
To promote national integration and social and economic development	To inform, entertain, sell— but chiefly to help discover truth and check government	To inform, entertain, sell—but chiefly to raise conflict to the plane of discussion	To provide avenues by which diverse opinions can be made public; to promote democracy in all social spheres, including the economic
Government has right to use for programs for public good	Anyone with economic means to do so	Everyone who has something to say	All citizens
Government and/or party control, and legal constraints	By "self-righting process of truth" in "free marketplace of ideas" and by courts	Community opinion, consumer action, professional ethics	Collective management and legal constraints
Challenge to authority; information damaging efforts for progress	Defamation, obscenity, indecency, wartime sedition	Serious invasion of recognized private rights and vital social interests	Undue interference with individual rights and other recognized social interests
Private or state; most often state	Chiefly private	Private unless government has to take over to ensure public service	Public (non-state), non-profit, and private (at this time)
Information is national resource to be used for developments; societal concerns more important than individual concerns	Instrument for checking government and meeting other needs of society	Media must assume obligation of social responsibility; if they do not, someone must see that they do	Media must not be unduly controlled by government, economic, or social interests

Tax went a long way toward fomenting the revolution. "Taxation without representation is tyranny," we shouted. Nevertheless, such stamp acts lasted until 1855 in England.

Seditious libel: criticism of the government, making it illegal to spread rumors about the crown, even if they were true.

Laws regarding *seditious libel* (criticism of the state) were a third form of control over the press. Seditious libel originated somewhere around the thirteenth century in England, when it became illegal to spread rumors about the crown and nobility. The crime was so serious that it was completely irrelevant whether the rumors were truthful or not. Until the eighteenth century, all a jury had to do was to ascertain whether the accused had indeed published the rumor or criticism. Because truthful criticism was more likely to stir up the angry crowds than readily disproved untruthful criticism, the general principle for many years was that *the greater the truth, the greater the libel.*

Even today, the majority of the world's people live under authoritarianism.

It would be wrong to discuss authoritarianism in the past only. Even today, most of the people of the world live under authoritarian rather than libertarian governmental systems. Despite the fact that for the past third of a century the United Nations' Universal Declaration of Human Rights has demanded that its members guarantee freedom of speech, only two-fifths of the world's people actually receive this and other parallel freedoms: freedom of assembly, freedom of religion, and freedom to petition their governments. Annual analyses of worldwide trends in freedom and control demonstrate the fragility of such assurances. In the mid-1970s, for example, the world's largest democracy, India, summarily abolished its constitutional guarantees of free speech, press, and assembly. The besieged Prime Minister, the late Indira Gandhi, called her action "The Emergency." She justified it on the basis of economic and political instability in her nation of a half billion people. A decade later, the troubled nation of South Africa imposed a series of emergency restrictions on journalists and others embroiled in antiapartheid movements. Television crews were barred from showing riots or police actions, and news accounts were officially censored. The political rationales of India and South Africa mirrored that of both revolutionaries and established dictators around the globe. All of them—from Albania to Zanzibar, and most certainly including Iran, Nicaragua, Greece, Venezuela, and Chile—are merely following the well-established political theory that holds that the less stable or secure the governors are, the less they tolerate dissent from the governed.

Communist Application

Communist media are part of the state, and reflect party decision making.

The Soviet-Communist theory of the press, logically, is an extension of authoritarianism, with one important exception. Authoritarian theory holds dear the need to control, but it still recognizes the press as an entity outside of government. In Communist theory as exemplified by the Soviet Union, the press for all practical purposes is a part of the state; it does not exist outside the state. Consequently, all media reflect the national policy or the party line, and all media efforts are pointed toward furthering the state's aim. There can be no repression of the press, for the press is the state.

Reprinted with permission from Heritage Features Syndicate.

Because the communist press has the responsibility of perpetuating and expanding the socialist system, it spends its time transmitting policy—already established truth, as it were—not searching for a nebulous truth that might emerge from a clash of ideas. As one might suspect, the content of the communist press is neither escapist nor entertainment oriented. Media are responsible for informing and indoctrinating society, and as such, media reflect what William Stephenson refers to as "social control" rather than "communication play."[5] Such is the nature of media functioning from the top down, operating as the voice of government.

Libertarianism

Karl Marx, whose ideas are the basis of communist thinking, felt that we had to improve society in order to improve individual men and women. Marx's views are diametrically opposed to the arguments of libertarians, who believe that we improve society by improving the individual.

During the Age of Reason, beginning in the seventeenth century, a period of unparalleled medical and scientific discovery, a new view took hold in which people began to be seen as rational and capable of making decisions. Philosophers from John Milton to Thomas Jefferson to John Stuart Mill defended the rationality of people and the ability of truth to withstand discussion.

Poet John Milton, as far back as 1644, argued in ringing terms before the British Parliament many of the basic tenets of libertarianism. His speech, "Appeal for the Liberty of Unlicensed Printing" (later reprinted as the poem *Areopagitica*), called for the "open marketplace of ideas" and the "self-righting process." Let all with something to say be free to express themselves, Milton argued. "Let truth and falsehood grapple; whoever knew truth to be worse in a free and open encounter. . . . Though all the winds of doctrine were let loose to play upon the earth, so truth be in the field, we do injuriously by licensing and prohibiting, to misdoubt her strength," he claimed.

Lest we be smitten by the ringing tones of *Areopagitica,* we should remember that Milton's motives were self-centered. He was expressing irritation over censorship of his own writings. He wanted serious-minded writers, such as himself, to be free to share their honest, although divergent, opinions. Ironically, history found Milton serving as a censor for the British government several years after the speech. Milton's motives in *Areopagitica* appear to have been somewhat narrow in that the freedom he was advocating was negative (freedom from licensing) rather than positive (freedom for certain ends). He still maintained that government had the obligation of prohibiting certain types of publications, particularly those that were blasphemous or seditious. Despite the narrowness and selfishness of Milton's arguments, they were powerful statements made about one century ahead of their time.

Thomas Jefferson's views on a free press were soundly libertarian in nature. He felt that although individual citizens may err in exercising their reason, the majority as a group would inevitably make sound decisions, as long as society consisted of educated and informed citizens. The way to help people avoid making errors of judgment, wrote Jefferson to a friend in 1787,

is to give them full information of their affairs through the channel of the public papers, and to contrive that those papers should penetrate the whole mass of the people. The basis of our government being the opinion of the people, the very first object should be to keep that right; and were it left to me to decide whether we should have a government without newspapers, or newspapers without a government, I should not hesitate a moment to prefer the latter.[6]

Editors, reporters, and other defenders of press freedom frequently quote the last sentence of the excerpt just cited, forgetting that Jefferson went on to qualify the seemingly anarchistic statement: "But I should mean that every man should receive those papers, and be capable of reading them." The qualification is significant, for it recognizes that even the philosophy of libertarianism is meaningless if the people being promised liberties have no means of utilizing them.

A third major contributor to libertarian theory was English philosopher John Stuart Mill. Liberty, to Mill, was the right of mature individuals to think and act as they pleased so long as they harmed no one else in the process. Mill's 1859 treatise *On Liberty* states that all human action should aim at creating the greatest happiness for the greatest number of people. According to Mill, the good society is one in which the greatest possible number of persons enjoy the greatest possible amount of happiness. And, unless they have the right to think and act for themselves, they will never reach this point. Indeed, he argued that if everyone but one individual had one opinion, society would be no more justified in silencing the individual than the individual, if he or she had the power, would be justified in silencing the rest of mankind.

Mill's philosophy is attractive to many Americans, especially those who are interested in the media. Mass communications in America are anchored in libertarian ideas, and America has, for the most part, endorsed a libertarian point of view. Whether or not America, or any country, approaches, or ever could approach, the fullest form of libertarianism is open to debate.

New legislation to control the effect of mass media and to protect their freedom is constantly advocated under social pressure. This has prompted a new theory or set of premises known as *social responsibility*. The theory was first enunciated by the Commission on Freedom of the Press, a group that was led by Robert Hutchins and met in the 1940s to assess the state of the press in America (see chapter 15).

Essentially, the theory of social responsibility is an extension of libertarianism in that it seeks to protect free expression. It requires the mass media to adequately represent all hues of the social spectrum. It seeks to make the mass media responsible for the quality of their offerings, print or broadcast. It seeks to inject truth in advertising and remove the concept of *caveat emptor* ("let the buyer beware"), which in the uncontrolled commercial world has seriously eroded a part of media credibility. Social responsibility charges the mass media with the development of, and enforcement of, ethics in the public interest. The theory is a formal recognition of the corporate organization, and institutionalized nature, of mass media. Much of the remainder of this chapter, and nearly all of chapter 15, considers the ramifications of this theory in contemporary America.

Social Responsibility

Social responsibility theory requires our mass media to adequately represent all hues of the social system. It tries to strike a balance between freedom and responsibility.

Earlier we mentioned that some scholars have chided educators and journalists for assuming that there are only four explanations of how the media operate—the authoritarian, communist, libertarian, and social responsibility theories. At least three additional theories or explanations of media systems have been offered; they are the revolutionary, developmental, and democratic socialist press theories.

Other Theories of the Press

Revolutionary Press In countries coping with revolutions against the existing government or foreign domination, the media appear to be in a transitional stage, divorced from normal state-media relationships.[7] The stage of unrest is initially fomented by the media, as we saw in the propaganda efforts of Samuel Adams and others in colonial America, by "underground journalists" in the restless 1960s and 1970s, and by many Third World nations today.

Because the press is managed by people who feel alienated from the government, there are certain risks involved in revolutionary publishing or broadcasting. Our own heritage is replete with journalistic martyrs, as are the heritages of Third World nations. If the work of the revolutionaries is successful, they are likely to become the established journalists once they replace their oppressors. The shape of the new government and its media system is determined by a number of factors, including the mood and philosophy of the victorious revolutionaries. For instance, the Soviet Communist media system emerged from Lenin's writings, whereas the American libertarian media system emerged from the philosophy of journalists such as Adams. Thus we can say that the revolutionary press theory shows indeterminant tendencies; it is not necessarily authoritarian nor libertarian in nature.

Developmental Press A natural offshoot of the revolutionary press theory, at least in twentieth century Third World nations, can be called the *developmental theory* of the press. It reflects modern nationalist and political independence movements, and draws upon socialist thought and developmental principles. According to one description, media under this theory are used to help promote social and economic development and achieve national integration. Like the revolutionary theory, this developmental theory is probably transitional, but seems to be longer lasting, as the nations move slowly toward modernization. And it is similar to the revolutionary theory in one more respect. Because of the vast number of variables that determine what form of government is going to emerge, it is practically impossible to judge whether the developing nations will move toward the libertarian or the authoritarian ends of the governmental-press continuum.[8]

One trend that can be pinpointed is the resentment many Third World countries feel toward the Western world. Their cries against cultural imperialism or communications colonialism, as we discussed in chapter 13, have resulted in calls for a *New World Communications Order,* or a *New World Information Order.* Such concerns are natural, given that most peoples of the Third World are colored, poor, ill-nourished, and illiterate, and resent the primarily Caucasian, affluent, well-fed, and literate West.[9] Given those vast disparities, it is unlikely the developing nations will gravitate naturally toward a media system created by and suited to the Western world.

Democratic Socialist Theory In some Western societies, especially those in Scandinavia and elsewhere in Europe, a modified form of social responsibility theory seems to be emerging. This *democratic socialism* has developed out of a combination of Marxist thought and the writings of classical

Revolutionary press encourages the development of a new political system. The system is neither authoritarian nor libertarian.

Developmental press, especially in Third World nations, promotes social and economic development.

libertarian philosophers. It recognizes the uniqueness of a fully developed media system that has been given free reign in a *laissez faire* marketplace, but has not adjusted to the fact that increased monopolies and concentration of media ownership may demand new forms of accountability and responsibility. Abuse by private owners, motivated by the economic bottom line, differs very little from abuse by despots, according to adherents of democratic socialism.

Like the social responsibility theory, those advocating democratic socialism suggest that the media be granted positive freedoms—freedoms they should be exercising to achieve certain goals. For instance, the media are expected to open their editorial columns to a wide diversity of opinions, and "to fuel the political and social debates necessary for the continued development of democratic governance."[10] If the press is not naturally inclined to open itself to these diverse views, someone or something will see that it happens. That "someone or something" could be a citizens group, a press council, or a governmental agency. Collective management, governmentally enforced ownership by political, social, and racial minorities, and a whole new body of legal constraints are logical outgrowths of this theory.

Libertarians do not take such suggestions lightly. They see little difference between the democratic socialism brand of media responsibility and the old-fashioned theory of authoritarianism. Indeed, they see signs that the Marxist philosophy of state ownership is inevitable once the public and government are given "control" over the press. They are uncomfortable with the premise that media become common carriers, subjected to regulatory pluralism. Nevertheless, they are forced to acknowledge that a totally free, unfettered, commercially motivated mass media system is highly unlikely to operate with the best interests of the citizenry in mind on all occasions.

All in all, the press theories described lead us to conclude that there are no pure or simple explanations of how the media operate within any political and economic system. As we have said, the general tendencies seem to be either in the direction of authoritarianism, with its central control over ideas, or libertarianism, with its general freedom from external constraints. The more complex the economic and political system, it seems, the more shades of gray are seen in the way mass media function. Modern American media are certainly not exempt from this idea.

Although the predominant view of media in America has been libertarian, no medium can operate with unlimited freedom. Government imposes restrictions despite the apparently absolutist nature of the First Amendment to the Constitution, which maintained that "Congress shall make no law . . . abridging the freedom of speech, or of the press."

Democratic socialism accepts a free press, but resists the freedom of unchecked economic growth; some external control is called for.

Limitations on the Media

Congress of the United States.

Begun and held at the City of New York,

on Wednesday the fourth of March one thousand seven hundred and eighty nine

---

The First Amendment to the U.S. Constitution

Congress shall make no law respecting an establishment of religion, or prohibiting the free exercise thereof; or abridging the freedom of speech, or of the press; or the right of the people peaceably to assemble, and to petition the Government for a redress of grievances.

A thoughtful analysis of such limitations has been offered by journalism historians who note:

Press freedom depends greatly on government's moods. Even a "free press" has limitations.

1. The extent of governmental control of the press depends on the nature of the relationship of the government to those subject to the government.
2. The area of freedom contracts and the enforcement of restraints increases as the stresses on the stability of the government and the structure of society increase.[11]
3. The more heterogeneous a society, the more freedom of expression it will tolerate.
4. The more developed a society, the more subtle will be the controls it exerts on expression.[12]

To see the validity of these four propositions, we will look to our media's rights to print, to criticize, and to report about government and society.

The first newspapers printed in America had to be approved by colonial officials, but before the middle of the eighteenth century, journalists were arguing for freedom from censure and prior restraint. Their independence was seen in their total disregard of the Stamp Act of 1765, which the Crown imposed in an attempt to bolster its depleted treasury, not to impose absolute prior restraint as in earlier cases in Britain.

American journalists have continued their opposition to any taxes that might be seen as discriminatory. Even in the twentieth century, when governments have attempted to place a tax on information, the press has successfully fought the taxes through the courts. In the most notable example, Governor Huey Long of Louisiana convinced the state legislature in 1934 to enact a 2 percent tax on the gross receipts of newspapers having circulations over twenty thousand per week. Because Long was feuding with all but one of the thirteen largest newspapers in the state, the United States Supreme Court unanimously declared the tax an unconstitutional misuse of governmental authority.

Media's ability to fight this type of taxation does not mean the media are immune from all taxes. Historically, they have been subjected to those ordinary forms of taxation that are considered legitimate support of their government. For instance, the Associated Press wire service lost a 1937 court case over taxation because the Supreme Court said the taxes in question were not threats to freedom of the press or freedom of speech, but were merely equitable and nondiscriminatory taxes on business.

The right to print is meaningless if the right to distribute is impeded. From time to time throughout its history, American government in its several forms has subverted the provision of the First Amendment for a free flow of information. The majority of court cases attempting to reconcile the conflict have involved limitations imposed upon the distribution of leaflets or pamphlets, usually those that are political, religious, or commercial in tone.

Additional control has been available to the government through the use of second-class mailing permits, which have helped the print media—especially magazines and promotional material—reach their audiences. Such permits were created by the government to encourage and subsidize distribution of knowledge, and thus became a form of licensing.

Some control over distribution continues to exist. By their simple ability to determine the rates to be charged, the postal authorities intimidate publishers operating at slim profit margins. Some mass circulation magazines have cut down their size and weight to offset rising postal costs. During the 1950s and on through the 1980s, others went out of business because they were unable to cope with distribution costs which were rising exponentially.

Limitations on Printing

American media oppose discriminatory "taxes on knowledge."

Limitations on Distribution

Government has several ways of controlling media distribution.

UPI

The New York Times

CITY EDITION

VOL CXX.... No. 41,431 — NEW YORK, THURSDAY, JULY 1, 1971 — 15 CENTS

SUPREME COURT, 6-3, UPHOLDS NEWSPAPERS ON PUBLICATION OF THE PENTAGON REPORT; TIMES RESUMES ITS SERIES, HALTED 15 DAYS

Nixon Says Turks Agree To Ban the Opium Poppy

PRESIDENT CALLS STEEL AND LABOR TO WHITE HOUSE

Pentagon Papers: Study Reports Kennedy Made 'Gamble' Into a 'Broad Commitment'

BURGER DISSENTS

First Amendment Rule Held to Block Most Prior Restraints

Soviet Starts an Inquiry Into 3 Astronauts' Deaths

U.S. and Diem's Overthrow: Step by Step

Prior Restraint

The Pentagon Papers and the
Progressive Magazine cases
indicate that prior restraint is
still one of government's
options.

We need to look no further than the Nixon administration and a perplexing 1979 case to see evidence that government has other ways of dealing with unpopular or dissident news and views. Sometimes, as in the 1971 Pentagon Papers case and the 1979 *Progressive Magazine* controversy, the government will attempt to curtail potentially troublesome journalism before it is printed or aired.

The most famous test case was probably not the most significant. But because it involved the Vietnam War, the Nixon administration, and the media giants (particularly the *New York Times* and the *Washington Post*), the 1971 case of the Pentagon Papers is viewed as a highly significant struggle over prior restraint.

In June 1971 the *New York Times* culminated a three-month effort of rewriting and editing a purloined forty-seven-volume secret study of the origins of the Vietnam War entitled "History of the U.S. Decision-Making Process on Vietnam Policy." It had been compiled by think-tank experts at RAND Corporation, one of whom was Daniel Ellsberg. Ellsberg was convinced that the American public had a right to know how it had gotten involved in the long and unpopular war. Although the story told by the *New York Times* was basically historical and revealed nothing that could significantly affect current United States military strategy, it was politically and diplomatically embarrassing to the Kennedy, Johnson, and Nixon administrations. The *Times* was convinced that the public's right to know overshadowed governmental embarrassment and the ethical questions surrounding the printing of top secret information. The Nixon administration disagreed. When the newspaper refused to honor Attorney General John Mitchell's request to suspend publication, the government successfully obtained a temporary restraining order—in effect, creating prior restraint—assuring that the series would not be continued until the issue was decided in the courts.

The case went quickly to the Supreme Court and two weeks later, after the justices had a relatively brief time to deliberate the particulars of the case, press attorneys gained a small victory when they convinced a six-member majority of the court that the government could not prove that anything in the Pentagon Papers actually endangered national security. Publication of the papers then continued, but the essential question about the right of government to impose absolute prior restraint went unanswered.[13]

In 1979, more of an answer was provided when the 40,000-circulation liberal magazine *The Progressive* found itself in a legal and ethical vortex after its editor assigned a free-lancer to write a story on the hydrogen bomb. The story, to put it mildly, was explosive. Writer Howard Morland, who had majored in economics in college and had taken a smattering of physics courses, did an excellent job of researching textbooks, the *Congressional Record,* encyclopedias, and scientific journals, and taking guided tours of atomic power plants—all sources readily available to anyone inclined to use them. The magazine scheduled Morland's piece for publication under the title, "The H-Bomb Secret: How We Got It and Why We Are Telling It." *Progressive* editor Erwin Knoll defended publication of the piece, claiming it did not disclose any new or classified information and that it served the important purpose of demonstrating laxity in governmental security over what is supposed to be one of government's best-kept secrets.

But the article was such a definitive treatise on how to build a hydrogen bomb that when *The Progressive* voluntarily but reluctantly sent it to officials to review for accuracy before publication, the government reacted swiftly and severely. It asked the magazine to delete some details in the story, but editor Knoll refused. The Justice Department then filed suit in a federal court and obtained a temporary restraining order against publication of the article. The government's case, logically, was that public dissemination of the information would increase the proliferation of nuclear weapons and thereby severely undercut the arms control and disarmament policies of the United States.[14]

Federal District Judge Robert Warren in Wisconsin agreed with the government, saying, "If a nuclear holocaust should result, our right to life becomes extinguished, and the right to publish becomes moot."[15]

The *Progressive* appealed Judge Warren's injunction, and was preparing to carry the case all the way to the United States Supreme Court. Some critics, including journalists, hoped it would not get that far, because it seemed likely the court would rule against the magazine.[16] As it turned out, the Burger court never ruled on the case, because nearly identical information about the construction of an H-bomb was printed in the *Madison Press Connection.* In the face of mounting pressure from the press and public, the federal government dropped its charges against *The Progressive,* which went ahead with its original story plus additional stories about its legal hassles over the bomb. As Jack Anderson commented before the case had been fully resolved, "regardless of the outcome of the investigation and the entire *Progressive* case, journalists and their sources will become more and more timid in their efforts to inform the public of things it has a right to know. Governmental harrassment," he said, "will result in self-censorship."[17]

Sedition

Governments in stress impose sedition laws; the United States did so in 1798 and during more recent war times.

The 1735 trial of New York publisher John Peter Zenger saw the jury revolting against the obviously antilibertarian tradition that truth was no defense against seditious libel charges. As noted in chapter 3, Zenger's acquittal helped establish the rights of journalists to criticize the government, but it did not settle the issue permanently. Such rights are tenuous. Proof came shortly after the new nation was established and the Constitution with its press freedoms was adopted. The Sedition Act, in effect from 1798 to 1801, was instigated by President John Adams' Federalist party. It was an effort to quell verbally abusive editorial criticism by newspapers under the control of rival anti-Federalists or Jeffersonians. The Sedition Act expressly outlawed false, scandalous, and malicious publications against the United States government, the president, and Congress. Fifteen people were prosecuted under the Sedition Act, some for offhanded witty remarks about President Adams. When Adams was replaced by Jefferson as president in 1800, the act expired, and Jefferson pardoned everyone who had been convicted under it.

American citizens and their courts must support the First Amendment before it becomes workable.

Of the lessons learned from the Sedition Act of 1798, media law scholars maintain that the most important one is the proposition that constitutional guarantees, in and of themselves, do nothing to assure freedom of expression; American citizens and their courts must support the First Amendment before it becomes workable.[18]

In the twentieth century we find further support of the proposition that press freedoms expand and contract in inverse relationship to the government's sense of security. Wartime tends to be a period of political repression, and modern wars are no exception. During World War I, for instance, while most Americans were demonstrating solidarity in the effort to keep the world safe for democracy, those who questioned the will of the majority were regarded with suspicion. Given the socially and politically homogeneous—and therefore intolerant—climate of the times, it is not surprising to find another Sedition Act being put on the books. The 1917 Espionage Act and the 1918 Sedition Act made it illegal to openly oppose the war effort, and resulted in the prosecution of two-thousand offenders. As a result the Supreme Court had to resolve the issues of what was and was not protected by the First Amendment.

In a 1919 test case, *Schenck v. United States,* Justice Oliver Wendell Holmes maintained that sedition occurs only when the words used create a "clear and present danger" to society.[19] For four decades courts attempted to interpret Holmes's test. Finally, a 1957 case found the court agreeing that merely advocating the overthrow of government was not seditious; rather, there had to be a "clear and probable danger" that the government would actually be overthrown. Abstract arguments were not a crime, but incitements to immediate action were.

Due to the difficulty of proving what constitutes actual incitement, the government has brought few prosecutions for sedition since the 1950s. Criticism of American government remains robust, but courts are spending relatively little energy in the attempt to balance the rights of government to remain secure and the rights of critics to complain. However, history's tendency to repeat itself has given us the political truism that an insecure government and a homogeneous populace might result once again in repression.

Seditious libel may have faded into the sunset, but civil libel certainly has not. Civil libel is a dispute between two private parties (individuals, corporations, associations, etc.) in which the government, through the court system, acts as a referee.

Recently American news media have been hit with some five hundred lawsuits a year, charging them with damaging the reputations of individuals or groups of people. At any given time today, there are some 2,500 libel suits working their way through the courts. At the trial level, in a recent three-year period, the media were losing 83 percent of those libel suits. In one year alone, juries awarded plaintiffs $1 million or more in twenty-two separate libel cases. Upon appeal, 71 percent of the libel cases were overturned or the amount of the award was reduced by judges who were more likely to take First Amendment issues into consideration than were the local juries. (For what it's worth: in nonlibel cases, during the same three-year period, juries voted for the plaintiffs only 51 percent of the time. While 83 percent of the journalists sued for libel were losing their cases, only 38 percent of manufacturers were losing their product liability suits, and only 33 percent of doctors were losing their malpractice suits.)

The conclusion to be drawn seems obvious. If you are a journalist, the men and women of your community are far less likely to side with you than if you are almost any other kind of a defendant. As the head lawyer for *Time* magazine put it, "Juries work out their passions against the press. They perceive it as one little guy up against the huge media."

Even when they win these civil libel suits, the media are likely to be spending small fortunes in legal fees. CBS, it is estimated, spent more than $10 million in legal fees defending itself against the $120 million libel suit filed by General William Westmoreland, who took issue with the "60 Minutes" claim that he had lied about enemy troop strength during the Vietnam

Libel

Libel laws give ordinary citizens a chance to hold their own against powerful media. The news media find libel cases an expensive and sometimes crippling form of intimidation.

In one of the most famous libel cases of all time, General William Westmoreland unsuccessfully sued CBS for $120 million, claiming the CBS show, "60 Minutes," defamed him when it discussed his role in the Vietnam War.

War. The *Pacific Daily News,* a small paper in the tiny island of Guam, spent $2 million over a seven-year period successfully defending a $50 million libel suit. The only reason it is still in business is that it had the backing of the powerful Gannett chain, to which it belongs. Had it been operating on its own, it never could have afforded the protracted lawsuit.

Meanwhile, for the months if not years spent in litigation, the individual news media suffer the obvious disruptions of daily reporting and editing that is entailed in being dragged into court. When they lose, not only do they pay out what are sometimes astronomical damage awards, but they also find themselves a little more timid and a little less likely to engage in robust investigative journalism.

Libel suits are seen as the public's way to balance the power struggle between themselves and the news media.

According to many Americans, it serves the press right. There are few ways for the celebrated and uncelebrated people in the street to hold their own against the powerful news media, so a libel suit—or even the threat of one—now and again may balance the power. As the laws of libel have developed, the common citizen gets a few considerations to offset the handicap of not being rich enough or politically powerful enough to go one-on-one against the press.

The best-known libel cases involve investigative reporting and editorial writing, but the *Associated Press Stylebook and Libel Manual* warns journalists that in 95 percent of libel cases, ordinary citizens claim that their reputations were injured by reporters who carelessly reported charges of crime, immorality, incompetence, or inefficiency. A study of nine-hundred libel and invasion of privacy suits between 1974 and 1984 showed a similar trend, although not as extreme. Researchers from the University of Iowa reported that only about 45 percent of libel suits resulted from stories that appeared on the front page of a newspaper.[20] The simple explanation: investigative stories are done with care, and usually checked by the news media's team of lawyers, whereas routine stories are given a quick once-over by the copy desk. When factual errors or inexact language crop up, people's reputations can be damaged. If the media report a drug arrest at 250 South Main Street and it actually occurred at 250 North Main Street, the upstanding citizens at the incorrect address are likely to be humiliated and embarrassed among their friends and acquaintances, and the media will have caused the problem.

Innocent-appearing stories damage more reputations than sensational investigative ones.

In order for the plaintiffs to prove that libel has occurred, they have to show the presence of four things: *defamation, publication, identification,* and *negligence.* Different states have different definitions of defamation, but most include the idea that defamation is communication that exposes persons to hatred, contempt, or ridicule; lowers them in the esteem of their peers; causes them to be shunned; or injures them in their business, trade, or profession. Examples of words and expressions that juries have found to be defamatory are those that imply that someone has done something illegal or has engaged in questionable sexual behavior, for example, or that causes others to shun a person's business because of things said about the person's business practices.

Plaintiffs must demonstrate defamation, publication, identification, and negligence.

A defamation is considered to have been published if it appears in a form in which another person, besides the writer and the person being defamed, is likely to see it. Anyone handling the copy during the publishing process can be hauled into court, although only those who can afford to pay damages are likely to be named in a suit.

To collect for libel, plaintiffs have to prove that they were singled out in some way that would leave no doubt in the minds of their associates that they were the subject of the attack, such as being shown in a picture or identified in context. Identification can occur indirectly, and someone can sue over misidentification. When the media do name individuals, they have to be careful not to injure other persons who have the same name. Full identification, including middle names, initials, ages, and addresses, is standard operating procedure in most newsrooms.

The fourth criterion, negligence—and its kissing cousin, malice—is fairly recent. Key Supreme Court cases of the 1960s—particularly one involving the *New York Times* and L. B. Sullivan, a city commissioner from Montgomery, Alabama—made it difficult for public officials and public figures to win libel suits unless they were able to prove *actual malice.* By that the court meant that defendants had known that their articles were false or had recklessly disregarded whether they were true or false.

In the 1970s and early 1980s, the tide began to turn somewhat against the media, as Supreme Court rulings made it easier for both private citizens and public officials to win libel suits. The legal questions in these cases centered around whether the media defendants were negligent in their journalistic practices. (The legal question in cases involving private individuals centered around whether the media defendants had been guilty of simple negligence; in cases involving public officials or public figures, the issue was whether the media had been guilty of negligence so extreme as to constitute actual malice.) Although it may be too early to tell, it seems that the tide may be shifting again, this time in the media's favor. Failure of General William Westmoreland to win his megasuit against CBS, and Israeli Defense Minister Ariel Sharon to win his $50 million suit against *Time* magazine in 1985, gave the media some confidence to continue pursuing risky and controversial investigative stories. Then, in 1986, the Supreme Court handed down two decisions that cheered journalists. One, *Jack Anderson v. Liberty Lobby,* makes it clear that public figures have a much harder time than private figures if they hope to beat the press. If public figures do not convince the judge, in pretrial hearings, that they have a good case against the news media, the judge is expected to use *summary judgment* and throw the case out of court prior to a jury trial. The second case, *Philadelphia Newspapers v. Hepps,* held that even private figures who have inadvertently found themselves the subject of a controversial news story of public concern, have the same burden of proof as public officials have long had—they all have to prove that the news items were false before the case will proceed. Media watchers expressed some surprise that a conservative Supreme Court would grant the news media such extra protection, but said it would be a while before the apparently libertarian decisions were put into practice.

Meanwhile, tradition holds that if plaintiffs are unable to convince juries that defamation, publication, identification, and negligence have all occurred, their case will probably be thrown out. Even if judge and jury agree with the plaintiffs that all four elements are present, victory for the aggrieved party is not yet assured. It is time for the defendants to build their cases, which they can do with any or a combination of *complete* or *partial defenses.*

The most common defenses are *truth, privilege,* and the *fair comment and criticism.* In some states, truth alone is not enough; the defendant must also prove lack of malice. Privilege is the right of the media to report activities of public officials acting in an official and public capacity. Impartial and fair abridgments of what transpires in open legislative and judicial proceedings are considered privileged and protected from charges of libel.

Media frequently get away with defamations because of the well-established defense known as fair comment and criticism. Generally it holds that it is permissible to publish defamatory material consisting of comment and opinion, rather than fact, about issues of public interest or importance. The editorial, review, or criticism must focus upon the public aspects of the subject at hand. A drama critic can harshly judge an actress's performance with impunity, but not her home life. A sportswriter can criticize a quarterback's ability to pass a football, but not his morals. Additionally, the opinions expressed should be those of the journalist, not just those of somebody being quoted about the person being criticized. Finally, responsible journalists should include factual data in their commentaries, so audiences have the opportunity to understand the basis of criticism being made.

As long as the editorial, commentary, or report about a public official or public figure is based on the above, journalists can probably win a libel case unless the plaintiff is able to prove actual malice. Carol Burnett's successful suit in 1981 against *The National Enquirer,* which falsely implied that the actress known for her anti-drug-abuse stance had been drunk in public, reminds us that fair comment and inaccurate reporting have their bounds. Another in a recent spate of similar suits found *Penthouse* magazine facing a multi-million dollar libel judgment after it printed a humorous article about the sexual exploits of an allegedly fictional Miss Wyoming. The real Miss Wyoming did not find the article amusing, and convinced a jury her reputation had been damaged.

The Westmoreland and Sharon trials, along with hundreds of others, indicate the media are expected to be accountable for investigative reports, fictional accounts, and even minor news items that disparage either private or public citizens. And that, most observers agree, is the way it should be. Other observers, meanwhile, insist that the whole system of libel law has gone awry, and that there should be other nonlegal means available for aggrieved individuals to salvage their reputations.[21] Even General Westmoreland, after the lengthy—and indecisive—legal battle that cost him several million dollars and CBS $10 million to wage, said he would have been happy to find some other means of letting the world know he felt his reputation had been damaged. Short of a massive change in our legal system, however, there seems little likelihood the law of civil libel will disappear.

Invasion of Privacy

Courts and state legislatures have developed another means of checking media excesses as they have put teeth into the vague concept known as the *right of privacy,* or the *right of the individual to be left alone.* Because of our national concern over dossiers (files of detailed information) compiled by credit bureaus, Uncle Sam's "super information bank" containing an incredible amount of private information about each citizen, the omnipresent closed-circuit television camera where we bank, and so on, Americans have gained a new awareness of this so-called right to be left alone.

Privacy laws protect a person's right to be left alone or to have peace of mind, while libel laws protect a person's reputation.

The "law of privacy" is relatively new, tracing its ancestry to an 1890 *Harvard Law Review* article by two young lawyers, Samuel D. Warren and Louis D. Brandeis. Offended by the snoopy Boston press, they argued that citizens should have legal redress from the prying, gossiping media. Since then, eleven states have passed privacy statutes, and courts in nearly all states have demonstrated sympathy with the basic principles.

Privacy lawsuits involve any of four different torts, or legal wrongs: *misappropriation* of a name or likeness for commercial gain, *unreasonable intrusion* into someone's seclusion, publication of *private information,* and putting someone in a *false light* by fictionalizing.

Privacy laws apply to appropriation, intrusion, and publication of private or false information.

America's first invasion of privacy suit involved the utilization of a picture of a young girl on a flour sack without her permission. This is known as *appropriation* of someone's likeness for commercial gain. After she lost

the case in 1902, the New York State Legislature produced the first state statute protecting people in such cases. Recently, plaintiffs have won damages for the right to control their own publicity and to make their own profits from it.

Intrusion into a person's solitude, including the use of microphones or cameras, is a second form of invasion of privacy. A classic case occurred in 1973 when a federal judge enjoined freelance photographer Ron Galella from his continual and bothersome photographing of Jacqueline Kennedy Onassis. He was told to stay at least twenty-four feet from Mrs. Onassis, and thirty feet from her children. He was also prohibited from blocking their movements; from doing anything that might put them in danger or that might harass, alarm, or frighten them; and from entering the children's play area at school.[22] After repeatedly violating the judge's orders, Galella in 1982 was found to be in contempt of court. He was told that he could photograph Mrs. Onassis in public places, but that his conduct had been so outrageous that the Onassis family deserved some privacy from his prying cameras.

Other intrusion cases center around plaintiffs' reasonable expectations of privacy—in their own homes, in a hospital bed, in a restaurant, etc. In addition, successful suits have been filed against media which took extremely embarrassing photos of persons in public places or which used hidden cameras or tape recorders to conduct investigations. Of course, we have all seen such photos and news stories appearing with impunity, so it is obvious that our nation's courts have held that intrusion is a matter of degree.

The original issue raised by Warren and Brandeis was the use of *private information,* which has involved the most controversy. To what degree does the press have the right to publish gossip, the substance of private conversations, and the details of private tragedies and illnesses, especially if the people in question do not want such information made public? Courts generally have agreed that if the public is interested in it, the press has the right to publish it. Only when good taste and good sense are totally abandoned will the press find itself in deep water.

The kinds of issues in question include the publishing of rape victims' names; digging up old information about a person and repeating it even though it may have questionable relevance today; reporting distressing details; or publishing articles or pictures that are in poor taste. While juries will be interested in the degree of embarrassment caused to the plaintiffs, the media seem to win more often than they lose. This is because courts have tended to accept journalists' definitions of what is or what is not newsworthy. The questions center not around what people should read about or be interested in, but what they actually do read about or show interest in. This explains the relative immunity of supermarket journals such as *National Enquirer, Midnight,* and *Star* from such lawsuits.

The fourth area of privacy to be considered is *false light,* which entails putting plaintiffs in a false position in the public eye. It has been a problem area for photographers misusing pictures, and for "new journalists" and "docu-drama" writers who tend to stretch the truth to make their stories more interesting.

A photograph not taken in a public place or a caption to a picture that creates a false impression of someone, even if the impression is not unfavorable, should not be used unless the subject has granted permission. Examples from court records show that if a newspaper prints a picture of a couple kissing in public or of an accident scene because it is newsworthy, then there is no problem; but if a newspaper reprints the same photo later to illustrate an article on a slightly different subject, it is guilty of false light invasion of privacy.

Fictionalization is a growing problem for some contemporary journalists and for script writers who frequently embellish their true stories with some small touches, such as descriptions of people or dialogue that might not be entirely accurate. To avoid being charged with invasion of privacy, such writers resort to using *composite characters* (giving a fictitious name to a person who represents characteristics of several people) or changing the names of central figures. If real persons are so clearly described in the stories that they can convince a jury their peace of mind has been disturbed despite the writer's attempts to disguise the identity, they may win their suits.

In summary, because privacy is a relatively new area of the law, it lacks the body of court decisions and state statutes which for centuries have defined the boundaries of libel. Therefore media do not yet know how far they can invade privacy in their quest to tell the public what it needs—or merely wants—to know. Most law texts and discussions of the issue conclude that if journalists exercise reasonable care and a sense of fair play, they will face few difficulties in this area.

Obscenity and Pornography

Obscenity and pornography seem to be related to audience size and duration.

There is an additional area of mass communication in which controversy rages, an area specifically dealing with media's entertainment function rather than their information function. This is the matter of obscenity and pornography as suitable content in the mass media. Here the question may well revolve around audience size and duration. *Tropic of Cancer* was unquestioned while restricted to the relatively small audience to which a hard-cover book appeals. It ran into trouble, however, once it was published in paperback and made available to far wider audiences. A range of specifically erotic movies playing to tiny audiences goes unchallenged, as does a considerable roster of girlie, stag, transvestite, homosexual, and even, in some circles, kiddie porn magazines of limited circulation. But these themes do not appear on national television, in the *Reader's Digest,* or the daily newspaper.

Is obscenity illegal because it is offensive, or because it is harmful? The jury is still out.

Some people contend obscenity ought to be illegal because it is offensive, not because it is harmful. They say it should be punished not because it is a crime, but because it is sinful. The Supreme Court, on the other hand, seems to believe pornography is offensive, rather than merely harmful, and could conceivably lead to a clear and present danger (or at least to a clear and probable danger) to the social order.

In the United States Supreme Court's 1957 *Roth v. U.S.* decision, the majority opinion, written by Justice William Brennan, made it clear that the court had always assumed that obscenity was not protected by the First Amendment. Implicit in the history of the First Amendment is the rejection of obscenity as being utterly void of redeeming social importance, Brennan noted. In that case the court held that the test for obscenity should be "whether to the *average* person, applying *contemporary community* standards, the *dominant theme* of the material, *taken as a whole,* appeals to prurient interest" (italics added).[23] The definition, as one can see, is loaded with qualification. Not surprisingly, the years following this liberal decision found a barrage of questionable matter being challenged in the courts on a wide variety of technicalities.

The 1973 Miller case checked the trend toward more liberal standards brought on by the 1957 Roth decision.

Because of the difficulty it had judging what constituted "redeeming social value," the Supreme Court attempted to escape its dilemma in 1973. The highly controversial *Miller v. California* case, involving mailing of sexually explicit brochures, found Chief Justice Warren E. Burger speaking for a new and more conservative High Court majority. After criticizing the "redeeming social value" test, Burger redefined pornography as that which, *taken as a whole, appeals to the prurient interest* in sex, which portrays sexual conduct in a *patently offensive way* in violation of *acceptable state laws,* and which, *taken as a whole, lacks serious literary, artistic, political, or scientific value* (italics added).[24] In addition, Burger stated and later decisions agreed, states are not constrained to judge obscenity in terms of a particular national standard. Juries were told to use their own ideas of what their own dominant community standards are when deciding what appeals to prurient interests and what is patently offensive.[25] Another case decided about the same time, *Jenkins v. Georgia,* warned censors that they could not use local community standards to be overly repressive.

Communities have a good deal of latitude in determining their own standards of obscenity.

The Miller decision threw an immediate scare into many people in the media, particularly film producers and book and magazine publishers. Their concern was that what was apparently acceptable to communities such as New York or Los Angeles, where much of their production was done, would be found completely illegal in America's hinterlands, where community values differed. Numerous test cases went through the lower courts—some arriving for final determination at the United States Supreme Court—over the ensuing years.

Hustler magazine publisher Larry Flynt, whose offices were in Columbus, Ohio, was tried and convicted in Cincinnati, 100 miles away. He was found guilty of pandering obscenity—specifically advertising that his magazine would appeal to lewd interests—and for violating Ohio's organized crime statutes. Flynt later carried his banner to other and even more unfriendly battlegrounds, including Lawrenceville, Georgia, where he was shot and permanently crippled during a subsequent pornography trial. From the accompanying front page essay in *Hustler* magazine of October, 1978, we can see how Flynt justifies continuing to wage the battle against obscenity laws and community standards. Obviously, not all would agree that libertarianism should extend to such media fare as *Hustler, Smut, Screw,* and *Deep Throat.*

Film Censorship Recently, films have been protected by the First Amendment, but that has not always been the case. A 1915 Supreme Court decision ruled that exhibition of films was a business rather than a form of literary expression deserving constitutional protection. Since 1952, however, films, even if considered obscene or sacrilegious by some elements of the community, have received at least limited benefits of First Amendment protection and usually get their day in court before arbitrary censorship and prior restraint can be applied.

That is not to say that censorship and prior restraint of films no longer occur. Indeed, even the Supreme Court has ruled that individual city ordinances that provide for prescreening and licensing of motion pictures prior to public viewing are constitutional. In Chicago this ordinance was upheld in 1961 by a 5 to 4 majority: "It shall be unlawful for any person to show or exhibit in a public place—any . . . motion picture . . . without first having secured a permit therefore from the superintendent of police."[26] In the ordinance's defense, Justice Tom Clark said, "It has never been held

Films receive First Amendment protection, but censorship and prior restraint remain.

Hustler Magazine,
October 1978, p. 5.
Reprinted by permission.

PUBLISHER'S STATEMENT

Glad to Be Back!

From the beginning HUSTLER was meant to be your magazine—created by the reader for the reader. For the most part I think we have succeeded in this goal; I believe that all of you out there who supported us and stuck by us through all the controversy and hardship of our development have made HUSTLER the success it is.

Any doubts I might have had about this were more than dispelled by the overwhelmingly generous outpouring of letters I received while recuperating from the almost-fatal shots fired at me last March. Because of these letters, I realized (more than ever before) that you, the readers, have shared the battles with me—enjoying the same victories and suffering the same defeats. In that sense it wasn't just me who was gunned down on the streets of Lawrenceville, Georgia; it was every American who holds dear the values that have made this country great.

I feel no regrets about my fight for free expression. Even though I have been crippled, I am not intimidated by the people who shot me, nor will I be intimidated by standing trial in Georgia again. I will continue to fight for what I believe in, even if I have to stop another assassin's bullet as a result.

I don't know who shot me and I don't want to know. The important thing is the knowledge that a particular element living in America feels so threatened by our ideas that they will subvert justice and one of God's commandments ("Thou shalt not kill") to stop us.

These people are the real victims of society. They have been victimized by sexual, religious and political repression; by the lack of love inherent in sexual repression; by the fear of honesty intrinsic to organized religion in order to maintain its control; and by the hypocrisy of politicians who are not only lying to the American taxpayers but also lining their own pockets with *our* hard-earned money.

This is HUSTLER's message. It is not a very different message from the one we started with; it is only more mature and more responsive to the needs of the people. HUSTLER will remain essentially what it has always been. The only difference is that I have a new set of values for myself, among them the desire to help eliminate discrimination against women.

HUSTLER will continue to be honest, sexually candid, outrageous and iconoclastic. We will continue to explore social and sexual taboos in the belief that an ongoing dialogue is the best hope for solving the problems that afflict our society. And it is you, our readers, who will still dictate our direction.

No single personality or ideology formed HUSTLER. It is a magazine for the average American, and it's put out by average citizens like you and me. I'm glad to be back.

Larry Flynt
**Publisher &
Chairman of the Board**

that liberty of speech is absolute. Nor has it been suggested that all previous restraints on speech are invalid."[27] Four years later, a similar law in Maryland was overruled because it did not assure theater owners of procedural safeguards if they wanted to question its prior restraint, and a Dallas ordinance was overturned for its vague references to guaranteeing suitability of films to children. Thus city and state censorship boards have been found acceptable to the Supreme Court, so long as they operate under clear, precise guidelines that do not undermine constitutional guarantees of due process.

Broadcast Obscenity For most media, the biggest thicket in obscenity is determining what are or are not appropriate reflections of community standards. In broadcasting, the issue is somewhat different, and for apparently logical reasons: the pervasive and intrusive nature of broadcasting, and the omnipresence of children in the broadcast audience.

Broadcasting receives special consideration in obscenity cases because it's a unique medium.

In a 1977 Federal Communications Commission decision concerning a George Carlin monologue ironically entitled "Seven Words You Can't Say On Radio," the commission noted that radio—and by extension, television—requires special treatment because of four important considerations:

1. children have access to radios and in many cases are unsupervised by adults;
2. radio receivers are in the home, a place where people's privacy interest is entitled to extra defense;
3. unconsenting adults may tune in a station without any warning that offensive language is being or will be broadcast; and
4. there is a scarcity of broadcast spectrum space, the use of which the government must license in the best public interest.[28]

Despite the fact that federal broadcast law prohibits censorship of programming, Congress granted the FCC special authority to outlaw obscene or indecent speech over the air.

In the Carlin case (*FCC v. Pacifica Foundation*) the Supreme Court split 5 to 4. The majority agreed that the FCC had the right to ban broadcasting of indecent speech from the airwaves during hours when children may be in the audience. In suggesting that Carlin's monologue was not essential to the exposition of an idea, the court rejected arguments presented in other cases that speech must be legally pornographic to be regulated.

Comedian George Carlin learned that there really are seven dirty words you can't use on the air.

Part of the controversy over this *Pacifica* case, which took five years to reach the Supreme Court, was the validity of the concept of "channeling" programming to periods when children would be unlikely to overhear it. One study cited showed that large numbers of children are in the broadcast audience until 1:30 A.M. and that the number of children does not fall below one million until 1 A.M. If true, how would it be possible for broadcasters to air legally permissible adult fare and reach only the audience they seek?[29] The FCC and the Supreme Court have not answered this question.

Obscenity on Cable If the main question raised about broadcast obscenity is the medium's ready availability to persons of all ages, what about cable television? This new medium, as we saw in chapter 10, now penetrates half of America's homes. It makes use of the same television set that receives over-the-air broadcasts which are subject to governmental control, but it differs in the simple fact that customers have to specifically subscribe to

Obscenity law affecting cable
TV falls somewhere between
broadcasting and film.

the service. It is not available to everyone, and even its customers have the
option of not subscribing to the pay channels on which racier material is
most likely to be seen. Therefore, there is some doubt as to whether the
Pacifica guidelines apply to cable.

In two major test cases, federal courts in the mid-eighties struck down
local ordinances and state statutes prohibiting cable companies from distributing indecent programs. A Miami, Florida ordinance and a Utah state
statute would have held local cable operators to blame for showing R-rated
films or programs that did not meet the *Miller* guidelines for obscenity but
which would have violated local community standards for decency—quite
a separate matter. The Utah act defined "indecent material" as the visual
or verbal depiction, display, representation or description of particular parts
of the body and particular sexual acts "which the average person applying
contemporary community standards . . . would find (were) presented in a
patently offensive way for the time, place, manner and context in which the
material is presented."

Cable companies successfully argued that cable television is considerably less intrusive and considerably more controlled by the viewer than
over-the-air broadcasting. For instance, parents can use a "lock box" to
shield their children from exposure to potentially objectionable materials.
They also convinced the court that the Utah statute was unconstitutionally
vague, that cable operators would be deterred from distributing protected
material because of uncertainty over whether a given program was or was
not indecent and offensive to community standards. Finally, the court held
that unlike radio, cable television is not an "uninvited intruder" into the
privacy of the home.[30]

Courts hold that cable TV is
invited into the privacy of the
home.

Audience acceptance
becomes the principal restraint
upon obscenity and
pornography.

Social Controls Because Supreme Court decisions are the law of the land,
they generally have been successful in preempting state and local regulations of obscenity in most mass media, leaving only self-regulation and social control to the state and local levels. This means that audience acceptance
becomes the principal restraint upon obscenity and pornography in the
media. Such social controls, while a form of restriction, generally work fairly
well in removing unsavory material from the mainstream while permitting
its availability to the smaller audiences who actively seek it. (Indeed, in
Stanley v. Georgia, the High Court ruled that citizens have the right to
possess and use obscene materials in the privacy of their own homes,[31] and
in 1970, the Post Office Department's "antipandering statute" said that
mail recipients could individually decide whether advertising materials sent
to them were arousing or sexually provocative, and could request that their
names be deleted from such mailing lists or the advertisers would face prosecution.)

From many of the Supreme Court rulings and from public application of them we see what might be called *variable obscenity,* indicating that
what is obscene for one person or group under one set of circumstances may

not be obscene for others under different circumstances. What is acceptable to adults may be unacceptable for children, and what is acceptable in private may be obscene if made public. Confusion over the issue provides still another indication that media take on the coloration of a given society at a given point in time.

A good indication of changing attitudes toward good taste is the two divergent views expressed by recent national panels on obscenity. In the late 1960s a federal Commission on Obscenity and Pornography appointed by Democratic President Lyndon Johnson spent three years and two million dollars researching the "puzzle of pornography." After wading through reams of social scientific evidence, it eventually concluded that because it could find no direct linkage between viewing obscenity and subsequent antisocial behavior, all laws regulating the consumption of obscenity should be abolished. Many scholars and social critics questioned the commission's findings, and 80 percent of Americans surveyed by the Harris and Gallup organizations in 1969 called for more stringent controls over pornography and obscenity. By 1970 the United States Senate and newly inaugurated Republican President Richard Nixon saw to it that the commission's findings were never implemented.

A decade and a half later, another pornography commission was called to action, this time by President Ronald Reagan and his Attorney General, Edwin Meese. The Meese Commission's objectives were "to determine the nature, extent, and impact on society of pornography in the United States, and to make specific recommendations to the Attorney General concerning more effective ways in which the spread of pornography could be contained, consistent with constitutional guarantees." Given that charge, and the fact that the eleven-member commission included several well-known conservatives who had been very outspoken about what they called pornography's evils, few were surprised in 1986 when the Meese Commission's 1,960-page report included ninety-two recommendations for legislatures, law enforcement agencies, and citizens' groups to prevent dissemination of sexually explicit material. The report called on public interest groups to file complaints, pressure prosecutors, monitor judges, and boycott stores selling pornography.

The commission claimed it found a direct cause-effect relationship between exposure to sexually explicit material depicting violence and aggressive behavior toward women. In addition, it found a link between exposure to nonviolent sexually explicit materials "depicting degradation, domination, subordination, or humiliation" and "the incidence of various nonviolent forms of discrimination against or subordination of women in our society."[32] However, the commissioners could not agree about how much harm is caused by nonviolent or nondegrading explicit sexual matter, or matter that merely contains nudity. In short, there was no consensus from the Meese Commission on the nature of matter that is merely indecent but not legally obscene—the same legal and semantic problem being faced by the designers of legislation affecting cable television.

"Variable obscenity" laws show that in such matters as pornography, it's different strokes for different folks.

Two national commissions on obscenity have arrived at two quite different sets of conclusions.

The 1986 Meese Commission report claimed there was a direct cause-effect relationship between sexually explicit material and inappropriate social behavior. However, numerous challenges have undermined the report's impact.

Attorney General Ed
Meese announces the
findings of his Obscenity
Commission in a news
conference staged
before an immodestly
draped statue of justice.

Attorney General Ed Meese announces the findings of his Obscenity Commission in a news conference staged before an immodestly draped statue of justice.

Because of the makeup of the panel, and the fact that it took only a year for the eleven commissioners, working in their spare time, to review selected evidence, the Meese Commission came under fire from civil libertarians and social scientists across the nation—including two of its own members. While Jerry Falwell, president of the conservative Liberty Foundation, called the commission's work "a good, healthy report that places the United States Government clearly in concert with grass roots America," a lawyer for the American Civil Liberties Union called it "little more than prudishness and moralizing masquerading behind social science jargon."[33]

In short, while supporters of the Meese Commission insisted the proposals would protect society, critics maintained they would undermine free speech. Attorney General Meese insisted the report would not lead to censorship. "This department, as long as I am Attorney General, is not going to engage in any censorship that violates the First Amendment," he promised.[34]

Meanwhile, the report itself has been accused of being pornographic. It contains numerous examples of sexually explicit materials, lists of publications and videotapes found in "adult" bookstores, and graphic descriptions of books and films. Those three-hundred pages of lists and descriptions have made the cumbersome report one of the hottest-selling governmental publications in memory.

We have shown how the First Amendment is supposed to protect media from governmental meddling. However, the government has made an impressive argument that without external regulation, there would be no freedom of broadcast communication and, ultimately, no broadcasting industry. Its argument is a simple one. Because the broadcast spectrum is finite, access to airwaves is limited; should private interests prevail, they could readily monopolize the marketplace of ideas. Additionally, because broadcast media enter the home directly, with little opportunity for preview by consumers, and because a massive percentage of the audience consists of children, broadcasting is said to need supervision lest it behave irresponsibly. Finally, the government reminds us, the infant broadcast industry of the 1920s took the unusual step of specifically requesting governmental regulation when it found it was unable to regulate itself.

In 1927 the Federal Radio Commission and, in 1934, the Federal Communications Commission, maintained that broadcasters had to operate in the "public interest, convenience, and necessity" if they expected to earn and retain their licenses. Thus, for more than half a century, Congress has maintained that the airwaves are public, not private, property; their use is contingent upon "responsible" performance. Interest, convenience, and necessity—an admittedly vague generalization—served the purpose of screening the worthy from the unworthy.

Broadcasters have long been expected by the FRC and FCC to operate in the "public interest, convenience, and necessity."

When Congress wrote the 1927 and 1934 acts, it specifically precluded the commissions from exerting control over programming. The acts concentrated on technological and economic concerns: how much power was to be used to broadcast signals; what areas were to be served on what frequencies and at what times; and what their call letters and locations were to be. A clause in the 1927 Radio Act, followed by a nearly identical statement in the 1934 Act, went so far as to state,

Nothing in this Act shall be understood or construed to give the licensing authority the power of censorship . . . and no regulation or condition shall be promulgated or fixed by the licensing authority which shall interfere with the right of free speech by means of radio communications.

However, whatever Congress can give, Congress can taketh away. Not only did these two acts permit the commissions the most severe form of subsequent punishment that a broadcaster could face—revocation of its license and therefore termination of its business—they even spelled out a few areas in which the commissions are granted specific censorship rights: obscenity, indecency, profanity, fraud, and lottery information. These criteria to broadcast in the "public interest, convenience, and necessity" do, in fact, constitute a form of program control because they describe the boundaries within which a station may program if it expects its license to be renewed. Broadcasters are reminded again and again that the airwaves belong to the people, and a broadcast license is merely a temporary lease of the facilities, implying no rights of ownership.

Airwaves belong to the public, and a broadcast license is merely a temporary lease of the facilities.

On the other hand, the regulatory agencies have made it clear that broadcast stations are not common carriers like telephone and telegraph companies, who, as monopolies, are obligated to accept business from anyone who wishes to use their services. Broadcasters have the right to refuse access to their facilities to various members of the public, should they so desire. And when they do make their facilities available, they can charge whatever prices traffic will allow. In that sense, at least, broadcasting is founded on the basis of free competition among licensees.

In the following discussions, keep in mind that the FCC historically has had no direct jurisdiction or control over network operations. Networks do not actually broadcast; they simply feed programming to their owned or affiliated stations. The FCC, of course, does exert considerable control over the owned and affiliated stations.

Licensing Procedures Congress established the broadcast regulatory agencies, and continues to legislate the industry's behavior. The Federal Communications Commission's primary responsibility is to grant and renew licenses—for radio stations, seven year licenses; and for television, five year licenses. The FCC's combined executive, legislative, and judiciary roles are handled by five members (prior to 1982 it was seven members), each of whom is appointed for a seven-year term by the United States President with the approval of the Senate. Under authority granted by Congress, the FCC and its large support staff write, administer, and interpret a vast array of rules and regulations. Most of them deal with technological questions concerning radio and television broadcasting, cable, satellite, telephone, and telegraph communications systems. However, the rules and regulations covering the license application and renewal process fill a good-sized book.

Applicants must disclose to the FCC any other broadcast interests they are involved in, because the commission vigorously enforces its standards concerning multiple ownership. An individual or corporation can own up to twelve television stations (as long as they do not reach more than 25 percent of the nation's television homes), twelve AM radio stations, and twelve FM stations. (Before the early 1980s the total was seven outlets in each category.) Group broadcasters who buy interests in stations more than half owned by minorities are able to own up to fourteen television, AM, or FM outlets, and are permitted to reach 30 percent of the nation's television households through their TVs as long as two of the stations in each service are controlled by minorities. Newspaper owners may no longer purchase broadcast properties in the market where they publish their papers, nor may radio station owners acquire television stations there, nor may television owners purchase local radio outlets. TV stations may no longer acquire cable TV franchises in the same city, and networks may not own cable systems at all.[35]

The Media Environment

Because most of the desirable broadcast frequencies have long been occupied, the government spends most of its time concerned with relicensing practices. Despite the constant record keeping for station operators and the concern over how well they are meeting their license expectations, most existing licenses have overwhelming odds of being renewed, even if they are challenged by another party wishing to take over the license. In its first forty years of license renewal practices, the FCC revoked or failed to renew only 100 of the 50,000 or so licenses that came up for review. With deregulation, it has become even less likely an existing licensee will be denied the right to continue in business.

Equal Opportunities The greatest abridgment of broadcast freedom has resulted from the application of two interrelated and sometimes confusing doctrines: those of *equal opportunities* (usually called *equal time*) and *fairness*.

Section 315 of the Communications Act of 1934 establishes the principle of opportunities. Note that the principle does not apply to print media, which are not required to give equal time. This concept applies only to candidates for political office and provides that all bona fide candidates for a given office must be accorded the same amount of air time on the same terms or at the same cost. Thus, if one mayoral candidate in a race is given an opportunity to appear on the air and make a political appeal, all twelve or fifteen other qualified candidates must be given the same amount of comparable time; if the time was purchased by the first candidate, none of the other candidates can be charged a higher rate. In no case can broadcasters charge more than their minimum or lowest unit rates to any of the candidates. Also, they cannot censor what the candidates say on the air, which at least protects the licensee from any subsequent libel suits. The original law said broadcasters were under no obligation to allow the use of their stations by *any* candidates, but a more recent act of Congress requires that stations accept advertising by federal political candidates.

Section 315 has been modified to exempt certain types of programming from adhering to the above provisions: bona fide newscasts, news interviews like "Meet the Press," news documentaries, on-the-spot news events, and, since 1975, press conferences. Coverage of face-to-face debates between two candidates has been a difficult issue. In 1960 it took an act of Congress to suspend the equal time requirement in order to permit the famous Kennedy-Nixon debates to be aired. The 1976 Carter-Ford debates and the 1980 Carter-Reagan debate occurred only because they were sponsored by an outside agency, the League of Women Voters, which was not controlled by the candidates or the broadcasters and was therefore exempt from the requirements. Currently, political debates are considered legitimate news events, and are therefore exempt from the equal time rule. This has made life easier for programmers, who otherwise would have to contend with a half-dozen or so minor presidential candidates seeking equal time

FCC's Section 315 says broadcasters must grant *equal opportunities* for air time to bona fide political candidates.

in the debate forum. Because few viewers would watch such a circus, such arrangements would have cost the networks millions of dollars in lost revenue.

A quirk of the equal time provision is that once individuals formally declare their candidacy, even entertainment programming brings Section 315 to life. Therefore, when Ronald Reagan was running for president, stations could not legitimately run his old movies without facing an equal time claim from opposing candidates.

Fairness Doctrine Unlike Section 315, which as part of the 1934 Communications Act has been passed by Congress, the fairness doctrine is the creature of the FCC, gradually evolving over the years in answer to specific problems. The fairness doctrine applies not to people or politicians but to items of controversy.

The fairness doctrine as it has developed since 1949 stated that stations had to devote a reasonable amount of time to controversial public issues, and in so doing, they had to actively encourage the presentation of all perspectives. They were similarly encouraged to editorialize on community issues and to make their station position known, provided they accorded a balance of views. If during the course of editorializing individuals or organizations were attacked, they had to be offered comparable free time in which to reply. News was exempt from all the foregoing.

Although the ground rules are now a little clearer, fairness is still a complex doctrine, subject to many strange innovations. The net result, far from encouraging broadcasters, still tends to discourage them from active participation in controversy.

The fairness doctrine was extended in 1967 to include broadcast advertising. For example, cigarette smoking was held to be a controversial issue, and stations were required by the FCC to give free time to smoking opponents to answer the barrage of cigarette commercials on the air. Such remedial advertising can be an expensive burden on a station. Although Congress subsequently resolved part of the problem by banning cigarette commercials entirely, the question of fairness in advertising remained.

In its 1969 Red Lion Broadcasting decision, concerning a Pennsylvania radio station that preached religion and right-wing politics and excluded opposing views, the United States Supreme Court reaffirmed that because broadcasting is a limited facility, the rights of the public, not the broadcaster, are paramount. Consequently, the court ruled that Congress and the FCC are not violating the First Amendment when they require a radio or television station to give reply time to answer personal attacks and political editorials.[36]

Confusion still exists concerning who determines what is a controversial issue of public interest, and even whether the Fairness Doctrine is constitutional. The FCC has advocated the elimination of the doctrine, but continues to support a means for meeting "the interest of the listening and

The *fairness doctrine* assures that controversial issues of public importance will be treated in a balanced way.

However, the quirks of the fairness doctrine tend to discourage broadcasters from tackling controversy.

The FCC has advocated the elimination of the fairness doctrine.

viewing public in obtaining access to diverse and antagonistic sources of information," and insists that the Fairness Doctrine is neither a necessary nor appropriate means by which to effectuate this interest.[37] In mid-1987, Congress, President Reagan, and public interest groups were fighting over removing the Fairness Doctrine from the FCC's province and enacting it into law.

Control of Cable In the 1960s the FCC began regulating cable television. Originally, the FCC had held that because cable TV did not operate on an interstate basis, it had no jurisdiction, and little interest in protecting local over-the-air broadcasters from the then insignificant upstart medium. By the mid-1960s it recognized the simple reality that cable indeed does deal in interstate signals, and that it was a commercial force to be reckoned with.

Cable had been retransmitting broadcasts from distant stations without paying for them, which meant that cable had a distinct advantage over local broadcasters. Local stations, which had to pay an arm and a leg for the same syndicated programming that cable was picking up free and delivering to customers for a monthly fee, feared for their economic survival. The argument local stations brought before the FCC was simple, if not altogether altruistic: if local stations went out of business the public would suffer.

Throughout the sixties and seventies, it became apparent that cable presented some unique regulatory problems. On the local, state, and federal levels, regulations were put into effect. Everything seemed to be affected, from the local communities' terms for installation and wiring and what programs and how many channels would be made available, to the FCC's 500-page book of regulations designed to guide cable development without doing significant harm to existing broadcasters. All in all, there was no peace in the industry.

Cable TV presents unique regulatory problems on the local, state, and federal levels.

In practical terms, for instance, the FCC rules maintained that cable systems had to carry the signals of all stations within a thirty-five-mile radius (the *must carry rule*), but they could not duplicate commercial programming within a specified period of time (the *non-duplication rule*). All systems had to carry educational stations and they had to grant local governments and interested citizens *public access* to a channel. The number of channels available for immediate or potential use depended on the size of the market or the size of the system. These variables also were considered in many of the other rules contained in the FCC's 1972 document.

Many of the rules have changed in recent years due to court decisions, the FCC spirit of deregulation, and broadcasters' reluctant acceptance of cable in the marketplace.

Since the late 1970s, many rules affecting cable TV have been dropped or softened. Deregulation is in the wind.

The 1978 revision of federal copyright laws at long last made cable operators responsible for paying royalties to program owners for the distant nonnetwork programs they carry. Restrictions on the use of distant signals were eased in 1979 when the FCC voted 6 to 1 to allow cable operators access to as many distant broadcast stations as they wished without having to obtain consent from the broadcasters whose signals they picked up. The only obligation of the cable operators thereafter was to pay royalties to program owners. Also in 1979 the Supreme Court struck down the FCC's *public-access rule* on a federal level. Because many groups had been violating the intent of the public-access proviso by filling the cable with dull and frequently obscene amateur programming, cable operators had protested.

In 1985 the courts struck down the *must-carry* rule as overly broad and unconstitutionally protective of local broadcasters and discriminatory against cable operators.[38] A year later, the FCC came up with a revised set of rules governing which types of cable systems have to carry which local stations. Another mid-decade decision granted cable increased First Amendment protection when it was ruled that a Los Angeles cable exclusive franchising plan was discriminatory.[39] The latter decision reinforced the spirit of the FCC's Cable Communications Policy Act of 1984, which sharply limited local governments' control over cable and permitted cable operators to set their own rates.

It has become obvious that the FCC, in cooperation with the federal courts, is interested in the gradual deregulation of cable on both the federal and local levels. As such it appears to represent a return to a laissez-faire philosophy of allowing natural marketplace forces to determine economic survival.

Broadcast Deregulation In keeping with its efforts to deregulate the trucking and airline industries, the Carter administration sought to minimize federal involvement in the broadcast arena. Between 1978 and 1980 several efforts were made in both the House of Representatives and the Senate to rewrite the 1934 Communications Act. Although Carter's Democratic administration was unable to affect significant change, the spirit of deregulation gained in intensity under the Republican, business-oriented administration of Ronald Reagan in the 1980s.

One reason for movement toward deregulation, as implied earlier, is a new recognition of the difficulties broadcast station licensees have had with the maze of FCC rules and regulations. If broadcasters could be freed from some of the paperwork, they would be able to devote more time to serving their audiences, the deregulators insisted.

Deregulators also maintain that the original reasons for many of the controls over broadcasting are irrelevant in today's world. There is no real band of scarcity any more, they say, pointing out the simple mathematic fact that there are more than ten-thousand radio and television stations operating in America, and fewer than seventeen hundred daily newspapers. With so many stations on the air, and more to come once Low Power TV, cable, and other new technologies peak out, owners should be able to do

Carter and Reagan administrations have attempted to allow marketplace forces, rather than federal law to control the broadcast industry.

Deregulators insist there is no "band of scarcity."

whatever they like—even broadcast commercials twenty-four hours a day, if they can keep an audience. Why, they ask, should broadcasters have a second class First Amendment? After all, there were only a handful of newspapers in America when the Constitution was signed, and no one insisted that they all be licensed and controlled the way broadcasting has been.

The reasoning is compatible with the basic economic philosophy of the Reagan administration, which is eager to encourage the growth of American business by allowing marketplace forces to prevail wherever possible. In order for stations to survive economically, they must give people what they want. Those that succeed will prosper; those that do not will go out of business.

Many of the bills brought before Congress support the movement to deregulate. Some have tried to adjust the troublesome concept of stations asked to operate in the "public interest, convenience, and necessity." Other bills would either eliminate or modify the equal time provision and fairness doctrine, giving radio more freedom than television.

During the 1980s the FCC itself extended the licensing period to seven years for radio and five years for television (it used to be three years for each), simplified the relicensing procedure, permitted individuals and groups to own more stations, eliminated ascertainment requirements, dropped news and public affairs programming rules along with suggested limits on advertising time, expanded the prime-time television programming hours, actively sought the elimination of the fairness doctrine, and watered down its regulations on children's programming.

Naturally, all these moves toward deregulation have not gone uncriticized. Consumer groups, minority groups, and public interest groups still believe in the traditional role of the FCC and governmental agencies; they should be watchdogs or trustees for the disenfranchised public, looking out for those unable to afford their own mass media. The critics look with alarm at the behavior of some recently deregulated broadcast stations, which have used the lack of specific program guidelines as an excuse to concentrate on mass merchandising of deodorant, cars, and beer. They maintain that stations should be promoting the democratic process through talk shows, newscasts, documentaries, and editorials rather than going for the easy buck. Radio, for instance, has cut back on its news programs over the past several years, and in some cases stations have eliminated news and public affairs broadcasts altogether.

Consumer groups, minority groups, and public interest groups are opposed to deregulation.

Obviously, the jury is still out. The trend seems to be in the direction of laissez-faire economics, as noted earlier. Optimists see the day when it does not matter whether a few stations cop out of their civic responsibility, because there are enough outlets available—at reasonable costs—for all the public interest groups to gain a hearing.

After all, there is no way ten-thousand broadcast outlets can survive on commercials and lowest-common-denominator music and escapist programming alone.

The Adversary Relationship

We have already investigated some of the formal means that government employs to minimize media excesses. But there are numerous other external controls over journalism, some of them dealing with the news reporting functions, others relating to the economic functions.

In today's sophisticated climate, instead of overt and obvious attempts to control the press, government generally tries manipulating or outmaneuvering it. Manipulation ranges from the devious *disinformation* campaigns (planting false stories and conducting other forms of psychological warfare through and upon the news media), to the more frequently used *classification* system.

On the federal level, classification involves certain kinds of information deemed in the national interest that is labeled as secret or top secret and withheld from the press. Dissemination of such material is punishable by law. No one reasonably questions the right or wisdom of government to withhold certain information from the public view that would seriously jeopardize the nation's position in its sensitive and sometimes perilous international dealings. However, in reality, this practice works differently. Classification is an invitation to government authorities to hide their shortcomings, errors, duplicity, and poor judgment in the guise of national security.

But the mass media are nearly as institutionalized as government itself. They are not without power to obtain some classified materials. Washington provides ample opportunities for the press to get information. Ambitious or disgruntled government insiders who are anxious to serve their own devious ends make it almost impossible to keep a secret or even a top secret for very long without the voluntary cooperation of the press. Using sources and other methods, the media probe and seek to uncover sensational and scandalous stories. They frequently cloak their investigations with references such as "it is the public's right to know how tax dollars are being spent," but media critics correctly argue that the media are looking for the kind of spicy news that will attract audiences in droves.

Regardless of the motives of the media, the adversary relationship heightens the competition with government for information. The greater the competition, the better the balance. If government were more forthright, the press would be less diligent; if the press were less alert, the government would be more secretive.

Freedom of Information

These conflicts have become institutionalized. On the national level since 1966 America has operated under a *Freedom of Information Act* (FOIA), and on the state and local levels public records and open meetings statutes have been enacted. FOIA gives journalists, authors, other investigators and the general public access to literally millions of pages of documents that the federal government is inclined to protect—information gathered by various federal administrative agencies such as executive departments, military departments, government corporations, government controlled corporations, other executive agencies, and independent regulatory agencies.

> "Knowledge will forever govern ignorance. And a people who mean to be their own governors, must arm themselves with the power knowledge gives. A popular government without popular information, or the means of acquiring it, is but a prologue to a farce or a tragedy, or perhaps both."
>
> —James Madison

> "But it is so difficult to draw a clear line of separation between the abuse and the wholesome use of the press, that as yet we have found it better to trust the public judgment, rather than the magistrate with the discrimination between truth and falsehood."
>
> —Thomas Jefferson

Records of the entire federal government, except for Congress and the courts, are theoretically included.[40] We say theoretically, because the FOIA by statute exempts much sensitive data from prying eyes, and bureaucrats make it difficult for even much nonsensitive data to be reported.[41]

Today every state, along with the District of Columbia, has open records and open meetings laws on the books. Such laws vary widely from state to state, of course, but most make the point that when business is carried out using taxpayers' funds, the taxpayers have a right to know what is going on. Exceptions to the general principle of openness occur in such delicate areas as personnel decisions and real estate negotiations. Such negotiations are usually permitted to occur behind closed doors. However, when it comes time to make formal decisions, state (and many local) governments have promised to abide by the general principle of opening the doors and letting the press and public know what is happening. And the records of such decisions are expected to be available for public scrutiny. When such meetings and records are improperly closed, the statutes provide legal remedies.

Because of government's inherent tendencies to prefer working in peace, out of the spotlight, and because of media's inherent tendencies to focus that spotlight on every speck of dirt, there are bound to be conflicts. A subtle system of checks and balances prevails between press and government when it comes to open meetings and open records. And, as often as not, journalists have found that the best routine for keeping channels of information open is to use common courtesy, rather than threats and intimidation. When journalists remind bureaucrats that both groups are there to serve the interests of the public rather than their own selfish interests, the system runs much more smoothly.

News Management

Techniques of news management include *backgrounders, off-the-record* and *not-for-attribution* interviews, and *trial balloons.*

Because the government cannot classify everything in Washington, those in power have devised other methods for controlling the dissemination of information. Such techniques come under the general heading of *news management.*

One technique favored by government is the *backgrounders.* The theory of backgrounding is commendable; it provides reporters a depth of perspective in sensitive situations with which they can better interpret the events they report. As some of the background data may be classified, reporters are asked not to reveal them.

In practice, however, this technique frequently has acted as either a kind of gag on journalists or as a perversion of its purposes. Backgrounders are of two types. Some are *off the record* and others are *not for attribution.* In an off-the-record statement or interview, an official talks quite candidly about a whole range of highly classified material. The press is free to listen but not to report. There is an informal honor system to this, for if off-the-record statements are reported, the journalists' sources of information may dry up. No one will talk to them again, not even to say good morning. Nor are reporters free to use the material from an off-the-record statement even if they get it from another source; they are effectively muzzled on that topic.

A second type of backgrounder involves not-for-attribution statements, whereby the press is quite free to use the background material as it sees fit, but it may not quote the sources nor specifically acknowledge where the information came from. Not-for-attribution backgrounders are responsible for those phrases so often encountered in the news: "a generally reliable administration source disclosed today . . ." or "a high Pentagon official has announced. . . ."

The not-for-attribution backgrounder is often used as a *trial balloon.* A projected government policy is released not-for-attribution, and public reaction to it is assessed. If public reaction is favorable, an appropriate spokesperson confirms the policy on record and takes the credit. If public reaction is unfavorable, there is no one to blame.

These practices are tolerated because backgrounders often do what they set out to do; they provide a realistic framework for interpretation.

Furthermore, out of the entire operation some scraps of significant information do come to the experienced player—information that is probably not obtainable in any other way.

The First Amendment guarantees press freedom—the public's right to know. As applied by the courts it upholds the common-law principle of a public trial in which the press is the representative of the people, not all of whom can crowd into the courtroom. The Sixth Amendment guarantees an individual a speedy and public trial by an impartial jury. A conflict arises when the press's preoccupation with the sensational, and its diligence in seeking it, results in a trial by press wherein defendants are often found guilty in public opinion before their judicial trial has commenced. For example, in the 1954 trial of Dr. Samuel Sheppard, who was accused of killing his wife, the Cleveland press found Sheppard guilty even before the jury went out to deliberate. Not surprisingly, Sheppard was convicted. Also not surprising was the 1966 Supreme Court decision that overturned Sheppard's conviction due to publicity. It is apparent that such instances make it increasingly difficult to form panels of objective jurors, particularly in sensational cases. This was an issue in choosing a jury for the trial of Jack Ruby after Ruby shot Lee Harvey Oswald on national television; when John Hinkley, Jr., shot President Reagan in full view of millions; and in other cases that have attracted national attention before and during the actual trial. Of course, the sensational trials are the exception. The overwhelming number of criminal cases get settled at the pretrial level, where plea bargaining and other legal machinations occur. But the fact that most of us can recall in vivid detail one or another famous trial means that rules need to be set up for such instances.

It says something about the nature of the controversy that journalists refer to it as a free-press/fair-trial issue, while members of the bar call it the fair-trial/free-press issue. Attempts to balance the constitutional provisions have had lengthy and sometimes nasty histories in this country, but there are signs that things may be getting better.

Is there a way to find a middle ground between the public's right to know and the defendants' rights to fair trials? Supreme Court Justice Felix Frankfurter expressed the need for such a balance:

A free press is not to be preferred to an independent judiciary, nor an independent judiciary to a free press. Neither has primacy over the other; both are indispensable to a free society. The freedom of the press in itself presupposes an independent judiciary through which that freedom may, if necessary, be vindicated. And one of the potent means for assuring judges their independence is a free press.[42]

A majority of the American states have developed voluntary press-bar guidelines with the purpose of balancing the needs of the news media, the judiciary, the defendants, and the public. While they differ from locale to locale, in scope and application, such guidelines generally tell media they

Conflict: The First and Sixth Amendments

Media's guarantee of access to information often clashes with defendants' rights to a fair trial. The search for a balance has been arduous.

Some of the pretrial coverage in the Sam Sheppard case.

Most states have voluntary *press-bar guidelines* for news coverage of court cases.

should not print: (1) confessions or stories about confessions; (2) results of lie detector tests or whether the defendant would or, in fact, did take such a test; (3) stories about a defendant's past criminal record; (4) stories that question credibility of witnesses or personal feelings of the judiciary; (5) stories about a defendant's character; and (6) stories that inflame the public mood against the defendant. Although the guidelines are intended to be voluntary, a judge in the state of Washington created an uproar among journalists in the early 1980s when he refused to let them attend and cover a trial until they pledged to obey every aspect of the guidelines. Some refused to sign as a matter of principle, regardless of whether they had intended to follow the guidelines. They did not like the idea of a judge making ethics mandatory. Ethical decisions, they argued, should remain matters of free choice. Generally, however, such guidelines work if only because they allow well-meaning journalists and members of the bar a framework for mutual understanding.

At times, the formal procedures allowed by law and the informal guidelines accepted by the press and the bar are not enough. In extreme cases, courts in the past decade have resorted to closing some court proceedings to the public, sealing some court records, and issuing restrictive orders, whereby trial judges lay out whatever rules they think necessary, and can get away with, to protect the sanctity of their courts.

Gag orders limit attorneys, court officials, journalists, and others from discussing particulars of a legal case.

Restrictive orders place limits on what attorneys and court officials can say to the news media and even what the media can share with their audiences. The latter constitute *gag orders,* which are legally binding dictates that reporters are forced to follow regardless of their constitutionality. When reporters violate gag orders, relying upon First Amendment guarantees, judges utilize their authority to hold the reporters in contempt of court. Reporters can be, and have been, sentenced to prison until the case is terminated or until they agree to comply with the judge's orders. Because most reporters find such contempt authority a repugnant violation of their constitutional rights, many have become media celebrities by serving out their sentences.

The pendulum has swung in favor of a qualified right of access to criminal proceedings.

A series of fascinating and sometimes conflicting Supreme Court decisions in the 1970s and 1980s seems to have resulted in the general understanding that reporters deserve to have access to both pretrial proceedings and actual trials. The battles for access have been acrimonious at times, with judges and attorneys barring press and public from various proceedings, and the press arguing vociferously that without media coverage the public's best interests would not be served. The infamous British Star Chamber, with its secret accusations, trials, and punishments, was brought up time and time again during the debates over the pros and cons of open trials. As of this writing, at least, the pendulum has swung in favor of a qualified First Amendment right of access to criminal proceedings—the judge can close the trials to press and public only when there is substantial probability that the defendant's right to a fair trial will be prejudiced by publicity.[43]

Ronny Zamora was found guilty of murder in a noted 1977 televised trial. Zamora's lawyer had argued that the boy was driven to murder by television-induced insanity.

One side note to the issue is the interesting history of attempts to open courts to film, video, and still photography. Since the popping flashbulbs and unwieldy cameras made a mockery of justice during the 1935 trial of Bruno Hauptmann, accused of kidnapping and murdering the infant son of national hero Charles Lindbergh, cameras in the courtroom have created discord between the press and the bar. Two years after Hauptmann was found guilty, the American Bar Association adopted Canon 35, calling for a complete ban on courtroom photography. As amended in 1952 and 1963, the ban also extended to television and radio broadcasting.

Cameras in the Courtroom

As photographic and broadcasting equipment has become more miniaturized and less conspicuous during ensuing decades, electronic journalists have fought for equal rights with their pencil-and-pen brethren. Experiments have met with varying results. Some of the least satisfactory experiments resulted in Supreme Court intervention.

Recent attempts in states experimenting with televised trials have proven to the satisfaction of some that televising court proceedings, if done inconspicuously, need not be detrimental to jurisprudence. In 1981 the Supreme Court ruled that an absolute constitutional ban on broadcast coverage of trials cannot be justified. Since that time, forty-three states have come to permit some form of extended media coverage—television, radio, or still photography. The Supreme Court under Chief Justice Warren Burger

The U.S. Supreme Court has agreed that cameras and electronic equipment do not necessarily jeopardize a fair trial.

vehemently opposed broadcasting or even photographing any of its proceedings, but it remains to be seen whether the new Chief Justice, William Renquist, will hold firm to tradition. (The mid-1980s saw another of the nation's most traditional bodies, the United States Senate, permitting television coverage of its proceedings, and to date there is little evidence of permanent damage to the institution. The only noticeable cost has been repeated exposure of primping and strutting solons—a small price to pay for higher civic awareness by the American public.)

Shield Laws

Shield laws protect journalists from having to reveal the identity of their confidential news sources.

One other point of contention between the law and the news has come to the fore: the question of *confidentiality,* or reporters' *shield laws.* It involves whether or not reporters can be forced by a grand jury, a court, or Congress to reveal their sources of information. There are those who hold that such confidentiality of sources is essential to press freedom. How else, they ask, can a reporter acquire privileged and sometimes dangerous information that serves the general welfare?

Journalistic investigations of the social, economic, and political environment have need for sources such as the government insider, called Deep Throat, who provided Bob Woodward with information about Watergate. Because many sources are willing to talk with reporters but fear being dragged into court, this kind of journalism has served as an investigative agency, which many see as having a highly significant role in the judicial process. Such investigations, media insist, must remain absolutely above and beyond official governmental investigations, lest the press be seen as merely another branch of government.

On the other hand, confidentiality of sources is subject to abuse by journalists themselves. Witness the infamous Janet Cooke case, in which the young *Washington Post* reporter refused to divulge to her editor—the same Bob Woodward of *Deep Throat* fame—the identity of an eight-year-old heroin addict about whom Cooke wrote a gut-wrenching story. Eventually, it became obvious that a prime reason for not identifying the boy was because he was a figment of Cooke's fertile imagination. America's journalists are taking years to recover the public confidence lost in that episode.

Meanwhile, the First Amendment argument for a shield is countered by a clause within the Sixth Amendment that states, "In all criminal prosecutions, the accused shall enjoy the right to be confronted with the witnesses against him; and have compulsory process for obtaining witnesses in his favor." Thus we are left with still another arena in which a delicate balancing must be made between the rights of the press and public to have full access to information of interest and value, and the rights of the governmental system to control that same information for what it maintains are essentially the same purposes—preservation of the just society.

THE FUNCTION OF THE PRESS IS VERY HIGH. IT IS ALMOST HOLY. IT OUGHT TO SERVE AS A FORUM FOR THE PEOPLE THROUGH WHICH THE PEOPLE MAY KNOW FREELY WHAT IS GOING ON. TO MISSTATE OR SUPPRESS THE NEWS IS A BREACH OF TRUST. —Supreme Court Justice Louis Brandeis

Student journalists are reminded of their freedoms and responsibilities.

Summary

The press operates according to four generally recognized theories: authoritarian, communist, libertarian, and social responsibility. Other explanations are made for the revolutionary, developmental, and democratic socialist press. The American system is best described as libertarian, but even under this theory controls are imposed on the mass media. The amount of control exerted by a government depends on the relationship between a government and the governed, the stability of the government, the heterogeneity of society, and the degree to which the society is developed.

Although many restrictions on media have only modest application to American society today, American citizens and the courts must support the First Amendment in order to make it work. Subtle but nonetheless influential controls are placed on journalists' capacity to gather, print, and distribute information and opinions. Laws dealing with sedition, libel, invasion of privacy, copyright, and obscenity must be obeyed by all citizens.

The government has maintained that the airwaves are finite and radio and television are intrusive, so through the FCC it controls the broadcast media. Most broadcasters who comply with FCC rules can expect to get their licenses renewed, although some challenges to existing licenses have succeeded in the courts. Equal Opportunity is a federal law dealing with access to airwaves by office seekers. The Fairness Doctrine is a complex set of rules created by the FCC that pertains to issues of controversy. In today's laissez faire climate, deregulation of the broadcast media is an agenda item.

An adversarial relationship between government and the press exists in the United States. This provides government with opportunities to try to manipulate the news, and motivates the press to outmaneuver the government. Freedom of information laws, open meetings laws, and public records laws have institutionalized the conflict. Meanwhile, backgrounders, off-the-record statements, and other subtle components of press-governmental relationships continue to prevail.

Occasionally conflicts arise between the First Amendment and the Sixth Amendment. To balance the rights of the press and of defendants, courts have resorted to issuing restrictive orders, which sometimes include "gagging" the press. Supreme Court decisions have opened pretrial and trial court proceedings to press and public, and have granted photographers and broadcasters access to most of the nation's judicial proceedings. Shield laws are supposed to protect the confidentiality of press sources, but like so many other controls, they are subject to some abuse.

All in all, an overview of media-governmental relations demonstrates the delicacy of the balance among competing interests—the media, government, and public, along with individual citizens subjected to scrutiny of "the system."

Notes

1. Frederick S. Siebert, Theodore Peterson, and Wilbur Schramm, *Four Theories of the Press* (Urbana, Ill.: University of Illinois Press, 1963), pp. 1-2.
2. Robert G. Picard, "Revisions of the *Four Theories of the Press* Model," *Mass Comm Review;* Winter/Spring, 1982/83, pp. 25-28.
3. John C. Merrill and Ralph L. Lowenstein, *Media, Messages, and Men* (New York: David McKay, 1971), p. 175.
4. Walter M. Brasch and Dana R. Ulloth, *The Press and the State: Sociohistorical and Contemporary Studies* (Lanham, MD: University Press of America, 1986), p. 3.
5. William Stephenson, *The Play Theory of Mass Communications* (Chicago: University of Chicago Press, 1967).
6. Edwin Emery and Michael Emery, *The Press and America: An Interpretative History of the Mass Media,* 4th ed. (Englewood Cliffs, N.J.: Prentice-Hall, 1978), p. 94.
7. William Hatchen, *The World News Prism: Changing Media, Clashing Ideologies* (Ames, Iowa: Iowa State University Press, 1981), p. 70.
8. Picard, "Revisions of the *Four Theories of the Press* Model," p. 27.
9. Hatchen, *The World News Prism* p. 74.
10. Picard, "Revisions of the *Four Theories of the Press* Model," p. 27.
11. Frederick S. Siebert, *Freedom of the Press in England, 1476-1776* (Urbana, Ill.: University of Illinois Press, 1952).
12. John D. Stevens, "Freedom of Expression: New Dimensions," in *Mass Media and the National Experience,* ed. Ronald T. Farrar and John D. Stevens (New York: Harper & Row, 1971), pp. 14-37.
13. *New York Times Co. v. United States,* 713 U.S. 403 (1971).
14. "Could the *Progressive* Help Make Possible a Nuclear Holocaust?" *Quill,* April 1979, p. 6.
15. Ibid.
16. James J. Kilpatrick, "Time to Announce, 'We've won'," syndicated column, *Deseret News,* 9 April 1979.

17. Jack Anderson, "FBI Pursues Author of H-bomb Story," syndicated column, *Logan* (UT) *Herald-Journal,* 29 May 1979.

18. Don R. Pember, *Mass Media Law,* 4th ed. (Dubuque, Iowa: Wm. C. Brown Company Publishers, 1987), p. 56.

19. *Schenck v. United States,* 249 U.S. 39 S. Ct. 247 (1919).

20. Randall P. Bezanson, Gilbert Cranberg, and John Soloski, "Libel Law and the Press: Setting the Record Straight," *Iowa Law Review,* October 1985, Vol. 71, No. 1; and *The 1985 Silha Lecture,* University of Minnesota, 15 May 1985.

21. Ibid.

22. *Galella v. Onassis,* 533 F. Supp. 1076 (1982).

23. *Roth v. United States,* 324 U.S. 476, 77 S. Ct. 1304 (1957).

24. *Miller v. California,* 413 U.S. 15 (1973).

25. Harold L. Nelson and Dwight L. Teeter, Jr., *Law of Mass Communications: Freedom and Control of Print and Broadcast Media,* 4th ed. (Mineola, N.Y.: The Foundation Press, 1982), pp. 313-18.

26. Ibid., p. 328.

27. Ibid.

28. *Pacifica Foundation v. FCC,* 556 F. 2d 9 (D.C. Cir. 1977).

29. *FCC v. Pacifica Foundation,* 488 U.S. 726 (1978).

30. "Cable Porn Law Unconstitutional," *The News Media & The Law,* Fall 1986, pp. 43-44.

31. *Stanley v. Georgia,* 394 U.S. 557 (1969).

32. "Meese Commission Releases Porn Report, Can't Agree on Harm of 'Indecent' Matter," *The News Media & The Law,* Fall 1986, pp. 40-42.

33. Ibid., and "Pornography report stirs expected storm," *FOI/FYI, A publication of the First Amendment Center,* August 1986, Vol. 2, No. 1, p. 6.

34. Ibid.

35. "A short course in broadcasting, 1987," *Broadcasting Cablecasting Yearbook, 1987.*

36. *Red Lion Broadcasting Co. v. FCC,* 395 U.S. 367 (1969).

37. Marc A. Franklin, *Cases and Materials on Mass Media Law,* 3d ed. (Mineola, N.Y.: The Foundation Press, Inc., 1987), pp. 818-19.

38. *Quincy Cable TV, Inc. v. FCC,* 768 F. 2d 1434 (D.C. Cir. 1985), certiorari denied, 106 S. Ct. 2889 (1986).

39. *City of Los Angeles v. Preferred Communications, Inc.,* 106 S. Ct. 2034 (1986).

40. Wayne Overbeck and Rick D. Pullen, *Major Principles of Media Law,* 2d ed. (New York: Holt, Rinehart and Winston, 1985), p. 229.

41. Steve Weinberg, "Trashing the FOIA," *Columbia Journalism Review,* January/February, 1985, pp. 21-28.

42. *Pennekamp v. Florida,* 328 U.S. 331 (1946).

43. "Public Access to Pretrial Hearing Protected by First Amendment," *News Media & The Law,* Summer, 1986, p. 9.

15

Ethics and Social Responsibility

The previous chapter demonstrated that within the libertarian framework, there has come a shift in emphasis from freedoms to responsibilities. We have seen that early in their history mass media fought for the right to gather and report news, to express opinions, and to operate in the marketplace of ideas and goods. Once those beachheads had been established, no matter how tentatively, the shift away from ideal libertarianism toward social responsibility followed quite naturally. Even the most eloquent spokespersons for a free press, Thomas Jefferson included, came to maintain that there was a need for some checks on extremely irresponsible media. The question is, which groups or organizations will prevent the media from issuing biased propaganda, misleading advertising, and antisocial entertainment and programming? Social responsibility theorists maintain that the government should guide the media along the proper path. Strict libertarians would rely upon public opinion and consumer response to do the job. Self-control on the media's part is another possible check. Sometimes self-control springs naturally from individuals attempting to serve society. More often, perhaps, it emerges as a defensive reaction when faced with real and perceived control from government or an outraged public.

How media practitioners have responded to ethical questions is the subject of this chapter. Throughout, we will use the terms *social responsibility* and *ethics* almost interchangeably, because it is our belief that members of institutions do have certain obligations to function in a socially responsible fashion and that, at base, ethics are manifestations of that social consciousness.

In April 1981 the Pulitzer Prize for feature writing was awarded to Janet Cooke, a reporter for the *Washington Post*. Shortly after the prize was announced, Cooke revealed that she had fabricated parts of her story about an eight-year-old drug addict. The child did not exist, the writer admitted before the *Post* returned the Pulitzer and asked her to resign. Why did she do it? "In my case, the temptation didn't derive from ambition," Cooke said. "I simply wanted to write a story that I had been working on so that I wouldn't have to go back and say, 'I cannot do it.' I did not want to fail."[1]

Ethics are moral guidelines for the resolution of difficult dilemmas. Indeed, there probably would be no need for ethics if there were no dilemmas, no tough choices to make. However, as can be seen in the Janet Cooke example, journalists and other media practitioners do have to make choices. Frequently the decisions involve ethics, complicated by conflicting guidelines—economic and political ones, and tyrannies of deadlines and media mechanics, for example. Put to a pragmatic test, journalistic ethical standards fall back too often in a fanfare of rationalization.

Behaviors institutionalized in news reporters, particularly investigative journalists, frequently involve questionable moral acts, and sometimes even prize-winning reporters have to make—or *do* make—decisions that many observers find morally repugnant.

Introduction

Who, or what institutions should keep the media socially responsible?

Ethical Dilemmas

Ethical choices are complicated by conflicting guidelines and everyday forces.

THE MIRAGE

A report on the 'fix' in Chicago

That man wearing a uniform is a Chicago Fire Department lieutenant. He has just taken a payoff for overlooking code violations in the Mirage, a small Chicago tavern. We know. The Mirage was our tavern. We had to make the payoff. It was a not-so-hidden cost of doing business. With a civic watchdog group, the Better Government Assn., The Sun-Times owned and operated the Mirage to uncover and document firsthand public and private corruption in the city that works—if you know how to work it. Inside is the full Sun-Times series, reports on the reaction and comment from the editors and readers.

The 1978 Distinguished Service Award winner of the Society of Professional Journalists—Sigma Delta Chi (SPJ/SDX)—for general reporting in newspapers was a *Chicago Sun-Times* investigative team that exposed shakedowns from city and state inspectors. The *Sun-Times* purchased a bar, named it "The Mirage," trained a couple of reporters as bartenders, and went into business. A pair of photographers was concealed in a loft above the bar, photographing the action below through a ventilation duct. Throughout the summer and fall of 1977 the reporters dealt with a variety of city and state inspectors who disregarded fire, health, and construction violations by the dozen in exchange for payoffs ranging from $10 to $100.[2]

In accepting their national SPJ/SDX award, *Sun-Times* reporters said they were concerned over the ethical questions involved in the investigation, but insisted that they had remained perfectly legal in every one of their activities; they engaged in no illegal entrapment, they used no hidden microphones, and they reported all bribes. Finally, they argued, the good they did for the city of Chicago and the state of Illinois more than justified any questionable behavior on their part.

The ingenuity demonstrated by the *Sun-Times* so impressed the nominating jury for the Pulitzer prizes that it submitted the Chicago paper's entry to the Pulitzer prize board, which makes the final selections. But no Pulitzer prize for local reporting went to the Windy City newspaper, because a majority of prize-board members expressed serious reservations about the *Sun-Times*' undercover methods of reporting.

Earlier Pulitzer prizes have been awarded to journalists posing as workers in nonjournalistic occupations. As ambulance drivers they exposed collusion between police and private ambulance companies, and as social workers they investigated social problems. But by the late 1970s the profession of journalism was growing leery of stepping across that narrow line separating "pretense" and "deception." Prize-board member James Reston of the *New York Times* said he had no problem with pretense, a passive act in which reporters allow someone to draw the wrong conclusions about a reporter's identity and level of information. But he was bothered by deception, in which reporters actively intend to mislead.

Is undercover reporting ethical if it results in benefits to society? The jury is split.

The issue was further heightened when another board member, Ben Bradlee, the editor of the *Washington Post* who had coordinated the *Post*'s Pulitzer prizewinning investigations of Watergate, asked:

How can newspapers fight for honesty and integrity when they themselves are less than honest in getting a story? Would you want a cop to pose as a newspaperman?[3]

(Bradlee, of course, had a bitter pill to swallow three years later when his young staffer, Janet Cooke, proved herself less than honest.)

Other board members said they were not convinced the *Sun-Times* had not engaged in entrapment, and still others criticized the newspaper for having taken the easy, theatrical approach to solving a civic problem.

Some pointed out that the reporters were engaged in illegal activities whenever they paid bribes to inspectors, regardless of the fact that each bribe was promptly reported to the Illinois Department of Law Enforcement.

In short, media professionals are divided over what constitutes honorable service and what constitutes unethical behavior. Several research studies, investigating why journalists react to ethical dilemmas as they do, indicate real confusion over values, recognition of the need for formalized codes of ethics, and belief that some journalists are more ethical than others.

One Chicago journalist adequately summed up journalistic concerns over ethical principles in light of the realities of reporting, editing, and gatekeeping under daily pressure:

> I believe all journalists with any talent and sense want to be ethical (fair, accurate, truthful) reporters but when you're faced with standing firm on a bunch of abstract principles you learned in college, or losing your job because you won't allow your publisher to review a story before it's printed, the decision becomes a very economic one. Sure it's unethical, but you can't get hired by presenting a resumé filled with short-term jobs you left because of "ethical principles."[4]

Other studies of journalism ethics reinforce the general conclusions noted here—that the profession has begun to articulate its ethical dilemmas, but has not reached consensus over which issues are most vexing, which ones create the greatest breach of the public trust, which ones cause the greatest disturbance to journalism's internal operations, or even which ones can be resolved by effective codes of ethics or some other means.[5] As one journalism scholar concluded, after a national survey of editors and publishers, "Newspaper people tend to be ethically confused. We tend to have two kinds of newspaper people: those who are totally insensitive to ethical matters and those who are hypersensitive. And one response is about as bad as the other."[6] A philosopher, after scouring the literature of mass media ethics, concluded simply that there is no such thing as the ethics of journalism, if by ethics we mean that those in the institution or organization agree on norms and principles.[7] Many would disagree, insisting that journalists tend to agree informally about the need for such principles as accuracy, fairness, restraint, and truth. But, like most abstractions, such principles are often difficult to pin down.

Junkets and Freebies

Journalists, particularly editors and reporters engaged in the gatekeeping function, frequently are offered *junkets* (expense-paid trips) in the hope that they will write glowing articles. To a lesser degree, theater critics, entertainment editors, and sportswriters get their share of *freebies* (free tickets and press invitations, passes to special events, copies of books, records, and even art). This is done, from the hosts' points of view, to call attention to newsworthy situations, to thank journalists for a given effort, or to gain favorable attention of the journalists who may find it difficult to be dispassionate in writing about their hosts.

When asked, journalists cannot agree over values, codes of ethics, or their own "professional" standards.

News sources frequently offer expense-paid trips (*junkets*) and *freebies* to gain favorable press coverage.

Newsroom ethics: How tough is enforcement? A survey of who gets suspended, who gets dismissed—and why.

A 1986 mail survey of 226 newspaper editors, by the American Society of Newspaper Editors, showed the following sometimes contradictory conclusions about how America's newspapers deal with ethical dilemmas:

More than one out of every three editors reported at least one ethics violation occurred at their papers in the past three years. A total of 240 ethics violations were reported by the 122 respondents who answered the question; 11 papers reported six or more violations in the past three years.

About one out of six editors said at least one newsroom employee had been dismissed because of ethics violations in the past three years. Another 11% said at least one employee had been suspended in the past three years for an ethics violation.

Slightly more than a third—37%—of the editors said they had a written code of ethics. More than half—54%—said they did not. Four percent said they were preparing one and 6% did not answer the question.

An overwhelming majority of editors—87%—at papers with written codes of ethics said the code included exceptions. Among them: Review tickets, tickets to sporting events, and freelance work approved by a supervisor.

Almost three out of 10 editors at papers with written codes said the code included penalty provisions.

Four percent of the editors said their papers require employees to sign a stock disclosure form.

Half of the editors said their code applied only to the newsroom. (The survey did not, however, ask whether the business side had a separate code of ethics.)

Five percent of the editors surveyed said they had a separate code of ethics for business reporters and editors.

Plagiarism, using unpublished information for financial gain, and discounts for personal purchases made from companies the reporter or editor handled generally were viewed as the most serious ethical violations of those listed.

Spouses owning stock in companies a reporter covers, and social relationships between newsroom personnel and newsmakers were viewed as minor or no ethical problems.

Social contacts between reporters, editors and newsmakers, as well as reporters who rewrote competitors' stories without verifying any information, were the most frequently encountered situations of those listed.

Source: ASNE Ethics Committee Report, "Newsroom Ethics: How Tough is Enforcement?", *Journal of Mass Media Ethics*, Fall/Winter 1986–87. By permission.

There is a balance to this, however; many newspapers and smaller media cannot afford to buy all the tickets involved, so the alternatives are either freebies or no coverage at all. The solution to this problem is not easy.

There are at least two primary issues involved with freebies:
(1) Acceptance of a freebie may consciously or unconsciously influence the
journalist's objectivity; and (2) the acceptance of a freebie, when known,
may give the public reason to distrust the news medium, even though the
journalists may not be (or think they are not) affected. Consequently, many
major media pay their own way to everything, even insisting that their reporters pay the government for travel on governmental aircraft or convoys
during political campaigns or junkets. Recently enacted codes of ethics in
many media clearly state that nothing of value is to be accepted by journalists. Other guidelines indicate that journalists can accept food or drink
as long as it can be eaten or drunk in ten minutes.

In response to a national survey of newspaper journalists, slightly more
than a third of those queried indicated that their newspaper had a stated
policy concerning the acceptance of freebies; a quarter said their newspaper
had no policy; and 35 percent said there was no written policy but that a
general understanding existed among staff members as to what was acceptable. Individual respondents showed a very low regard of their colleagues' ability to remain objective when offered freebies and junkets, even
though only 15 percent of those queried admitted to having been so influenced themselves. The lower the dollar value of the freebie, the less likely
the journalists were to feel compromised. In other words, there is a parallel
between ethical practices and monetary enticements.[8]

In sum, the national study on freebies indicated a strong regard for individual professionalism, a general distaste for freebies, a rejection of freebies as essential, and a general support for codes of ethics, particularly the code of the Society of Professional Journalists, Sigma Delta Chi, which is the nation's largest professional journalism group.[9]

This does not mean that the issue of freebies and junkets has gone away. In late 1986, some 10,600 people from American and foreign publishing and broadcast businesses attended a free bash thrown by Disney World to celebrate its fifteenth anniversary. The classic four-day, all expenses paid pseudoevent included appearances by former United States Chief Justice Warren Burger and Nicholas Daniloff (an American journalist who had just been released from a brief captivity in the Soviet Union); the premiere of a new movie by Michael Jackson; and dozens of press conferences with entertainers and Disney celebrities. Disney, the airlines, and area hotels and resorts spent an estimated $8 million on the junket. It was, according to some, a pittance in comparison with the free publicity Disney was expecting to garner in return.[10] Of course, the ethics of the junket were so widely discussed in journalism trade magazines that journalists are far less likely to accept, and justify, such junkets in the future. Whether the general public will even know if subsequent news coverage about Disney World is legitimate or "payback" is another story.

Bribes

Commercial influences frequently have been seen as primary tests of media ethics. In the late 1950s there was a great network scandal about "The $64,000 Question." Glued to their television chairs, Americans rooted week after week for the young father with the fantastic memory. Would he win the $64,000? Then, disillusionment came. The public learned that he had been coached by the network. Producers of "The $64,000 Question" and other prime-time network quiz programs had been seeking to maintain audience interest for the commercial sponsors by making sure that some attractive contestants did not make mistakes that would eliminate them from competition. When challenged in this unethical practice, producers justified their behavior with the argument that they were producing entertainment programs, not genuine intellectual contests. Few people supported them in this argument, and big-money quiz shows disappeared from the schedule for several years. After the Federal Communications Commission added a Section 509 clause, making it illegal to assist any contestant in a quiz program with the intent of deceiving the listening public, and after networks made a great deal of publicity out of their claims that henceforth they, and not the commercial sponsors, would determine rules and procedures of quiz and game shows, such programming returned to the air.[11]

While quiz show scandals were going on in the 1950s, the practice of *plugola* was in full operation. Publicity agents representing manufacturers of certain products would pay a program's writers and director for giving on-camera exposure to their products. Many people in the broadcast industry were on the take, including performers, whenever they slipped in

Quiz show scandals, *plugola*, and *payola* are examples of commercial influences over TV and radio programming.

the free boosts or ads, called *plugs*. Nowadays, the practice has become far more refined, particularly in the film industry. Producers seeking authenticity in their movies make extensive use of easily recognized brand name products—cars, soft drinks, beer, cigarettes, junk food, etc.—for which they are paid a fee by the product manufacturers and advertisers. The practice has become so widespread that agents openly advertise their "product placement" services in trade journals such as *Advertising Age*.

The most widespread and publicized use of plugola has been found in the recording industry in the form of *payola* to disc jockeys. In days past, record companies quite openly put disc jockeys on the payroll to play their records on the air, because such exposure from a popular personality often guaranteed sales. Many DJs were making more from payola than they were from their salaries, and some refused to play anything without payola. Payola was essentially advertising for which the stations received no income and which, because it was not labeled as advertising, deceived the public. Despite congressional hearings and attempts by the FCC and broadcasters to terminate the practice, reports of payola continue to surface within the industry. Lately, payola has consisted of drugs in lieu of cash. Obviously the law is circumvented, and the issue remains a legal as well as ethical one.

Commercial Influences

Magazine and newspaper editors, particularly in the fields of fashion, travel, and real estate, often trade a certain amount of editorial coverage in return for significant volumes of advertising. An advertiser can, in effect, double or sometimes triple the product's exposure in a single issue at the same cost.[12] Frequently the editorial content of these sections of newspapers is simply an excuse for the advertising volume-filler material to meet the sixty-forty advertising-editorial ratio. More often than not, this filler material is supplied by publicists for advertisers, as can be seen by even casual readers of newspaper "progress editions" or boat show, home show, county fair, and other sections. When business offices formally or informally contract to combine editorial and advertising coverage, they issue "business office musts" or BOM's, telling the editorial department that given PR blurbs must appear on the news pages. Not surprisingly, the practice offends most reporters.

There seems to be a kind of double damage here. Not only are the magazine's or newspaper's subscribers reading puffery, and assuming it is fact, but preferential treatment is being shown to certain advertisers who are acquiring greater space at lesser cost. Other advertisers, ignorant or unable to take advantage of such arrangements, have legitimate cause for complaint. Ultimately, of course, the greatest damage is probably to an unmeasurable commodity: credibility. It is only natural for credibility to suffer when audiences expect media to be serving the information function while in fact they are serving the advertising or persuasion function. As a result of all of this, BOMs and puffery appear to be falling into disfavor

Print media sometimes exchange editorial coverage for advertising. The price to be paid: credibility.

> Let us be as journalists, then. And like all good journalists, we shall present our facts in an order that will satisfy the famous five *Ws:* wow, whoopee, wahoo, why-not and whew.
>
> —Former journalist and novelist Tom Robbins, *Even Cowgirls Get the Blues* (New York: Bantam Books, 1977), p. 144.

among most responsible media. Even advertising departments are coming to recognize that their advertising message carries a greater weight when it is in a publication that readers perceive as independent and credible.

There are many who, looking at the enormous corporate power of the mass media and their commercial advertisers, fear misuse. A case might well be made, however, that there are some practical reasons why these fears may be overemphasized.

Two-thirds of the nation's daily newspapers are chain owned and generally are controlled financially by corporations that are far removed from the daily news-reporting and editorial activities. Although the primary commitments of chain owners are probably to net profits, occasionally abuse of ownership becomes blatant. In 1977 Panax Corporation told its eight daily newspapers and forty weeklies to publish a pair of front page articles critical of President Jimmy Carter. Two editors refused and were subsequently dismissed. The furor that followed became national news in trade journals and on network television, and assuredly warned other chains that such meddling would not go unnoticed.

In theory, of course, the trend to corporate ownership could spell the end of local autonomy and diversity of opinions. Conglomerates have the power to control the content of media, and when they do, editors and reporters have few defenses. It is an issue that many media observers take very seriously, if only in the abstract.

By the same token, the relationship between advertisers and media decision making is a concern. At one time in its history, the American news industry was heavily influenced by advertisers who assumed that because they paid two-thirds of the cost of print media and nearly all the cost of electronic media, they could call all the shots. Advertiser ideology was omnipresent, creeping into news and editorial functions.

Today it has become more and more obvious that the mass media in larger markets are greater and more powerful than their advertisers. They do not need to jeopardize their audience position to satisfy advertisers anymore. Whereas advertisers once enforced their will by threatening to withhold or cancel their advertising, in general they can no longer do so. The truth is that they need the media to market their goods as much as the media need them for financial support. To cancel advertising is self-defeating in the long run. Institutionalized media demonstrate time and time again that neither party can long exploit the other without destroying the delicate relationship.

Owners and Advertisers

Editorial abuse by media owners and advertisers is possible, but several factors tend to limit manipulation.

Perquisites

Some "star" journalists have many privileges associated with their jobs, but such *perks* can be abused.

A good many columnists and "star reporters" supplement their incomes by taking advantage of their unique positions as influential gatekeepers. We might call these advantages perquisites (meaning privileges) or *perks*. Some go so far as to include favorable (or unfavorable) mention of individuals or companies for pay. Others, especially syndicated columnists, have booked themselves with talent agencies and joined the lecture circuit, where they receive tidy sums per visit to college assemblies or other groups. Jack Anderson, for one, makes so much from his lectures that he takes no income from his widely syndicated column. America's highest paid speakers include Walter Cronkite, Dan Rather, Paul Harvey (at $20,000 per speech), and Barbara Walters (at $15,000).

One of the problems associated with media stardom cropped up recently when Barbara Walters became embroiled in an international diplomatic incident. Walters arranged a television interview with arms dealer Manucher Ghorbanifar, a shadowy figure involved in the controversial sale of American weapons to Iran in exchange for the release of hostages. At the conclusion of the interview Ghorbanifar handed Walters some secret documents pertaining to the arms sale, and asked Walters to deliver them to the White House. Walters said she reluctantly did what Ghorbanifar asked, and then was roundly attacked in media circles for blurring the line between being a reporter and being a maker of the news—even of being a governmental agent.

On another front, financial editors of newspapers and magazines are often in a position to glean inside knowledge of proposed transactions. These financial journalists have a double-barreled opportunity. They can enhance a financial situation (or detract from it) through mention in their pages, and they are in a position to personally profit from privy information by speculation. Because of the sensitivity of financial transactions, distorted news read as fact can adversely affect readers where it hurts, in their pocketbooks.

A *Wall Street Journal* reporter learned the price of sharing "inside" information.

The most famous abuser of these perks is R. Foster Winans, the former *Wall Street Journal* reporter who was fired, jailed, and fined for leaking information about upcoming *WSJ* columns to stockbrokers. Winans had access to tidbits scheduled to appear in the *Journal*'s "Heard on the Street" column, an influential "gossip" column that could make or break stocks. He struck a deal with a pair of greedy stockbrokers to leak "Heard" items far enough in advance for them to manipulate the stock market—a practice known on Wall Street as "insider trading." He also conspired to write about stocks his associates had already purchased in order to influence the stock's value. The Securities and Exchange Commission brought a civil suit against Winans and others in 1985, and the young journalist was found guilty of numerous counts of securities, wire, and mail fraud and conspiracy.

Ironically, Winans maintained in court that the *Wall Street Journal* never explicitly told him that such practices would violate both the law and the newspaper's code of ethics. In a book about the case, Winans admitted that what he had done was "technically unethical" for a journalist; but he said that "Since I knew that I wasn't letting my investment alter my judgment at work in any way, the ethical question was purely one of appearances."[13] His editors and coworkers, of course, were outraged. They suffered a double blow when the Securities and Exchange Commission attempted to use the courts to enforce the newspaper's ethics code upon its employees, to "protect" the newspaper's good reputation. The ensuing legal and ethical morass led some to suggest that the newspaper would have been better off if it had never created an ethics code in the first place or if it had never bothered developing a reputation worth protecting.[14]

It is important to remember that the Winans case—like the Janet Cooke case at the *Washington Post*—is famous because it is such an extreme departure from the norm. If nothing else, the cases are classical negative role models; they teach us in clear black-and-white terms what is unacceptable.

Formal surveys and anecdotal insights from journalists have suggested recently that the public may be losing faith in their news media. The reasons for—and the consequences of—the credibility dilemma are complex.

Credibility Problems

Reasons for—and consequences of—the media's credibility dilemma are complex.

Los Angeles Times media critic David Shaw's short but distressing list of credibility problems shows public dissatisfaction with news media's invasion of privacy; general insensitivity to audiences (with no formal vehicles available by which most media can respond to complaints); a suspicion that journalists are members of the liberal elite class; journalists' reliance upon unnamed sources in their stories; inaccuracy (and a reluctance to admit to their inaccuracies); and a general reluctance to write about journalism and media with the same insights and skepticism employed in covering almost all other institutions. Shaw, who writes a regular column on media for the *Los Angeles Times,* said his list emerged from a *Times* survey of three thousand readers.[15]

Earlier, another metropolitan newspaper ombudsman devised his own list, based on listening to twenty phone calls a day from disgruntled readers of the *Kansas City Star* and *Times*. According to Donald C. Jones, readers simply do not trust journalists for their inaccuracy, arrogance, unfairness, disregard of privacy of ordinary citizens, contempt for local community areas, insensitivity to race, religion, and sex, glorification of the criminal and the bizarre, and overall bad writing and editing.[16]

Meanwhile, several nationwide surveys of media audiences have shed conflicting lights on the credibility issue. The American Society of Newspaper Editors interviewed sixteen hundred American adults in 1984 and 1985 about newspaper credibility, with results that disturbed many journalists and media observers. ASNE's data showed that newspapers had high credibility with only 32 percent of the public; 84 percent said they had "some respect" or "great respect" for newspapers; and 76 percent said they thought the press helps keep public officials honest. People who had little faith in newspapers also tended to be hostile to the concept of press freedom, a finding which reinforced a suspicion many editors had developed through informal contacts and letters to the editor. For instance, 42 percent agreed with the statement "Sometimes there's too much freedom of the press," and 39 percent agreed that "the media abuse their constitutional guarantee of a free press." Those most critical of the press were members of two diverse socioeconomic groups: the poorest and least educated, and the richest and best educated.[17] Some ASNE members took that insight to mean that the American people do not really want to see life as it exists, while others suggested the greatest significance of the findings is that public pressure groups would jump on the antipress sentiment and mount propaganda campaigns against the bearers of bad news.[18]

A year later, journalists were told to relax, that the public really did not hate them. History's most massive—and expensive—study of public attitudes toward the mass media, conducted by the Gallup organization for the Times Mirror Corporation in 1986, concluded that there really is no credibility problem. The controversial study said the three thousand Americans who were interviewed at length overwhelmingly believe the news they get, and actually have a higher opinion of the print and broadcast media than of President Reagan, Congress, big business, or the military. Times Mirror said that if credibility is defined as believability, then credibility is, in fact, one of the media's strongest suits. However, the survey showed the public was still critical of the media in the areas of independence, political bias, negativism, intrusiveness, and unwillingness to admit to mistakes.[19]

A review of the ASNE and Times Mirror studies, along with several others produced by academic and professional groups over the past several years, demonstrates the fragility of the relationship between the mass media and the American public. The studies also suggest that the "credibility glass" might be half-empty or half-full, depending on how one cares to ask the questions and exploit the findings.

A fifty-year review of America's continuing love-hate relationship with its news media, published by the Gannett Center for Media Studies at Columbia University, concluded that America tends to give the news media about the same unenthusiastic votes of confidence it gives other institutions. The public is concerned about the news media's arrogance and power just as it is concerned about those traits in other powerful institutions. However, Americans are quite sensitive to the ethics of journalists, particularly as they see journalists treating private citizens cavalierly, and they really do wish there were not so much bad news in the papers and on the air. The Gannett Center study also concluded that the public does not pay enough attention to government-press relations, press freedom, and the media's autonomy in dealing with pressure groups and economic influences.[20]

The relationship between the mass media and their audiences is fragile.

> It's the presumption of omniscience, of having the last word, in the very style of journalism that ticks people off. Every day every newspaper is filled with mistakes and errors of omission, no matter how hard it tries. But we ought to make it clear—in every lead, in every caption that we write—that we are an imperfect institution. That what we are giving the readers is not the last word but a collection of clues.
>
> —Philip Foisie, foreign editor of *Washington Post,* quoted by Michael T. Malloy, "Journalism Ethics: Not Black, Not White, But . . . . A Rainbow of Gray," *National Observer,* July 25, 1975.

When summarizing the entire arena of credibility studies, Gannett Center Director Everette Dennis has lamented what he sees as a problem in public education. He noted that whereas the mass media have emerged as a major American institution in recent years, nowhere in our educational system is there a coherent effort to introduce students to the news media, "to provide not only an understanding of how the media work, but also the overarching issues of freedom of expression and its role in our constitutional scheme." Dennis suggested that people are not enthusiastic about freedom of the press largely because they have never been exposed to a thoughtful explanation of the Constitution and the role of information in a democratic society.[21] If Dennis is right, there is no end in sight for the credibility problem until journalists settle the issues inherent within their profession and the public gets a better lesson in civics.

The public may need to be educated about media freedoms and responsibilities.

Codes of Conduct

Mass media ethics are pretty much a product of the twentieth century. During the early years of the American republic, the press was dominated by political interests that subsidized many newspapers. But political control of the press diminished with the success of the penny press in the 1830s and the growth of advertising. Throughout the rest of the century the press reflected the spirit of the time—freewheeling days of national expansion, the winning of the West, and social Darwinistic ideas of the survival of the fittest.

This climate lasted until the turn of the century when the muckrakers, individual journalists, discovered their considerable influence and touched a public nerve with their crusade against business corruption and government collusion. For a decade, they attacked social injustice through books, magazines, and newspapers. This period may have been the highwater mark of journalistic ethics, paralleled perhaps only by post-Watergate self-analysis.

A great deal of media self-criticism existed in the 1920s, as thoughtful journalists recognized public disaffection with media excesses, and called for commitments to journalism's professional status.[22] Schools of journalism, most offering courses in journalism ethics, attempted to place journalism in an academic framework. The ethics literature of the period demonstrated a strong moral responsibility to one's professional community, ethics being construed as right action toward one's fellows.

THE JOURNALIST'S Creed

I believe

IN THE PROFESSION OF

JOURNALISM.

I BELIEVE THAT THE PUBLIC JOURNAL IS A PUBLIC TRUST; THAT ALL CONNECTED WITH IT ARE, TO THE FULL MEASURE OF THEIR RESPONSIBILITY, TRUSTEES FOR THE PUBLIC; THAT ACCEPTANCE OF A LESSER SERVICE THAN THE PUBLIC SERVICE IS BETRAYAL OF THIS TRUST.

I BELIEVE THAT CLEAR THINKING AND CLEAR STATEMENT, ACCURACY, AND FAIRNESS, ARE FUNDAMENTAL TO GOOD JOURNALISM.

I BELIEVE THAT A JOURNALIST SHOULD WRITE ONLY WHAT HE HOLDS IN HIS HEART TO BE TRUE.

I BELIEVE THAT SUPPRESSION OF THE NEWS, FOR ANY CONSIDERATION OTHER THAN THE WELFARE OF SOCIETY, IS INDEFENSIBLE.

I BELIEVE THAT NO ONE SHOULD WRITE AS A JOURNALIST WHAT HE WOULD NOT SAY AS A GENTLEMAN; THAT BRIBERY BY ONE'S OWN POCKETBOOK IS AS MUCH TO BE AVOIDED AS BRIBERY BY THE POCKETBOOK OF ANOTHER; THAT INDIVIDUAL RESPONSIBILITY MAY NOT BE ESCAPED BY PLEADING ANOTHER'S INSTRUCTIONS OR ANOTHER'S DIVIDENDS.

I BELIEVE THAT ADVERTISING, NEWS AND EDITORIAL COLUMNS SHOULD ALIKE SERVE THE BEST INTERESTS OF READERS; THAT A SINGLE STANDARD OF HELPFUL TRUTH AND CLEANNESS SHOULD PREVAIL FOR ALL; THAT THE SUPREME TEST OF GOOD JOURNALISM IS THE MEASURE OF ITS PUBLIC SERVICE.

I BELIEVE THAT THE JOURNALISM WHICH SUCCEEDS BEST—AND BEST DESERVES SUCCESS—FEARS GOD AND HONORS MAN; IS STOUTLY INDEPENDENT, UNMOVED BY PRIDE OF OPINION OR GREED OF POWER, CONSTRUCTIVE, TOLERANT BUT NEVER CARELESS, SELF-CONTROLLED, PATIENT, ALWAYS RESPECTFUL OF ITS READERS BUT ALWAYS UNAFRAID, IS QUICKLY INDIGNANT AT INJUSTICE; IS UNSWAYED BY THE APPEAL OF PRIVILEGE OR THE CLAMOR OF THE MOB; SEEKS TO GIVE EVERY MAN A CHANCE, AND, AS FAR AS LAW AND HONEST WAGE AND RECOGNITION OF HUMAN BROTHERHOOD CAN MAKE IT SO, AN EQUAL CHANCE; IS PROFOUNDLY PATRIOTIC WHILE SINCERELY PROMOTING INTERNATIONAL GOOD WILL AND CEMENTING WORLD-COMRADESHIP; IS A JOURNALISM OF HUMANITY, OF AND FOR TODAY'S WORLD.

Walter Williams

DEAN SCHOOL OF JOURNALISM, UNIVERSITY OF MISSOURI, 1908-1935

Codes of ethics and courses in media ethics emerging from this period displayed a concern for common values of a shared culture. One of the earliest was the journalist's creed, written in 1908 by Walter Williams, the first dean at the University of Missouri's School of Journalism. Williams emphasized honor amid the search for truth. "I believe that no one should write as a journalist what he would not say as a gentleman," he wrote in the creed. This concept of journalism ethics as "moral responsibility to my fellows" faded from view after the beginning of the 1930s. Into the vacuum moved a new understanding of what constituted ethical journalism, the *cult of objectivity*, which went unchallenged for the next several decades. Ethically responsible journalism was seen as achievable only through factual accuracy and verification of details, devoid of analysis or interpretation. Readers were left to draw their own conclusions, which oftentimes was an

Early codes of ethics were supplanted by the *cult of objectivity*.

impossible task especially when they were not provided the framework within which an event took place.

An example of this occurred during the tyranny that surrounded the late Senator Joseph McCarthy in the early 1950s when McCarthy labeled prominent State Department officials as Communists and homosexuals. The press duly reported this factually—the fact of his utterance. It did not report that this was a familiar tactic used by McCarthy, nor that many of his past allegations had proven erroneous. Journalists did not report this because it would have been an expression of editorial opinion contrary to the accepted tenets of objective reporting.

The environment that gave rise to objectivity led to debates about media morality. Starting in the 1920s, first print journalists and then broadcasters formulated codes of ethics, attempts to define acceptable and unacceptable professional behavior.

Newspaper Codes

In the past sixty years the press has developed numerous codes of ethics, some prepared by editors and others by reporters. The codes share many elements, differing primarily in the perspectives of the two groups.

American Society of Newspaper Editors Code Social responsibility as a press theory probably goes back to the Canons of Journalism adopted by the American Society of Newspaper Editors (ASNE) in 1923. It should be observed that members of the society were the working editors, not the publishers, and that their doctrine of press responsibility was adopted in the heyday of jazz journalism. In essence, they reflected their own individual repugnancies and consciences.

As originally written, the 1923 Canons of Journalism (revised in 1975) reflected the social concerns, the group ethics typical of the times. In their preamble, ASNE members maintained that:

The primary function of newspapers is to communicate to the human race what its members do, feel, and think. Journalism, therefore, demands of its practitioners the widest range of intelligence, of knowledge, and of experience, as well as natural and trained powers of observation and reasoning. To its opportunities as a chronicle are indissolubly linked its obligations as teacher and interpreter.

Many saw the Canons as an interesting but toothless collection of lofty thoughts suitable for framing on the newsroom wall. Indeed, early in its history the ASNE tried to expel one of its members, *Denver Post* editor Fred G. Bonfils, who was accused of blackmailing oil millionaire Harry Sinclair in connection with the Teapot Dome Scandal, the Watergate of its day.[23] It could not agree to exercise its self-policing powers, however, so the ASNE appeared to many as a less than adequate means of guaranteeing media responsibility.

The ASNE code stresses social ethics, but it lacks the power of enforcement.

It was no coincidence that the ASNE Canons were created three years after author Upton Sinclair had turned his muckraking attention toward the press, equating it to a house of prostitution. In a book he published privately, because established publishing houses were not willing to take the chance, Sinclair said journalists constantly violated the public trust by taking the "brass check," the medium of exchange in brothels.

The Brass Check is found in your pay-envelopes each week—you who write and print and distribute our newspapers and magazines. The Brass Check is the price of your shame—you who take the fair body of truth and sell it in the market-place, who betray the virgin hopes of mankind into the loathsome brothel of Big Business.[24]

Society of Professional Journalists/Sigma Delta Chi Code Quite similar to the ASNE Canons, both in justification for existence and tone, is the code of ethics adopted by Sigma Delta Chi (SDX) in 1926. SDX consisted of reporters and journalism students, as distinct from ASNE's management and editorial constituency. As such, SDX reflected the concerns of the bulk of working reporters, gathered together into an organization dedicated to their freedoms and professionalism. In the 1970s SDX, the largest group of journalists in America, renamed itself the Society of Professional Journalists, Sigma Delta Chi, and revised its code of ethics to reflect contemporary issues.

The SPJ/SDX code maintains that the duty of journalists is to serve the truth, that public enlightenment is the forerunner of justice, and that the Constitutional role to seek the truth is part of the public's right to know the truth. "We believe those responsibilities carry obligations that require journalists to perform with intelligence, objectivity, accuracy, and fairness," the code says; it then goes on to list a bill of ethical particulars. Like the ASNE code, it has been subject to considerable debate. Detractors insist that it artificially ascribes to journalism a false sense of professionalism, and they inquire skeptically why the organization has been unable—or unwilling—to implement enforcement procedures on the local and national levels.[25] Adherents insist it helps remind journalists of their responsibilities, holds them morally accountable for their behavior, and increases news media credibility in the eyes of the public.[26]

Broadcast Codes

There is a major distinction between the earliest codes drawn up by the newspaper industry and those drawn up by broadcasters. In a phrase, it is the difference between positive and negative liberties. Newspapers, with their tradition over the centuries of fighting for freedoms, expressed in their codes a belief in the self-righting process and the general rationality of audiences, and, secondarily, a commitment to garner and maintain the public's trust.

The codes of the radio industry (1937) and its successor, television (1952), along with the Motion Picture Production Code (1930), differ markedly in intent and scope. All are patent efforts to forestall new governmental regulation or to deal with existing regulation, and as such primarily view responsibility in negative terms—"Thou shalt not"—rather than in positive terms stating the scope of responsibility to the various publics that support them. The codes reflect the electronic media's emphasis on entertainment, although they do give some attention to educational functions. Displaying a shift from rationalism prevalent under libertarian philosophy, the codes appear to depict audiences as readily corruptible. Page after page of the codes tell practitioners how to treat issues of sex, religion, manners, and morals so as to operate in the public interest, in entertainment, advertising, and news.

The Radio Code Between 1937 and the early 1980s the Radio Code of the National Association of Broadcasters (NAB) went through some twenty revisions, until it filled a thirty-page pamphlet. By 1982, when the code came under fire from the federal government for its antitrust violations, its index alone listed nearly one hundred different topics, ranging from *adult themes* to *vulgarity*. (Ironically, a federal appeals court found fault with the code's limitations on the numbers of minutes per hour radio and television stations could carry advertising. The NAB's best intentions, it seemed, were contrary to public interest.)

The Radio Code contained several pages devoted to regulations and procedures, outlining the functions of the Code Authority of the NAB, the organization to which broadcasting stations belong if they wish to carry the prestigious "seal of good practice." Revoking a station's subscription to the code was the most serious direct impact the Code Authority could have on a violator. Obviously, if a station wished to violate the standards of the code and cared not whether its membership in the NAB meant anything to its audiences and advertisers, there was nothing to stop it from doing so. Significantly, only one-third of the nation's radio stations were code subscribers. In hard times or in highly competitive markets where a couple of extra minutes of advertising per hour or some questionable programming practices could bring solvency, the owners could opt for profits and forgo the public relations.

The Television Code Essentially similar to the Radio Code was the NAB's Television Code, which also went through some twenty revisions following its inception in 1952. It ended up being even longer and more specific than the Radio Code, due in part to the pervasiveness of the medium and the controversial nature of much of its programming, especially its programming to children. Modifications in the Television Code came on the heels of public and political outcries against violence, deceptive advertising on programs for children, and news practices. In 1975 the code was amended in light of Senate hearings on nutrition and human needs and public outcry

Sex, violence, children's advertising, and news practices have been central to the TV code.

against television advertising's exploitation of children; its utilization of real and cartoon characters from the regular programs to sell products; its reliance upon peer pressure and fear appeals to foster buying habits; and its appeal to violent, dangerous, or otherwise antisocial behavior.

Many of television's "Special Program Standards" codes were nearly identical to those in the Radio Code. Some of the areas treated were narcotics, gambling, betting, the handicapped, sex, quiz programs, payola, lotteries, fictionalization of the news, and advertising. In almost every case the standards were expressed negatively—shall not, should avoid, should be de-emphasized, is prohibited, and so forth. Likewise the Television Code's statements on the treatment of news and public events, controversial public issues, political telecasts, and religious programs read much like those in the Radio Code, differing mainly in statements regarding visual impact.

After the NAB Code Authority closed down in 1982, most of the individual networks and stations—both television and radio—implemented their own codes, which had much the same tone as the NAB's. In addition, on the news front, a stringent set of guidelines laid down by the Radio Television News Directors Association (RTNDA), first passed in 1966, took up much of the slack left by the NAB code's demise. Implementation of its code has occupied much of RTNDA's attention of late.

Meanwhile, the film industry, as discussed in chapter 9, has struggled with its own voluntary code and rating system.

> The NAB code was dropped in 1982, but RTNDA has picked up much of the slack.

Codes—A Conclusion

Good and honorable people, disturbed by the appearance or practice of social irresponsibility, have labored long hours to develop equitable means of improving both the image and reality of their businesses. In recent years a growing proportion of journalists and other media personnel have joined professional organizations—one list consisting primarily of working news reporters and editors includes 119 such groups—[27] that have devoted more and more of their meeting time and publications content to ethical dilemmas and questions about heightening levels of performance and status. Academic conferences, bull sessions, formal publications, talk shows, and even court cases make constant reference to the need for codes of ethics in the mass media. (More than one libel plaintiff has insisted that all he or she expected was for the media to conform to their own written codes.) If the 1980s has seen a boomlet of interest in mass media ethics and codes of behavior, it has also seen frustration over the scope and implementation of those codes.

A general summary of media industry codes has led us to the present conclusion that overall the codes seem to have certain elements in common. They have few if any teeth; they are both unenforced and unenforceable. They are incumbent upon members only, and the only sanction that can be applied against a member is expulsion from membership, sometimes a small penalty. The codes tend to be bland statements drawn up in response to public disenchantment with media operations. At best, they are a stopgap

> Noble in tone, most codes are unenforceable efforts to mold ethical behavior.

of semiserious self-regulation in the hope that somehow their platitudes will satisfy both public critics and government's temptation to regulate.

By their existence, they are an attempt to organize, to standardize, and to codify ethics. In a sense, they circumvent individual moral development by imposing corporate responsibility. This is a large order and, in the complexity of institutionalized mass media, comprises really only a halfway measure toward the regulation that media are trying to avoid. Codes arise in response to public demand, but they are framed to cause the least commotion.

Commission on Freedom of the Press

We have already seen that calls for social responsibility in all the mass media have been in existence from almost the beginning. Various codes of ethics, whether created spontaneously or as a defensive response to external pressures, have been common in mass media for more than half a century. Insofar as the news media were concerned, the codes probably had minimal effect on the daily performance of journalists. It may be that hanging a code of ethics on the newsroom wall was sufficient; when challenged on their daily performance, media practitioners could point to their code and say, "We're trying."

But when an influential external group comes along and seriously questions a craft's basic operating premises, something more than a feeble "We're trying" seems in order. Such has been the case on several occasions, but never so significantly as in 1947 when the Commission on the Freedom of the Press, a sort of blue-ribbon public conscience, issued its broad-ranging report.[28] The commission had been created during World War II (1942) with University of Chicago Chancellor Robert M. Hutchins as chairman. It was funded by *Time* magazine's Henry Luce and the *Encyclopaedia Britannica.*

The influential Hutchins Commission, set up as a public conscience, told the press how to be both free and responsible.

As seen by the Hutchins Commission, freedom of the press is a moral and not a natural right. Therefore, the right can be lost or forfeited when abused. Underlying the basic social responsibility theory postulated by the commission are two very fundamental propositions:

1. *Whoever enjoys freedom has certain obligations to society,* and because the media enjoy a libertarian heritage, they have the concomitant responsibility to use those freedoms to serve the welfare of society as a whole.

With freedom comes
responsibility. Society's welfare
is entrusted to the press and
protected by the First
Amendment.

2. *Society's welfare becomes the most overriding concern.* Essentially, this demonstrates a shift in the theoretical foundation of press freedom from the individual to society. Individual rights to speak out are balanced by group rights to be free from invasion of privacy, or libel; personal rights to free expression are described in terms of public access to the media, or the "public's right to know."

Based largely on testimony from journalists and from the literature of journalism organizations, the Hutchins Commission settled on a list of five basic requirements, or expectations, society believes the press should fulfill. Those requirements have been the subject of a great deal of debate since the time they were outlined. Few media organizations accepted the commission's recommendations at the time, but as the decades have passed, it is interesting to note how those recommendations continue to serve as a valid framework for press criticism.

Meaningful News

Media were told to provide a "truthful, comprehensive, and intelligent account of the day's events, in a context which gives them meaning." Objectivity was not enough.

The first requirement was that the media provide "a truthful, comprehensive, and intelligent account of the day's events, in a context which gives them meaning." It is a call for accuracy in news, but beyond that, it is a call for the clear separation of fact from opinion. Recognizing the value of objective, value-free journalistic reporting, the commission also said the media fail society if they fail to place the news in perspective. "It is no longer enough to report the *fact* truthfully. It is now necessary to report *the truth about the fact*," the commission concluded. In an ever more complex society, people need to know more than the basic "who said what to whom, when and where." They also need to know the "how" and "why" of news. False objectivity, as in merely reporting that Senator McCarthy has announced the existence of communists in the state department, is socially harmful, even though it may be true that the senator did say so. What is needed, the commission said, is to report the greater truth, which is the question of whether, and just how many, communists there really are in the state department, what are the motives and special interests of the senator, and what was the overall political climate in which the statement was made. Many people may not recognize or appreciate the attempts to add socially responsible interpretation to the news, especially because much interpretive news comes across as manipulative or editorialized reportage. But despite shortrun problems with credibility, more and more news media have begun to recognize the value of making the effort, spending the extra dollars on news staffs trained in complex political, economic, and social issues that will enable them to give that truthful, comprehensive account of the news.

Access for Comment and Criticism

Second on the Hutchins Commission's list of requirements was that the media should serve as "a forum for the exchange of comment and criticism." Because ownership and control of the news media were falling into fewer and fewer hands, the commission said there was a need for those

Media should serve as ''a forum for the exchange of comment and criticism,'' especially because media were falling into the hands of fewer and fewer owners.

media to serve as "common carriers" of controversial ideas and views that individual and corporate gatekeepers may not otherwise allow into the news flow. While the media should not be seen as common carriers in the sense that railroads have legal obligations to carry diverse freight, and they need not give space and time to everyone's ideas, they should as a matter of policy carry views contrary to their own, according to the commission.

Within the past two decades this question of public access to the media has grown in intensity, and many attempts to resolve the question have emerged. Most newspapers have put into practice their own voluntary fairness doctrines, opening up their editorial pages for divergent political views. *Op-ed* pages (opposite the editorial pages) have become commonplace, serving as cafeterias of opinions written by a broad spectrum of society. This valuable newspaper space emerged as a logical outgrowth of the socioeconomic turmoil of the 1960s, when disenfranchised minorities, disgruntled youth, and others realized they could not be heard by the general public by merely grabbing a bullhorn or cranking out a few hundred copies of their underground newspapers.

A Representative Picture of Society

The third requirement of the press laid down by the commission was that the media project "a representative picture of the constituent groups in society." Media were asked to take into consideration the values and aspirations, and the weaknesses and vices, of different social groups, and to avoid stereotyping the different groups in news coverage and entertainment shows.

Media have no reason to disagree with this request, except perhaps to wonder how they can possibly meet a precise "quota system" for minority coverage and depiction. Indeed, through their various codes of ethics they had already placed the same general responsibility on themselves. But whether they have met that responsibility is entirely another matter.

Of all the complaints against the media that surfaced during the 1960s and 1970s, those by minorities were perhaps the most vehement. Tired of

Media were asked to project ''a representative picture of the constituent groups in society.'' Some progress has been seen in news, entertainment, and advertising dealing with minorities.

being depicted as shiftless, rhythmic, happy-go-luckies in films, or as news-worthy only in newspaper sports and crime stories, black Americans began asserting their own identities in media. Likewise women, long seen as either sex symbols or domestics, fought for and began to win greater representation in gatekeeper roles. Other groups, including Chicanos, the elderly, religious minorities, and others long overlooked, have slowly cut away at culturally chauvinistic media organizations. Although there has been substantial progress, by no means have the battles over stereotyping in news, entertainment, and persuasive media been won.

Clarifying Goals and Values

The press was also asked to improve its "presentation and clarification of the goals and values of the society." Being the national educator is difficult.

Fourth on the list of requirements is that the press be responsible for "the presentation and clarification of the goals and values of the society." Here the media's educational function is called for, with the Hutchins Commission telling media to assume a responsibility like that of educators in stating and clarifying the ideals toward which the community should strive. Rather than appealing to the lowest common denominator of taste and value, media were asked to elevate public interests.

While agreeing in principle with this dictate, the press has found it difficult to fulfill. Different subgroups within society have widely divergent goals and values, and to carry all of those goals and values to the satisfaction of every group is impossible, especially because so many of them are contradictory.

The Right to Know

Finally, Hutchins asked the press to provide the public with "full access to the day's intelligence." Serious questions arise over the public's "right to know."

Fifth on the Hutchins Commission's list of requirements for the press is that the media should provide "full access to the day's intelligence." Recognizing the exponential nature of our information explosion, coupled with what appears to be an exponential increase in individuals' need to know how to cope, this becomes an awesome requirement.

Federally mandated freedom of information laws have followed vigorous efforts by the Society of Professional Journalists and other news organizations to eliminate or at least curtail unnecessary governmental secrecy, the classification system, closed meetings, and inaccessible records. As we saw in chapter 14, this social responsibility mandate flies in the face of government's natural inclinations, and the issue is far from being resolved. Recent court decisions have held that the media do not have rights of access to news events exceeding those of the general public, and some seriously question whether the public really needs to know all that the press says it does.

Ultimately, one might argue that the burden for being fully informed falls back squarely on the shoulders of the individual citizen, as it theoretically did under libertarianism. Such a burden is just what the Hutchins Commission was attempting to avoid, because it pictured the citizenry as being far more lethargic than did the promulgators of libertarian theory.

Perhaps what bothered the media the most about the Hutchins group was the shift in liberty being suggested. For two centuries American journalists had operated on the basis of negative liberty, or freedom from external restraints. Suddenly, however, the thrust was on positive freedom, a freedom for pursuit of some predetermined goals. This new social responsibility seems grounded on a school of thought that sees negative liberty as insufficient and ineffective, somewhat like telling people they are free to walk without first making sure they are not crippled. As one scholar noted:

To be real, freedom must be effective. It is not enough to tell a man that he is free to achieve his goals; one must provide him with the appropriate means of attaining those goals.[29]

Who better than government is able to provide humanity with the "appropriate means"? Government, even a democratic government, is seen by social responsibility adherents as the only force strong enough to guarantee effective operation of freedom.

This does not set well with observers who find social responsibility to be only a slightly disguised version of authoritarianism. Media philosopher John Merrill has been one of the most outspoken critics of the theory. To him the proposition that pluralism of ideas should be governmentally mandated is ludicrous. Journalists, he says, must retain their freedom to make their own news and editorial judgments. Even well-intentioned attempts by outside groups seeking media improvement are self-serving, and inevitably lessen the autonomy of journalists, Merrill maintains. So long as journalism voluntarily gives up its capacity to think for itself—its editorial self-determinism—it is in "grave danger of becoming one vast, gray, bland, monotonous, conformist spokesman for some collectivity of society."[30] Merrill's views seem to be in the minority, however, as moves increase toward social responsibility and corporate accountability in all public areas—press, business, government, and so forth.

Social responsibility stresses positive freedom—freedom to pursue predetermined goals; libertarianism stresses negative freedom—freedom from restraints. The shift in freedoms is vexing to many; they call it authoritarianism.

When the Commission on Freedom of the Press suggested that the media engage in vigorous mutual criticism and that an independent agency for evaluating the media be established, it was calling for new vehicles for media improvement. Also novel were its recommendations that academic-professional centers of advanced study, research, and publication in communications be established and that journalism schools guarantee their students broader liberal training. Since that 1947 report, activities have taken place on all these fronts. Whether such activities are entirely beneficial to the media and society may be debatable. Let us look at some evidence of change in each of the areas posed by the Hutchins Commission.

Improving the Media

Self-criticism and academic analysis of the press were suggested by Hutchins. Since 1947 much has been done in this arena.

Journalism Reviews Criticism of the mass media, most notably that of self-criticism by practitioners seeking improvement in their product, has taken on new dimensions since the Hutchins Commission report. Only in the past

(a)　　　　　　　　(b)

Columbia Journalism Review, Washington Journalism Review, Chicago Journalism Review, and two dozen others have actively criticized the press.

two decades have practitioners collected their observations and published journalism or media reviews on a regularly scheduled basis.

One of the earliest *j-reviews* is the *Columbia Journalism Review;* the professionally written and edited, almost staid journal we have cited several times in this text. *CJR* appeared first in the fall of 1961 and was published quarterly until the 1970s, at which time it was issued bimonthly from Columbia University's graduate school of journalism. Underwritten by private donors, foundations, subscriptions, and, only recently, advertisers, *CJR*'s policy, as stated in its first editorial and repeated in each subsequent issue, is "To assess the performance of journalism in all its forms, to call attention to its shortcomings and strengths, and to help define—or redefine—standards of honest, responsible service . . . to help stimulate continuing improvement in the profession and to speak out for what is right, fair, and decent."

In *CJR*'s pages can be found ongoing debates over media behavior. "Darts" and "laurels" are handed out to media seen deserving of criticism or praise. Editorials, book and article reviews, original research on media performance, and scholarly articles (some of them perhaps overly stuffy and sometimes establishmentarian in tone) about ethics, economics, and press-government relations are there for *CJR*'s steadily growing readership.

Sharing space on the nation's newsracks with *CJR* is the *Washington Journalism Review.* Breezier in tone and slicker in appearance than *CJR,* the *Washington Journalism Review* has great appeal to the nation's growing cadre of media watchers who delight in inside information about news economics, personalities, and trends.

Among the best-known j-reviews put out by reporters was the *Chicago Journalism Review* (1968–1975), a feisty journal established "to fight news management, news manipulation, and assaults on the integrity of the working press." The review emerged to criticize the media coverage of the 1968 Democratic National Convention in Chicago. Many reporters had felt as though they had been manipulated by their editors, and their editors by local officials—especially Mayor Richard Daley—in covering the confrontations between Chicago police and demonstrators during that convention. Demonstrators in full view of network cameras had chanted "The whole world is watching," and a sufficient number of Chicago journalists became so disillusioned at the distorted pictures conveyed of police-demonstrator confrontations that they risked their jobs to tell what they said was the whole story. That whole story included police brutality—unprovoked clubbings of reporters—condoned by the mayor's office and covered up by editors, according to many reporters.

Over the next several years the *Chicago Journalism Review,* with its front cover always an editorial cartoon by Pulitzer prize winner Bill Mauldin, served as an alternative to the news media in that city. However, its circulation never did reach a level sufficient to become a valid alternative for mainstream audiences. Like most reviews, it frequently served as a forum for internal complaints; journalists used it as a vehicle to talk to themselves and to pinprick their bosses.

The same could be said about most of the two dozen or so other local reviews that appeared during the late 1960s and 1970s. At this writing, the most solidly entrenched and influential of them is the *St. Louis Journalism Review.* There may be some question over whether the on-again, off-again journalism review movement is primarily an effort to improve media for the public or to improve media as places of employment. Whatever, it has displayed many of the trappings of the social responsibility theory as envisioned by the Hutchins Commission.

Press Councils Of all the Hutchins Commission's recommendations for improving the press, the one that seemed most heretical to journalists was the call for an independent public agency for evaluating the media. However, even prior to the Hutchins challenge, such efforts were already under way. The first semblance of press councils in America may have been the 1946 advisory councils consisting of community leaders in Redwood City, California, and Littleton, Colorado.

Additional efforts at local newspaper councils and citizens' advisory groups came in 1967, aided in part by a $40,000 grant from the Mellett Fund for a Free and Responsible Press earmarked to "encourage responsible press performance without infringing on First Amendment Freedoms."[31] The four newspaper councils (Redwood City, California; Bend, Oregon; and Sparta and Cairo, Illinois) and broader-based media councils in St. Louis and Seattle met with mixed success.

Over the next several years, similar media-community councils were established in Honolulu, Louisville, and Boston; and America's first statewide newspaper-press council was begun in Minnesota in 1971. The latter uses a grievance hearing panel, which seeks particulars from complainants

Outside agencies charged with reviewing media performance have met with mixed success.

who unsuccessfully attempted to seek satisfaction directly through the "offending" newspaper. If the hearing panel decides that the case should be adjudicated, complainants are asked to waive the right to sue the newspaper in a court of law. This serves to protect the newspaper from double jeopardy. The Minnesota press council has continued to rule against the state's newspapers as often as it has ruled for them.[32]

Press councils have no legal authority, and their only real power surfaces if constituent media cooperate by submitting evidence and publicizing the councils' findings—the "cleansing light of publicity." They serve an intermediate function between the routine, informal settlement of a reader-editor conflict and a formal court action. Those in support of the press council concept argue that the councils aid media by resolving press-audience differences in-house. Because the councils provide their services free of charge to those who complain, and because complainants frequently are more interested in "moral vindication" than in going to court, news councils can and should result in fewer lawsuits being filed against the media. Indeed, General William Westmoreland has indicated that he received bad advice when told to sue CBS and "60 Minutes" for libel. After the expensive, protracted, no-win libel suit, he said all he wanted was the kind of open airing of his complaint and the opportunity for moral vindication that a news council provides.

In short, press councils are modified forms of self-regulation, similar to the Better Business Bureau, National Advertising Review Board, Motion Picture Association of America, and other trade groups. The self-initiated efforts to monitor media performance would seem to be far preferable to externally imposed controls from legislatures, courts, or powerful public interest groups. Such efforts are in sync with the version of press freedom and accountability laid down by Thomas Jefferson and other social libertarians. A press council is little more than an open marketplace of ideas, a "tournament of reason."[33]

National News Council But not all agree with this premise. When there were calls in the early 1970s for a national news council in America, several major news organizations strongly opposed the proposals. The *New York Times,* for one, maintained that such "voluntary regulation" would diminish press freedom. The late John S. Knight, of the influential Knight-Ridder empire, maintained that "Any self-respecting editor who subscribes to meddling by the National News Council is simply eroding his own freedoms. Editors are accountable to their readers, not to a group of busybodies with time on their hands."[34] That argument hearkens back to the type of libertarianism espoused by James Madison, who thought the press should remain free from as many external constraints as possible—including the kinds of pressures a voluntary group of public-spirited advisors might impose.[35]

Nevertheless, there were strong forces in favor of establishing a national council, and those forces prevailed. Their Jeffersonian experiment lasted a little over a decade, from 1973 to 1984, with funding coming from foundations (42%), media organizations (29%), corporations (25%), and individuals (4%). As things developed, several major media companies continued to oppose the council. Despite the fact that former CBS News President Richard Salant was at one time president of the council, his influential colleague Walter Cronkite did not agree with the council concept. Neither did ABC; Time, Inc., and NBC were never enthusiastic, and the *New York Times* would not cooperate. The NNC's visibility, funding, and subsequent impact on the nation's major media remained at such a low level that in 1984 it died a quiet death.[36]

The always-controversial, never-influential National News Council folded in 1984. It should not rest in peace.

The council had investigated 242 complaints in its eleven years, of which eighty-two were found to be warranted. The complaints included seventy-eight made against the television networks, sixty-eight against newspapers, twenty-seven against news syndicates, nineteen against AP and UPI, and the remainder against nationally distributed magazines and other media. Many of the complaints were less than earthshaking in nature; most were immediately dismissed as being unwarranted or lacking in specifics, or were readily remedied by NNC staffers before being brought to the full council. All in all, the National News Council's findings were hardly threatening to press freedom in America.

Upon its death, the NNC's complete records were turned over to the University of Minnesota School of Journalism, where hopes remain that the time will come when the Jeffersonian forces for a meaningful national news council will once again be mustered.

Ombudsmen About the same time a few American communities began experimenting with press councils, another concept of media accountability was initiated: the ombudsman system. *News ombudsmen* are neutral arbitrators usually employed by a given medium to accept and investigate audience complaints and to publish corrections and explanations of media policy.

Ombudsmen are neutral, in-house arbitrators, acting on public complaints about the media.

Ombudsmen (or "ombuds practitioners," if we prefer the nonsexist but awkward term) are found in many different organizations—corporations, foundations, international bodies, school systems, governmental offices, professional associations, prisons, colleges, and so forth. Their general functions are:

To give a personal and confidential hearing, to defuse rage, to provide a
    caring presence to those in grief about a dispute;
To provide (and sometimes to receive) information on a one-to-one basis;
To counsel people (confidentially) on how to help themselves, by helping
    to develop new options through problem-solving and role-playing;
To conciliate (as in shuttle diplomacy);
To mediate (bringing people together face-to-face);

To investigate formally or informally, either with or without presenting
    recommendations;

To arbitrate or adjudicate;

To facilitate systems change, by recommending "generic" solutions, by
    upward feedback and "management consulting" within institutions,
    by public reports, by recommendation to legislatures, and by
    supporting education and training.[37]

According to one scholar, the classic description of the ombudsmen's
role is: "They may not make, or change, or set aside any law or policy or
management decision; theirs is the power of reason and persuasion."[38]

At last count, there were forty-one members of the Organization of
News Ombudsmen, or "ONO." ("Oh! No!" is supposedly the most commonly heard expression when the ombudsman comes walking through the
newsroom, because it is usually a sign that he or she is bearing bad tidings
for some reporter or editor.) All but ten of them are from metropolitan
American newspapers which have circulations above 100,000; two are from
ABC and CBS network news, five are from Canadian newspapers, and the
final three are from England, Israel, and Japan. Several of them go by the
title "readers representative." Few of them have identical job descriptions.
Altogether, however, they serve the roles described here.

The first ombudsman on an American paper was appointed for the
Louisville Courier-Journal and *Times* in 1967. He was John Herchenroeder, the assistant to the executive editor, who had been city editor of the
Courier-Journal for twenty years. At the time of his retirement in 1980,
he was receiving and acting upon approximately four thousand complaints
a year dealing with news coverage, advertising, and circulation problems.
Corrections were run under the standing headline, "Beg Your Pardon," and
regular columns by Herchenroeder explained in some detail how the newspapers operated, so readers could get a greater understanding of the mechanics and thought processes involved in daily production.

Many of the current "ONO" members write regular columns for their
newspapers' readers. Highlights of their columns and activities are sum-
marized in *Quill* magazine, the monthly publication of the Society of
Professional Journalists. As one might expect, those columns are filled with
provocative and sometimes juicy ethical dilemmas.

Since 1970 at least a half dozen staff members of the *Washington
Post* have served at different times as ombudsmen. Some of the turmoil
over the ombudsmen's functions at the *Post* have been echoed elsewhere in
the country, as other papers experimented with in-house critics who were
also serving as public watchdogs. In some cases, staffers seemed resentful
of having their newspaper actually pay someone to criticize them, and won-
dered why it was necessary to hang their dirty laundry out in public. The
Washington Post's ombudsman in 1986, Joe Laitin, offered one retort for
reporters or editors who have experienced the sting of the ombudsman. After
Laitin had been rather critical of a *Post* news story, the wary reporter asked

Janet's World

The Story of a Child Who Never Existed — How and Why It Came to Be Published

By Bill Green
Washington Post Ombudsman
© 1981, Washington Post Co.

"Jimmy" never existed, but his story convulsed the city and humiliated The Washington Post—proud house of Watergate investigations.

The story was a lie and, after all its celebrated achievements, The Post owes its readers an accounting of its spectacular failure.

How did it all happen? Why?

This account was prepared from 47 interviews, primarily with members of The Post's staff. It was written by The Post's ombudsman, who is the fifth person to occupy the position of reader representative since it was created by the newspaper in 1970.

This is essentially a story of the failure of a system that, in another industry, might be called "quality control." On newspapers, it is called editing.

The fabrication of Janet Cooke's story eluded all of The Post's filters that are set up to challenge every detail in every news story the paper publishes. From the time she applied for a job, questions were not asked. Editors abandoned their professional skepticism.

This narrative reconstruction suggests that "Jimmy" moved through the cycle of news reporting and editing like an alien creature, unimpeded by ordinary security devices. This account has available the marvelous tool called hindsight.

It is also the story of a young and talented reporter, flaming with ambition, who showed irresistible promise of achievement. The Post accelerated her success, and may have thereby hastened her failure.

Janet Cooke declined to be interviewed for this report, so her version of events is not represented. Where Cooke is quoted directly, the remarks are usually attributable to those in conversation with her.

For Cooke, it was personal tragedy. For The Post, it was inexcusable.

For instance, none of the editors pressed Cooke for confidential details on the identity of "Jimmy" or his family.

Reporters with doubts about the story discussed their skepticism among themselves rather than taking it up the line to editors who might have taken action.

Some Post staff members, however, did express suspicions to editors who weren't listening.

While some reporters and editors were talking up their doubts, The Post's top news executives, unaware of newsroom anxieties, were nominating "Jimmy's World" for journalism's highest award, the Pulitzer Prize.

At the center of all this was Janet Cooke. Whether diligence in the newsroom would have prevented her hoax no one can say. Nevertheless, The Post didn't work hard enough.

The Ombudsman's Report

- **Warning signals were ignored.**
- **Senior editors were uninformed.**
- **Competition for prizes clouded good judgment.**

Washington Post ombudsman Bill Green took his own newspaper to task for its handling of the Janet Cooke episode in 1981.

him, "Is this the beginning of my persecution?" Laitin replied, "No, I am not a persecutor, I am a coroner. I just did an autopsy, and once should be enough."[39]

With ombudsmen like Laitin on the job, it is little wonder staffers are uncomfortable!

Generally, when supported by management, those news media experimenting with ombudsmen have sensed growing public acceptance, manifest in increased credibility.

The acceptance of ombudsmen by management and journalists should lead to increased credibility in the eyes of the public.

Toward Professionalism?

To hear Max Lerner describe it, journalism was but one of many professional organizations attempting to resolve new issues created by emergent conflicting roles in the final quarter of the twentieth century. Each of these organizations was caught in something like an identity crisis:

> But all of them together share with the larger society something of the anguish of the moral crisis of our time . . . The social structures that held the professions in balance internally have broken down. They were personal structures, of family, church, small community, face-to-face contracts, training by apprenticeship. They have been replaced by impersonal, massive aggregates, which may be good schools for developing skills but not for developing character.[40]

Given this identity crisis, it can be argued that journalism must transcend bureaucratic tendencies and concentrate on its original ethic—the ethic of fraternity and common values.[41] Codes of ethics could be a start, given their potential for promoting fraternity, and bringing status to the profession. However, in another sense such codes can be extremely limiting, if they serve to set journalism apart from the very public it serves, insulating journalists from public scrutiny.

In a provocative statement about the place of journalism in society, Association for Education in Journalism President James Carey warned journalists of the dangers of professionalism.[42] To appreciate Carey's concerns over professionalism, we should avoid loose rhetorical usage of the term and consider the sociological foundations of the professions. Then we will recall that the professional:

Questions are raised over whether journalism is, *or ever* should be, *a profession.*

1. works in a closed group and shares certain values and norms with co-workers;
2. works primarily for the satisfaction derived from the work and secondarily for the monetary rewards;
3. unlike the nonprofessional, places high value on prestige, honors, and other types of recognition;
4. places allegiance to the profession rather than to the immediate organization, and
5. wants autonomy in daily work.[43]

Additionally, professionals are recognized by adherence to enforceable canons of performance, by high social status, and by extensive and specialized bodies of knowledge capped by examinations leading to entrance into the profession.[44]

Recognizing these standards for professional status, Carey questioned whether it is proper for journalism to attempt to join the elites. He recognized that the rise of journalism education was a response to the need to train professionals, to establish consistent practices, techniques, and standards by which practitioners could be identified, judged, and ultimately controlled. An increase in legitimacy, according to Carey, came to journalism only once it became taught in the universities, yet the industry's motivations for support of journalism education may be suspect. The business itself stood to gain in public acceptance once it allied itself with the universities instead of print-shop apprenticeships for its products, and the business could assert control over its reporters by instituting standards of writing, reporting, and ethical behavior through professional education.

Journalism thus joined engineering, teaching, and even medicine and law in using universities to train new practitioners. As the professions gained an academic toehold during the twentieth century, they assumed new and more important roles in society. Professional people came to be looked upon as being a cut above average, as though they had answers to complex moral, social, and political questions the laypeople could not answer.

As professions specialize, they codify their own behavior and frequently reinterpret morality so it fits their peculiar needs. For example, secrecy and privilege among doctors and patients, lawyers and clients, and journalists and news sources may lead to morally defective relationships in which the interests of the patients, clients, and news sources, rather than those of the general public, are served.

Professionals codify their morality and gain elite educations. Is that appropriate for journalists? Or are they more concerned about each other than about the public?

Professionals have the tendency to be far more concerned with the way they are viewed by their colleagues than with the way they are viewed by their clients. In journalistic terms, this can be seen time and again when editors and publishers and reporters attend conferences, accepting mutually congratulatory applauses and awards from their peers, frequently not knowing—and perhaps not caring—what their readers or viewers *really* think of them. Such behavior can further isolate professional journalists from their publics. The irony pointed out by Carey is that by publicizing their awards professionals are encouraging the view that their members are people of unusual ability who belong to an elite group and that the public lacks the power to make effective evaluations and criticisms of their work.

"We would, in short, all be better served if professionals, including journalists, were to see themselves less as subject to the demands of their profession and more to the demands of the general moral and intellectual point of view," Carey suggested.[45]

The issue thus returns to the universities, where the general moral and intellectual points of view supposedly are inculcated. The American Council on Education in Journalism and Mass Communication (ACEJMC), which accredits academic journalism programs, long has held to the policy that journalism students should take no more than one-quarter of their

courses within the mass media disciplines, and the remaining three-quarters in liberal arts. In its argument that the business of journalism should be peopled with broadly educated citizens sensitive to the needs of the public, ACEJMC has echoed some of Carey's concerns. However, journalism educators long have recognized that the students who get hired into business are frequently the ones who spend more of the collegiate waking hours pursuing immediate journalism interests, working on the student newspapers or other media where it appears they are getting exactly the kind of training needed to enter the pragmatic marketplace. A traditional liberal education appears less attractive to educators and media students than an applied curriculum infused with *relevance*. Since the 1960s, journalism enrollments have grown exponentially—from 10,664 majors in 1958 to nearly 25,000 in 1967 to over 92,000 in the mid-eighties.

Meanwhile, the voices of Carey and many others are heard crying, "Relevant for what?"

Summary

We have used the complex terms *ethics* and *responsibility* interchangeably. Ethical dilemmas faced by journalists, and research studies on the subject, indicate that there are distinct patterns of performance deemed acceptable by some and unacceptable by others in the business. Junkets and freebies, and payola and plugola, have plagued journalists and broadcasters. For some journalists, such as columnists, opportunities abound to take advantage of the perquisites of the position. Other ethical problems arise when advertisers are given special editorial attention or when public relations firms submit material to newspapers that is printed as news.

Various codes have been created since the 1920s to guide the behavior of media people. These codes are, for the most, harmless statements of ideal conditions, unenforced and unenforceable, drafted by the accused as blandly as possible as a hopeful stopgap to regulation and control, to gain public credibility, or to encourage internal regulation. We could argue that controls appear inevitable once ethics have deteriorated to the point that codified guidance is demanded. Therefore, the codes of the various media industries can be seen as way stations on the road to controls.

Our discussion of these responsibilities of the mass media revolves around the concerns and pronouncements uttered by the Hutchins Commission on Freedom of the Press. Social responsibility is a protective doctrine labeling humanity as lethargic. As such it has authoritarian overtones, because someone—the government, the media, or organizations of the public—is called upon to see that the lethargic populace is prodded and served.

Gradually, though reluctantly, the media have accepted many of the 1947 Hutchins Commission challenges. Information media have broadened their bases, offering more meaningful news and access for comment and criticism. They now offer a somewhat more representative picture of society, and through interpretive reporting they clarify social goals and values. Granted, there is much room for improvement if one sees cries for social responsibility to be calls for improvement in media. The public's right to know is increasingly an agenda item, and numerous efforts to improve media through journalism reviews, press councils, and ombudsmen are seen as thrusts in that direction.

Finally, we raise disturbing questions about the media's general credibility, and the institutionalization and professionalization of journalism. A plethora of studies and anecdotal evidence suggests the general public feels disenfranchised from the media; citizens increasingly are critical of media ethics, insensitivity, inaccuracy, and arrogance. Our concluding question is whether professionalism is a goal to be sought by media practitioners if the price of such professionalism is alienation from the very publics they should be serving and upon whom they depend for their own survival.

Notes

1. From transcript of interview with Phil Donahue on NBC's "Today," 1-2 February 1982.
2. "Winners of This Year's SDX Distinguished Service Awards," *Quill,* April 1979, pp. 14-15.
3. Steve Robinson, "Pulitzers: Was the Mirage a Deception?" *Columbia Journalism Review,* July-August 1979, p. 14.
4. David Gordon, "Chicago Journalists and Ethical Principles," *Mass Comm Review;* Fall 1979, pp. 17-20.
5. See, for instance, the annual *Journalism Ethics Report* of the Society of Professional Journalists/Sigma Delta Chi, in which large scale surveys of SPJ members and other journalists demonstrate disagreement about the seriousness of and solutions to ethical dilemmas.
6. Philip Meyer, summarizing his American Society of Newspaper Editors survey, at Association for Education in Journalism and Mass Communications national convention, Gainesville, Florida, August 1984.

7. Fred Korn, "What's Right? There's no consensus," *1984–85 Journalism Ethics Report* of the Society of Professional Journalists / Sigma Delta Chi, pp. 36–37.

8. Keith P. Sanders and Won H. Chang, "Freebies: Achille's Heel of Journalism Ethics?" (Unpublished paper presented to the Association for Education in Journalism, Mass Communication and Society Division, Logan, Utah, 30 March 1979); and "Codes: The Ethical Free-for-All: A Survey of Journalists' Opinions About Freebies," *Freedom of Information Foundation Series, No. 7,* March 1977.

9. Ibid.

10. Alan Prendergast, "Mickey Mouse Journalism," *Washington Journalism Review,* January-February 1987, pp. 32–35; and Mike Moore, "Disneygate," *Quill,* November 1986, p. 2.

11. Harrison B. Summers, Robert E. Summers, and John H. Pennybacker, *Broadcasting and the Public,* 2d ed. (Belmont, Calif.: Wadsworth Publishing Company, 1978), p. 403 n.

12. John Hohenberg, *The News Media: A Journalist Looks at His Profession* (New York: Holt, Rinehart & Winston, 1968).

13. R. Foster Winans, *Trading Secrets* (New York: St. Martins Press, 1986).

14. Charles Koshetz, "Winans case spotlights ethics policies," *1985–86 Journalism Ethics Report* of the Society of Professional Journalists / Sigma Delta Chi, p. 9.

15. David Shaw, in panel discussion at the Washington Journalism Center and the Organization of News Ombudsmen annual meeting, Washington, D.C., May 1986.

16. Thomas Griffith, "Why Readers Mistrust Newspapers," *Time,* 9 May 1983, p. 94.

17. American Society of Newspaper Editors, *Newspaper Credibility: Building Reader Trust,* a study conducted by MORI Research, Inc., Minneapolis, Minn., April 1985.

18. Alex S. Jones, "Public Views Newspapers With Mixture of Faith and Mistrust, Poll Finds," *New York Times,* 13 April 1985.

19. Times Mirror, *The People & The Press: An investigation of public attitudes toward the news media,* conducted by the Gallup Organization, Los Angeles, January 1986.

20. D. Charles Whitney, *The Media and the People—Americans' Experience with the News Media: A Fifty-Year Review* (New York: Gannett Center for Media Studies, November 1985).

21. Everette Dennis, "The Politics of Media Credibility," (unpublished lecture delivered at Journalism Ethics Institute, Washington and Lee University, Lexington, Va., 28 March 1986).

22. Clifford G. Christians, "Mass Media Ethics: An Analysis of the Academic Literature," (unpublished paper presented to Association for Education in Journalism, College Park, Maryland, August 1976).

23. Edwin Emery and Michael Emery, *The Press and America: An Interpretative History of the Mass Media,* 4th ed. (Englewood Cliffs, N.J.: Prentice-Hall, 1978), p. 511.

24. Upton Sinclair, *The Brass Check: A Study of American Journalism* (Pasadena, Calif.: Published by the author, 1920), p. 436.

25. See annual *Journalism Ethics Reports,* published by Society of Professional Journalists/Sigma Delta Chi, throughout the 1980s. Also see the special theme issue on mass media codes of ethics in the *Journal of Mass Media Ethics,* Vol. 1, No. 1, Fall/Winter 1985-86. In particular note Clifford Christians, "Enforcing Media Codes," pp. 14-21; Deni Elliott, "A Conceptual Analysis of Ethics Codes," pp. 22-26; and Jay Black and Ralph Barney, "The Case Against Mass Media Codes of Ethics," pp. 27-36.

26. Casey Bukro, "The SPJ Code's Double-Edged Sword: Accountability and Credibility," *Journal of Mass Media Ethics,* Vol. 1, No. 1, Fall/Winter 1985-86, pp. 10-13, and the *1986-87 Journalism Ethics Report* of the Society of Professional Journalists, Sigma Delta Chi.

27. Warren K. Agee, "The Joining of Journalists," *Quill,* November 1978

28. Commission on Freedom of the Press, *A Free and Responsible Press* (Chicago: University of Chicago Press, 1947).

29. Theodore Peterson, "The Social Responsibility Theory of the Press," *Four Theories of the Press* by Fred S. Siebert, Theodore Peterson, and Wilbur Schramm (Urbana, Ill.: University of Illinois Press, 1963), pp. 93-94.

30. John C. Merrill, *The Imperative of Freedom: A Philosophy of Journalistic Autonomy* (New York: Hastings House, 1974), p. 3.

31. William L. Rivers, et al., *Backtalk: Press Councils in America* (San Francisco: Canfield Press, 1972), p. 14.

32. Donald M. Gillmor, "Press Councils in America," in *Enduring Issues in Mass Communication,* ed. Everette E. Dennis, Arnold H. Ismach, and Donald M. Gillmor (St. Paul, Minn.: West Publishing Co., 1978), p. 342.

33. J. Herbert Altschull, *Agents of Power: The Role of the News Media in Human Affairs* (New York: Longman, 1984).

34. Creed C. Black, "A flawed idea whose time had not come," *ASNE Bulletin,* August 1984, p. 23.

35. Robert A. Logan, "Jefferson's and Madison's Legacy: The Death of the National News Council," *Journal of Mass Media Ethics,* Vol. 1, No. 1, Fall/Winter 1985-86, pp. 68-77.

36. Andrew Rodolf, "National News Council Folds," *Editor & Publisher,* 31 March 1984, pp. 9, 28; "News Council to Fold," *Broadcasting,* 26 March 1984, p. 30; Thomas Griffith, "Watchdog Without a Bite," *Time,* 9 April 1984, p. 103.

37. Mary P. Rowe, "Notes on the Ombudsman in the U.S., 1986," (unpublished paper presented to the Washington Journalism Center and the Organization of News Ombudsmen annual meeting, Washington, D.C., May 1986).

38. Ibid.

39. Anecdote by Joe Laitin, at the Washington Journalism Center and the Organization of News Ombudsmen annual meeting, Washington, D.C., May 1986.

40. Max Lerner, "The Shame of the Professions," *Saturday Review,* 1 November 1975, p. 10.

41. Clifford Christians, "Mass Media Ethics."

42. James W. Carey, "A Plea for the University Tradition" (AEJ Presidential Address, Seattle, Washington, August 1978), *Journalism Quarterly,* Winter 1978, pp. 846-55.

43. Everette Dennis, *The Media Society,* p. 98.

44. See theme issue on mass media professionalism, *Journal of Mass Media Ethics,* Vol. 1, No. 2, Spring/Summer 1986, esp. Marianne Allison, "A Literature Review of Approaches to the Professionalism of Journalists," pp. 5-19; William F. May, "Professional Ethics, the University, and the Journalist," pp. 20-31; Louis W. Hodges, "The Journalist and Professionalism," pp. 32-36; Douglas Birkhead, "News Media Ethics and the Management of Professionals," pp. 37-46; Arthur J. Kaul, "The Proletarian Journalist: A Critique of Professionalism," pp. 47-55; and John C. Merrill, "Journalistic Professionalization: Danger to Freedom and Pluralism," pp. 56-60.

45. James Carey, "A Plea for the University Tradition," p. 853.

Glossary

a

AAAA
American Association of Advertising Agencies; national trade organization that has developed extensive standards of good practice to guide agencies in producing ethical advertising

ABC
American Broadcasting Company

account executives
ad agency personnel who maintain a constant liaison with clients, mediating between clients and the rest of the ad agency personnel

A. C. Nielson Company
the largest commercial ratings service used by broadcasters; makes extensive use of audimeters, diaries, and people meters; its primary work is with networks

adjacencies
spot announcements before, during, and after network programming

advertising
any paid form of nonpersonal presentation of ideas, goods, or services by an identified sponsor

affiliate
radio or television station under contract with one of the three commercial broadcast networks (ABC, CBS, NBC); receives but does not have to air programming from its particular network

AM
amplitude modulation

ANPA
American Newspaper Publishers Association

Arbitron Company (ARB)
commercial ratings service used by broadcasters; makes extensive use of diaries and "telephone coincidental surveys"; its primary work is with individual stations

ascertainment
FCC regulations demanding that broadcasters seeking licenses determine the needs, interests, and problems of the communities they hope to reach

ASNE
American Society of Newspaper Editors

Associated Press (AP)
oldest American news service; a cooperative with members

audiences
large, anonymous, and heterogeneous masses of individuals attending to mass communications

audimeters
tuning meters with recording circuits attached to in-home TV sets to gauge amount and type of TV viewing

audion tube
a vacuum tube that made voice transmission possible; invented by Lee De Forest in 1906

authoritarianism
political system under which communication flows from the top down; centralized control of media is inherent

b

blacklisting
practice in Hollywood in the late 1940s and early 1950s, during "Red Scare" period, by which filmmakers, performers, and others who were accused of left-wing or communistic leanings were ostracized by Hollywood and unable to work in media

black weeks
periods during the year when no ratings are made of TV shows; during these weeks stations and networks tend to air public affairs shows and other programs that do not draw large audiences

blind bidding
film marketing system that demands that exhibitors sign contracts for films before the films are released or, sometimes, prior to production

block booking
film marketing system in which film exhibitors are forced to contract for a number of motion pictures from a studio, including the studio's B-grade films, in order to get the studio's better films for exhibition; the practice has been held to be illegal

brand image
advertisers' attempt to establish a distinction among products on the basis of snob appeal

brand name advertising
one of the earliest advertising techniques, attempting to establish the brand name (Coca-Cola, Kleenex, etc.) as synonymous with the generic product or with quality

budget
news services' suggested top stories to be transmitted to clients/members during each news cycle

c

Canons of Journalism
code of ethics developed by the American Society of Newspaper Editors in 1923 and revised in 1975, stressing responsibility of journalists

CATV or community antenna television
cable television system by which a town or area wired with coaxial cable receives distant TV signals boosted by a local cable company

CBS
Columbia Broadcasting System

chain ownership
two or more newspapers published in different communities but owned by the same company; term used interchangeably with group ownership

channel noise
interference within or external to the medium; physical or mechanical barriers to effective communication

classification
certain kinds of information the government has labeled secret or top secret and withholds from the press and public

classified advertising
sales or merely informational notices, generally nonillustrated, concentrated in one section of a newspaper or magazine

clipping book
collection of articles from newspapers reflecting favorably upon a public relations person's client; used as an unscientific measure of public relations effectiveness

codex
a system by which sheets of papyrus or parchment were tied by cords between wood boards; invented by the Romans in the fourth century A.D.

cold type
see offset printing

commission system
method used by advertising agencies, who retain about 15 percent of a client's payment to the media in exchange for producing and placing the ads

communication
a stimulus-response process involving a source and human receiver

communications
the means—tools and mechanics—with which the communication process takes place

consent decree
agreement between United States Department of Justice and Hollywood film studios in which studios consented to give up control over film distribution to their own theaters, thereby breaking up studios' production and distribution monopoly

controlled circulation
print media, generally magazines or newsletters, distributed free of charge to preselected groups and underwritten by advertisers or sponsors

copyright
body of law protecting authors, publishers, and other media producers from having their words illegally reproduced

corrective advertising
means by which advertisers found guilty of false or misleading advertising are required by the FTC to use a certain percentage of their future advertising to acknowledge and rectify the earlier misrepresentation

cost per thousand (CPM)
the cost to a mass medium or advertiser of reaching each one thousand members of the audience

CPB
Corporation for Public Broadcasting

CPM
see cost per thousand

d

democratic socialism
a modified form of social responsibility theory emerging from a combination of Marxist thought and the writings of classical libertarian philosophers

demographic breakouts
specialized sections or editions of newspapers or magazines intended to appeal to particular geographic, ethnic, or economic parts of each medium's audience

demographics
readily measurable characteristics of audiences such as age, sex, race, income, and level of education

diary
rating system in which audiences keep written records of their TV and radio usage

Direct Broadcast Satellite (DBS)
communications satellites emitting powerful signals that can be received on small and relatively inexpensive rooftop antennas

display advertising
showcase advertisements occupying considerable space and distributed throughout a newspaper or magazine

documentation
FTC requirement that advertisers fully document the claims made in their advertising

dummy
a diagram or mock-up of a print-medium page, indicating where illustrations, advertising, news and so forth are to be placed

e

electromagnetic waves
electrical impulses used to transmit radio and TV signals

electronic newsroom
contemporary newspaper offices employing video display terminals and computers for reporting, editing, and producing the newspaper

ENG
electronic newsgathering equipment, such as mini-cams and microwave relays

equal time
FCC's Section 315 of the 1934 Communications Act providing that broadcasters provide the same amount of air time on the same terms or at the same cost for all bona fide candidates for political office

f

fairness doctrine
FCC provisions calling for airing of both sides of controversial issues of public importance

family viewing time
provisions of National Association of Broadcasters (NAB) calling for eliminations of violent or sexually oriented programming for the early evening hours when children are most likely to be in the audience

feature syndicates
packagers of entertainment—in verbal and pictorial form—primarily for newspaper clients

Federal Communications Commission (FCC)
broadcast regulatory agency established in 1934 to oversee licensing and operation of all wire and radio communication and to assure stations operated in the "public interest, convenience, and necessity"

Federal Radio Commission (FRC)
broadcast regulatory agency established in 1927, predecessor to FCC

feedback
reactions, either immediate or delayed, of communication receivers; feedback conditions future communication by limiting options of participants

fiber optics
tiny flexible strands of glass capable of carrying beams of light or electromagnetic energy around curves and corners; may revolutionize cable television because of its size, cost, information-carrying capacity, and versatility

First Amendment
Constitutional guarantee of press, speech, and religious freedom; it reads: Congress shall make no law respecting an establishment of religion, or prohibiting the free exercise thereof; or abridging the freedom of speech, or of the press; or the right of the people peaceably to assemble, and to petition the Government for a redress of grievances

FM
frequency modulation; broadcasts over line-of-sight signals

four-walling
film marketing system in which distributors rent theaters directly from theater owners and keep all the box-office receipts for the limited engagements of their films; such films are generally heavily promoted on television and released simultaneously in carefully selected areas

freebies
complimentary tickets, passes, books, records, and so forth offered to journalists by news sources who hope to receive favorable press treatment

free press/fair trial
controversies centering around the media's first amendment rights to gather and report news and the sixth amendment's guarantee of a fair trial unbiased by journalistic coverage

FTC
Federal Trade Commission; national regulatory agency overseeing such trading issues as truth in packaging, lending, and advertising

functions of communication
information, entertainment, persuasion, and transmission of culture

g

gag orders
court-ordered limitation to what attorneys and court officials can say to the news media and, in turn, what the media can publish or broadcast

galley proof
the first printed version of a book, magazine, or newspaper article, which is checked for errors before final production

gatekeepers
people who determine what will be printed, broadcast, produced, or consumed in the mass media

group ownership
see chain ownership

h

hard news
accounts of "significant" news events, intended to inform and educate audiences about the "real world"

HomeComCen
home communications center, which in its fuller form today might consist of microcomputer, interactive cable television, VCR and videodisc systems, home satellite-receiving dish, telephone interface, and a facsimile printer

hot type
see letterpress

Hutchins Commission
informal but influential 1940s Commission on the Freedom of the Press, headed by Robert M. Hutchins, whose review of journalism enunciated the social responsibility theory of the press

hype
heavy promoting of products or performers, frequently by press agents who are seeking media and public attention

i

iconoscope
camera tube in Zworykin's original television system

institutional advertising
public relations advertising, in which corporate or governmental agencies attempt to improve their image at the same time they promote their products

International News Service (INS)
news service founded by William Randolph Hearst in 1909, noted for its lively, sensational, and international news coverage. INS in 1958 merged with UP to form UPI

invasion of privacy
published or broadcast information that violates an individual's right to be let alone or damages one's peace of mind

inverted pyramid
a style of organizing a news report so that the essential information appears at the outset, with information of decreasing importance following in descending order; sometimes called "journalese"

j

jazz journalism
early twentieth-century newspaper practices aimed at increasing circulation; gawdy pictures, tabloid format, splashy and sensationalistic writing

joint operating agreement
practice by which two newspapers in the same city share printing presses and production equipment in order to reduce costs

journalese
see inverted pyramid

journalism review
formalized media, usually magazines, for self-criticism and public criticism of journalism

junkets
free trips offered to journalists, with the expenses picked up by news sources who hope to receive favorable press treatment

k

kinescope
(1) Zworykin's cathode ray tube television receiver carrying an image consisting of thirty horizontal lines; (2) in the 1950s, a means of taking films of TV programs directly from the TV screen

kinetoscope
Edison's earliest motion picture projector system, invented in the 1880s

l

least objectionable programming (LOP)
a theory that holds that a TV program need not be good or even highly popular to succeed, but that it should be less objectionable—and therefore draw more viewers—than the competition

letterpress
"hot type" typesetting process involving raised and indented lettering and pictures whose images are transferred directly onto paper during the printing run

libel
published or broadcast information that damages an individual's reputation

libertarianism
political system under which communications flows in all directions, without central control of media

licensing
means by which, under authoritarian government, only those media operating in accord with governmental policies will receive permission to publish or broadcast

lowest common denominator (LCD)
the largest audience likely to consume a given media production; implication is that demographic and psychographic characteristics of such an audience result in the LCD accepting rather bland and inoffensive programming

low power television (LPTV)
ruling recently approved by FCC by which LPTV or "drop-in TV" channels are to serve special interest or geographically limited audiences

m

mass communication
The process whereby mass-produced messages are transmitted to large, anonymous, and heterogeneous masses of receivers. Used in the collective singular, the term refers to broad, theoretical considerations, whereas mass communications (with an *s*) refers to the mechanics or media involved in achieving the process

mass communications
the mechanical means whereby mass-produced messages are transmitted to large, anonymous, and heterogeneous masses of receivers; the term is synonymous with mass media

mass media
see mass communications

modem
electronic devices permitting computers to communicate over telephone lines

MOR
middle-of-the-road radio music formats, more recently called "adult contemporary"

motivational research (MR)
highly Freudian sociopsychological advertising techniques, based on the idea that people frequently purchase and use products to satiate needs and desires they themselves do not fully understand; the motivations are most frequently sexual in nature

MPAA
Motion Picture Association of America, established in 1945 to replace the MPPDA; its 1968 Code classifies movies into G, PG, PG–13, R, and X categories

MPPDA
Motion Picture Producers and Distributors of America, a national trade organization begun in the 1930s to assure that Hollywood films were socially acceptable; it drew up the Code of the Motion Picture Industry

muckrakers
journalists, magazine writers, and book authors who uncovered corruption and questionable practices of businesses and government

Multipoint Distribution Service (MDS)
a variety of over-the-air TV that uses a microwave signal to transmit pay programs, generally to a master antenna serving a motel or apartment building or, occasionally, an entire city

n

NAB
National Association of Broadcasters; voluntary trade organization

narrowcasting
news, entertainment, advertising, and other programming produced for a small, self-selective segment of the electronic media's audience

NBC
National Broadcasting Company

negative option
marketing technique whereby books, records, or tapes are automatically sent to members unless they specifically decline each offer

neovideo
a generic term to describe new electronic communication technology

news hole
the amount of space remaining in the newspaper after advertising and all the standard feature (nonnews) material has been allocated

news management
general term describing the various techniques governments have developed to control access of reporters and the public to news about governmental activities; backgrounders, off-the-record interviews, not-for-attribution statements, and trial balloons are among the techniques utilized

newspaper
a regularly issued, geographically limited print medium, which serves the general interest of a specific community; printed on unbound newsprint, it commonly contains news, comment, features, photographs, and advertising

nickelodeon
early film-exhibition halls or theaters; admission cost was a nickel

NPR
National Public Radio, network made up of noncommercial stations by the Corporation for Public Broadcasting

o

objectivity
journalistic goal of reporting news without bias or color, staying with the literal, provable truth even if further independent investigation would suggest the reporter should give some interpretation of the news

offset printing
"cold type" reproduction system by which copy is reproduced on a smooth, photographic plate; the printing process occurs when the plate passes over a series of rollers and ink is transferred onto paper indirectly from the plate; it tends to be cleaner printing than the older letterpress or raised-lettering processes

ombudsmen
in-house journalism critics, employed by the newspapers or broadcast stations, who mediate between public and journalists when the press is criticized

op-ed pages
the page opposite the editorial page where guest opinions usually appear

owned-and-operated (O & O)
broadcast outlets owned and operated by the networks; by FCC rules, each network is allowed only twelve O & Os, no more than seven of which can be the generally more profitable VHF (very high frequency) stations

p

participating spots
isolated commercial messages inserted into broadcast programs by national or local advertisers

pay cable
services such as Home Box Office or sports programs for which cable TV viewers pay an extra fee, usually on a monthly basis

payola
offering of money or other bribes to disc jockeys to encourage them to play a particular recording

pay TV
generic term for pay-cable TV, subscription TV, and Multipoint Distribution Services—all services for which television viewers pay an extra fee

PBS
Public Broadcasting Service

penny press
American newspapers, beginning in the 1830s, that lowered their street sales price to one cent, attracting such large circulations that advertisers were willing to underwrite most of the cost of producing the paper to reach those masses

people meter
state of the art television ratings service; a remote control handset records each family member's viewing habits

persistence of vision
psychological principle upon which the motion picture is based, that is, that the eye retains an image fleetingly after it is gone

playlist
list of popular songs being played by radio stations

plugola
free boosts or promotional "plugs" made by program writers, hosts, or guests who are frequently paid by manufacturers who wish to gain exposure for their products

positioning
recent advertising techniques of aiming ads, and products, at specific demographic—and psychographic subgroups within the larger mass audience

press agentry
a form of public relations whose goal is to gain free publicity in the news media for a particular client, often a performer or artist

press councils
external advisory boards that review press performance and serve as a sounding board for public complaints against the press

prior restraint
censorship, through licensing, taxation, or another means, by which media are limited prior to being printed or distributed

PRSA
Public Relations Society of America; professional trade association established to elevate and professionalize the practice of PR

PSA
public-service announcements, print ads, or broadcast commercials for nonprofit institutions, usually carried free of charge

pseudoevent
term coined by historian Daniel Boorstin to describe newsworthy events that have been purposively created to be filmed or reported by journalists; such events would not occur if the media were not there to report them

psychographics
internal characteristics of audiences such as values, needs, beliefs, and interests

public relations
the broad "umbrella" of mass communications including publicity, opinion research, promotion, press agentry, advertising, lobbying, and political campaigning; frequently behind-the-scenes, subtle communications

puffery
gratuitous exaggerations about individuals or products, found in advertising, public relations releases, and news stories

Pulitzer prizes
journalistic and literary awards given annually to the nation's most outstanding writers; prizes were established by publisher Joseph Pulitzer

r

ratings
a measure of what percentage of all 86 million American TV households happen to be watching a particular show at a given time in comparison with anything else people might choose to be doing

RCA
Radio Corporation of America, parent company to National Broadcasting Company (NBC)

rip 'n' read
practice of radio newscasters who take wire-service news directly from the teletype and read it over the air, without rehearsing or localizing the news

s

Satellite Master Antenna TV (SMATV)
hotels, motels, and apartment or condominium complexes use a large satellite dish to receive nationally distributed TV signals; the signals are split to serve tenants, and additional over-the-air broadcast stations are included in the package

scatter plan
advertisers' practice of purchasing time for their commercials to appear during many different programs, at different times on different networks

sedition
criticism of the state or rulers

selective contract adjustments
system by which theater owners who cooperate with distributors in promoting and running their films are given preferential treatment

semantic noise
interference within the communication process, within the human sources and receivers; psychological or language barriers

sequestration
locking up a jury or otherwise shielding jurors from potentially manipulative information about the court case they are hearing

share
a measure of what percentage of all the TV sets that are turned on at a given time happen to be tuned to each show; thus, the "share of the audience" at any point in the day

shield laws
laws that protect journalists from having to tell law enforcement agencies the identity of their confidential sources

slogans
catchy capsule summaries of given products or political candidates, used in advertising or political campaigns

social responsibility
twentieth-century political theory holding that instruments of business or communications have responsibilities as well as freedoms

soft news
accounts of "insignificant" events or ideas, frequently intended to amuse or lend generalized insights into the human condition

Soviet totalitarianism
political theory holding that media are branches of the government, aiding the control of information and propaganda

SPJ/SDX
The Society of Professional Journalists, Sigma Delta Chi; largest national organization of working reporters and editors

split run
press run in which half of a given newspaper or magazine edition carries one form of an advertisement, the other half a different form so relative appeal of the ads can be gauged

star system
Hollywood's method of promoting individual actors and actresses to assure an eager audience for their film, regardless of the quality of a given film

storyboards
static graphic depictions of television commercials, produced by ad agency creative staff

subliminal advertising
advertising intended to motivate consumers' unconscious minds; ad techniques include "embeds," hidden messages in verbal and written ads

subscription TV (STV)
over-the-air pay television, using satellite signals, for which viewers pay an additional fee each time they watch a special program, film or sports event

subsequent punishment
a form of control over the media, subjecting them to regulations after publication; laws of libel, for instance, are used to remedy damages that media may have caused by publishing and distributing harmful news and commentary

summary judgment
a judgment granted to a party in a legal dispute when the judge feels there is no significant factual basis for a jury trial

superstation
television station that distributes its signals to a wide area by means of satellites

t

tabloid
technically, the half-sized magazine style newspaper; in common terms, *tabloid* refers to splashy, sensationalistic journalism

Taffies
technologically advanced families, who have access to a variety of neovideo: home computers, VCRs, etc.

talking head
derogatory term to describe dull TV shows, usually on public television, filled with discussion and no action

teletext
information system encoded on an unused space on ordinary TV signal; offers a series of indexed information to viewers who must make selection with a control panel, but does not allow true interactive or consumer-generated information transfer

tie-in
marketing technique by which one medium (book, TV show, film, etc.) is produced and heavily promoted in conjunction with another medium

trade book
books produced for mass or general readership, usually sold through bookstores

u

UHF
Ultrahigh frequency allocations on the TV band; channels 14 through 70

unique selling proposition (USP)
advertising strategy based on the principle that advertising itself must establish in customers' minds whatever difference there is between essentially identical products. Usually the USP is a dramatization or exploitation of some characteristic of the product, even though other products share that characteristic

United Press (UP)
private news service noted for human interest, personalized journalism founded by E. W. Scripps in 1906; in 1958 it merged with Hearst's International News Service to become the UPI

United Press International (UPI)
formed in 1958 from United Press and International News Service; a private news service with clients

USIA
United States Information Agency; American government's official propaganda and information agency dealing with United States image abroad

v

value-added theory
Martin Mayer's theory that advertising adds a vague dimension to a product, such as raising the consumer's hopes or granting self-fulfillment

variable obscenity
what is obscene for one person or group under one set of circumstances may not necessarily be obscene for others under different circumstances

VDT
video display terminals; typewriter keyboards and cathode ray tube screens used by reporters and editors; part of the electronic phototypesetting system

VHF
Very high frequency allocations on the TV band; channels 2 through 13

videocassette recorder (VCR)
telecommunications equipment using videotape to record and play back live scenes or programs from television

videodiscs
prerecorded discs that, when played over a videodisc player and shown on a standard television set, allow a variety of viewer controls, such as slow motion, stop action, instant search, and reviewability

videotex
information system providing for consumer control over interactive cable or telephone lines; can be used for information storage and retrieval, in-home banking and shopping, and other interactive communication

w

wireless
system developed by Marconi for transmitting signals over the airwaves, without use of wires; early radios were called "wireless"

y

yellow journalism
sensationalistic journalism in the 1880 to 1900 period, typified by the circulation-building gimmicks of Hearst and Pulitzer

Bibliography

a

"ABC Audits *USA Today*." *Publishers Auxiliary*, 2 July 1984.

Abel, Elie, ed. *What's New: The Media in American Society*. San Francisco: Institute for Contemporary Studies, 1981.

A. C. Nielsen Company. "Audience Research." In *TV Guide Almanac*, edited by L. Craig, T. Norback and Peter G. Norback. New York: Ballantine Books, 1980.

Adams, Russ. "Cashing in on electronic mail systems." *Business Software*, June 1984.

Adler, Richard, ed. *Understanding Television: Essays on Television as a Social and Cultural Force*. New York: Praeger, 1980.

Agee, Warren K. "The Joining of Journalists." *Quill*, November 1978.

Agee, Warren K.; Ault, Phillip; and Emery, Edwin, eds. *Perspectives on Mass Communications*. New York: Harper & Row, 1982.

A Look at USA Today, A Changing Newspaper Committee Report. A Report to the Associated Press Managing Editors Association, November 1983.

Altheide, David L. *Media Power*. Beverly Hills: SAGE Publications Inc., 1985.

Altschull, J. Herbert. *Agents of Power: The Role of the News Media in Human Affairs*. New York: Longman, 1984.

American Newspaper Publishers Association, *Facts About Newspapers, '87*. Washington, D.C.: American Newspaper Publishers Association, May 1987.

American Newspaper Publishers Association News Research Report No. 13, 12 July 1978.

American Press Institute, *The Public Perception of Newspapers. Examining Credibility*. Reston, Va.: American Press Institute, 1984.

American Society of Newspaper Editors and American Newspaper Publishers Association Foundation. *Free Press & Free Trial*. Washington, D.C.: The Newspaper Center, 1982.

Anderson, Jack. "FBI Pursues Author of H-bomb Story." Syndicated column, *Logan* (Utah) *Herald Journal*, 29 May 1979.

Andrews, Peter. "The Birth of the Talkies." *Saturday Review*, 12 November 1977.

Armstrong, David. *A Trumpet to Arms: The Alternative Press in America*. Los Angeles: Torcher, 1981.

Aronson, Steven M. L. *Hype*. New York: William Morrow and Company, Inc., 1983.

"A short course in broadcasting, 1987." *Broadcasting Cablecasting Yearbook, 1987*.

"AT&T Prophesies a $15 billion market for videotex." *Marketing News*, 26 September 1984.

Atlas, James. "Beyond Demographics: How Madison Avenue knows who you are and what you want." *The Atlantic Monthly*, October 1984.

Atwan, Robert; Orton, Barry; and Vesterman, William, eds. *American Mass Media: Industries and Issues*. New York: Random House, 1978.

b

Bagdikian, Ben H. *The Effete Conspiracy and Other Crimes by the Press*. New York: Harper & Row, 1972.

Bagdikian, Ben H. "Fast-food news: a week's diet." *Columbia Journalism Review*, March/April 1983.

Bagdikian, Ben H. *The Information Machines: Their Impact on Men and the Media*. New York: Harper & Row, 1971.

Bagdikian, Ben H. "Newspaper Mergers—The Final Phase." *Columbia Journalism Review*, March/April 1977.

Baird, Kathleen Hunt. "P.M. to A.M.: Is a Trend Building?" *Presstime* (journal of the American Newspaper Publishers Association), December 1979.

Baker, John F. "1984: The Year in Review." *Publishers Weekly*, 15 March 1985.

Baker, Kent. "Cable TV Reshapes Sports." *The Press*, June 1982.

Baldwin, Brad. "Dow Jones News/Retrieval." *Infoworld*, 30 April 1984.

Barnouw, Erik. *The Golden Web: A History of Broadcasting in the United States from 1933 to 1952*. New York: Oxford University Press, 1968.

Barnouw, Erik. *The Image Empire: A History of Broadcasting in the United States Since 1953*. New York: Oxford University Press, 1970.

Barnouw, Erik. *A Tower in Babel: A History of Broadcasting in the United States to 1933*. New York: Oxford University Press, 1966.

Barnouw, Erik. *Tube of Plenty: The Evolution of American Television*. New York: Oxford University Press, 1975.

Bauer, Raymond A., and Bauer, Alice H., "America, 'Mass Society,' and Mass Media." *Journal of Social Issues*, Vol. 16, 1960.

Baus, Herbert M., and Ross, William B. *Politics Battle Plan*. New York: Macmillan, 1968.

Bell, D. *The Coming of Post-industrial Society*. New York: Basic Books, 1976.

Bell, D., ed. *Toward the Year 2000: Work in Progress*. Boston: Beacon Press, 1967.

Berelson, Bernard. "Communication and Public Opinion." In Wilbur Schramm, ed. *Mass Communication*. Urbana: University of Illinois Press, 1949.

Bergreen, Laurence. "Just Don't Get Booked After the Animal Act . . ." *TV Guide*, 17 March 1979.

Bernays, Edward. *Crystallizing Public Opinion*. New York: Liveright, 1961.

Bernstein, Carl, and Woodward, Bob. *All the President's Men*. New York: Simon & Schuster, 1974.

Berstein, Peter W. "Psychographics is still an issue on Madison Avenue." *Fortune*, 16 January 1978.

Bettinghaus, Erwin P. *Persuasive Communication*. New York: Holt, Rinehart & Winston, 1973.

Bezanson, Randall P.; Cranberg, Gilbert; and Soloski, John. "Libel Law and the Press: Setting the Record Straight." *Iowa Law Review*, October 1985, Vol. 71, No. 1; and *The 1985 Silha Lecture*, University of Minnesota, 15 May 1985.

Bittner, John R. *Mass Communication: An Introduction.* 4th ed. Englewood Cliffs, N.J.: Prentice-Hall, 1986.

Bittner, John R. *Professional Broadcasting: An Introduction.* Englewood Cliffs, N.J.: Prentice-Hall, 1981.

Black, Creed C. "A flawed idea whose time had not come." *ASNE Bulletin,* August 1984.

Blake, Reed H., and Haroldsen, Edwin O. *A Taxonomy of Concepts in Communication.* New York: Hastings House, 1975.

Blumer, Herbert. "Elementary Collective Groupings." In Robert E. Park, ed., *An Outline of the Principles of Sociology.* New York: Barnes and Nobel, 1939.

Blumer, Jay G. "Information overload: is there a problem?" In Eberhard Witte, ed., *Human Aspects of Telecommunication.* New York: Springer-Verlag, 1980.

Blumer, Jay G., and Katz, Elihu, eds. *The Uses of Mass Communications: Current Perspectives on Gratification Research.* Beverly Hills, Calif.: SAGE, 1974.

Bogart, Lee. *Press and Public: Who Reads What, When, Where, and Why in American Newspapers.* Hillsdale, N.J.: Lawrence Erlbaum Associates, Publishers, 1981.

Boorstin, Daniel J. *The Image: A Guide to Pseudo-Events in America.* New York: Harper & Row, Harper Colophon Books, 1964.

Bowen, Ezra. "A Debate Over 'Dumbing Down.' " *Time,* 3 December 1984.

Bower, Robert T. *Television and the Public.* New York: Holt, Rinehart & Winston, 1973.

Boyd-Barrett, Oliver. *International News Agencies.* Beverly Hills, Calif.: SAGE 1980.

Brasch, Walter M., and Ulloth, Dana R. *The Press and the State: Socio-historical and Contemporary Studies.* Lanham, Md.: University Press of America, 1986.

Bronson, Gail. "Videotapes give Hollywood a second shot at success." *US News & World Report,* 4 March 1985.

Brown, Charlene; Brown, Trevor; and Rivers, William. *The Media and the People.* New York: Holt, Rinehart & Winston, 1978.

Brown, J. A. C. *Techniques of Persuasion: From Propaganda to Brainwashing.* Harmondsworth, Middlesex, England: Penguin Books, 1963.

Brown, Les. *Keeping Your Eye on Television.* New York: The Pilgrim Press, 1979.

Bukro, Casey. "The SPJ Code's Double-Edged Sword: Accountability and Credibility." *Journal of Mass Media Ethics,* Vol. 1, No. 1, Fall/Winter 1985–86; and the *1986–87 Journalism Ethics Report* of the Society of Professional Journalists, Sigma Delta Chi.

Bullion, James Stuart. "Truth, Freedom and Responsibility: Seeking Common Ethical Ground in International News Work." *Journal of Mass Media Ethics,* Vol. 1, No. 1, Spring/Summer 1986.

Burgess, Anthony. "TV is Debasing Your Lives." *TV Guide,* 18 September 1982.

Busby, Linda, and Parker, Donald. *The Art and Science of Radio.* Boston: Allyn and Bacon, Inc., 1984.

C

"Cable Advertising in First Quarter: WTBS Gets Half." *Broadcasting,* 10 May 1982.

"Cable Opportunities Beyond the Horizon of Entertainment." *Broadcasting,* 10 May 1982.

"Cable Porn Law Unconstitutional." *The News Media & The Law,* Fall 1986.

"Cable Restraints Dropped." *NAB Radio/ TV Highlights,* 28 July 1980.

"Cable Throws a Party in Las Vegas." *Broadcasting,* 10 May 1982.

Canape, Charlene. "The Chase Is On: Can *TV-Cable Week* Catch *TV Guide?*" *Washington Journalism Review,* June 1983.

Cantor, Muriel G. *Prime-Time Television: Content and Control.* Beverly Hills, Calif.: SAGE 1980.

Cantril, Hadley. *The Invasion From Mars: A Study in the Psychology of Panic.* Princeton, N.J.: Princeton University Press, 1940.

Carey, James W. "A Plea for the University Tradition." Association for Education in Journalism presidential address, Seattle, August 1978, published in *Journalism Quarterly,* Winter 1978.

Cassata, Mary B., and Asante, Molefi K., eds. *Mass Communication: Principles and Practices.* New York: Macmillan, 1979.

Castro, Janice. "Reuters' $1.5 Billion Bonanza." *Time,* 6 February 1984.

Castro, Janice. "Selling Off A Magazine Empire." *Time,* 3 December 1984, pp. 62–63.

Charren, Peggy, and Krock, Robert. "Program-Length Commercials Turn Kiddies' TV Into Toy Store." *Variety,* January 1986.

Chin, Kathy. "Keyfax service to hit Chicago." *Infoworld,* 8 October 1984.

Chiu, Tony. "MSN, MPC and OPT." *Panorama,* March 1980.

Christians, Clifford G. "Mass Media Ethics: An Analysis of the Academic Literature." Unpublished paper presented to Association for Education in Journalism, College Park, Md., August 1976.

Cirino, Robert. *Don't Blame the People: How the News Media Use Bias, Distortion, and Censorship to Manipulate Public Opinion.* Los Angeles: Diversity Press, 1971.

City of Los Angeles v. Preferred Communications, Inc., 106 S. Ct. 2034 (1986).

Clark, Ruth. *Relating to Readers in the '80s.* Washington, D.C.: American Society of Newspaper Editors, 1984.

Cocks, Jay. "Come On, Let's Get Banglesized!" *Time,* 26 December 1983.

Cocks, Jay. "Sing a song of seeing." *Time,* 26 December 1983, pp. 54–55.

"Codes: The Ethical Free-for-All: A Survey of Journalists' Opinions About Freebies." *Freedom of Information Foundation Series, No. 7,* March 1977.

Cohen, Bernard. *The Press and Foreign Policy.* Princeton, N.J.: Princeton University Press, 1963.

Cole, Barry, ed. *Television Today: A Close-up View; Readings from TV Guide.* New York: Oxford University Press, 1981.

"Collect royalties on video-recorder sales?" *US News & World Report,* 13 February 1984.

Commission on the Freedom of the Press. *A Free and Responsible Press.* Chicago: University of Chicago Press, 1947.

Compaine, Benjamin M., ed. *Anatomy of the Communications Industry: Who Owns the Media?* White Plains, N.Y.: Knowledge Industry Publications, Inc., 1982.

Compaine, Benjamin M. *The Book Industry in Transition: An Economic Analysis of Book Distribution and Marketing.* White Plains, N.Y.: Knowledge Industry Publications, 1978.

Compaine, Benjamin M., ed. *Who Owns the Media? Concentration of Ownership in the Mass Communications Industry.* New York: Harmony Books, 1979.

"Computers provide book market." United Press International. *Logan* (Utah) *Herald Journal,* 5 March 1986.

Comstock, George. *Television in America.* Beverly Hills, Calif.: SAGE, 1980.

Comstock, G.; Chaffee, S.; Katzman, N.; McCombs, M.; and Roberts, R. *Television and Human Behavior.* New York: Columbia University Press, 1978.

Cook, Anthony. "The Peculiar Economics of Television." *TV Guide,* 14 June 1980.

Cook, Bruce. "Los Angeles: The Monthly That Wants Respect." *Washington Journalism Review,* April 1984.

Corliss, Richard. "Backing into the future." *Time,* 3 February 1986.

"Could the Progressive Help Make Possible a Nuclear Holocaust?" *Quill,* April 1979.

Cowan, Geoffrey. *See No Evil.* New York: Simon & Schuster, 1979.

Crouse, Timothy. *The Boys on the Bus.* New York: Random House, 1973.

Cullen, Maurice R., Jr. *Mass Media and the First Amendment: An Introduction to the Issues, Problems, and Practices.* Dubuque, Iowa: Wm. C. Brown, 1981.

"Curl Up with a Good Movie." *Deseret News,* 30 July 1979.

Cutlip, Scott M., and Center, Allen H. *Effective Public Relations,* 6th ed. Englewood Cliffs, N.J.: Prentice-Hall, 1985.

d

Danna, Sammy R. "The Press-Radio War." *Freedom of Information Center Report No. 213,* Columbia, Mo.: University of Missouri School of Journalism, December 1968.

Davis, Dennis K., and Baran, Stanley J. *Mass Communication in Everyday Life.* Belmont, Calif.: Wadsworth, 1980.

Davison, E. Phillips, and Yu, Frederick T. C. *Mass Communication Research: Major Issues and Future Directions.* New York: Praeger, 1974.

DeFleur, Melvin L. *Milestones in Mass Communication Research.* New York: Longman, 1983.

DeFleur, Melvin L., and Ball-Rokeach, Sandra. *Theories of Mass Communication,* 4th ed. New York: Longman, 1982.

DeFleur, Melvin L., and Dennis, Everette E. *Understanding Mass Communication,* 2d ed. Boston: Houghton Mifflin, 1985.

De Forest, Lee. *Television: Today and Tomorrow.* New York: Dial Press, 1942.

Denisoff, R. Serge. *Solid Gold: The Popular Record Industry.* New York: Transaction Books, 1975.

Dennis, Everette E. *The Media Society.* Dubuque, Iowa: Wm. C. Brown, 1978.

Dennis, Everette E. "The Politics of Media Credibility." Unpublished lecture delivered at Journalism Ethics Institute, Washington and Lee University, Lexington, Va., 28 March 1986.

Dennis, Everette E., and Ismach, Arnold. *Reporting Processes and Practices.* Belmont, Calif.: Wadsworth, 1978.

Dennis, Everette E.; Ismach, Arnold; and Gillmor, Donald, eds. *Enduring Issues in Mass Communication.* St. Paul, Minn.: West, 1978.

Desmond, Robert W. *The Information Process: World News Reporting to the Twentieth Century.* Iowa City: University of Iowa Press, 1978.

Dessauer, John P. *Book Publishing: What It Is, What It Does.* New York: Bowker, 1974.

Deutsch, Linda. "Publishers Finance Lavish Book Promotions." *The Salt Lake Tribune,* 28 April 1978.

"Direct Mail campaigns pinpoint customers." United Press International. *Washington Times,* 30 May 1985.

Dizard, Wilson P., Jr. *The Coming Information Age: An Overview of Technology, Economics, and Politics.* New York: Longman, 1985.

"Don't judge tomorrow's videotex by today's data, expert says." *Presstime,* November 1984.

Drummond, William J. "Local TV News: Today the Bay Area, Tomorrow the World." *San Jose Mercury News West,* 16 February 1986.

e

Early, Steven C. *An Introduction to American Movies.* New York: New American Library, 1979.

Edwards, John. "Exploring private message systems." *Online Today,* Vol. 3, No. 6, June 1984.

Edwards, Kenneth. "Information Without Limit Electronically." In *Readings in Mass Communications,* 4th ed., edited by Michael Emery and Ted Curtis Smythe. Dubuque, Iowa: Wm. C. Brown, 1980.

Edwards, Kenneth. "Teletext Broadcasting in U.S. Endorsed by FCC." *Editor & Publisher,* 18 November 1978.

Efron, Edith. *The News Twisters.* Los Angeles: Nash, 1971.

Ellul, Jacques. *Propaganda: The Formation of Men's Attitudes.* New York: Alfred A. Knopf, 1965.

Ellul, Jacques. *The Technological Society.* New York: Alfred A. Knopf, 1964.

Emery, Edwin, and Emery, Michael. *The Press and America: An Interpretative History of the Mass Media,* 5th ed. Englewood Cliffs, N.J.: Prentice-Hall, 1984.

Emery, Michael, and Smythe, Ted Curtis, eds. *Readings in Mass Communication,* 6th ed. Dubuque, Iowa: Wm. C. Brown, 1986.

Emery, Walter B. *Broadcasting and Government.* East Lansing, Mich.: Michigan State University Press, 1971.

Engel, Jack. *Advertising: The Process and Practice.* New York: McGraw-Hill, 1980.

Epstein, Edward Jay. *News from Nowhere.* New York: Random House, 1973.

Evans, Harold. *Pictures on a Page.* Belmont, Calif.: Wadsworth, 1979.

f

Fadiman, William. *Hollywood Now.* London: Thames and Hudson, 1973.

Fang, Irving E. *Television News, Radio News.* Minneapolis, Minn.: Rada Press, 1980.

Farrar, Ronald T., and Stevens, John D., eds. *Mass Media and the National Experience.* New York: Harper & Row, 1971.

Fascell, Dante B., ed. *International News: Freedom Under Attack.* Beverly Hills, Calif.: SAGE, 1979.

FCC v. Pacifica Foundation, 488 U.S. 726 (1978).

Fearing, F. "Influence of the Movies on Attitudes and Behavior." *Annals of the American Academy of Political and Social Sciences,* 1947.

Fields, Howard. "Forum Predicts Future of Books, Software." *Publishers Weekly,* 8 June 1984.

Fields, Howard. "Survey Finds Eight Million New Readers in Five Years." *Publishers Weekly,* 27 April 1984.

Fields, Howard. "The View From Washington." *Publishers Weekly,* 24 May 1985.

"Fifth Estate's $30 billion-plus year." *Broadcasting,* 30 December 1985.

Finley, John A. "Special Report: Reading." *Presstime,* September 1984.

"First U.S. anarchist radio show." *Logan* (Utah) *Herald Journal,* 31 October 1984.

Fischer, Heinz-Dietrich, and Merrill, John C., eds. *International and Intercultural Communication.* New York: Hastings House, 1976.

Flander, Judy. "TV's Top Dollar Talent." *Washington Journalism Review,* March 1986.

Fletcher, Alan D. "City Magazines Find a Niche in the Media Marketplace." *Journalism Quarterly,* Winter 1977.

"FM Growth Continues." *Broadcasting,* 16 June 1980.

Foltz, Kim. "Radio's Wacky Road to Profit." *Newsweek,* 25 March 1985.

Foster, Eugene S. *Understanding Broadcasting.* Reading, Mass.: Addison-Wesley, 1978.

Francois, William E. *Mass Media Law and Regulation,* 3d ed. Columbus, Ohio: Grid, 1982.

Franklin, Marc A. *Cases and Materials on Mass Media Law,* 3d ed. Mineola, N.Y.: The Foundation Press, Inc., 1987.

Franklin, Marc A. *The First Amendment and the Fourth Estate.* Mineola, N.Y.: Foundation Press, 1981.

Freberg, Stan. "The Freberg Part-time Television Plan." In *Mass Media in a Free Society,* edited by Warren K. Agee. Lawrence, Kansas: The University Press of Kansas, 1969.

Fricke, David. "The Rock and Roll Hall of Fame." *Rolling Stone,* 13 February 1986.

Friedman, Barbara J. "Videotex languor claims victims." *Presstime,* April 1986.

Friedman, Barbara J. "Viewtron alters market goals." *Presstime,* February 1986.

Friedman, Jack. "Radio, Magazines are Examples for Cable Advertising to Follow, Says Warner's Schneider." *Broadcasting,* 11 August 1980.

Friedman, Jack. "Suddenly Cable Is Sexy." *TV Guide,* 19 July 1980.

FTC v. Warner-Lambert Co., 435 U.S. 950 (review denied).

Furlong, William Barry. "The Monster Is Lurking Just Over the Hill." *Panorama,* March 1980.

g

Galella v. Onassis, 533 F. Supp. 1076 (1982).

Gans, Herbert J. *Deciding What's News: A Study of CBS Evening News, NBC Nightly News, Newsweek and Time.* New York: Pantheon Books, 1979.

Garfield, Robert. "New book publishers look for concept, then author to write it." *USA Today,* 2 October 1984.

Genovese, Margaret. "Transmission by Satellite: Pages Today, Wire News Tomorrow, Ads in the Future." *Presstime,* October 1980.

Genovese, Margaret. "UPI." *Presstime,* December 1986.

Gerbner, George, ed. *Mass Media Policies in Changing Cultures.* New York: Wiley, 1977.

Gerbner, George. Testifying before the National Commission on the Causes and Preventions of Violence, 1969.

Gillmor, Donald M. "Press Councils in America." In *Enduring Issues in Mass Communication,* edited by Everette E. Dennis, Arnold H. Ismach, and Donald M. Gillmor. St. Paul, Minn.: West, 1978.

Ginsburg, Douglas H. *Regulation of Broadcasting: Law and Policy Towards Radio, Television, and Cable Communications.* St. Paul, Minn.: West, 1979.

Goldsen, Rose. "Why Television Advertising is Deceptive and Unfair." *ETC.,* Winter 1978.

Goldsmith, Alfred N., and Lescarboura, Austin C. *This Thing Called Broadcasting.* New York: Henry Holt, 1930.

Goldstein, Seth. "Cable and Pay-TV." *Panorama,* April 1980.

Goldstein, Tom. *The News at Any Cost: How Journalists Compromise Their Ethics to Shape the News.* New York: Simon and Schuster, 1985.

Goodman, Ellen. "Author's hoax makes important point." Syndicated column, *Logan* (Utah) *Herald Journal,* 2 October 1984.

Goodman, Ellen. "Indiana Jones' spoils PG film rating." *Logan* (Utah) *Herald Journal,* 26 June 1984.

Gordon, David. "Chicago Journalists and Ethical Principles." *Mass Comm Review,* Fall 1979.

Gordon, Robbie. "God's Medium." *Washington Journalism Review,* April 1986.

Gramling, Oliver. *AP: The Story of News.* New York: Farrar and Rinehart, 1940.

Grannis, Chandler B. "Title Output and Average Prices: 1984 Final Figures." *Publishers Weekly,* 23 August 1985.

Gray, Richard. "Implications of the new information technology for democracy." In John W. Alhauser, ed., *Electronic Home News Delivery: Journalistic and Public Policy Implications.* Bloomington, Ind.: School of Journalism and Center for New Communications, Indiana University, 1981.

Greenfield, Jeff. "Don't Blame TV." *TV Guide,* 18 January 1986.

Greenfield, Jeff. "TV is Not the World." *Columbia Journalism Review,* May/June 1978.

Griffith, Thomas. "Watchdog Without a Bite." *Time,* 9 April 1984.

Griffith, Thomas. "Why Readers Mistrust Newspapers." *Time,* 9 May 1983.

Groop, Marvin M. "Now You See It! Now You Don't!" *Magazine Newsletter of Research No. 44,* March 1984.

Gross, Lynne Schafer. *The New Television Technologies,* 2d ed. Dubuque, Iowa: Wm. C. Brown, 1986.

Gross, Lynne Schafer. *Telecommunications: An Introduction to Radio, Television, and Other Electronic Media,* 2d ed. Dubuque, Iowa: Wm. C. Brown, 1986.

Grossman, Michael B., and Kumar, Martha J. *Portraying the President.* Baltimore: Johns Hopkins University Press, 1981.

"Group Says Children's TV Programming Suffers." United Press International. *Logan* (Utah) *Herald Journal,* 23 May 1985.

Grunig, James, and Hunt, Todd. *Managing Public Relations.* New York: Holt, Rinehart & Winston. 1984.

Guback, Thomas. "Theatrical Film." In Benjamin M. Compaine, Christopher Sterling, Thomas Guback, and J. Kendrick Noble, Jr., *Anatomy of the Communications Industry: Who Owns the Media?* White Plains, N.Y.: Knowledge Industry Publications, Inc., 1982.

Gumpert, Gary, and Cathcard, Robert, eds. *Inter/Media: Interpersonal Communication in a Media World,* 2d ed. New York: Oxford University Press, 1982.

Gunther, Marc. "Pacifica: Radio's Outlet for the Outrageous." *Washington Journalism Review,* May 1983.

Gunther, Marc, and Gunther, Noel. "And now a word from pay radio." *Washington Journalism Review,* July/August 1983.

Gunther, Marc, and Gunther, Noel. "Black Radio: Playing for the Big Time." *Washington Journalism Review,* October 1983.

Gutman, John. "A marriage of sight and sound, video rock is on a roll." *Salt Lake Tribune,* 27 July 1984.

h

Hafer, W. Keith, and White, Gordon E. *Advertising Writing.* St. Paul, Minn.: West, 1977.

Halberstam, David. *The Powers That Be.* New York: Alfred A. Knopf, 1979.

Hannegan v. Esquire, 327 U.S. 146 (1946).

Harmetz, Aljean. "With return of Indy Jones comes debate over ratings." *Salt Lake Tribune,* 22 May 1984.

Harris, Louis. "Public Prefers News to Pablum." *Deseret News,* 1 Jan 1978.

Hass, Alan D. "Some Businesses are Sheer Murder." *Family Weekly,* 18 November 1984.

Hatchen, William. *The World News Prism: Changing Media, Clashing Ideologies.* Ames, Iowa: Iowa State University Press, 1981.

Head, Sydney W. *Broadcasting in America.* Boston: Houghton Mifflin, 1978.

Heller, Karen. "Hot On the Press: Magazines That Are Making a Mint." *Washington Journalism Review,* April 1984.

Hennessy, Bernard. *Public Opinion,* 3d ed. Scituate, Mass.: Duxbury Press, 1975.

Henry, William A., III, and Bruns, Richard. "Esquire at Mid-Century." *Time,* 21 November 1983.

Henry, William A., III. "Reuter's Hot Financial Flash." *Time,* 11 June 1984.

Henry, William A., III. "The Ten Best U.S. Dailies." *Time,* 30 April 1984.

Hess, Stephen. *The Washington Reporters.* Washington, D.C.: Brookings Institute, 1981.

Hickey, Neil. "Do You Know Why Miniseries Usually Start On Sunday?" *TV Guide,* 7 December 1985.

Hickey, Neil. "Goodbye '70s, Hello '80s." *TV Guide,* 5 January 1980.

Hickey, Neil. "The Newsroom Power Plays That Viewers Never See." *TV Guide,* 18 May 1985.

Hickey, Neil. "Read Any Good Television Lately?" *TV Guide,* 16 February 1980.

Hickey, Neil. "Turn On Your TV, And Aim For A Diploma." *TV Guide,* 17 November 1984.

Hickey, Neil. "Viewers Beware: Your People Meter Will Be 'Watching' Your Every Move." *TV Guide,* 15 December 1984.

Hiebert, Ray E., and Spitzer, Carlton, eds. *The Voice of Government.* New York: Wiley, 1968.

"High tech and the home." *Newsweek,* 17 February 1986.

Hill, Doug. "Will the Latest in Home Video Empty the Movie Theaters?" *TV Guide,* 20 March 1982.

Hohenberg, John. *The News Media: A Journalist Looks at His Profession.* New York: Holt, Rinehart & Winston, 1968.

Holder, Dennis. "Local Coverage On Ku." *Washington Journalism Review,* October 1985.

Holder, Dennis. "Mixing Public Radio With Private Enterprise." *Washington Journalism Review,* June 1984.

Hollinger, Hy. "Proliferation of videocassettes and barter TV syndication made a tumultuous 1985 for film biz." *Variety,* 8 January 1986.

Holm, Wilton R. "Management Looks at the Future." In *The Movie Business: American Film Industry Practice,* edited by A. W. Bleum and J. E. Squire. New York: Hastings House, 1972.

"How the best-seller lists are made." *USA Today,* 13 September 1983.

"How the U.S. Uses Radio Today." *Beyond The Ratings,* Vol. 7, No. 5, May 1984.

Hulteng, John L. *The Messenger's Motives: Ethical Theory in the Mass Media,* 2d ed. Englewood Cliffs, N.J.: Prentice-Hall, 1985.

Hulteng, John L. *The News Media: What Makes Them Tick?* Englewood Cliffs, N.J.: Prentice-Hall, 1979.

Hutchinson, Thomas H. *Here is Television: Your Window to the World.* New York: Dial Press, 1946.

Hynds, Ernest C. *American Newspapers in the 1980s.* New York: Hastings House, 1980.

i

Institute for Propaganda Analysis. "How to Detect Propaganda." *Propaganda Analysis,* 1 November 1937.

j

Jacob, Miriam. *"Reader's Digest:* Who's in Charge?" *Columbia Journalism Review,* July/August 1984.

"Jankowski foresees shakeout in radio." *Broadcasting,* 5 November 1984.

Janowitz, Morris, and Hirsch, Paul, eds. *Reader in Public Opinion and Mass Communication,* 3d ed. New York: Free Press, 1981.

Johnson, Harriet C. "Radio refuses to roll over and die." *USA Today,* 10 October 1984.

Johnstone, John W. C.; Slawski, E. J.; and Bowman, William W. *The News People: A Sociological Portrait of Journalists and Their Work.* Urbana, Ill.: University of Illinois Press, 1976.

Jones, Alex S. "Public Views Newspapers With Mixture of Faith and Mistrust, Poll Finds." *New York Times,* 13 April 1985.

Jory, Tom. "PBS and Corporate Underwriting." *Salt Lake Tribune,* 28 May 1978.

Josephson, Sheree. "Uses and Gratifications." Unpublished manuscript, April 1984.

"Journalism in 1985: Bolder and Wiser." *Broadcasting,* 16 December 1985.

Jowett, Garth. *Film: The Democratic Art.* Boston: Little, Brown, 1976.

Jowett, Garth, and Linton, James M. *Movies as Mass Communication.* Beverly Hills, Calif.: SAGE, 1980.

k

Kahn, Frank J., ed. *Documents of American Broadcasting,* 4th ed. Englewood Cliffs, N.J.: Prentice-Hall, 1984.

Kaiser, Charles, and Karlen, Neal. "Time Inc.'s $47 Million Mistake." *Newsweek,* 26 September 1983.

Katz, Elihu; Blumer, Jay G.; and Gurevitch, Michael. "Utilization of Mass Communication by the Individual." In Blumer and Katz, eds., *The Uses of Mass Communications on Gratification Research.* Beverly Hills, Calif.: SAGE, 1974.

Katz, Elihu; Gurevitch, Michael; and Hass, H. "On the Use of Mass Media for Important Things." *American Sociological Review,* Vol. 24, 1973.

Katz, S. N., ed. *A Brief Narrative of the Case and Trial of John Peter Zenger.* Cambridge: Harvard University Press, 1963.

Kaye, Elizabeth. "Sam Phillips: The *Rolling Stone* interview." *Rolling Stone,* 13 February 1986.

Key, Wilson Bryan. *The Clam-Plate Orgy and other Subliminal Techniques For Manipulating Your Behavior.* New York: Signet, 1981.

Key, Wilson Bryan. *Media Sexploitation.* New York: Signet, 1976.

Key, Wilson Bryan. *Subliminal Seduction: Ad Media's Manipulation of a Not So Innocent America.* New York: Signet, 1974.

Kidder, Rushworth M. "Halting the Billion-Dollar March Of War Toys and War Cartoons." *Christian Science Monitor,* 16 December 1985.

Kidder, Rushworth M. "Video culture." Reprinted from the *Christian Science Monitor,* 10–14 June 1985.

Kilpatrick, James J. "Time to Announce, 'We've Won.' " Syndicated column, *Deseret News,* 9 April 1979.

Kilpatrick, James J. *Washington Star* syndicated column, 16 May 1978.

Kizol, Jonathan. "A Nation's Wealth." *Publishers Weekly,* 24 May 1985.

Kleppner, Otto; Russell, Thomas; and Verrill, Glenn. *Advertising Procedure,* 8th ed. Englewood Cliffs, N.J.: Prentice-Hall, 1983.

Knight, Arthur. *The Liveliest Art: A Panoramic History of the Movies,* rev. ed. New York: New American Library, 1979.

"Knight-Ridder's cutbacks at Viewtron show videotex revolution is faltering." *Wall Street Journal,* 7 October 1984.

Korn, Fred. "What's Right? There's no consensus." *1984–85 Journalism Ethics Report* of the Society of Professional Journalists/Sigma Delta Chi.

Koshetz, Charles. "Winans case spotlights ethics policies." *1985–86 Journalism Ethics Report* of the Society of Professional Journalists/Sigma Delta Chi.

Kowet, Don. "UPI's Prince With a Mysterious Past." *Washington Times Insight,* 16 December 1985.

Kowinski, William. "Talk Back to Television." *Penthouse,* February 1979.

Kraus, Sidney, and Davis, Dennis. *The Effects of Mass Communication on Political Behavior.* University Park, Pa.: Penn State Press, 1976.

Krieghbaum, Hillier. *Pressures on the Press.* New York: Crowell, 1972.

l

LaBrie, Henry G., III, ed. *Perspectives on the Black Press.* Kennebunkport, Me.: Mercer House, 1974.

Lachenbruch, David. "A Buyer's Guide to Videodisc Players." *TV Guide,* 6 March 1982.

Lachenbruch, David. "The Coming Videodisc Battle." *Panorama,* April 1980.

Lanier, Robin. "The Interactive Videodisc in Action." *E-ITV, The Techniques Magazine for Professional Video,* April 1986.

Lasswell, Harold D., "The Structure and Function of Communication in Society." In L. Bryson, *The Communication of Ideas.* New York: Harper, 1948.

Lazarsfeld, Paul F.; Berelson, Bernard; and Gaudet, Helen. *The People's Choice.* New York: Duell, Sloan & Pearce, 1944.

"Learning to Live with TV." *Time,* 28 May 1979.

Lenhart, Maria. "The Author as Peddler." *Deseret News,* 4 August 1979.

Lerner, Max. "The Shame of the Professions." *Saturday Review,* 1 November 1975.

Lesly, Philip, ed. *Public Relations Handbook.* Englewood Cliffs, N.J.: Prentice-Hall, 1978.

Levy, Leonard *Freedom of the Press from Zenger to Jefferson.* Indianapolis: Bobbs-Merrill, 1966.

Levy, Leonard. *Legacy of Suppression: Freedom of Speech and Press in Early American History.* Cambridge, Mass.: Harvard University Press, 1960.

Levy, Mark R. "Home Video Recorders and TV Program Preference." Paper presented to the Mass Communications and Society Division, Association for Education in Journalism Annual Conference, Boston, Mass., August 1980.

Levy, Mark R. "The Audience Experience with Television News." *Journalism Monographs,* no. 55, April 1978.

Lichter, S. Robert, and Rothman, Stanley. "The Media Elite." *Public Opinion,* October/November 1981.

Lichty, Lawrence H., and Topping, Malachi C. *American Broadcasting: A Sourcebook on the History of Radio and Television.* New York: Hastings House, 1975.

Liebling, A. J. *The Press.* New York: Ballantine, 1964.

Liebling, A. J. *The Wayward Pressman.* New York: Doubleday, 1948.

Lippmann, Walter. *Public Opinion.* New York: Harcourt, Brace, 1922.

Lockwood, Russ. "Electronic Oracle." *A+ Magazine,* May 1984.

Loder, Kurt. "The music that changed the world." *Rolling Stone,* 13 February 1986.

Lodge, Sally A. "Paperback Top Sellers." *Publishers Weekly,* 15 March 1985.

Logan, Robert A. "Jefferson's and Madison's Legacy: The Death of the National News Council." *Journal of Mass Media Ethics,* Vol. 1, No. 1, Fall/ Winter 1985–86.

Logan, Robert A. *"USA Today's* Innovations and their Impact on Journalism Ethics," *Journal of Mass Media Ethics,* Vol. 1, No. 2, Spring/ Summer 1986, pp. 74–87.

London, Michael. "Movie exhibitors hear new video war strategy." *Salt Lake Tribune,* 15 March 1984.

Lowery, Shearon, and DeFleur, Melvin. *Milestones in Mass Communication Research: Media Effects.* New York: Longman, 1983.

"Low Power Television Pioneers Begin a New Broadcast Era." *Broadcasting,* 17 May 1982.

Lynes, Russell. "The Electronic Express." *TV Guide,* 24 February 1968.

m

Mabry, Drake. "Editors vs. Syndicates." *Presstime,* January 1982.

MacDonald, J. Fred. *Don't Touch That Dial! Radio Programming in American Life 1920–1960.* Chicago: Nelson Hall, 1979.

MacDougall, A. Kent. "Magazines: Fighting for a Place in the Market." *Los Angeles Times,* 9 April 1978.

MacDougall, William. "Why People Are Turned Off By Television." *US News & World Report,* 13 February 1984.

MacNeil, Robert. "Is Television Shortening Our Attention Spans?" *New York University Education Quarterly,* Vol. 14, No. 2, Winter 1983.

MacNeil, Robert. "The Mass Media and Public Trust." *Occasional Paper,* No. 1. New York: Gannett Center for Media Studies, 1985.

Madison, Charles A. *Book Publishing in America.* New York: McGraw-Hill, 1966.

Malloy, Michael T. "Newspapers May Some Day Let You Pick News You Want." *National Observer,* 21 February 1976.

Mankekar, D. R. *One Way Free Flow; Neo-colonialism Via News Media.* New Delhi: Clarion Books, 1978.

Mankiewicz, Frank, and Swerdlow, Joel. *Remote Control: Television and the Manipulation of American Life.* New York: Ballantine Books, 1978.

Martin, Richard, and Innerst, Carol. "Textbooks Besieged for Ducking 2Rs— Religion and Relevance." *Washington Times Insight,* 23 October 1985.

Marzolf, Marion. *Up From the Footnote: A History of Women Journalists.* New York: Hastings House, 1977.

Maslow, Abraham. *Motivation and Personality.* New York: Harper & Row, 1970.

Maslow, Abraham. *Toward a Psychology of Being,* 2d ed. New York: D. Van Nostrand, 1968.

Masmoudi, M. "The New World Information Order." *Journal of Communication,* Spring 1979.

Massing, Michael. "CBS: Sauterizing the News." *Columbia Journalism Review,* March/April 1986.

Mast, Gerald. *A Short History of the Movies.* New York: Bobbs-Merrill, 1976.

Mathews, Jack. "Vivid images, visceral action." *USA Today,* 14 November 1983.

Matusow, Barbara. "Station Identification: Network Affiliates Loosen The Apron Strings." *Washington Journalism Review,* April 1985.

Matusow, Barbara. *The Evening Stars: The Making Of The Network News Anchor.* New York: Ballantine Books, 1983.

Mayer, Martin. *About Television.* New York: Harper & Row, 1972.

Mayer, Martin. *Madison Avenue U.S.A.* New York: Harper & Brothers, 1958.

McCombs, Maxwell E., and Becker, Lee. *Using Mass Communication Theory.* Englewood Cliffs, N.J.: Prentice-Hall, 1979.

McCombs, Maxwell E., and Shaw, Donald L. "The Agenda-Setting Function of Mass Media." *Public Opinion Quarterly,* 1972.

McGinniss, Joe. *The Selling of the President.* New York: Trident, 1969.

McGuigan, Cathleen. "Rock music goes to Hollywood." *Time,* 11 March 1985.

McKenna, George, ed. *Media Voices: Debating Critical Issues in Mass Media.* Guilford, Conn.: Dushkin Publishing Group, 1982.

McLeod, Jack; Becker, Lee; and Byrnes, J. "Another Look at the Agenda-Setting Function of the Press." *Communication Research,* 1974.

McLuhan, Marshall. *The Gutenberg Galaxy.* Toronto: The University of Toronto Press, 1967.

McLuhan, Marshall. *Understanding Media: The Extensions of Man.* New York: McGraw-Hill, 1965.

McQuail, Dennis. *Towards a Sociology of Mass Communication.* London: Collier-Macmillan, 1969.

McQuail, Dennis, and Gurevitch, Michael. "Explaining Audience Behavior: Three Approaches Considered." In Jay G. Blumer and Elihu Katz, eds., *The Uses of Mass Communication: Current Perspectives on Gratifications Research.* Beverly Hills, Calif.: SAGE, 1974.

"Media Tie-ins." *Publishers Weekly,* 11 April 1980.

"Meese Commission Releases Porn Report, Can't Agree on Harm of 'Indecent' Matter." *The News Media & The Law,* Fall 1986.

Melanson, James. "Homevideo attains mass-medium status." *Variety,* 8 January 1986.

Merrill, John C. *The Imperative of Freedom: A Philosophy of Journalistic Autonomy.* New York: Hastings House, 1974.

Merrill, John C., and Fisher, Harold A. *The World's Great Dailies: Profiles of 50 Newspapers.* New York: Hastings House, 1980.

Merrill, John C., and Lowenstein, Ralph L. *Media, Messages, and Men,* 2d ed. New York: Longman, 1979.

Meyer, Philip. *Editors, Publishers and Newspaper Ethics: A Report to the American Society of Newspaper Editors.* Washington, D.C.: ASNE, The Newspaper Center, 1983.

Meyer, Philip. *Ethical Journalism.* New York: Longman, 1987.

Miller, Jim. "Rock's new women." *Time,* 4 March 1985.

Miller v. California, 413 U.S. 15 (1973).

Minnow, Newton. Addressing the National Association of Broadcasters, 9 May 1961.

Mogel, Leonard. *The Magazine: Everything You Need to Know to Make It in the Magazine Business.* Englewood Cliffs, N.J.: Prentice-Hall, 1979.

Monaco, James. *How to Read a Film: The Art, Technology, Language, History, and Theory of Film and Media,* 2d ed. New York: Oxford University Press, 1981.

Moore, Mike. "Disneygate." *Quill,* November 1986.

Morris, Joe Alex. *Deadline Every Minute: The Story of the United Press.* New York: Doubleday, 1957.

Morton, John. *"USA Today's* Ad-versity." *Washington Journalism Review,* September 1984.

"Movie studios put more emphasis on home video pay-TV markets." *The Wall Street Journal,* 1 May 1984.

Murray, Bob L. "Long Live the Little Magazine." *St. Louis Journalism Review,* September 1985.

Murrow, Edward R. Addressing the Radio Television News Directors Association, 15 October 1958.

n

Nadler, Eric. "Guiding TV to the Right." *Mother Jones,* April 1984.

Naisbitt, John. *Megatrends: Ten New Directions Transforming Our Lives.* New York: Warner Books, Inc., 1982.

"NCTA Report on Local Cable Programming." *Broadcasting,* 1 September 1980.

Near v. Minnesota, 383 U.S. 697 (1931).

Nelson, Harold L. *Freedom of the Press from Hamilton to the Warren Court.* Indianapolis: Bobbs-Merrill, 1966.

Nelson, Harold L., and Teeter, Dwight L. *Law of Mass Communications,* 5th ed. Mineola, N.Y.: Foundation Press, 1986.

Nelson, Mark. "Newspaper Ethics Codes and the NLRB." *Freedom of Information Center Report No. 353.* Columbia, Mo.: University of Missouri School of Journalism, May 1976.

Nelson, Roy Paul, and Hulteng, John. *The Fourth Estate: An Informal Appraisal of the News and Information Media,* 2d ed. New York: Harper & Row, 1983.

Newcomb, Horace, ed. *Television: The Critical View.* New York: Oxford University Press, 1982.

"New Owners Take Over Reins at UPI." United Press International. *Salt Lake Tribune,* 12 June 1985.

"News Council to Fold." *Broadcasting,* 26 March 1984.

"News mixed for radio." *Broadcasting,* 16 December 1985.

Newsom, Clark. "The Beat Goes on for P.M.S to A.M.S" *Presstime,* December 1980.

Newsom, Doug, and Scott, Alan. *This Is PR: The Realities of Public Relations.* Belmont, Calif.: Wadsworth, 1981.

Newsom, Doug, and Seigfried, Tom. *Writing in Public Relations Practice.* Belmont, Calif.: Wadsworth, 1981.

"Newspaper Credibility. Building Reader Trust." American Society of Newspaper Editors. A study conducted by MORI Research, Inc., Minneapolis, Minn., April 1985.

"Newspapers Are Public's Favorite Ad Medium." *Editor & Publisher,* 29 March 1975.

New York Times Co. v. United States, 713 U.S. 403 (1971).

Nimmo, Dan. *Political Communication and Public Opinion in America.* Santa Monica, Calif.: Goodyear Publishing, 1978.

Nobel, J. Kendrick, Jr. "Book Publishing." In Benjamin M. Compaine; Christopher H. Sterling; Thomas Guback; and J. Kendrick Noble, Jr., *Anatomy of the Communications Industry: Who Owns the Media?* White Plains, N.Y.: Knowledge Industry Publications, Inc., 1982.

Noelle-Neumann, Elisabeth. "Return to the Concept of Powerful Mass Media." In H. Eguchi, and K. Sata, eds., *Studies of Broadcasting: An International Annual of Broadcasting Science.* Tokyo: The Nippon Hoso Kyokai, 1973.

o

Oates, William R.; Ghorpade, Shailendra; and Brown, Jane D. "Media technology consumers: demographics and psychographics of 'Taffies.'" Paper presented to the Mass Communications and Society Division, Association for Education in Journalism and Mass Communications Annual Convention, Norman, Oklahoma, August 1986.

Ogilvy, David. *Confessions of an Advertising Man.* New York: Atheneum, 1963.

Olasky, Marvin N. "Ministers or Panderers: Issues Raised by the Public Relations Society Code of Standards." *Journal of Mass Media Ethics,* Vol. 1, No. 1, Fall/Winter 1985–86.

O'Toole, John. "Madison Ave.: Selling ideas, not products." *U.S. News and World Report.* 10 March 1986.

Overbeck, Wayne and Pullen, Rick D. *Major Principles of Media Law,* 2d ed. New York: Holt, Rinehart and Winston, 1985.

Owen, Jan. *Understanding Computer Information Networks.* Sherman Oaks, Calif.: Alfred Publishing Co., 1984.

p

Pacifica Foundation v. FCC, 556 F. 2d 9 (D.C. Cir. 1977).

Packard, Vance. *The Hidden Persuaders.* New York: Pocket Books, 1968.

Panitt, Merrill. "Do the Networks Have a Death Wish?" *TV Guide,* 18 September 1982.

Patterson, Thomas E., and McClure, Robert D. *The Unseeing Eye: The Myth of Television Power in National Politics.* New York: G. P. Putnam's Sons, 1976.

Pearce, Alan. "How the Networks Have Turned News Into Dollars." *TV Guide,* 23 August 1980.

Pember, Don R. *Mass Media in America,* 5th ed. Chicago: SRA, 1987.

Pember, Don R. *Mass Media Law,* 4th ed. Dubuque, Iowa: Wm. C. Brown, 1987.

Pennekamp v. Florida, 3328 U.S. 331 (1946).

"Perils and prospects over the electronic horizon." *Broadcasting.* In readings in *Mass Communication: Concepts and Issues in the Mass Media,* 4th ed. Michael Emery and Ted Curtis Smythe, eds. Dubuque, Iowa: Wm. C. Brown, 1980.

Petersen, Clarence. *The Bantam Story: Thirty Years of Paperback Publishing,* 2d ed. New York: Bantam Books, 1975.

Peterson, Theodore. *Magazines in the Twentieth Century.* Urbana, Ill.: University of Illinois Press, 1956.

Peterson, Theodore. "The Social Responsibility of the Press." In *Four Theories of the Press,* by Fred S. Siebert, Theodore Peterson, and Wilbur Schramm. Urbana, Ill.: University of Illinois Press, 1963.

Picard, Robert G. "Revisions of the *Four Theories of the Press* Model." *Mass Comm Review,* Winter/Spring, 1982/83.

Picard, Robert G. *The Press and the Decline of Democracy: The Democratic Socialist Response in Public Policy.* Westport, CT: Greenwood Press, 1986.

Polskin, Howard. "Acting: awful; scripts: weak; sales: terrific." *TV Guide,* 30 June 1984.

Pope, Leroy. "Pay TV, After a Shaky Infancy, Is Off and Rolling." *Deseret News,* 1 August 1979.

"Pornography report stirs expected storm." *FOI/FYI, A publication of the First Amendment Center,* Vol. 2, No. 1, August 1986.

Postman, Neil. *Amusing Ourselves to Death: Public Discourse in the Age of Show Business.* New York: Elisabeth Sifton Books, Viking, 1985.

Postman, Neil. *Teaching as a Conserving Activity.* New York: Delacorte Press, 1979.

Postman, Neil. *The Disappearance of Childhood.* New York: Delacorte Press, 1982.

Powers, Ron. *The Newscasters.* New York: St. Martin's, 1977.

Powers, Ron. "Where Have the News Analysts Gone?" *TV Guide,* 8 October 1980.

"Predictions: Changes to Conjure With." *Publishers Weekly,* 6 August 1979.

Prendergast, Alan. "Mickey Mouse Journalism." *Washington Journalism Review,* January-February 1987.

Prescott, Eileen. "Author Tours: The Bloom is Off the Rose." *Publishers Weekly,* 3 August 1984.

Price, Jonathan. *The Best Thing on TV: Commercials.* New York: Penguin Books, 1978.

"Public Access to Pretrial Hearing Protected by First Amendment." *News Media & The Law.* Summer 1986.

"Public TV Study Says VCRs Don't Detract from Viewing Levels." *Broadcasting,* 21 July 1980.

"Publishers Go Electronic." *Business Week,* 11 June 1984, pp. 84–97.

Publishers Information Bureau. "Revenues & Pages Set Records in 1983." New York: Magazine Center, 1984.

"Pulling Wires." *Time,* 6 May 1985.

"Putting Social Trends to Use in Cable." *Broadcasting,* 10 May 1982.

q

Qualter, Terrence H. *Propaganda and Psychological Warfare.* New York: Random House, 1962.

Quincy Cable TV, Inc. v. FCC 768 F. 2d 1434 (D.C. Cir. 1985), certiorari denied, 106 S. Ct. 2889 (1986).

r

"Radio." *Broadcasting,* 31 December 1984.

"Radio psychologists: Misleading or helpful?" *Logan* (Utah) *Herald Journal,* 3 January 1984.

Radolf, Andrew. "Move into cable is not a way to protect newspapers." *Editor & Publisher,* 29 October 1983.

Rambo, C. David. "In the Race for News, Technology Leads at the Wires." *Presstime,* August 1981.

Randall, Richard S. "Censorship: From *The Miracle to Deep Throat.*" In *The American Film Industry,* edited by Tino Balio. Madison, Wisc.: The University of Wisconsin Press, 1976.

"Reagan bringing radio back as political tool." *Broadcasting,* 29 October 1984.

"Recording Industry a 4-Billion-Dollar Hit." *U.S. News & World Report,* 30 April 1979.

Red Lion Broadcasting Co. v. FCC, 395 U.S. 367 (1969).

Reed, J. D. "New rock on a red-hot roll." *Time,* 18 July 1983.

Reeves, Richard. "Lucas and Spielberg Strike Back for America." *Logan* (Utah) *Herald Journal,* 17 August 1982.

Reston, James. *The Artillery of the Press.* New York: Harper & Row, 1967.

Righter, Rosemary. *Whose News? Politics, the Press and the Third World.* London: Times Books, 1978.

Rivers, William L., and Dennis, Everette E. *Other Voices: The New Journalism in America.* San Francisco: Canfield, 1974.

Rivers, William L., and Nyhan, Michael J., eds. *Aspen Notebook on Government and Media.* New York: Praeger, 1973.

Rivers, William L; Blakenburg, William B.; Starck, Kenneth; and Reeves, Earl. *Backtalk: Press Councils in America.* San Francisco: Canfield Press, 1972.

Rivers, William L.; Schramm, Wilbur; and Christians, Clifford G. *Responsibility in Mass Communication.* New York: Harper & Row, 1980.

Robinson, G. O. *Communications for Tomorrow.* New York: Praeger, 1978.

Robinson, John P., and Levy, Mark R. *The Main Source: Learning from Television News.* Beverly Hills: SAGE, 1986.

Robinson, Steve. "Pulitzers: Was the Mirage a Deception?" *Columbia Journalism Review,* July–August 1979.

Rodman, George, ed. *Mass Media Issues: Analysis and Debate.* Chicago: SRA, 1982.

Rokeach, Milton. *Beliefs, Attitudes, and Values.* San Francisco: Jossey-Bass, 1968.

Rokeach, Milton. "Images of the Consumer's Mind On and Off Madison Avenue." *ETC,* September 1964.

Rokeach, Milton. *The Nature of Human Values.* New York: Free Press, 1973.

Rokeach, Milton. *The Open and Closed Mind.* New York: Basic Books, 1960.

Rose, Ernest D. "How the U.S. Heard About Pearl Harbor." *Journal of Broadcasting,* Vol. 5, no. 4, Fall 1961.

Rosenberg, Bernard, and White, David M., eds. *Mass Culture Revisited.* Princeton, N.J.: Van Nostrand Reinhold, 1971.

Rosenthal, Sharon. "$77,000 For 30 Minutes? *Gimme a Break!" TV Guide,* 8 December 1984.

Rosenthal, Sharon. "Here's How They're Selling Alka-Seltzer to Yuppies." *TV Guide,* 22 December 1984.

Rothman, Robert. "Public broadcasting bill shot down twice." *Washington Times,* 2 November 1984.

Rouch, Matt. "Liberty's Fourth: 5-day ABC-TV fest." *USA Today,* 22 May 1986.

Rowe, Mary P. "Notes on the Ombudsman in the U.S., 1986." Unpublished paper presented to the Washington Journalism Center and the Organization of New Ombudsmen annual meeting, Washington, D.C., May 1986.

Ruth, Marcia. "Newspapers Are Asking for Less Exclusivity." *Presstime,* November 1986.

Ruth, Marcia. "Syndicates Serve Spreading Smorgasbord." *Presstime,* November 1986.

S

Salisbury, David. "The Third 'Industrial Revolution': Robot Factories and Electronic Offices." *Christian Science Monitor,* 8 October 1980.

Sandage, Charles H.; Fryburger, Vernon; and Rotzoll, Kim. *Advertising Theory and Practice.* Homewood, Ill.: Irwin, 1979.

Sanders, Keith P. "What Are Daily Newspapers Doing to be Responsive to Readers' Criticisms? A Survey of U.S. Daily Newspaper Accountability Systems." In *News Research for Better Newspapers,* edited by Galen Rarick, American Newspaper Publishers Association Foundation. Washington, D.C., 1975.

Sanders, Keith P., and Chang, Won H. "Codes: The Ethical Free-for-All; A Survey of Journalists' Opinions About Freebies." *Freedom of Information Foundation Series,* No. 7, March 1977.

Sanders, Keith P., and Chang, Won H. "Freebies: Achille's Heel of Journalism Ethics?" Unpublished paper presented to Association for Education in Journalism, Mass Communication and Society Division, Logan, Utah, 30 March 1979.

Sandman, Peter; Rubin, David; and Sachsman, David. *Media: An Introductory Analysis of American Mass Communications,* 3d ed. Englewood Cliffs, N.J.: Prentice-Hall, Inc., 1982.

Schemenaur, P. J., ed. *Writer's Market.* Cincinnati: Writer's Digest Books, 1986.

Schenck v. United States, 249 U.S. 39 S. Ct. 247 (1919).

Schramm, Wilbur, and Lerner, Daniel, eds. *Communication and Change: The Last Ten Years—and the Next.* Honolulu: University of Hawaii Press, 1976.

Schramm, Wilbur, and Porter, William E. *Men, Women, Messages, and Media: Understanding Human Communication,* 2d ed. New York: Harper & Row, 1982.

Schramm, Wilbur, and Roberts, Donald F., eds. *The Process and Effects of Mass Communication.* Urbana, Ill.: University of Illinois Press, 1971.

Schramm, Wilbur; Lyle, J.; and Parker, E. *Television in the Lives of Our Children.* Stanford, Calif.: Stanford University Press, 1961.

Schultze, Quentin. "Comments on the History of Ethical Codes in the Advertising Business." Paper presented to Association for Education in Journalism, Mass Communications and Society Division, Logan, Utah, 30 March 1979.

Scigliano, Eric. "Public Radio Airs a Feud." *Washington Journalism Review,* December 1982.

Seitel, Fraser P. *The Practice of Public Relations.* New York: Charles E. Merrill, 1980.

Seldes, Gilbert. *The New Mass Media: Challenge to a Free Society.* Washington, D.C.: American Association of University Women, 1957.

Selig, Robert W. "Exhibs Think Big As Filmgoers Have Small Home Screens." *Variety,* 8 January 1986.

Sellers, Leonard, and Rivers, William. *Mass Media Issues.* Englewood Cliffs, N.J.: Prentice-Hall, 1977.

Severin, Werner, and Tankard, James, Jr. *Communication Theories: Origins, Methods, Uses.* New York: Hastings House, 1979.

Shaw, David. "Book Biz Best-Sellers—Are They Really? Laziness and Chicanery Play Major Roles." *Los Angeles Times,* 24 October 1976.

Shaw, David. In panel discussion at the Washington Journalism Center and the Organization of News Ombudsmen annual meeting, Washington, D.C., May 1986.

Shaw, David. *Journalism Today.* New York: Harper's College Press, 1977.

Shaw, David. "Newspapers Challenged as Never Before." *Los Angeles Times,* 26 November 1976.

Shaw, Donald L. "Technology: Freedom for What?" In *Mass Media and the National Experience,* edited by Ronald T. Farar and John D. Stevens. New York: Harper & Row, 1971.

Siebert, Fredrick S. *Freedom of the Press in England 1476–1776.* Urbana, Ill.: University of Illinois Press, 1952.

Siebert, Fredrick S.; Peterson, Theodore; and Schramm, Wilbur. *Four Theories of the Press.* Urbana, Ill.: University of Illinois Press, 1963.

Sigel, Efrem. "Videotex: into the cruel world." *Datamation,* 15 September 1984.

Simon, Raymond. *Public Relations: Concepts and Practices,* 2d ed. Columbus, Ohio: Grid Publishing, 1980.

Sinclair, Upton. *The Brass Check: A Study of American Journalism.* Pasadena, Calif.: published by the author, 1920.

Siskel, Gene. "The new PG-13 film rating: Is it effective?" *Salt Lake Tribune,* 29 August 1984.

Sklar, Robert. *Prime Time America: Life On and Behind the Television Screen.* New York: Oxford University Press, 1981.

Skornia, Harry J. *Television and Society: An Inquest and Agenda for Improvement.* New York: McGraw-Hill, 1965.

Skylar, David. "Why Newspapers Die." *Quill,* July/August 1984.

Small, William. *To Kill a Messenger.* New York: Hastings House, 1972.

Smith, Anthony. *Goodbye Gutenberg: The Newspaper Revolution of the 1980s.* New York: Oxford University Press, 1980.

Smith, Anthony. *The Newspaper: An International History.* London: Thames and Hudson, 1979.

Smith, Desmond. "Is The Sun Setting On Network Nightly News?" *Washington Journalism Review,* January 1986.

Smith, Desmond. "What Is America's Secret Weapon in the Energy Crisis? Your Television Set." *Panorama,* April 1980.

Smith, F. Leslie. *Perspectives on Radio and Television: An Introduction to Broadcasting in the United States.* New York: Harper & Row, 1979.

Smith, F. Leslie. "The 'Praise the Lord' Television Network." BM/E (Broadcast Management/Engineering) March 1980.

Smythe, Ted C., and Mastroianni, George A. *Issues in Broadcasting: Radio, Television, and Cable.* Palo Alto, Calif.: Mayfield, 1975.

Sonderling Broadcasting Corporation, Station WGLD-FM, 27 R.R. 2nd 285, 11 April 1973.

"Special Report: The World of TV Programming." *Broadcasting,* 22 October 1984.

Stakelin, Bill. "Beating the bushes for AM resurgence." *Broadcasting,* 13 January 1986.

Stakelin, Bill. "Radio at 60: It's still a fun business." *Variety,* 8 January 1986.

Stanley, Robert H., and Ramsey, Ruth G. "Television News: Format as a Form of Censorship," *ETC.,* Winter 1978.

Stanley v. Georgia, 394 U.S. 557 (1969).

Steigerwald, B. "Videodisc: Ultimate Weapon or Video Revolution?" *Los Angeles Times,* 24 March 1981.

Steinberg, Charles. *The Communicative Arts.* New York: Hastings House, 1970.

Steiner, Gary A. *The People Look at Television.* New York: Alfred A. Knopf, 1963.

Stempel, Guido H., and Westley, Bruce H., eds. *Research Methods in Mass Communication.* Englewood Cliffs, N.J.: Prentice-Hall, 1981.

Stengel, Richard. "The Sound of Quality." *Time,* 5 November 1984.

Stephenson, William. *The Play Theory of Mass Communications.* Chicago: University of Chicago Press, 1967.

Steritt, "Movies Try New Tactics in Battle with TV." *USA Today,* 14 November 1983.

Sterling, Christopher H. "Television and Radio Broadcasting." In *Who Owns the Media? Concentration of Ownership in the Mass Communication Industry,* edited by Benjamin M. Compaine. New York: Harmony Books, 1979.

Sterling, Christopher H. "Trends in Daily Newspaper and Broadcast Ownership, 1922–1970." *Journalism Quarterly,* Summer 1975.

Sterling, Christopher H., and Haight, Timothy R. "Characteristics of Newspaper Readers." Tables in *The Mass Media: Aspen Institute Guide to Communication Industry Trends.* New York: Praeger, 1978.

Sterling, Christopher H., and Kittross, John M. *Stay Tuned: A Concise History of American Broadcasting.* Belmont, Calif.: Wadsworth, 1978.

Stern, Gerald. E., ed. *McLuhan: Hot and Cool.* New York: New American Library, 1969.

Sterritt, David. "Movie Ratings—From G to X: Are They Out of Focus?" *Christian Science Monitor,* 16 September 1982.

Sterritt, David. "Movies Try New Tactics in Battle With TV." *Christian Science Monitor,* 23 August 1982.

Stevens, John D. "Freedom of Expression: New Dimensions." In *Mass Media and the National Experience,* edited by Ronald T. Farrar and John D. Stevens. New York: Harper & Row, 1971.

Stevenson, Robert L., and White, Kathryn P. "The Cumulative Audience of Network Television News." *Journalism Quarterly,* Autumn 1980.

Stewart, Sally Ann. "Cosby: The Undisputed King Of TV." *USA Today,* 5 December 1985.

Summers, Harrison B.; Summers, Robert E.; and Pennybacker, John H. *Broadcasting and the Public,* 2d ed. Belmont, Calif.: Wadsworth, 1978.

Swain, Bruce. *Reporter's Ethics.* Ames, Iowa: Iowa State University Press, 1978.

t

"Talking Back to the Tube." *Newsweek,* 25 February 1980.

Tannenbaum, Percy H., ed. *The Entertainment Functions of Television.* Hillsdale, N.J.: Lawrence Erlbaum Associates, 1981.

Tebbel, John. *A History of Book Publishing in the United States.* New York: Bowker. Four Volumes: 1972, 1975, 1978, and 1981.

Tebbel, John. *Compact History of the American Newspaper.* New York: Hawthorne, 1969.

Tebbel, John. *The American Magazine: A Compact History.* New York: Hawthorne, 1969.

Tebbel, John. *The Media in America.* New York: Crowell, 1975.

Terry, Sara. "Ultrovox singer sees a new age of responsibility in rock music." *Christian Science Monitor,* 13 February 1986.

Thayer, Lee, ed. *Ethics, Morality and the Media.* New York: Hastings House, 1980.

— "(The) New Face of TV News." *Time,* 25 February 1980.

The People and the Press: A Times Mirror Investigation of Public Attitudes Toward The News Media. Los Angeles: Times Mirror Co., 1986.

(The) Roper Organization, Inc. *Public Perceptions of Television and Other Mass Media: A Twenty-year Review,* 1959–1978. New York: Television Information Office, 1979.

"(The) Second 50 Years Of The Fifth Estate." *Broadcasting,* 30 December 1985.

Thomas, Bill. "What Does A Man Want?" *Washington Journalism Review,* December, 1984.

Tuchman, Gay. *Making News.* New York: Free Press, 1978.

"Turn on Your Neighbor." *Panorama,* April 1980, p. 21.

u

Udell, Jon G., ed. *The Economics of the American Newspaper.* New York: Hastings House, 1978.

United Press International. "Computers Make Many Functionally Illiterate." *Logan* (Utah) *Herald Journal,* 24 September 1980.

United Press International. "Curl Up With a Good Movie." *Deseret News,* 30 July 1979.

"UPI is Sold to Media News Corp." *Logan* (Utah) *Herald Journal,* 3 June 1982.

"UPI owner says news service to expand." *Logan* (Utah) *Herald Journal,* 15 June 1986.

"Up, Up and Away for Radio Networking." *Broadcasting,* 17 March 1980.

U.S. Government. *The Global 2000 Report to the President: Entering the Twenty-first Century* (Vol. 1). Washington, D.C.: Government Printing Office, 1980.

v

Valenti, Jack. "The Movie Rating System." In *Mass Communication: Principles and Practices,* edited by Mary B. Cassata and Molefi K. Asante. New York: Macmillan, 1979.

Valenti, Jack. "The Politics of Cable." *Media Digest,* Spring 1980.

Valenti, Jack. "The system works as guide for parents." *USA Today,* 14 November 1983.

"Videodisc." *Media Digest,* Spring 1980.

"Videodiscs catch on in military, business training." Associated Press. *Washington Times,* 30 October 1984.

"Videos making big changes in pop culture." United Press International. *Logan* (Utah) *Herald Journal,* 14 March 1985.

w

Wantzel, Michael. *"World Press Review* helps cure the 'all-American' outlook." *Baltimore Evening Sun,* 30 September 1983.

Waples, D.; Berelson, B.; and Bradshaw, F. R. *What Reading Does to People.* Chicago: University of Chicago Press, 1940.

Waters, Harry F. "What TV Does to Kids." *Newsweek,* 21 February 1977.

Waters, Harry F., and Uehling, Mark D. "Tuning In On The Viewer." *Newsweek,* 4 March 1985.

Weaver, David. *Videotex Journalism: Teletext, Viewdata, and the News.* Hillsdale, N.J.: Lawrence Erlbaum Associates, 1983.

Weaver, David H., and Wilhoit, G. Cleveland. *The American Journalist: A Portrait of U.S. News People and Their Work.* Bloomington, Indiana: Indiana University Press, 1986.

Weber, Ronald, ed. *The Reporter as Artist.* New York: Hastings House, 1974.

Weinberg, Steve. "Trashing the FOIA." *Columbia Journalism Review,* January/February, 1985.

Weisman, John. "Network News Today: Which Counts More—Journalism or Profits?" *TV Guide,* 26 October 1985.

Wells, Alan, ed. *Mass Media and Society.* Palo Alto, Calif.: Mayfield, 1979.

West, Woody. "Public Broadcasting: Pull the Plug." *Washington Times Insight,* 17 February 1986.

Westin, Av. *Newswatch: How TV Decides The News.* New York: Simon and Schuster, 1982.

Whetmore, Edward Jay. *Mediamerica: Form, Content, and Consequences of Mass Communication,* 3d ed. Belmont, Calif.: Wadsworth, 1985.

"White House, Congress at odds over public broadcasting funding." *Broadcasting,* 23 December 1985.

Whitney, D. Charles. *The Media and the People—Americans' Experience with the News Media: A Fifty-Year Review.* New York: Gannett Center for Media Studies, November 1985.

Wicker, Tom. *On Press.* New York: Viking, 1978.

Wicklein, John. "National news: public TV and the NPR example." *Columbia Journalism Review,* January/February 1986.

Wicklein, John. "The Assault On Public Television." *Columbia Journalism Review.* January/February 1986.

Wilkins, Lee. Review of Robert Picard, *The Press and the Decline of Democracy.* In *Journal of Mass Media Ethics,* Vol. 1, No. 2, Spring/Summer 1986.

Williams, Frederick. *The Communications Revolution.* Beverly Hills, Calif.: SAGE, 1982.

Williams, Lynne. *Medium or Message?* Woodbury, N.Y.: Barron's Educational Series Inc., 24 May 1971.

Williams, Martin. *TV: The Casual Art.* New York: Oxford University Press, 1982.

Wilson, Sloan. *The Man in the Gray Flannel Suit.* New York: Pocket Books, 1967.

Winans, R. Foster. *Trading Secrets.* New York: St. Martins Press, 1986.

Winick, Mariann Pezzella, and Winick, Charles. *The Television Experience: What Children See.* Beverly Hills, Calif.: SAGE, 1979.

Winn, Marie. *The Plug-In Drug.* New York: Bantam Books, 1977.

"Winners of This Year's SDX Distinguished Service Awards." *Quill,* April 1979.

Withey, Stephen, and Abeles, Ronald P., eds. *Television and Social Behavior.* Hillsdale, N.J.: Lawrence Erlbaum Associates, 1981.

Wolfe, Tom, ed. *The New Journalism.* New York: Harper & Row, 1973.

Wolseley, Roland E. *The Black Press, USA.* Ames, Iowa: Iowa State University Press, 1971.

Wolseley, Roland E. *The Changing Magazine: Trends in Readership and Management.* New York: Hastings House, 1973.

Wolseley, Roland E. *Understanding Magazines.* Ames, Iowa: Iowa State University Press, 1969.

Wood, R. Kent. "The Utah State University Videodisc Innovations Project." *Educational and Industrial Television,* May 1979.

Wright, Charles R. "Functional Analysis and Mass Communication." *Public Opinion Quarterly,* 1960.

Wright, Charles R. *Mass Communication: A Sociological Perspective,* 3d ed. New York: Random House, 1986.

Wright, Donald K. "Premises for Professionalism: Testing the Contributions of PRSA Accreditation." Paper presented to the Public Relations Division, Association for Education in Journalism Annual Convention, Houston, Texas, August 1979.

Wright, John W., ed. *The Commercial Connection: Advertising and the American Mass Media.* New York: Dell/Delta, 1979.

Z

Zimmerman, David. "Rock's hall-of-famers will never fade away." *USA Today,* 23 January 1986.

Zoglin, Richard. "Covering The Awful Unexpected." *Time,* 10 February 1986.

Zoglin, Richard. "Gremlins in the Rating System." *Time,* 25 June 1984.

Zoglin, Richard. "VCRs: Coming on strong." *Time,* 24 December 1984.

Credits

Chapter 1
Page 7: top, © Cleo Freelance Photography, **bottom left,** © Tom Pantages, **bottom right,** © Robert Eckert/EKM-Nepenthe; **page 19: top left,** © Robert Eckert/EKM-Nepenthe, **top right,** © James L. Reynolds, **bottom left,** © Mark Antman/Image Works, Inc., **bottom right,** © James Brey; **page 29:** © Stephen Ferry/Gamma Liaison.

Chapter 2
Page 38: top, AP/Wide World Photos, **bottom left,** © 1985 National Broadcasting Company, Inc., **bottom right,** AP/Wide World Photos; **page 46:** Courtesy of CBS; **page 57:** © Michael Hayman/Black Star.

Chapter 3
Pages 83, 88: Bettmann Archive, Inc.; **page 91: top,** AP/Wide World Photos, **bottom,** Bettmann Archive, Inc.; **page 94:** Springer/Bettmann Archive, Inc.; **page 100:** © Michael Siluk/EKM-Nepenthe; **page 104:** © Jay Black; **page 107:** © Bob Coyle; **page 109:** © Jill Cannefax/EKM-Nepenthe; **page 112:** © Jean-Claude Lejeune; **page 115: left,** © James L. Reynolds, **right,** © Mark Antman/Image Works, Inc.

Chapter 4
Page 127: © Bob Coyle; **page 133: left,** Reprinted from *The Saturday Evening Post.* © 1956 The Curtis Publishing Company, **right,** Rex Hardy, Jr./*Life Magazine,* © 1938 Time, Inc.; **page 139:** Reprinted with permission from *TV GUIDE® Magazine.* Copyright © 1953 by Triangle Publications, Inc. Radnor, PA; **page 140:** AP/Wide World Photos; **page 148:** © James Shaffer; **page 152: left,** Copyright © 1967, American Psychological Association, **right,** Copyright © 1987 American Psychological Association; **page 155:** Courtesy of *World Press Review.* Published by The Stanley Foundation; **page 156:** Courtesy of Meridian Publishing, Inc.

Chapter 5
Page 167: © Bob Adelman/Magnum Photos; **page 170:** Bettmann Archive, Inc.; **page 175:** © David Strickler/The Picture Cube; **page 177:** © Robert Eckert/EKM-Nepenthe; **page 184:** Courtesy of William Morrow & Company, Inc.; **page 189:** © Jay Black; **page 190:** © Tom Ballard/EKM-Nepenthe; **page 195:** © Robert Eckert/EKM-Nepenthe.

Chapter 6
Pages 206, 208, 213: AP/Wide World Photos; **pages 214, 218:** Bettmann Archive, Inc.; **page 222:** © Tim Jewett/EKM-Nepenthe; **page 230:** © Tyrone Hall/Stock, Boston; **page 233:** Courtesy of National Public Radio, Photo by Stan Barouh; **page 239:** © Bob Coyle.

Chapter 7
Page 247: Bettmann Archive, Inc.; **page 250:** AP/Wide World Photos; **page 251:** © 1984 Dan Sheehan/Black Star; **page 256:** Bettmann Archive, Inc.; **page 257:** AP/Wide World Photos; **page 259:** Springer/Bettmann Film Archive; **page 261:** UPI/Bettmann Newsphotos; **page 266:** Hitachi America, Ltd.

Chapter 8
Page 270: © Cleo Freelance Photography; **page 273:** AP/Wide World Photos; **page 274:** Bettmann Archive, Inc.; **page 283:** Courtesy of Nielsen Media Research; **page 286:** © National Broadcasting Company, Inc., Alan Singer, photographer; **page 294:** © 1986 Doug Wilson/Black Star; **page 297:** © Steve Schapiro/Gamma Liaison; **page 302:** © James L. Reynolds; **page 305:** Turner Broadcasting Systems, Inc.; **page 313: (all)** Brent Herridge/KSL-TV, Salt Lake City; **page 318:** NASA; **page 321:** © Martha Stewart/The Picture Cube.

Chapter 9
Page 328: © James Brey; **page 330:** From the Collection of the Library of Congress; **page 332:** Springer/Bettmann Film Archive; **page 334:** Bettmann Archive, Inc.; **page 335:** UPI/Bettmann Newsphotos; **page 336:** Burton Holmes Collection/EKM-Nepenthe; **page 337:** AP/Wide World Photos; **page 339:** Courtesy of The Academy of Motion Picture Arts and Sciences; **page 342:** AP/Wide World Photos; **page 343:** Courtesy of The Academy of Motion Picture Arts and Sciences; **page 345:** © James Shaffer; **pages 350, 352:** Courtesy of The Academy of Motion Picture Arts and Sciences; **page 355:** UPI/Bettmann Newsphotos; **page 358:** AP/Wide World Photos; **page 360:** Courtesy of The Academy of Motion Picture Arts and Sciences; **page 365: (both)** Courtesy of Caesar's Palace, Las Vegas, NV.

Chapter 10
Page 373: left, © James Shaffer, **right,** Courtesy of National Captioning Institute; **pages 375, 378:** © Robert Eckert/EKM-Nepenthe; **page 406:** Sears, Roebuck and Company; **page 409:** © James Shaffer; **page 413:** AT&T Bell Laboratories.

Chapter 11
Page 424: © Bob Coyle; **page 427:** Bettmann Archive, Inc.; **page 434:** © Tom Pantages; **page 438:** © James L. Reynolds; **page 440:** © Bob Coyle; **page 444:** © James Brey; **page 456:** © Bob Coyle.

Chapter 12

Page 468: Bettmann Archive, Inc.; **page 471:** AP/Wide World Photos; **page 480:** UPI/Bettmann Newsphotos; **page 486:** AP/Wide World Photos; **pages 487, 489:** © Alan Carey/Image Works, Inc.

Chapter 13

Pages 500, 502: Bettmann Archive, Inc.; **page 509:** © Bob Coyle; **page 514:** © Susan Meiselas/Magnum Photos; **page 516:** © Robert Eckert/EKM-Nepenthe; **page 519:** AP/Wide World Photos.

Chapter 14

Page 544: AP/Wide World Photos; **page 548:** © Tim Carlson/Stock, Boston; **page 552:** *Hustler Magazine,* October 1978, page 5. Reprinted by permission; **page 553:** AP/Wide World Photos; **page 556:** UPI/Bettmann Newsphotos; **page 569:** AP/Wide World Photos; **page 571:** © Jay Black.

Chapter 15

Page 576: Reprinted with permission of *Chicago Sun-Times;* **page 580:** Springer/Bettmann Archives, Inc.; **page 598: left,** Courtesy of *Columbia Journalism Review,* Edward Sorel, artisit, **right,** *Washington Journalism Review;* **page 606:** © Jay Black.

Index

First Amendment to the U.S. Constitution, 85, 119, 212, 537, 542–43, 550–51, 557, 563, 567–68, 594
Fisher, Eddie, 255
Fitzgerald, F. Scott, 153, 255
Five theories of the press, table, 530–31
Fleming, John Ambrose, 206
Fletcher, Alan, 144
Flush, 188
Flynt, Larry, 551–52
FM, 31, 223–24
Focus, 142
Foisie, Philip, 587
folk music, 252
folk rock, 261
Fonda, Jane, 480
Food and Drug Administration (FDA), 454
Forbes, 145
Ford, Gerald, 476, 559
Ford Motor Company, 136
Fortune, Inc., 145
Foster, Jodi, 43
Four Seasons, The, 258
Four Theories of the Press, 528
Fourdrinier, 89
Fourteenth Century English Mystic Magazine, 138
fox trot, 254
Frankfurter, Felix, 567
Frankie Goes to Hollywood, 37
Franklin, Benjamin, 84, 128
Franklin, James, 84
Freberg, Stan, 458
Freddie and the Dreamers, 259
freebies, 578–81
Freed, Alan, 256–57
Freedom of Information Act (FOIA), 564–65, 596
Freedom's Journal, 111
free-lancers, 157
Freneau, Philip, 128
Freud, Sigmund, 438–39, 443, 472
Friedman, Milton, 484
Front Page, 93
Front Page Woman, 94
"Frugal Gourmet, The," 185
functional analysis, 53–55
Future Shock, 194

g

gag orders, 568
Galella, Ron, 548
Gallup, George, 119, 472, 555, 586
Gandhi, Indira, 532
Gannett, 97, 106–8, 544, 586–87
Gannett Center for Media Studies, 586–87
Garrett, Paul, 472, 483
gatekeeper, 16–18, 28, 31, 41, 99–100, 102, 140, 252–53, 511–15, 584, 596
Gaye, Marvin, 258
Geissler, William, 504
Geldorf, Bob, 262
General Electric, 207, 210
General Magazine and Historical Chronicle, 128

General Motors Corporation, 472
general semantics, 25–27
geodemographics, 442
"Georgics," 466
Ghorbanifar, Manucher, 584
Gleason's Pictorial, 129
global village, 71
Godey's Lady's Book, 128
Goetz, Bernhard, 38
go-go, 261
Golden Notebook, The, 186
Goodbye Mr. Chips, 479
Goodman, Benny, 214, 255
Grain Information News, 517
gramophone, 248
Grand Funk Railroad, 260
graphophone, 247
Grass Is Always Greener Over The Septic Tank, The, 184
Grateful Dead, 260
Greeley, Horace, 90
Green, Bill, 603
Greenberger, Martin, 196
Gremlins, 352
Grey Advertising, 474
Grinning Idiot, 138
group ownership, 95–98
Grunig, James, 483–84
Guccione, Bob, 147
Gulf & Western, 173
Gutenberg Galaxy, The, 70
Gutenberg, Johann, 76, 168–69, 467
Guthrie, Woody, 260

h

Hadden, Briton, 140
Hall, Edward T., 195
Halsey, William, 132
Hamilton, Alexander, 86, 128, 468
Hamilton, Andrew, 85
Hancock, John, 128
"Hands Across America," 478, 480–81
Hannah, Mark, 485
Harcourt Brace Jovanovich Publications, Inc., 145
Hard Pressed, 138
Harlequin romances, 176–77
Harper's Bazaar, 161
Harper's New Monthly Magazine, 129
Harper's Weekly, 129
Harris, Benjamin, 82
Harris, Louis, 116–18, 555
Harrison, William Henry, 485–86
Harshe-Rotman & Druck, 474
Harte-Hanks, 121
Hartford (Conn.) Courant, 566
Harvard Law Review, 547
Harvey, Paul, 584
Hauptmann, Bruno, 238, 569
Hawthorne, Nathaniel, 92, 184
"Heard on the Street," 584
Hearst Corporation, 150
Hearst, William Randolph, 91–92, 130, 150, 502–3, 515
"Heartbreak Hotel," 258
heavy metal, 261

Hefner, Hugh, 147–48
Hemingway, Ernest, 153
Hendrix, Jimi, 260
Henry VIII, 83, 539
Herchenroeder, John, 602
Herman, Woody, 255
Herman's Hermits, 259
Herrold, Charles, 206
Hertz, Heinrich, 205
Hidden Persuaders, The, 52, 436
High Times, 147
"Highway to Hell," 43
Hill and Knowlton, 474
Hinkley, Jr., John W., 43, 567
History of Standard Oil, 470
Hitler, Adolf, 218
Holmes, Oliver Wendell, 543
Holt, Rinehart & Winston, 173
Home Box Office, 148
Hoover, Herbert, 209, 211, 218
Hope, Bob, 481
Horst, Gerald Ter, 476
hospital PR, 491
house organs, 145–46
Houseman, John, 216
Houston Living, 142
Houston, Whitney, 481
"How Much is that Doggie in the Window?," 255
Howells, William Dean, 171
Hughes, Charles Evans, 206
Hunt, Todd, 483–84
Hustler, 551–52
Hutchins Commission on Freedom of the Press, 593–604
Hutchins, Robert M., 535, 593
hype, 253–54, 478–82
hypodermic theory, 42–43

i

"I Can't Get No Satisfaction," 259
"I Spy," 46
"I Want to Hold Your Hand," 258, 480
Iacocca, Lee, 481, 484
IBM, 172
Illiterate America, 169
I'm OK, You're OK, 52
In Cold Blood, 187
Indiana Jones and the Temple of Doom, 352–53
inkjet printing, 120
Innis, Harold, 71
institutional advertising, 446–47
Intelsat IV, 141
Internal Revenue Service, 454
International Association of Business Communication (IABC), 494
International News Service (INS), 91, 217–18, 503
international news services, 512–15
International Organization of Journalists (IOJ), 514
inverted pyramid, 15, 501
Ismach, Arnold, 101
ITT, 172